S0-BRA-492

Sensor Networks for Sustainable Development

TK
7872
D48S4435
2014
WEB

Sensor Networks for Sustainable Development

EDITED BY

Mohammad Ilyas • Sami S. Alwakeel
Mohammed M. Alwakeel
el-Hadi M. Aggoune

CRC Press is an imprint of the
Taylor & Francis Group, an informa business

CRC Press
Taylor & Francis Group
6000 Broken Sound Parkway NW, Suite 300
Boca Raton, FL 33487-2742

© 2014 by Taylor & Francis Group, LLC
CRC Press is an imprint of Taylor & Francis Group, an Informa business

No claim to original U.S. Government works

Printed on acid-free paper
Version Date: 20140402

International Standard Book Number-13: 978-1-4665-8206-4 (Hardback)

This book contains information obtained from authentic and highly regarded sources. Reasonable efforts have been made to publish reliable data and information, but the author and publisher cannot assume responsibility for the validity of all materials or the consequences of their use. The authors and publishers have attempted to trace the copyright holders of all material reproduced in this publication and apologize to copyright holders if permission to publish in this form has not been obtained. If any copyright material has not been acknowledged please write and let us know so we may rectify in any future reprint.

Except as permitted under U.S. Copyright Law, no part of this book may be reprinted, reproduced, transmitted, or utilized in any form by any electronic, mechanical, or other means, now known or hereafter invented, including photocopying, microfilming, and recording, or in any information storage or retrieval system, without written permission from the publishers.

For permission to photocopy or use material electronically from this work, please access www.copyright.com (http://www.copyright.com/) or contact the Copyright Clearance Center, Inc. (CCC), 222 Rosewood Drive, Danvers, MA 01923, 978-750-8400. CCC is a not-for-profit organization that provides licenses and registration for a variety of users. For organizations that have been granted a photocopy license by the CCC, a separate system of payment has been arranged.

Trademark Notice: Product or corporate names may be trademarks or registered trademarks, and are used only for identification and explanation without intent to infringe.

Library of Congress Cataloging-in-Publication Data

Sensor networks for sustainable development / editors, Mohammad Ilyas, Sami S. Al-Wakeel, Mohammed M. Alwakeel, el-Hadi M. Aggoune.
pages cm
Summary: "Sensor networks are becoming an integral part of our society to meet some common needs such as healthcare, and smart energy use. The success of any enterprise depends upon how fast it can deal with its operations and optimize performance. This handbook is designed for anyone who deals with or intends to become involved with the field of sensor networks. "-- Provided by publisher.
Includes bibliographical references and index.
ISBN 978-1-4665-8206-4 (hardback)
1. Sensor networks--Industrial applications. 2. Sustainable engineering. I. Ilyas, Mohammad, 1953- editor of compilation.

TK7872.D48S4435 2014
681'.75--dc23 2014007236

Visit the Taylor & Francis Web site at
http://www.taylorandfrancis.com

and the CRC Press Web site at
http://www.crcpress.com

Contents

SECTION III Energy

SECTION IV Healthcare

Preface

The handbook *Sensor Networks for Sustainable Development* explores the great contribution made by sensor network technologies on sustainable development. The term *sustainable development* aims toward allowing human progress and the development of society to meet human needs, both in present times and for generations to come, while at the same time ensuring the sustainability of the environment and of natural resources.

The handbook is organized as a collection of articles authored by experts in the field and is valuable for anyone who deals with or intends to become involved with the field of sensor networks. The overarching goal of these articles is to provide the reader with a spectrum of applications where sensor networks are used to yield support for sustainable development through optimal use of resources and efforts while achieving the intended objective of quality of life for everyone now and in the future.

From the technical point of view, sensor nodes are small electronic components capable of sensing many types of information from their surroundings, including temperature, light, humidity, radiation, the presence or nature of biological organisms, geological features, seismic vibrations, specific types of computer data, and more. In its rudimentary form, a sensor network consists of a large number of sensor nodes that may be randomly and densely deployed for applications such as agriculture, environment, energy, healthcare, transportation, and disaster management. Recent advances have made it possible to make these components small, powerful, and energy efficient, and they can now be manufactured cost-effectively in large quantities for specialized telecommunications applications. Sensor nodes are usually very small and are capable of gathering, processing, and communicating information to other nodes and to the outside world. Based on the information-handling capabilities and compact size of the sensor nodes, sensor networks are often referred to as *smart dust* as they may have the capabilities of adaptation, self-awareness, and self-organization.

Sensor network research and development rely on many concepts and protocols derived from distributed communication networks. Based on their applications in various development fields, there are numerous ways of exploring the contribution of sensor networks for sustainable development. This handbook consists of a total of 21 chapters that capture the current state of sensor networks and their applications in six representative sustainable development application areas. The areas covered in these chapters range from basic concepts to research grade material including future directions as follows:

1. Sensor networks for smart agriculture
2. Sensor networks for smart environment
3. Sensor networks for smart energy use
4. Sensor networks for smart healthcare
5. Sensor networks for smart transportation
6. Sensor networks for public safety and disaster management

Smart agriculture is an application where sensors and actuators are used to ensure the optimal growth of a crop through a watering protocol that nearly eliminates waste. Smart grid is an application where sensor networks are used to ensure that the power generated and delivered adhere to preset quality and quantity criteria established to optimize the use of energy. Use of sensor networks for healthcare provides added benefit to healthcare professionals as well as to patients by giving them a better platform for monitoring health aspects remotely and by allowing them to make quick, accurate, and informed decisions. Smart transportation allows vehicles (and passengers) to share information about traffic conditions and provide a better platform for law enforcement, ensuring safety for everyone on the road. These are just a few of the many potential applications of sensor networks.

Accordingly, this handbook provides scientific tutorials and technical information about sensor networks and their applications in various areas of broad interest that are of concern to sustainability. It can also serve as a comprehensive source of reference material on sensor networks and their applications in various sustainable development areas. The targeted audience for the handbook therefore may include professionals who are designers and/or planners for emerging telecommunication networks, researchers working on sensor networks applied fields, and those (faculty members and graduate students) who would like to learn about the impact of sensor networks on sustainable development.

In conclusion, it is important to note that the field of sensor networks (wired and wireless) continues to evolve rapidly. Clearly, sensor networks are becoming an integral part of our society. They are in our homes, vehicles, the roads by which we travel, and in the farms from where we get our food. The applications of sensor networks for sustainable development will therefore continue to grow as well. This handbook will thus serve as a valuable collection and reference for readers who look for articles that focus on the applications of sensor networks that motivate sustainable development.

Mohammad Ilyas
Boca Raton, Florida

Sami S. Al-Wakeel
Riyadh, Kingdom of Saudi Arabia

Mohammed M. Al-Wakeel
Tabuk, Kingdom of Saudi Arabia

el-Hadi M. Aggoune
Tabuk, Kingdom of Saudi Arabia

Acknowledgments

Many people have contributed to this handbook in their unique ways. First and foremost, we would like to thank the group of highly talented and skilled researchers who have contributed 21 chapters to this handbook for their dedication, cooperation, and professionalism. The generous support from the leadership of the Sensor Networks and Cellular Systems (SNCS) Research Center at the University of Tabuk is highly appreciated. It has also been a pleasure to work with Nora Konopka and Jessica Vakili of Taylor & Francis Group/CRC Press. We are extremely grateful for their support and professionalism through all the stages of this handbook. Finally, we would like to acknowledge the unconditional support extended by our families during this project; they all deserve very special thanks.

Editors

Dr. Mohammad Ilyas is dean and professor in the College of Engineering and Computer Science at Florida Atlantic University, Boca Raton, Florida. He received his BSc in electrical engineering from the University of Engineering and Technology, Lahore, Pakistan, in 1976. From March 1977 to September 1978, he worked for the Water and Power Development Authority, Pakistan. In 1978, he was awarded a scholarship for his graduate studies. He completed his MS in electrical and electronic engineering in June 1980 at Shiraz University, Shiraz, Iran. In September 1980, Dr. Ilyas joined the doctoral program at Queen's University in Kingston, Ontario, Canada. He completed his PhD in 1983. His doctoral research was about switching and flow control techniques in computer communication networks. Since September 1983, he has been with the College of Engineering and Computer Science at Florida Atlantic University. From 1994 to 2000, he was chair of the Department of Computer Science and Engineering. From July 2004 to September 2005, he served as interim associate vice president for research and graduate studies. During the academic year 1993–1994, he was on a sabbatical from the Department of Computer Engineering, King Saud University, Riyadh, Saudi Arabia.

Dr. Ilyas has conducted successful research in various areas, including traffic management and congestion control in broadband/high-speed communication networks, traffic characterization, wireless communication networks, performance modeling, and simulation. He has published 1 book, 25 handbooks, and over 170 research articles. He has supervised 11 PhD dissertations and more than 38 MS theses to completion. He has been a consultant to several national and international organizations. Dr. Ilyas is an active participant in several IEEE technical committees and activities. He is also a senior member of the IEEE and a member of the ASEE.

Dr. Sami S. Alwakeel is a resourceful and innovative professional with more than 25 years of academic experience providing powerful learning opportunities for graduate and undergraduate university students, Saudi national educational facilities, and corporations. He is recognized for teaching excellence and expertise in developing computer education programs, for his contributions in information technology (IT) research, and in providing technical solutions and consultations for various IT sectors in Saudi society.

Dr. Alwakeel received his BSc in electrical engineering with honors from King Saud University, Riyadh, Saudi Arabia, and his MSc and PhD in electrical engineering from Stanford University, California. Upon his return from his graduate studies, he joined a small team of staff members who founded the College of Computer and Information Sciences at King Saud University. He has served in the university in various academic and administration positions, including as dean of the College of Computer and Information Science and as a member of King Saud University Council from 2003 to 2009.

In the general education sector, Dr. Alwakeel was the chairman and member of the National Computer Education Committee, Ministry of Education; author of a textbook on local area networking; author and coauthor of six books for secondary school; and author of textbooks on IT and computer science for the Ministry of Education, Saudi Arabia. He has also authored three books and several articles in social sciences and many articles on computers and IT in newspapers and culture magazines.

Dr. Alwakeel is also known as an expert in providing computer network solutions and consultations to individuals and corporations and in pioneering IT awareness and IT national planning. He was a founder of the Saudi Computer Society and chairman of the society's board of directors; chairman of the Committee for E-learning and Distant Learning Bylaws Development, E-learning Center, Ministry of Higher Education; and a member of the team that developed the "Saudi IT National Plan," 2002–2003, which was approved by the Saudi Council of Ministers in 2006. He was also a member of various organizational committees of many national computer conferences and symposiums held in Saudi Arabia and served as a consultant for various government and public agencies as well as business units in Saudi Arabia.

Dr. Alwakeel was a visiting scholar to various international organizations and universities. He was a research visitor at GMD and FHG Research Institutes; Siemens Research Labs in Germany; the Department of Computer Science and Engineering at Florida Atlantic University, Boca Raton, Florida; General Electric (GE) Corporation; and Space Labs, United States. He was a scientific visitor at the Center of International Cooperation for Computerization in Japan. He also supervised and examined more than 30 MS and PhD theses, holds two US patents, and has contributed to more than 60 scientific papers in research journals and conferences in the fields of computer science and engineering.

Dr. Mohammed M. Alwakeel received his BS in computer engineering and his MS in electrical engineering, both from King Saud University, Riyadh, Saudi Arabia, in 1993 and 1998, respectively, and his PhD in electrical engineering from Florida Atlantic University, Boca Raton, Florida, in 2005. From 1994 to 1998, he was a communications network manager at the National Information Center, Saudi Arabia. From 1999 to 2001, Dr. Alwakeel was with King Abdulaziz University. From 2009 to 2010, he was an assistant professor and the dean at the College of Computers and Information Technology at the University of Tabuk. He is now the vice-rector for development and quality at the University of Tabuk. His current research interests include teletraffic analysis, mobile satellite communications, sensor networks, and cellular systems.

Dr. el-Hadi M. Aggoune received his MS and PhD in electrical engineering from the University of Washington (UW), Seattle, Washington. He is a professional engineer registered in the State of Washington and senior member of the IEEE. His highest academic and administrative positions were endowed chair professor and vice president and provost. He was the founder and director of a major research and development laboratory that won the Boeing Supplier Excellence Award. Dr. Aggoune was also the winner of the IEEE Professor of the Year Award, UW

Branch. He is listed as inventor in a major patent assigned to the Boeing Company. His research work is referred to in many patents, including patents assigned to ABB, Switzerland, and EPRI, United States. He currently serves as a professor and director of the Sensor Networks and Cellular Systems Research Center, University of Tabuk, Tabuk, Saudi Arabia. His research interests include modeling and simulation of large-scale networks, sensors and actuators, scientific visualization, and control and energy systems.

Contributors

Ahsan Abdullah
Department of Information Technology
Faculty of Computing and Information Technology
King Abdulaziz University
Jeddah, Saudi Arabia

el-Hadi M. Aggoune
Sensor Network and Cellular System Network Research Center
University of Tabuk
Tabuk, Kingdom of Saudi Arabia

Hamid Aghajan
Ambient Intelligence Research Lab
Stanford University
Stanford, California

and

iMinds, Gent University
Ghent, Belgium

Rana E. Ahmed
Department of Computer Science and Engineering
American University of Sharjah
Sharjah, United Arab Emirates

Rashid Al Hammadi
Department of Computer Science and Engineering
American University of Sharjah
Sharjah, United Arab Emirates

Mohammed M. Alwakeel
Sensor Network and Cellular System Network Research Center
University of Tabuk
Tabuk, Kingdom of Saudi Arabia

Sami S. Alwakeel
Department of Computer Engineering
College of Computer and Information Sciences
King Saud University
Riyadh, Kingdom of Saudi Arabia

and

Sensor Network and Cellular System Network Research Center
University of Tabuk
Tabuk, Kingdom of Saudi Arabia

Arny Ambrose
Department of Computer and Electrical Engineering and Computer Science
Florida Atlantic University
Boca Raton, Florida

Mohammad Ammad-Uddin
Sensor Network and Cellular System Network Research Center
University of Tabuk
Tabuk, Kingdom of Saudi Arabia

Giuseppe Anastasi
Department of Information Engineering
Università di Pisa
Pisa, Italy

Angelos Antonopoulos
Communication Technologies Division
SMARTECH Department
Telecommunications Technological Centre of Catalonia (CTTC)
and
Department of Signal Theory and Communications
Technical University of Catalonia (UPC)
Barcelona, Spain

Luis Felipe Artigas
Laboratory of Oceanology and Geosciences (UMR CNRS 8187)
University of Littoral Côte d'Opale; MREN-ULCO
Wimereux, France

Ricardo J.M. Batista
MobiHealth BV
Enschede, the Netherlands

Carlo Alberto Boano
Institute for Technical Informatics
Graz University of Technology
Graz, Austria

Richard G.A. Bults
Telemedicine Group, Biomedical Signals and Systems
Faculty of Electrical Engineering, Mathematics and Computer Science
University of Twente
Enschede, the Netherlands

Mihaela Cardei
Department of Computer and Electrical Engineering and Computer Science
Florida Atlantic University
Boca Raton, Florida

Wei Chang
Department of Computer and Information Sciences
Temple University
Philadelphia, Pennsylvania

Francesco Chiti
Department of Information Engineering
University of Florence
Florence, Italy

Diane J. Cook
School of Electrical Engineering and Computer Science
Washington State University
Pullman, Washington

Sajal K. Das
Department of Computer Science
Missouri University of Science and Technology
Rolla, Missouri

Alessandra De Paola
Networking and Distributed Systems Laboratory
University of Palermo
Palermo, Italy

Rene A. de Wijk
Wageningen University Research Center/Food & Biobased Research
Wageningen, the Netherlands

Jantine Duit
Noldus Information Technology BV
Wageningen, the Netherlands

Melike Erol-Kantarci
School of Electrical Engineering and Computer Science
University of Ottawa
Ottawa, Ontario, Canada

Romano Fantacci
Department of Information Engineering
University of Florence
Florence, Italy

Hao Guo
School of Electrical and Electronic Engineering
Nanyang Technological University
Singapore, Singapore

Denis Hamad
Laboratoire d'Informatique Signal et Image de la Côte d'Opale
Université du Littoral Côte d'Opale
Calais, France

Peter Harliman
Compilers and Advanced Computer Systems Laboratory
School of Electrical Engineering
Korea University
Seoul, Republic of Korea

Hermie Hermens
Telemedicine Group, Biomedical Signals and Systems
Faculty of Electrical Engineering, Mathematics and Computer Science
University of Twente
and
Roessingh Research and Development
Enschede, the Netherlands

Loc Ho
Department of Computer Science
San Jose State University
San Jose, California

Farhan Hurmoodi
Department of Computer Science and Engineering
American University of Sharjah
Sharjah, United Arab Emirates

Antonio Iacoviello
Research Consortium on Intelligent Software Agent Technologies (CORISA)
University of Salerno
Salerno, Italy

Nada Ibrahim
Department of Computer Science and Engineering
American University of Sharjah
Sharjah, United Arab Emirates

Sozo Inoue
Library
Kyushu University
Fukuoka, Japan

Kyong Jin Jo
Compilers and Advanced Computer Systems Laboratory
School of Electrical Engineering
Korea University
Seoul, Republic of Korea

Valerie M. Jones
Telemedicine Group, Biomedical Signals and Systems
Faculty of Electrical Engineering, Mathematics and Computer Science
University of Twente
Enschede, the Netherlands

Elli Kartsakli
Department of Signal Theory and Communications
Technical University of Catalonia (UPC)
Barcelona, Spain

Seon Wook Kim
Compilers and Advanced Computer Systems Laboratory
School of Electrical Engineering
Korea University
Seoul, Republic of Korea

Isabelle Leblond
Laboratory of Sciences and Technologies of Information and Communication (Lab STICC)
National Engineering School of Advanced Techniques in Brittany
Brest, France

Joo-Ho Lee
College of Information Science and Engineering
Ritsumeikan University
Shiga-Ken, Japan

Joon Goo Lee
Compilers and Advanced Computer Systems Laboratory
School of Electrical Engineering
Korea University
Seoul, Republic of Korea

Giuseppe Lo Re
Networking and Distributed Systems Laboratory
University of Palermo
Palermo, Italy

Kay-Soon Low
School of Electrical and Electronic Engineering
Nanyang Technological University
Singapore, Singapore

Ali Mansour
Laboratoire Sciences et Technologies de l'information et de la Communication
École Nationale Supérieure de Techniques Avancées Bretagne
Brest, France

Melody Moh
Department of Computer Science
San Jose State University
San Jose, California

Teng-Sheng Moh
Department of Computer Science
San Jose State University
San Jose, California

Hussein T. Mouftah
School of Electrical Engineering and Computer Science
University of Ottawa
Ottawa, Ontario, Canada

Lucas P.J.J. Noldus
Noldus Information Technology BV
Wageningen, the Netherlands

Felix Jonathan Oppermann
Institute for Technical Informatics
Graz University of Technology
Graz, Austria

Marco Ortolani
Networking and Distributed Systems Laboratory
University of Palermo
Palermo, Italy

JongSeung Park
Graduate School of Information Science and Engineering
Ritsumeikan University
Shiga-Ken, Japan

Tommaso Pecorella
Department of Information Engineering
University of Florence
Florence, Italy

Marjan Popov
Faculty of Electrical Engineering, Mathematics and Computer Science
Delft University of Technology
Delft, the Netherlands

Agung B. Prasetijo
Faculty of Engineering
Department of Computer Engineering
Diponegoro University
Semarang, Indonesia

Kay Römer
Institute for Technical Informatics
Graz University of Technology
Graz, Austria

Akihito Sonoda
DNP LSI
Japan Co., Ltd.
Tokyo, Japan

Alfredo Vaccaro
Department of Engineering
University of Sannio
Benevento, Italy

Christos Verikoukis
Communication Technologies Division
SMARTECH Department
Telecommunications Technological
Centre of Catalonia (CTTC)
Barcelona, Spain

Zachary Walker
Department of Computer Science
San Jose State University
San Jose, California

Ing A. Widya
Telemedicine Group, Biomedical
Signals and Systems
Faculty of Electrical Engineering,
Mathematics and Computer Science
University of Twente
Enschede, the Netherlands

Kevin Bing-Yung Wong
Ambient Intelligence Research Lab
Stanford University
Stanford, California

Jie Wu
Department of Computer and
Information Sciences
Temple University
Philadelphia, Pennsylvania

Hiroto Yasuura
Department of Computer Science
Kyushu University
Fukuoka, Japan

Tongda Zhang
Ambient Intelligence Research Lab
Stanford University
Stanford, California

Imran A. Zualkernan
Department of Computer Science
and Engineering
American University of Sharjah
Sharjah, United Arab Emirates

Section I

Agriculture

1 A Review of Applications of Sensor Networks in Smart Agriculture

Ahsan Abdullah

CONTENTS

ABSTRACT

The world population is likely to double by 2050; however, neither the area under cultivation nor the availability of water is likely to double to meet the challenge. There are other challenges too, such as reduction in the number of farms and in the number of agriculture workforce. Climate change is expected to further worsen the existing situation. Therefore, in order for humanity to survive, agriculture has to become smart—one way is by integrating Wireless Sensor Networks (WSNs) in different types of traditional agriculture and that too at different levels. In this chapter, we will discuss different types of sensors, their market and different types of WSNs with their application in smart agriculture.

1.1 INTRODUCTION

During the last six decades, world agriculture has become significantly effective. There has been an enhancement in production systems, availability of different types of fertilizers, high yield varieties, etc. Effective livestock management and breeding programs have also contributed. Thus, there has been an agriculture revolution and a doubling of food production. However, there are challenges too, such as scarcity of water, diversity in pest attacks, and decline in agricultural workforce, to name a few. As per FAO (2011) Report, climate change is expected to further worsen the existing challenges that agriculture faces. Population changes over time, it is influenced by many factors, such as births, deaths, and migration into or out of the area. Global population has grown slowly for most of the human history; however, during the last 500 years, population has risen at an unprecedented rate, as shown in Figure 1.1 (Anonymous 2014a). As per Burney et al. (2010) and Bruinsma (2009), it is estimated that by 2050, the current world population is likely to grow from 6.7 to 9 billion, with most of the increase occurring in sub-Saharan Africa and South Asia. Considering the changes in the level of consumption and composition associated with growing household incomes, it is projected that feeding the world in 2050 will require a 70% increase in total agricultural production.

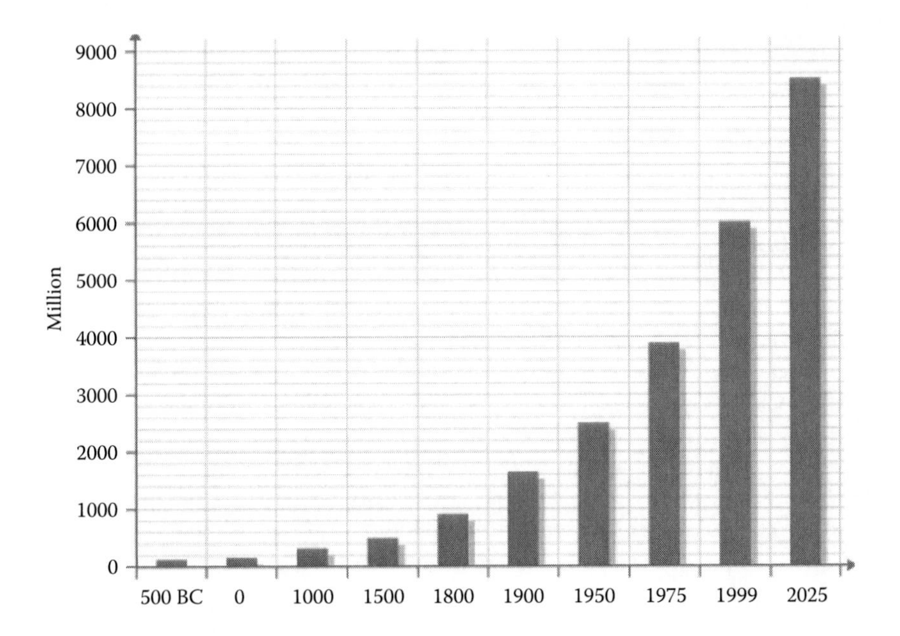

FIGURE 1.1 World population growth 500 BC–2025.

1.1.1 Objectives of Smart Agriculture

The objectives of smart agriculture are as follows (Anonymous 2012):

1. Using a combination of sensors such as temperature, light, and humidity, so as to spot the risk of frost, possible plant diseases and establish watering requirements based on soil dampness.
2. Manage cultivation of crop and to monitor the exact conditions in which the plants are growing from the comfort of your own home.
3. Control conditions in green houses, nurseries, and closely monitor performance of sensitive crops, such as vines or tropical fruit, where the smallest amount of change in climate can affect the final outcome.
4. Determine the best conditions for each crop, by comparing the data obtained during the best harvests.

1.1.2 Why Smart Agriculture?

As per Wark et al. (2007), farming has traditionally been a labor-intensive human activity, which involves tending plants and animals on an almost discrete basis. However, after the industrial revolution, modern agriculture became mechanized and automated. This has resulted in large farms per farmer and the subsequent disappearance of small farms. For example, in the United Kingdom, 200,000 farms vanished between 1966 and 1995, and 17,000 farmers and farmworkers abandoned farming in 2003. In the United States, between 1950 and 1999, the number of farms reduced by 64%. Furthermore, the global demographic shift in farm labor and reduced employment of young workers has created an aging farming population and an imminent labor shortage.

Agricultural production is also threatened by climate change. In different parts of the world with agricultural production already on a decline, the means of coping with adversarial events are limited. Climate change is expected to reduce yields to even lower levels and make it even more erratic (Fischer et al. 2002; Stern 2006). Because of climate change, long-term changes in the patterns of temperature and precipitation are expected. These changes are likely to shift production seasons, patterns of pests, and diseases. These changes will subsequently alter the set of practicable crops affecting production, prices, incomes, and eventually, livelihoods and lives. As per Tercek (2012), between now and 2050, agriculture has to convert much less territory, increase yields on existing farms and prairies and use water and other resources more effectively. Fertilizers and pesticides have to be used in a way so as to minimize pollution. All this has to be done while adjusting to the shifting weather patterns and a more erratic climate. Clearly this strengthening of agriculture has to be sustainable. Therefore, agriculture has to become smart in order to cope with these challenges, for example, by sensor-based greenhouse cultivation, precision agriculture (Section 1.5.1).

1.1.3 Why Wireless Sensor Networks?

Wireless is not truly a new technology. Since early 1950s wireless signals have been used in transmitting AM and FM to television and radios. For decades military has been using *line of sight* micro-wave towers. A vast majority of the world knows about the most common wireless technology, that is, cellular phones. As per the MIT Technology Review (2003), Wireless Sensor Networks (WSNs) are among the 10 emerging technologies that will change the world. WSNs are being used in varied and important applications such as military, agriculture, healthcare, and industrial process monitoring. WSN is an intelligent private network consisting of a large number of sensor nodes having explicit functions. Wireless transmission allows the placement of sensors at far-flung, dangerous, and tough environments. Ideally, the WSNs have a number of advantages, such as low power consumption, cost-effectiveness, easy installation, and small footprint. Wireless sensors can be deployed almost anyplace (including underwater and underground) and that too at a far lower cost as compared to a wired system. As per Thusu (2010), because of the recent developments in wireless technology and embedded systems, the cost of hardware used in WSN has not only reduced, but has also become widely available. These devices also comply with industry standards such as the IEEE 802.15.4 for radio communication hardware and the emerging ZigBee and WirelessHART for networking of devices resulting in an increase of end-user adoption.

The rest of the chapter is organized as follows. In Section 1.2, classification of sensors is presented along with their current market, in Section 1.3, five types of WSN are discussed, in Section 1.4, some recent applications of sensor networks in smart agriculture are discussed, which is followed by Section 1.5 with brief discussion of challenges of WSNs.

1.2 WHY SENSORS?

One of the major challenges of agriculture is field data collection and consequent action. Significant time and effort is involved in data collection from a large field or plantation. For example, to analyze soil, along with the environment temperature, humidity and other field parameters also need to be recorded. Depending on the nature of crop, field trips could be required every day, or in some cases, several times per day in order to collect samples or perform site monitoring. This is not only time consuming, but also requires trained manpower, resulting in a corresponding increase in expenses. Sensors are therefore a solution of choice for agriculture data collection and monitoring, especially from the comfort of the farmer's home (Figure 1.6). As per Miskowicz (2005), over the last few years, event-based sampling and control have received special attention by the researchers from the domain of networked control systems and WSNs. This is because event-based strategies reduce the exchange of information between sensor nodes, controllers, and actuators. This reduction in exchange of information translates into extension in the battery life of the wireless sensors. Reducing the computational load of the embedded devices also reduces the network contention.

1.2.1 Sensor Classification

As there are literally hundreds and thousands of types of sensors, so are there numerous sensor classifications. For example, some sensor classifications (White 1987) are based on (1) measurement, (2) technology, (3) means of detection, (4) conversion phenomenon, (5) sensor material, and (6) field of application. There is also a sensor classification by the W3C Semantic Sensor Network Incubator Group that classifies sensors into six classes based on 69 different principles. For example, humidity can be measured using neutron probe, lysimeter, tensiometer, etc. In this chapter, agriculture sensors are classified by broadly dividing them into three classes based on their properties, which are (1) Physical, (2) Mechanical, and (3) Chemical. Further breakdown of these sensor types is shown in Figure 1.2, with some of the smart agriculture applications of these sensors discussed in Section 1.4. Note that there can also be other agriculture sensor classifications, such as (1) soil sensor, (2) plant sensor, and (3) weather sensor, and further sensor classifications can based on monitoring vs. tracking and so on; however, discussion of these classifications is beyond the scope of this chapter.

A sensor network node (or mote) usually consists of several parts. However, the main parts being (1) a radio transceiver with a connection to an external antenna or an internal antenna, (2) a CPU, (3) an electronic circuit or actuator for interfacing with the sensors, and (4) an energy source. The energy source could be a battery or an implanted energy harvester (solar, thermal, etc.). A sensor node can vary in size from that of a shoebox down to the size of a speck of dust (Park and Chou 2006). A biosensor mote is a special miniaturized analytical device, which consists of a biological or biologically derived sensing element either connected or integrated within a physicochemical transducer. There are different types of biosensors, such as electrochemical, potentiometric, amperometric, optical, to name a few. A food quality

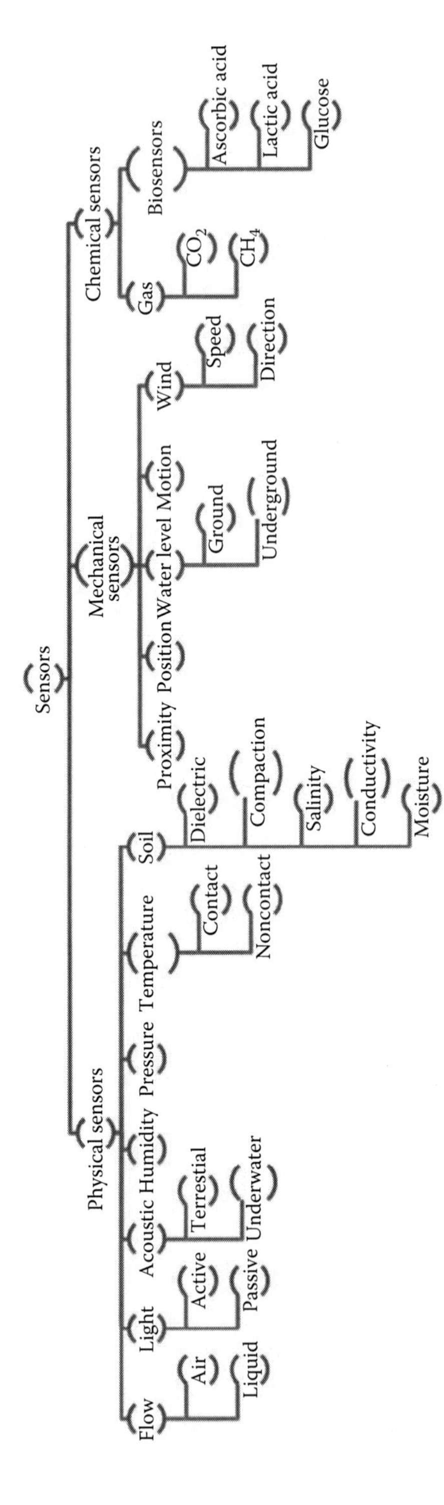

FIGURE 1.2 Nonexhaustive hierarchy of sensor classification based on properties.

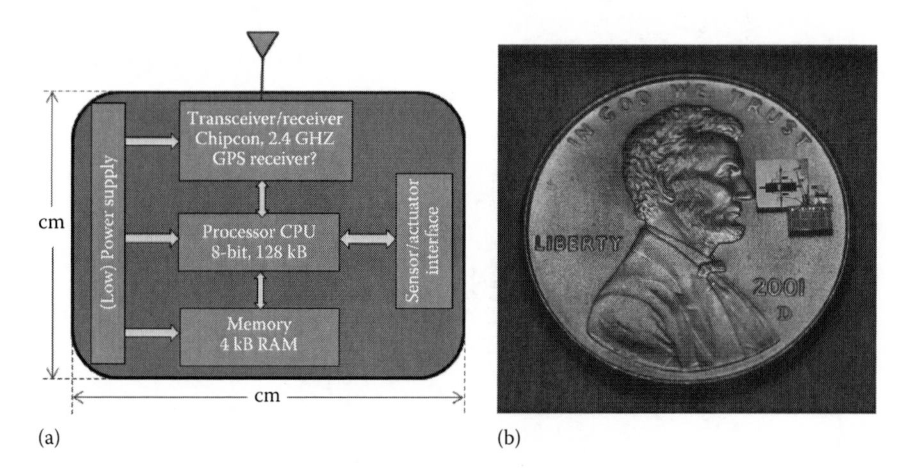

FIGURE 1.3 Wireless sensor network node: (a) sensor architecture and (b) smart dust.

biosensor is a special type of biosensor, which can react to some property or properties of food and transform the reaction(s) into measurable electric signals.

Sensors function within the network and typically fulfill one of the two roles: (1) perform data-logging, processing (and/or transmitting) sensor data collected from the environment or (2) acting as a gateway in the ad hoc wireless network formed by all the sensors. The gateway passes data back to a collection or unique data sink point. Figure 1.3a shows the architecture of a conventional wireless sensor node along with typical ratings of its different parts (Chien et al. 2011). Figure 1.3b shows a 11.7 mm^3 mock-up of Golem Dust system, showing a 0.25 μm CMOS ASIC, solar power array, accelerometer, and CCR, each on separate die (Warneke et al. 2003).

1.2.2 Sensor Market

The global environmental sensor and monitoring market was grown exponentially over the last decade (Anonymous 2013), and in 2010, it was valued at US$11.1 billion. This market is expected to reach US$15.3 billion in 2016, that is, a compound annual growth rate of 6.5% between 2011 and 2016. The market for the terrestrial category is expected to increase at a 5.3% compound annual growth rate to reach US$3.7 billion in 2016. Figure 1.4 shows the industrial wireless sensor market percent revenue for different types of sensors during 2009.

Figure 1.5 shows the adoption of WSNs. As the cost of producing WSNs is dropping, their sales are on the rise. This industry trend is expected to continue as the new technology matures and sensor applications grow.

1.3 TYPES OF SENSOR NETWORKS

The developments in the domain of sensor networks are supported by electronics miniaturization, availability of massive data storage, increase in computational capacity, and the Internet. Sensor networks allow the capability of distributed

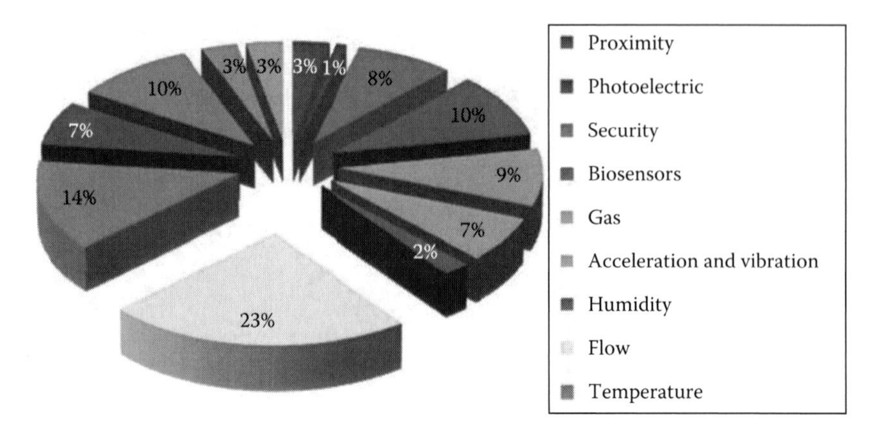

FIGURE 1.4 Industrial wireless sensor market percent revenue during 2009. (From Frost & Sullivan report by Thusu, R., Wireless sensor use is expanding in industrial applications, 2010, http://www.sensorsmag.com/networking-communications/wireless-sensor/wireless-sensor-use-is-expanding-industrial-applications-7212, accessed December 9, 2012.)

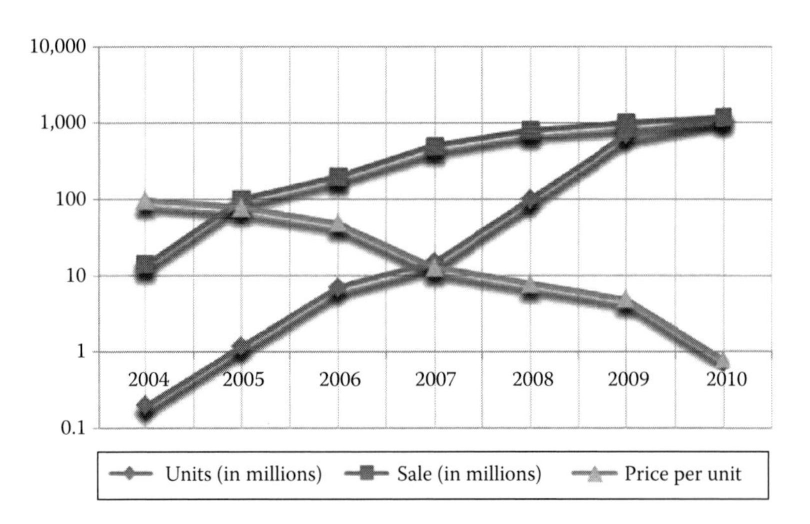

FIGURE 1.5 Wireless sensor network adoption. (Courtesy of Harbor Research, Boulder, CO.)

sensing, real-time data visualization, and analysis. Sensor networks also allow integration with adjacent networks and remote sensing data streams. WSN-based applications are diverse and involve one or a combination of different types of sensor networks. From the analysis of different WSNs around the world, five types or classes of WSNs can be identified (Meratnia et al. 2010). In this section, we will briefly discuss these different types of WSNs with reference to smart agriculture applications.

1.3.1 Environmental Sensor Networks

Environmental sensor networks (ESNs) are the forerunners of WSNs. Usually, environmental sensor networks have been exclusively deployed for data collection and monitoring. Environmental sensor networks are often static, nondense, large scale, and are deployed in harsh and unsupervised settings. ESNs are subject to harsh environmental elements that cause swift device and sensor failure. Heterogeneity of ESNs is more with regard to different types of sensor nodes (e.g., resource-limited nodes, gateways, routers) rather than types of sensors deployed. The state-of-the-art ESNs expand their single-hop communications capability to also support multihop communications and are likely to have more than one sink node. *A recent example of ESN is monitoring of dangerous gases in manure storage facilities* (*Murphy et al. 2012*)*; this is discussed in Section 1.4.*

1.3.2 Body Sensor Networks

Body Sensor Networks (BSNs) usually consist of very few wireless sensor nodes attached to a living body unified with one or more powerful personal device (e.g., smart phone). BSNs typically monitor vital signs, tracking, and data collection, this being the main purpose of using these sensor networks. Usage of BSNs has normally been centered on off-line analysis of collected data by experts and providing feedback mainly in the field of health and wellness; however, this has changed lately. The terms Body Sensor Networks (BSNs) and Body Area Network (BAN) have often been used in similar applications; sometimes interchangeably. Generally BAN refers to the network aspect of wearable sensing applications, whereas the BSN mainly refers to the infrastructure of wearable and sometimes implantable sensing applications. As compared to ESNs, BSNs are small scale, heterogeneous (in terms of different types of sensors) and require single-hop communication. Since different types of personal information can be collected by BSNs, both security and privacy are major concerns. Unlike ESNs, in static BSNs, energy consumption is not a major issue, because dependable data processing and well-timed feedback are of higher importance; however, the same is not true for mobile BSNs. *A recent example of BSNs is body temperature monitoring of farm workers to prevent thermal exhaustion* (*Kandel et al. 2012*)*; this is discussed in Section 1.4.*

1.3.3 Structure Sensor Networks

Structure sensor networks (SSNs) consist of medium to large number of wireless nodes usually installed in specific sites such as industrial locations or attached to structures (e.g., bridges) or buildings (e.g., office) or infrastructure (e.g., rails). As compared to ESNs, which are almost always deployed outdoors, SSNs may be installed both indoors and outdoors and can combine several environments simultaneously, including restricting access to buildings. In terms of security, intrinsically SSNs are often more security centric than ESNs and require protection mechanisms against attacks and corrosive effects of the surroundings. Due to their deployment complexity and difficulty and energy efficiency, long network lifetime is very important. Similar to most traditional environmental sensor networks, SSNs are usually static. Structure sensor networks may be both single and multihop (depending on the scale of the network) and are often heterogeneous (in terms of both type of sensors sensor nodes and functionality). *As mentioned in Section 1.3.2, monitoring of dangerous manure gases is one of the examples of SSN.*

1.3.4 Transport and Logistics Sensor Network

Recently numerous efforts have been focused toward wireless communication and networking between transportation vehicles such as cars, trucks, and trains. This has resulted in the development of a number of communication standards for vehicle to vehicle communication, such as IEEE 802.11p. Each individual vehicle can be considered to be a sensor node, which locally observes its own status while monitoring its surroundings too. In the context of this chapter, we are interested in Transport and logistics Sensor Network (TSN) for transport of food, that is, sensors attached to food boxes with sensors actively monitoring the transport conditions of food items and unlike passive RFID, reporting the measurements. This allows reacting to changes in the transport conditions early and appropriately (Becker et al. 2009) and ensuring food quality. For example, contrary to the currently employed First Expire First Out (FEFO) strategy that uses static best before dates, dynamic FEFO strategy takes the real best before dates into account using information acquired during storage and transportation (Becker et al. 2010). Depending on the application, TSNs may be either in the form of a network of vehicles or a combination of vehicle networks with multiple sensors attached per vehicle. *A recent example of TSN is food tagging by FoodLogiQ (www.foodlogiq.com) whereby the produce is tagged and tracked from growers to the retail outlet* (Anonymous 2014b).

1.3.5 Participatory Sensor Network or M2M

In its most basic form, M2M (Machine to Machine) involves devices that communicate independently, that is, without human intervention. Under M2M everyday objects are locatable, addressable, recognizable, readable, and controllable through the Internet (Ward 2012). In fact, M2M is now synonymous with the *Internet of things* (Anonymous 2010). Current advances in mobile technology have stretched Participatory Sensor Network (PSN) functionality to the level that making and receiving phone calls are considered rather rudimentary tasks. More and more mobile phones are now supplied with sensors (e.g., GPS, accelerometer, gyroscope, camera) and different types of connectivity mediums (bluetooth, wifi, GSM, etc.). This combination makes the mobile phone and in fact people carrying them a valuable source of gathering and transmitting data. *A recent example of M2M is Zebra Net (Zhang et al. 2004) where the Zebras were tagged with wireless sensors, and the information was used for monitoring their activities, details in Section 1.5.3.*

For each type of sensor network discussed, there are a large set of hidden conditions and functionalities offered. Although it may seem difficult to clearly distinguish

TABLE 1.1
Comparison between Different Sensor Network Types

	Covered Area			Lifetime		Mobility		Density		Diversity	
	Large	**Medium**	**Small**	**Long**	**Short**	**Mobile**	**Static**	**Low**	**High**	**Homo.**	**Hetro.**
1. ESN	✓			✓			✓	✓		✓	
2. BSN			✓		✓	✓		✓			✓
3. SSN	✓	✓	✓	✓			✓		✓		✓
4. TSN	✓	✓			✓	✓	✓		✓	✓	✓
5. PSN	✓				✓	✓		✓	✓		✓

between different types of sensor networks with reference to different types of agriculture, we will do a comparison based on some important factors such as (1) covered area, (2) network lifetime, (3) mobility, (4) density, and (5) diversity. These factors are briefly described as follows.

Network *lifetime* is possibly the most important metric for the assessment of sensor networks. Obviously, in a resource-constrained environment, the depletion of a limited resource must be taken into account. It could be impossible or problematic to recharge the battery in the remote location, and therefore, the critical requirement is to extend the network lifetime. For network *coverage* in WSNs, each sensor node attains a certain *view* of the environment. A sensor's coverage of the environment is limited by its accuracy and range; it can only cover a limited physical area of the environment with missing details being interpolated. High sensor *density* is required to assure measurement redundancy and to provide a deeper understanding of the variation of the parameter monitored, and to also monitor at the individual object level. For example, in a cattle-monitoring WSN, a sensor is attached to every animal, so sensor density is high. Sensor *diversity* is based on sensor nodes having diverse capabilities in terms of energy supply, storage space, communication, computation, reliability, and other aspects. Observe that sensors, gateway, base station also result in heterogeneity of a WSN. Sensor *mobility* is categorized by the node additions and failures as well as physical movement of the nodes. Physical mobility is caused by the controlled movement of objects (vehicles) or *random* movement of animals with sensor nodes attached to them. Table 1.1 presents a comparison between different network types using these parameters (here Homo. is Homogeneous and Hetro. is Heterogeneous).

1.4 APPLICATIONS OF SENSOR NETWORKS IN SMART AGRICULTURE

In this section we will discuss some recent applications of sensor networks in smart agriculture. The applications considered consist of some of the comparatively more established ones such as Precision Farming (PF) and Greenhouse, along with some of the more upcoming applications such as cattle net, food quality monitoring/tagging, and agriculture worker safety to name a few. A general challenge for most of the

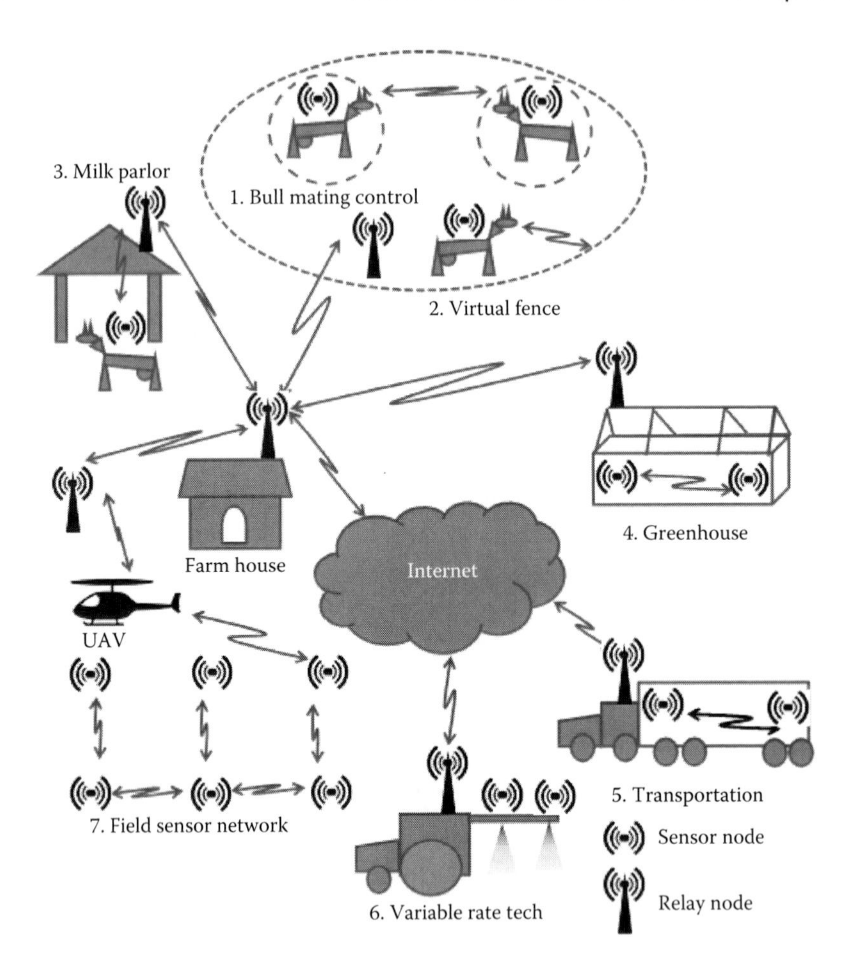

FIGURE 1.6 Wireless sensor networks in smart agriculture.

WSN applications is a requirement of interdisciplinary cooperation. Integrating IT with application domain experts is one of the potential hurdles to the growth of WSN.

Figure 1.6 shows a smart farm with multiple nested levels of WSN consisting of the relay nodes and the sensor nodes. Note that the relay nodes are used to expand the WSN coverage throughout the smart farm. The first level consists of the *wireless cloud* that connects broadband Internet to the farm along with connectivity with far-flung TSN. The second level of network is the farm-wide WSN with both monitoring and actuation nodes. The third level of nested network is a WSN at the field level where it is used to monitor the field parameters. The fourth nested level is the PSN or M2M between the cattle being reared on the farm/ranch. The fifth level is the hovering UAV (Unmanned Aerial Vehicle), which is a mobile wireless node; UAV is used to collect data from the sensors while it passes over them and subsequently transmits it to the farmhouse. The sixth and possibly the last level is the agriculture worker net without a relay node.

1.4.1 Precision Farming

Although PF started as a technology-driven development, it is not just another name for yield mapping and variable rate technology (VRT) for managing spatial variability within a field. Instead, PF integrates analysis and synthesis for crop production with the goal to reduce decision uncertainty. This is achieved by using a data-driven approach, through better understanding and management of natural variation. In PF, expertise from many disciplines is utilized, including information technology, telecommunication engineering, and GIS. Thus, the coverage of PF is diverse, which includes, but is not limited to, crop growth monitoring, global positioning systems, remote sensing, soil fertility, electronic equipment, global information systems, computer models, decision support systems, variable-rate technology, yield mapping, and accurate record-keeping. PF potentially leads us to *Push Button Agriculture.* Discussion of all applications of PF is beyond the scope of the chapter; therefore, in this section, we will only discuss a few recent applications.

1.4.1.1 Tea Farming

Tea tastes good when the tea leaves have high concentration of theanine, while it tastes bad when the tea leaves have high concentration of catechin. Theanine in new tea leaves changes to catechin through solar illumination. In Arai (2011), a tea leaf quality and quantity monitoring system along with tea tree vitality assessment has been discussed. The system uses network cameras and ASTER imagery data for the tea estates of Japan. Tea tree vitality was found to be highly correlated with the quality of new tea leaves (nitrogen content); this was measured using ground-based infrared cameras. The estimated nitrogen content was subsequently displayed on geographical map image using a GIS system. This allowed new tea leaves quality assessment using visible to near infrared radiometer data onboard remote-sensing satellites. Using the proposed method of nitrogen content and ground-based infrared camera network, it was also possible to determine the most appropriate time for harvesting new tea leaves.

1.4.1.2 Vineyard Sensors

A disease, a deficiency, pest, or other harmful agents can cause morphological or physiological changes in the vine plant. Therefore, the symptoms can be visually spotted because of the change of the green color and/or appearance of some color stains on the leaf. The proposed wireless sensor network by Lloret et al. (2011) is based on a collection of nodes capable of capturing images, processing them locally and generating corresponding responses based on its decision, that is, whether or not a bad (infected/damaged) leaf was found. This results in a distributed system, where each node makes its own decision, thus avoiding network contention and overwhelming the central server. The proposed system however was not able to distinguish between deficiency, pest, disease, or other harmful agents. Some of the problems faced were (1) misinterpretation of bad leaves with the ground because of similarity of color and (2) variation in size of the same leaves based on images taken at different distances.

1.4.1.3 Nongrain Yield Mapping

Crop yield maps are a valuable means to quantify yield variations within the field, and to provide a scientific basis for implementing site-specific crop management strategies. As per Maharlouie et al. (2012), cereal grain yield monitors are well established. Experimental methods developed by researchers have been commercialized, due to the availability of proven combined grain yield monitors. However, up to now, limited commercial methods have been developed for nongrain crops. This is mainly due to different methods of harvesting and significant variations in the physical and mechanical properties of the crops. In Maharlouie et al. (2012), the feasibility of developing a new yield monitoring system is explored that uses a mass flow sensor. The sensor utilized the momentum of chopped material impacting a spring loaded pivoted plate; the plate was installed at the end of a chopper discharge spout.

1.4.1.4 Water Conservation in Paddy Field

In a paddy crop field, the land under cultivation has to be fully watered. After some period of farming, it is not required to fully irrigate the field. Thus, subsequent irrigation depends on the soil type, the contours or the level of the land and as per need basis. Therefore, using wireless sensor network, the paddy crop field area can be monitored with little or no manpower (Nirmal Kumar et al. 2011). Nirmal Kumar et al. (2011) proposed a ZigBee wireless sensor network that was used for monitoring the crop field area by deploying water sensors in the land to detect and identify regions of low water level. Humidity sensors were used to sense the weather; from this, the farmer could get an idea about the climate. If there was any chance of rainfall, the farmer need not irrigate the field.

1.4.2 Greenhouse

Historically farmers have experienced huge financial losses because of wrong weather predictions and incorrect irrigation timings (Chaudhary et al. 2011). As per the FAO (2011) Report, climate change is expected to further aggravate the existing challenges that agriculture faces. In this context, with the evolution in wireless sensor technologies and miniaturized sensor devices, it is possible to use WSN for automatic environment monitoring and controlling the parameters in a confined environment, that is, a greenhouse. Greenhouses are an upcoming technology in PF. Rich sensor applications in greenhouse (Figure 1.7) help farmers grow a high quality crop or multiple cycles of a crop in a controlled and closely monitored environment. The main objective is to monitor and efficiently control the green house environment as per the crop requirements. Multitude of sensors and actuators are used, which are reconfigurable as per the stages of crop growth and take into consideration dynamic changes in the targeted area, soil temperature, climate, season, and type of crop. Some of the climatic control variables affecting a greenhouse are shown in Figure 1.7.

Smart WSN applications are expected to control acuter like pump, valve, carton slider, and fans, etc. (Chaudhary et al. 2011). The greenhouse control can be achieved through WSN via Ethernet connected to the central PC of a remote network. Bluetooth technology can be used for some applications to collect environmental data from a

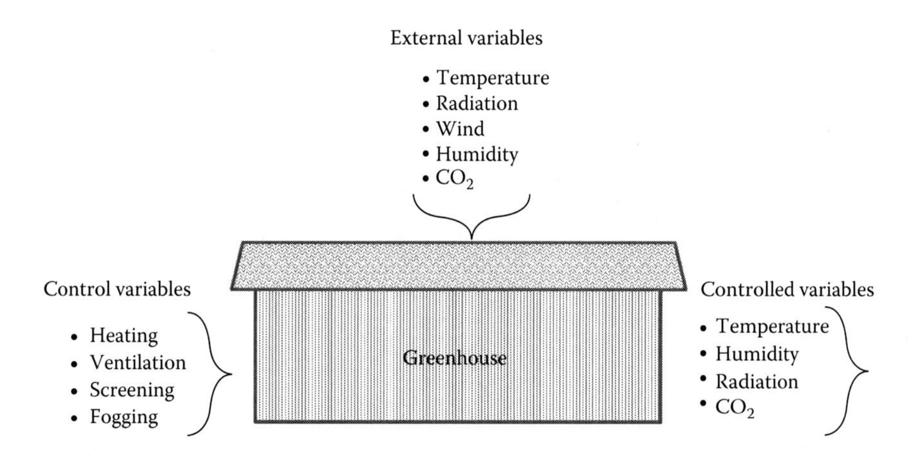

FIGURE 1.7 Climatic control variables in a greenhouse.

sensor network and transmit to a central control system. This type of remote control technology significantly improves productivity and reduces the labor cost.

1.4.2.1 Dew Control

Crop growth in a greenhouse is effected by dew condensation on the leaf surface of crops that can stimulate diseases caused by bacteria and fungus. Park and Park (2011) have presented a WSN-based automatic monitoring system solution that prevents dew condensation in a greenhouse environment. The system consisted of sensor nodes for collecting data, base nodes for processing collected data, relay nodes for driving devices for adjusting the environment inside the greenhouse, and an environment server for data storage and processing. Barenbrug formula was used for calculating the dew point on the leaves; the system was able to prevent dew condensation phenomena on the crop's surface for prevention of diseases.

1.4.2.2 Adjusting Shipping Timing

Lee et al. (2010) have discussed a *Paprika Green House System* (PGHS) that consists of a network of sensors measuring temperature, humidity, illuminance, and other relevant parameters. The PGHS also controls ventilators, humidifiers, lightings, and video-processing through a GUI application; this is achieved by analyzing the measured data. The system also maintains the best environment in the cultivation facility by using biometric data and creates optimum conditions at paprika root zone. The system optimizes the management of production elements and reduces energy, fertilizer, and water loss resulting in the reduction of production costs. To overcome the issue of separate conversion/control module for each sensor characteristic, an integrated sensor module was developed. This module integrates various sensors required for cultivation of the crop into a single node. LED used in PGHS prevented vermin and adjusted the growth rate and crop coloring so that shipping timings could be adjusted. The *greenhouse environment integrated management system* enabled the monitoring of PGHS in real-time through the Internet.

1.4.3 Cattle Monitoring

Usually a farmer relies on a combination of visual observation, experience, and instinct while making management decisions about the herd, but these data-intensive decisions made on intuition are likely to be far from optimal and also labor intensive. Agricultural community has a continued interest of using sensor networks to monitor and control livestock. The use of sensor networks with livestock allows farmers to monitor and control the herd even when the farmer is far from the field. This monitoring and control can help detect illness-related behavior, enhance land management, and develop animal behavior models. Mobile sensor network nodes can thus be used to influence and monitor animals' positions in the pastures and grasslands (Wark et al. 2007).

1.4.3.1 ZebraNet

One of the first major applications of WSN for animal monitoring was tracking of zebras reported by Zhang et al. (2004) as part of the ZebraNet project. ZebraNet used the zebra's GPS position data that was taken every few minutes. This data was subsequently hopped in a peer-to-peer fashion (as in PSN) to other zebras when they came in range. Subject to the amount of storage capacity of each device, a user could then download historical position data of multiple zebras by approaching a single zebra. Researchers have since proposed advanced systems for ad hoc routing of data through large networks of mobile cattle nodes (Radenkovic and Wietrzyk 2006).

1.4.3.2 Virtual Fence

Both Butler et al. (2004) and Lee et al. (2008) have used similar stimuli but achieved different goals using different algorithms. Butler et al. (2004) developed a *moving virtual fence* algorithm for herding cows. Each animal in the herd was given a smart collar consisting of a sound amplifier, a PDA, a GPS, and a radio unit (WLAN). The animal's location was determined using the GPS and confirmed through a measurement of vicinity of the cow relative to the fence boundary. When approaching the perimeter, the animal was given auditory stimuli, which pushed the animal away from the fence. Lee et al. (2008) have examined the potential of controlling the bulls during mating using mild electric shocks delivered through radio-controlled collars. Cows were assigned to the bulls; the nonassigned bull received a mild electric shock on approaching either the unassigned cow or another bull. Nonassigned bulls were sometimes observed avoiding the cow despite a change in its location. This suggests that the bull associated the electric shock with the cow and not with the location in which they received the electric shock.

1.4.3.3 Remote Temperature Monitoring

The US dairy industry is merging rapidly. Today fewer dairy operations are family operated, therefore in larger, more mechanized dairies the attention and care given to each dairy cow has consequently reduced. As a result, mortality rates and production loss from sickness are on the rise in these large dairy operations (Anonymous 2014c). A possible solution to these problems is the Dairy Monitoring System (Anonymous 2014c) that consists of (1) a bolus, equipped with a passive RFID chip, an embedded temperature sensor and (2) a reader for collecting temperature data from the bolus. Once the bolus is orally administered, it settles into one of the compartments of the cow's stomachs. The density of the bolus prevents it from exiting the digestive system. As the dairy cow passes between two panels fixed at the entry to the milking parlor, a small electrical charge is induced inside the bolus by a magnetic field. This charge is sufficient to energize the transmitter. The bolus responds to this query signal by transmitting its unique ID and temperature on a coded radio frequency. As the dairy cow enters the milking parlor two to three times per day, the system automatically monitors each cow's temperature and ID. This allows the dairy operator to receive a 24–48 h advance notice on any outbreaks or illnesses before visible signs appear, thus reducing treatment costs and mortality rates.

1.4.3.4 Issue of Big Data

According to Wang et al. (2006), unless the structure and processes are in place to take advantage of the massive amount of data generated by the wireless sensors, the data has all the potential to overwhelm the system while providing limited value. For example, a Dutch start-up called Sparked has used the M2M technology for monitoring the health of livestock. The vital signs of the animal are measured by a sensor implanted in a cow's ear. On average, each cow generates about 200 MB a year (Anonymous 2010). As per the latest figures of the US Department of Agriculture, as of January 1, 2012, there were about 91 million cattle and calves in the United States. Just a million of those cattle tagged with sensors would generate 200 TB in a year. Thus, there is a need for a Data Warehouse for managing and utilizing this vast amount of data for running ad hoc queries and performing data mining for knowledge discovery.

1.4.4 Safety of Agriculture Workers

Agriculture is considered to be one of the most harmful industries for youth workers in the United States. Recent regulations and training programs have facilitated the safety of youth workers, limiting the amount of injuries to less than 1% of the workers. However, considering the strength of youth workforce to be over 1,000,000 persons in farming alone, this apparently small percentile correlates with the deaths of 695 and injuries of 16,851 youth workers (Hill et al. 2000).

1.4.4.1 Monitoring Worker Position and Temperature

Because of insufficient training and little or no experience when first starting a new job, other than environmental dangers, young workers are also at a high risk of injury from equipment and runovers. Hyperthermia or heat stress, is defined as the condition when an individual's core body temperature spirals above the body's ideal operating temperature. As per Bouchama and Knochel (2002), this is also called the hypothalamic set point and results in severe central nervous system dysfunction. Kandel et al. (2012) have discussed the working of a temperature sensor circuit consisting of thermopile and thermistor that accurately estimated the worker core body temperature. XBee radio frequency modules were used to transmit information wirelessly via a portable easy-to-use device inserted in the worker's ear and communicating via the ZigBee communication protocol. Using wireless transmission properties, the Received Signal Strength Indication (RSSI) was utilized to approximate the distance between the signal receiving and transmitting devices. The accuracy of the worker position provided was comparable to that of GPS methods and that too without the line of sight to sky requirement.

1.4.4.2 Manure Storage Gas

Another hazardous work area for farm workers is confined space where manure is stored. For working in such an environment, gas detection equipment needs to be used so as to determine concentrations of harmful gases and oxygen levels. This should be done prior to and during entry in the premises, as extremely toxic gases often accumulate from decomposing manure and also have unsafe oxygen levels. A gas monitor with remote sampling enables measurements to be taken by workers located safely outside the storage facility. These measurements can also be used to establish air circulation times before workers enter the manure storage premises. Different types of gas sensors are available in the market. For example, those with one-time use based on changing color of the badges worn by workers costing about $140 for a box of 10. At the other end are the fixed sensors that can also monitor oxygen levels costing $2000 to $4000+ (Murphy et al. 2012).

1.4.5 Food Quality Monitoring and Tagging

The selection of a particular foodstuff by a consumer is influenced by many sensory observations, one of them being taste. Taste is affected by different factors including sweetness, saltiness, bitterness, and acidity, these being the most important factors. Texture is another key sensory observation and is influenced by many factors including moisture content, carbohydrate, and protein levels (Szczesniak 2006) in the foodstuff. Other important sensory observations include the shape, scent, and color of the foodstuff. Therefore, rapid, portable, and accurate sensors are required for automatic assessment of food quality and physiological state. Due to their numerous attributes, biosensors potentially offer a fast, accurate, relatively cheap, portable, stable, and user-friendly mechanism for on spot monitoring of fruit maturity and quality (Rana et al. 2010).

1.4.5.1 Freshness and Quality Monitoring

As per Kriz et al. (2002), monitoring of fruit quality is one of the major worries within the food industry. There is a growing need to develop analytical devices, which can provide quality monitoring for the entire food processing operation, that is, starting from the materials to the end products (Whitaker 1993). Sensors incorporated into food packages can potentially benefit consumers, by guaranteeing freshness and quality, and at the same time allowing retail industry to more effectively manage food stocks and product genuineness.

Meats and fish are prone to rapid growth of bacteria. Bacterial growth on meats and fish typically results in the release of nitrogen- and sulfur-containing compound, accompanied by strong and foul smell. Smits et al. (2012) used smart radio-frequency labels with sensors enabling measurement of temperature, humidity, and the presence of volatile amine compounds. The labels were made by means of high quality screen printing on low cost foils using lamination technologies. This was combined with pick and place technology. For data processing, commercially available cost-effective MSP430G microprocessors from Texas Instruments were used.

1.4.5.2 Hay Moisture Monitoring

Hay harvesting is a precise process with a significant bearing on the success of a farm's feeding program. Hay is used for the nourishment of livestock; the harvested hay to be of top quality so that it fetches a good price too. Hay is at its best through a particular week-long period during its maturation. Beyond this point, it becomes coarse and dry losing most of its nutritional value. Tracking the moisture levels in hay bales is one of the most important activities of the entire hay baling process. Any bale harvested below 12% moisture content has experienced field losses, resulting in low feeding value and yield losses. A bale with moisture content above 20% is at the risk of spontaneous combustion due to increased heat levels during the respiration process (www.thingmagic.com).

In the smart solution (www.thingmagic.com), a thin RFID tag is attached to one of the coils holding the hay together. On this small tag, all essential data is stored. With an RFID reader, anyone can check the field of origin of the hay with exact location of harvest, harvest date, average and high moisture levels, temperature, weight, amount of preservatives used, and a unique ID number. With this useful information, farmers can dispense hay to their livestock and ensure its consistent quality. The remaining hay can be sold for a good price since the buyer is aware of the exact nutritional value of each bale. Bales with too much moisture are removed from a stack to avoid contamination of the remaining bales. These bales are subsequently used to feed cows or put to other uses where high quality is not a priority.

1.5 CHALLENGES

Adoption, utility, and applications of WSNs in agriculture are not without a multitude of challenges and the requirement of addressing difficult research problems. Some of those challenges are mentioned in this section.

1.5.1 Battery Life

The biggest problem faced by WSN is energy. When a sensor is drained of energy, it can no longer accomplish its role unless the source of energy is replenished. Therefore, it is normally accepted that a wireless sensor dies when its battery runs out. Even when not in use, portable energy sources like batteries will experience current leakages that ultimately drain the resource; furthermore, any defects in the packaging due to long-term wear and tear can result in environmental issues (Seah et al. 2009). Therefore, it is urgently needed to increase the battery life or decrease its discharge rate. One possible solution being pursued is energy harvesting. *Energy harvesters* are small devices that take ambient energy and convert that into electrical energy to power the wireless sensor. The aim is a 20-year lifetime or more. Photovoltaics are most commonly used, but there is work underway on other types of energy harvesters too, such as piezoelectrics (harvesting energy from vibrations), electrodynamics (similar to bicycle dynamo), thermoelectrics (harvesting energy from a heat gradient), and more (Das 2011). However, these are not the major sources for harvesting energy in a typical agriculture environment.

1.5.2 Cost per Unit

A major obstacle to wider embracing of WSNs is the cost of motes. At $99–$300 apiece (Madan and Reddy 2012), motes are currently too costly for many of the applications its inventors envisioned, such as extensive use in agriculture. One idea calls for wireless sensors that you *peel, stick and forget* (Smits et al. 2012). The radio frequency identification (RFID) tag industry possibly has reached a cost as low as about $0.20 per tag and seeks to reach in a decade, the price of $0.05 per tag for inventory tracking purposes (Homs et al. 2004).

1.5.3 Signal Attenuation

Radio waves are attenuated while transmitted from the sender to the receiver. The degree of attenuation is dependent on the medium between the receiver and the transmitter. According to Becker et al. (2009), in case of food, major attenuation occurs because of water, as water constitutes main part of the food having propagation attenuation worse than air. According to Gabriel et al. (1996), attenuation is also caused by the animal body, which is highest for the fat and lowest for the blood. Consider the case of using wireless sensors for monitoring cattle health, since the cattle are generally fed in herd, this results in massive increase in the surface area, which seriously affects the radio signals. The sensor communication system is required to minimizing the impact of radio attenuation through food and animal body.

1.5.4 Authenticity and Security

Severe environmental operating conditions, hazards of physical compromise and unpredictable data transfer rates are some of the challenges for WSN. WSNs are usually deployed in an unattended environment, thus the sensors could be compromised for malicious reasons, such as intentional falsification of sensor data (Yu et al. 2010). In such a case, garbage in would obviously result in garbage out. The traditional solutions of data authenticity and confidentiality are there; however, due to the resource constraints of the sensors, the existing protocols cannot be used for WSN. There are other constraints and obstacles too that need to be resolved while designing a security protocol for WSNs. Some of these constraints being limited memory, storage, and processing capabilities, these limitations differentiate WSN security architecture design requirements from traditional network design requirements.

REFERENCES

Anonymous. 2010. Augmented business. http://www.economist.com/node/17388392, accessed February 10, 2014.

Anonymous. 2012. 50 Sensor Applications for a Smarter World. http://www.libelium.com/top_50_iot_sensor_applications_ranking/, accessed November 15, 2012.

Anonymous. 2013. Environmental Sensing and Monitoring Technologies: Global Markets. http://www.bccresearch.com/market-research/instrumentation-and-sensors/environmental-sensing-monitoring-technologies-ias030b.html, accessed November 10, 2013.

Anonymous. 2014a. Population change and structure. http://www.bbc.co.uk/schools/gcsebitesize/geography/population/population_change_structure_rev1.shtml, accessed February 10, 2014.

Anonymous. 2014b. Produce Traceability Solutions. https://pti.foodlogiq.com/, accessed February 10, 2014.

Anonymous. 2014c. Animal health: Automatic ID and cow temperature monitoring. http://www.phaseivengr.com/Solutions/WirelessSensingSolutionsInDepth/AnimalHealthandIdentification.aspx#sthash.nZQCFSAg.dpuf, accessed February 10, 2014.

Arai, K. 2011. Wireless sensor network for tea estate monitoring in complementally usage with satellite imagery data based on Geographic Information System (GIS). *International Journal of Ubiquitous Computing (IJUC)*, 1(2), 12–21.

Becker, M., Wenning, B.-L., Görg, C., Jedermann, R., and Timm-Giel, A. 2010. Logistic applications with wireless sensor networks. In *Proceedings of the Sixth Workshop on Hot Topics in Embedded Networked Sensors*, Killarney, Ireland, pp. 6–10.

Becker, M., Yuan, S., Jedermann, R., Timm-Giel, A., Lang, W., and Görg, C. 2009. Challenges of applying wireless sensor networks in logistics. In *Proceedings of CEWIT 2009. Wireless and IT Driving Healthcare, Energy and Infrastructure Transformation*, Marriott Islandia, NY.

Bouchama, A. and Knochel, J. P. 2002. Heat stroke. *New England Journal of Medicine*, 346(25), 1978–1988.

Bruinsma, J. 2009. The resource outlook to 2050: By how much do land, water and crop yields need to increase by 2050? Paper presented at the *FAO Expert Meeting on How to Feed the World in 2050*, June 24–26, 2009, Rome, Italy. Rome, Italy: Food and Agriculture Organization of the United Nations, Economic and Social Development Department.

Burney, J. A., Davis, S. J., and Lobell, D. B. 2010. Greenhouse gas mitigation by agricultural intensification. *Proceedings of the National Academy of Sciences*, 107(26), 12052–12057.

Butler, Z., Corke, P., Peterson, R., and Rus, D. 2004. Virtual fences for controlling cows. In *Proceedings of the 2004 IEEE International Conference on Robotics and Automation (ICRA'04)*, New Orleans, LA, Vol. 5, pp. 4429–4436. IEEE.

By the Editors of Technology Review. 2003. 10 Emerging technologies that will change the world. *Technology Review Magazine (MIT)*. http://www2.technologyreview.com/featured-story/401775/10-emerging-technologies-that-will-change-the/2/, accessed February 10, 2014.

Chaudhary, D. D., Nayse, S. P., and Waghmare, L. M. 2011. Application of wireless sensor networks for greenhouse parameter control in precision agriculture. *International Journal of Wireless & Mobile Networks (IJWMN)*, 3(1), 140–149.

Chien, T. V., Chan, H. N., and Huu, T. N. 2011. A comparative study on hardware platforms for wireless sensor networks. *International Journal on Advanced Science, Engineering and Information Technology*, 2(1), 70–74.

Das, R. 2011. Wireless sensor networks: The challenges and opportunities. http://www.mpdigest.com/issue/Articles/2011/sept/idtech/Default.asp, accessed February 10, 2014.

Fischer, G., Shah, M., and van Velthuizen, H. 2002. *Climate Change and Agricultural Vulnerability*. Laxenburg, Austria: International Institute for Applied Systems Analysis (IIASA).

Gabriel, S., Lau, R. W., and Gabriel, C. 1996. The dielectric properties of biological tissues: III. Parametric models for the dielectric spectrum of tissues. *Physics in Medicine and Biology*, 41(11), 2271.

Hill, J., Szewczyk, R., Woo, A., Hollar, S., Culler, D., and Pister, K. 2000. System architecture directions for networked sensors. In *Proceedings of the Ninth International Conference on Architectural Support for Programming Languages and Operating Systems*, Cambridge, MA. ACM Press, New York, pp. 93–104.

Homs, C., Metcalfe, D., and Takahashi, S. 2004. *Exposing the Myth of the 5-Cent RFID Tag*. Cambridge, MA: Forrester Research, Inc.

Kandel, M. K., Grisso, R., Diller, T. E., and Wicks, A. L. 2012. Monitor system to detect heat stress and position of youth lawn care workers. *Journal of Mechatronics*, 1, 1–9.

Kriz, K., Kraft, L., Krook, M., and Kriz, D. 2002. Amperometric determination of L-lactate based on entrapment of lactate oxidase on a transducer surface with a semi-permeable membrane using a sire technology based biosensor. Application: Tomato paste and baby food. *Journal of Agricultural and Food Chemistry*, 50(12), 3419–3424.

Lee, C., Prayaga, K. C., Fisher, A. D., and Henshall, J. M. 2008. Behavioral aspects of electronic bull separation and mate allocation in multiple-sire mating paddocks. *Journal of Animal Science*, 86(7), 1690–1696.

Lee, J.-W., Shin, C., and Yoe, H. 2010. An implementation of paprika green house system using wireless sensor networks. *International Journal of Smart Home*, 4(3), 57–68.

Lloret, J., Bosch, I., Sendra, S., and Serrano, A. 2011. A wireless sensor network for vineyard monitoring that uses image processing. *Sensors*, 11(6), 6165–6196.

Madan, V. and Reddy, S. R. N. 2012. Review of wireless sensor mote platforms. *VSRD-IJEECE*, 2(2), 50–55.

Maharlouie, M. M., Loghavi, M., and Kamgar, S. 2012. Feasibility study of a pivoted-plate sensor for silage corn yield monitoring. *International Journal of Agriculture Sciences*, 4(2), 190–195.

Meratnia, N., Van der Zwaag, B. J., Van Dijk, H. W., Bijwaard, D. J., and Havinga, P. J. 2010. Sensor networks in the low lands. *Sensors*, 10(9), 8504–8525.

Miskowicz, M. September 2005. Sampling of signals in energy domain. In *10th IEEE Conference on Emerging Technologies and Factory Automation (ETFA'2005)*, Catania, Italy, Vol. 1, 4pp. IEEE.

Murphy, D. J., Manbeck, H. B., and Steel, J. S. 2012. Reducing risk of entry into confined space manure storages. In *Proceedings of Got Manure? Conference & Tradeshow*, pp. 200–208.

Nirmal Kumar, K., Ranjith, P., and Prabakaran, R. 2011. Real time paddy crop field monitoring using Zigbee network. In *2011 International Conference on Emerging Trends in Electrical and Computer Technology (ICETECT)*, Nagercoil, India, pp. 1136–1140. IEEE.

Park, C. and Chou, P. H. 2006. Eco: Ultra-wearable and expandable wireless sensor platform. In *International Workshop on Wearable and Implantable Body Sensor Networks (BSN'06)*, Cambridge, MA, pp. 162–165.

Park, D. H. and Park, J. W. 2011. Wireless sensor network-based greenhouse environment monitoring and automatic control system for dew condensation prevention. *Sensors*, 11(4), 3640–3651.

Radenkovic, M. and Wietrzyk, B. 2006. Wireless mobile ad-hoc sensor networks for very large scale cattle monitoring. In *Proceedings of the Sixth International Workshop on Applications and Services in Wireless Networks (ASWN'06)*, Berlin, Germany, pp. 47–58.

Rana, J. S., Jindal, J., Beniwal, V., and Chhokar, V. 2010. Utility biosensors for applications in agriculture—A review. *Journal of American Science*, 6(9), 353–375.

Seah, W. K., Eu, Z. A., and Tan, H. P. 2009. Wireless sensor networks powered by ambient energy harvesting (WSN-HEAP)—Survey and challenges. In *1st International Conference on Wireless Communication, Vehicular Technology, Information Theory and Aerospace & Electronic Systems Technology, 2009 (Wireless VITAE 2009)*, Aalborg, Denmark, pp. 1–5. IEEE.

Semantic Sensor Network XG Final Report, W3C Incubator Group Report June 28, 2011. http://www.w3.org/2005/Incubator/ssn/wiki/Agriculture_Meteorology_Sensor_Network, accessed February 10, 2014.

Smits, E., Schram, J., Nagelkerke, M., Kusters, R., van Heck, G., van Acht, V., and Gerlinck, G. 2012. *Development of Printed RFID Sensor Tags for Smart Food Packaging*. In *IMCS 2012—The 14th International Meeting on Chemical Sensors*, Tagungsband, pp. 403–406.

Stern, N. H. 2006. *Stern Review: The Economics of Climate Change*, Vol. 30. London, U.K.: HM Treasury.

Szczesniak, A. S. 2006. Classification of textural characteristics. *Journal of Food Science*, 28(4), 385–389.

Tercek, M. 2012. Feeding the world through smarter agriculture. http://www.huffingtonpost.com/mark-tercek/world-hunger_b_1459961.html, accessed February 10, 2014.

Thusu, R. 2010. Wireless sensor use is expanding in industrial applications. http://www.sensorsmag.com/networking-communications/wireless-sensor/wireless-sensor-use-is-expanding-industrial-applications-7212, accessed December 9, 2012.

Wang, N., Zhang, N., and Wang, M. 2006. Wireless sensors in agriculture and food industry—Recent development and future perspective. *Computers and Electronics in Agriculture*, 50(1), 1–14.

Ward, J. 2012. The Promise of M2M. http://www.ntca.org/new-edge/epapers/the-promise-of-m2m, accessed February 10, 2014.

Wark, T., Corke, P., Sikka, P., Klingbeil, L., Guo, Y., Crossman, C., Valencia, P., Swain, D., and Bishop-Hurley, G. 2007. Transforming agriculture through pervasive wireless sensor networks. *Pervasive Computing, IEEE*, 6(2), 50–57.

Warneke, B., Scott, M., and Leibowitz, B., and Pister, K. S. J., 2003. Ultra-low energy circuits for distributed sensor networks (smart dust). http://www.eecs.berkeley.edu/XRG/Summary/Old.summaries/03abstracts/abstracts.KSP.html, accessed February 10, 2014.

Whitaker, J. R. 1993. The need for biosensors in the food industry and food research. In *Food Science and Technology*, Wagner, G. and Guilbault, G. G. (eds.). New York: Marcel Dekker, pp. 13–30.

White, R. M. 1987. A sensor classification scheme. *IEEE Transactions on Ultrasonics, Ferroelectrics and Frequency Control*, 34(2), 124–126.

Yu, C. M., Chen, C. Y., Lu, C. S., Kuo, S. Y., and Chao, H. C. 2010. Acquiring authentic data in unattended wireless sensor networks. *Sensors*, 10(4), 2770–2792.

Zhang, P., Sadler, C. M., Lyon, S. A., and Martonosi, M. 2004. Hardware design experiences in ZebraNet. In *Proceedings of the Second International Conference on Embedded Networked Sensor Systems*, Baltimore, MD, pp. 227–238.

2 Wireless Sensor Networks with Dynamic Nodes for Water and Crop Health Management

el-Hadi M. Aggoune, Sami S. Alwakeel, Mohammed M. Alwakeel, and Mohammad Ammad-Uddin

CONTENTS

ABSTRACT

Recent advances in microelectronic and microelectromechanical systems have produced new battery-powered sensor devices that have capabilities for detecting and processing physical information. These devices (nodes) can be connected to form a wireless sensor network (WSN) that performs a variety of operations. WSNs provide sensing accuracy and fault tolerance and can be deployed in harsh environments to provide continuous monitoring and processing capabilities.

WSNs collect various types of data from a monitored area. Depending on the application, parameters sensed may include moisture, temperature, nutrients, and pollutants. Sensed information is carried over multihop from node to node to a base station (BS) for further processing and action taking.

Given the numerous benefits WSNs offer, a case study was developed for their potential implementation in the farming sector in Saudi Arabia. Water utilization in Saudi is very critical as there is little permanent storage for it such as reservoirs or dams. At the same time, the Saudi land is fertile and has the potential to produce both quantity and quality crops such as wheat, dates, fruits, vegetables, flowers, and alfalfa. This case study focuses on WSNs to control water used for irrigation as well as for monitoring the quantity and quality of crops.

This study is motivated by both the lack of water and the premise that for the majority of crops, an excess of water may have an equally negative effect as does a deficit. Hence, the need for a technology as an aid to optimally dispense the appropriate amount of water for optimal crop quantity and quality. It is believed that WSNs provide an answer.

2.1 INTRODUCTION TO WSN

Wireless sensor networks (WSNs) consist of a large number of nodes [1,2] that are connected in the form of a wireless network. Each node consists of devices that sense physical parameters, process information, and broadcast messages to each other or to a BS [3,4]. Sensor nodes are characterized by their small size, limited battery power, low data rate transmission, and limited memory capacity. The data transmission occurs over a transmission channel that is characterized by a low data rate and a relatively short transmission range.

Proper usage of WSNs requires a deployment plan. Upon deployment, WSNs can operate with very little or no human supervision. WSNs are usually deployed in remote farms where weather and geographical conditions vary significantly. Sensor deployment problems have been investigated in a variety of applications, and many solutions have been proposed [5–7].

Node energy. An important consideration in WSN farming applications is the node battery life and storage capacity to support the computational and communication capabilities. Recently, research and development efforts have focused on devising techniques for increasing network lifetime through increasing battery lifetime using a variety of energy-harvesting techniques.

Memory and storage. The limited memory and storage capacity are caused mainly by the small physical size and low cost of the node. This, however, is addressed to a certain extent by networking the nodes and using a BS.

Routing protocols. When a sensor node initiates a message exchange with the BS in a large area, a routing operation is necessary [2,3]. The conventional on-demand routing protocols perform a network-wide flooding to discover routes. Flooding-based route discovery, however, causes a large routing overhead and high battery power consumption [2–4]. As a result, limiting the number of transient nodes (number of hops) that are involved in data transmission toward the BS limits the number of transmitted packets, which in turn reduces power consumption and thus increases the network lifetime.

Fault tolerance. Nodes in WSNs are prone to failure due to, among other things, energy depletion, hardware failure, or communication link errors. Fault tolerance is addressed through node placement, topology control, event detection, and data gathering and aggregation.

Scalability. Scalability deals with whether the information from the WSN remains accurate in the face of a growing number of nodes. Work in recent years has led to reliable protocols for WSN deployment, data routing, and congestion control to address the issue of scalability.

In the following section, Section 2.2, a case study of the Saudi agriculture will be laid out; in Section 2.3, the integration of WSNs and unmanned aerial vehicles (UAVs) will be proposed and explained; in Section 2.4, agricultural parcels are defined; in Section 2.5, the implantation of WSNs is discussed; in Section 2.6, major WSN hardware considerations are listed; in Section 2.7, efficient irrigation is discussed; in Section 2.8, conclusions and recommendations are offered; and finally, acknowledgements are given.

2.2 SAUDI ARABIA AGRICULTURE: CASE STUDY

The total area of the Kingdom of Saudi Arabia (Saudi) is 2,149,690 km^2. About 1.6% of it is urban area, and about 80% (1,736,250 km^2) is desert of which only 1.6% is arable land [8,9]. The biggest hurdles for cultivation are shortage of water, spread of land, and adverse weather and atmospheric conditions.

Saudi is a desert country with virtually no permanent rivers or lakes and with only limited bursts of rainfall during a typical year. Actually, the rain bursts, at times, cause far more damage than benefit. Additionally, there is an ever-increasing demand for water to suit the population of a typical fast-developing country in terms of construction, industry, and lifestyle [10–12].

Crops are grown in dispersed circular- or rectangular-shaped parcels of land having limited water resources and exposed to harsh environmental conditions including excessive heat or cold and sandstorms. Furthermore, the farming parcels have limited or no communication infrastructures. Most common crops include dates, seasonal fruits and vegetables, olives, wheat, and alfalfa. It is worth mentioning that wheat growing is receding because of its water requirements.

There are 200 dams that can collect an estimated 450 MCM of runoff annually. Some of the famous dams in Saudi are shown in Figure 2.1. There are also 25 sea water desalination plants. Some of the desalination plants are shown in Figure 2.2. Water demand and supply trends are shown in Figure 2.3. In 2010, the total water demand was 20,000 million cubic meters (MCM) while the total water available from the different resources was 20,100 MCM. There is a narrow difference suggesting a low reserve margin warranting a state of alert.

To cope with the scarcity of water, there is a need to equip the agricultural sector with modern tools and scientific approaches to agriculture and farming such as those suggested by the fast-developing precision agriculture and smart agriculture that rely on WSNs to achieve sustainability. More recently, with the advent of unmanned vehicles and the accompanying progress in research and development in ad hoc and vehicular communication, WSNs are positioned to gain further functionality as some of the nodes can become dynamic (carried by UAVs) facilitating both data

FIGURE 2.1 Dams in Kingdom of Saudi Arabia (KSA). Bisha Dam in Bisha city is the biggest dam in Saudi. It has a capacity of 325 MCM. Bisha city is 1002 km from Riyadh.

FIGURE 2.2 Desalination plants in Saudi. Jubail water desalination plant in Jubail city is the world largest plant in the world. Jubail is an industrial zone located 481 km from the capital Riyadh. It supplies 70% of the country's drinking water.

collection and wireless communication in areas that are not equipped with fixed communication infrastructures.

Sensors are normally planted in strategic locations forming disjoint networks and subnetworks in individual parcels. Sensors and actuators are placed where they are most effective depending on the information to be collected. For example, sensors can be placed on roots, trunk, branches, twigs, leaves, and even buds of trees. Some of the sensors can be coupled with actuators that control water valves, pesticide dispensers, sound-emitting devices to repel bugs, and even cameras. The objective is to collect and store vital data and information relative to the environment and crop health, allowing farmers to make informed decisions.

The data can be collected from the individual networks using unmanned vehicles that have the ability to loiter and hover at certain collection points. Data collected by the UAVs are then redirected to a supervisory control and data acquisition system (SCADA) residing in a control center. The control center serves as a hub for data and a hanger for UAV maintenance, including battery recharging and trajectory planning software upgrades.

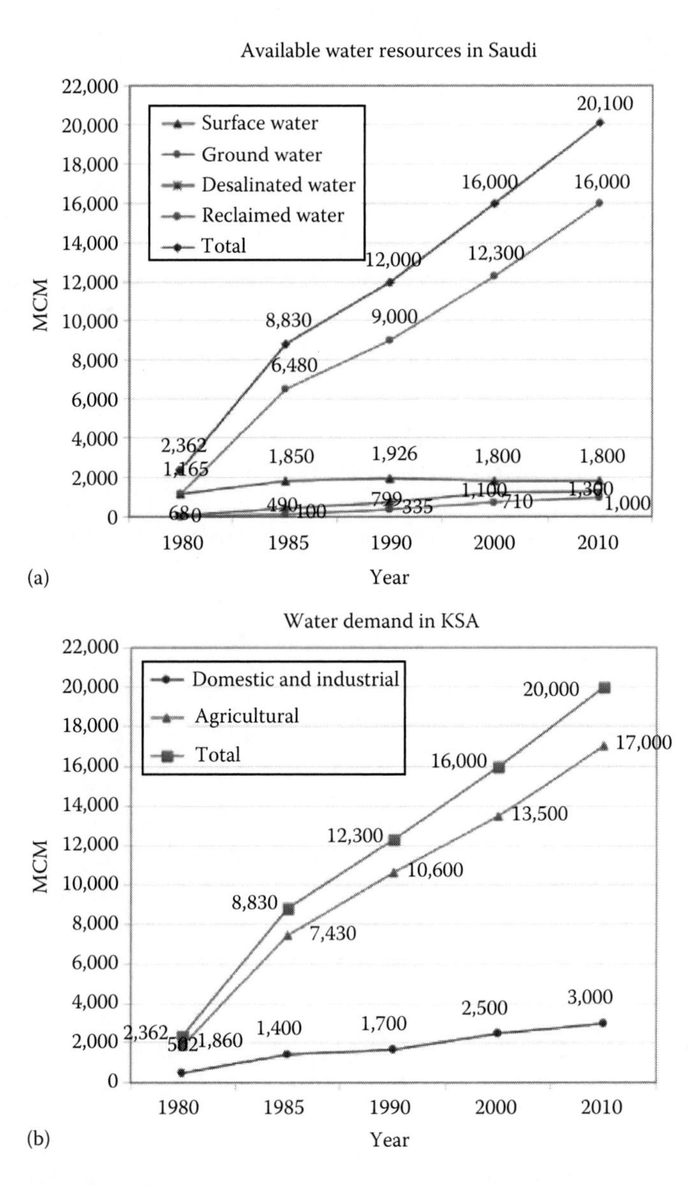

FIGURE 2.3 Water supply (a) and demand trend (b) in Saudi. There is a narrow difference between water supply and water demand suggesting a low reserve margin warranting a state of alert.

2.3 INTEGRATION OF UNMANNED AERIAL VEHICLES WITH WIRELESS SENSOR NETWORKS

UAVs have gained a lot of attention in recent years. Their use nowadays is not limited to defense. Rather, civilian applications have taken advantage of the advances scored in the defense sector. One can find very successful applications in such areas as forest, ocean, environment, and weather monitoring; topography; rescue and safety; and farming. Lately, suggestions on the use of UAVs have included airplane inspection.

The success of UAVs is due to their versatility. They can be very small, carry a customizable payload, and may not necessarily require takeoff or landing strips. Furthermore, they are becoming very affordable to the point where a group of them can be used as a swarm in a coordinated structure to take on a variety of participatory tasks or serve for redundancy and backup.

UAV command and control, and condition and capability in terms of self-awareness, situational awareness, self-organization, reconfiguration, and adaptation are well-established concepts that have been demonstrated and implemented successfully. It is worth mentioning that today, a simple UAV toy can exhibit many of these capabilities with added features such as the controllability through wearable computing devices such as smart phones.

UAV command and control interfaces along with trajectory planning options are also available. One can select a group of UAVs on a computer screen, designate a mission, specify the payload, assign a path or a destination, and launch a real-time scenario.

In [13], UAVs are used in order to minimize the excessive usage of pesticides and fertilizers in agricultural areas. The process of applying the chemicals is controlled by means of the feedback obtained from the WSN planted in the crop field. The UAV trajectory is adjusted based on the feedback from the sensors. In [14], UAVs are used to serve as a relay network to eliminate the disconnection of parts of a WSN and guarantee the delivery of data. In [15], cooperative multiple input multiple output (MIMO) techniques are used to support communication among static sensors in a sparse WSN and a relay network composed of UAVs in order to keep the WSN connected. In [16], a customizable virtual environment to display condition and capabilities of unmanned vehicles is disclosed. The virtual environment is highly customizable and can be used to generate mission scenarios as well as edit them.

2.4 FARMING PARCELS

Section 2.3 clearly suggests that UAVs can be used in a meaningful way to augment WSNs capabilities. Furthermore, we postulate the following with regard to farming:

1. The agricultural parcels are disjointed and separated by roads and right of ways for utilities such as power transmission and communication lines.
2. There is one or more WSNs in each parcel. Some of the nodes are coupled with actuators that control the flow of water, pesticide, or fertilizer.
3. The communication between WSNs is either nonexistent or very limited to be useful.

The suggested use of UAVs is mainly to provide additional capability for the following:

1. Gather data from the WSNs (Figure 2.4).
2. Relay the information to SCADA computers residing in a control center.

FIGURE 2.4 Data collection by UAVs.

3. Enable at least limited communication among the WSNs.
4. There can be any number of UAVs per parcel. They can be
   a. Autonomous following preplanned trajectories with a level of adaptation through commands from sensors.
   b. Semiautonomous with the intervention of an operator.
   c. Reconfigurable based on tasks or capabilities (payload).

One mode of operation is to associate one or more UAVs with a given parcel according to certain criteria such as the density of sensors and actuators per parcel. Thus, the UAVs associated with a given parcel can be considered as dynamic gateways for data collection and information processing; the WSNs include airborne nodes.

2.5 SENSORS FOR AGRICULTURE

Sensors traditionally have four layers, including the sensor layer, communication layer, control layer, and power layer. Their main functions are to detect, monitor, and measure physical parameters such as temperature, brightness, relative humidity, precipitation, sunshine, soil fertilizer, soil moisture, speed and direction of wind, and fruit and stem sizes.

On trees, sensors may be planted to monitor vital environmental and health parameters (Figure 2.5). The data collected can be accumulated by dedicated gateway nodes according to a routing protocol. Data can then be collected from all gateway nodes using a set of UAVs. The collected data can then be transmitted to the SCADA system.

During a collection session, data are transmitted from node to node to the gateway nodes according to a preset protocol and as triggered by a certain event such as a signal from a UAV. While great advances have been made in the development

FIGURE 2.5 UAVs in route to collect data.

of routing protocols, the issue of power consumption in nodes, especially during activity (transmission, reception, and processing of information), is still an ongoing challenge. Several remedies have been suggested in literature; some rely on hardware implementations and others on software. Some have suggested the organization of sensors into cluster patterns.

Since the agricultural land in Saudi is made up of disjoined parcels of land, clustering sensors according to location, task, or proximity makes sense. For example, sensors planted on a given tree may themselves form a cluster. This clustering scheme facilitates data collection, organization, and mining [17,18].

2.6 HARDWARE CONSIDERATIONS

A sensor node consists of a main board with slots for communication, sensing, data storage, and power supply. A single board may accommodate several sensors as shown in Figure 2.6.

Sensors can be divided into two categories:

1. *Central Located Sensors (CLS).* A single sensor (Figure 2.7) can monitor several parameters such as pressure, humidity, temperature, luminosity, and solar and ultraviolet radiation.
2. *Distributed Field Sensors (DFS).* Many sensors need to be installed to monitor the parameters at different locations (Figure 2.7). These sensors monitor parameters such as leaf wetness, soil moisture, soil temperature, trunk diameter, stem diameter sensor, and fruit diameter.

UAVs equipped with data collection, storage, and wireless communication facilities scan the parcels in a predefined pattern and collect data from the gateway nodes. Figure 2.8 shows a scenario of data collection. Sensors are organized in subnets on parcels. Data are relayed from node to node to a gateway, which in turn relays them

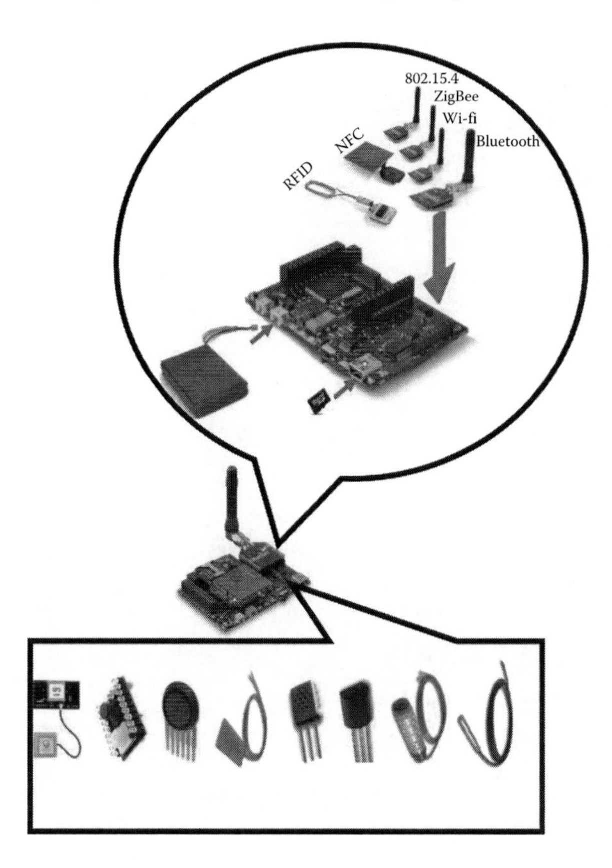

FIGURE 2.6 Sensor node and associated sensors.

FIGURE 2.7 Gateway node planted on a tree.

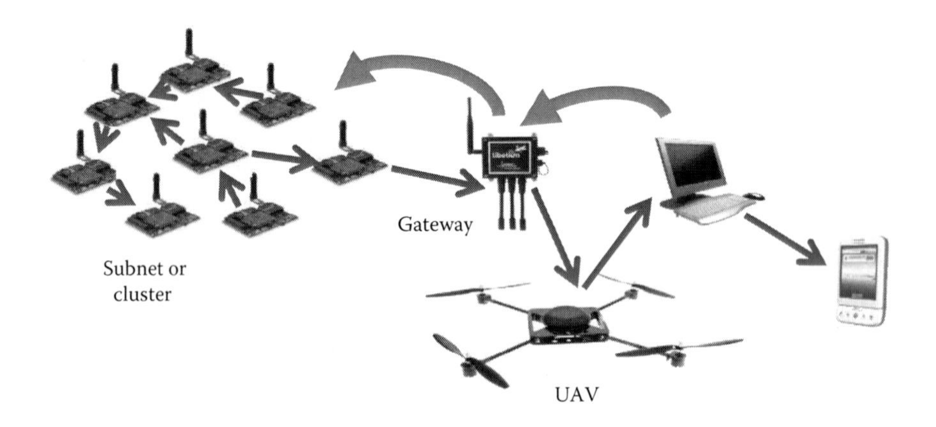

FIGURE 2.8 A scenario of data collection.

to the UAV. The UAV then transmits them to SCADA. To further facilitate data collection, subnets can also be organized into clusters as suggested earlier.

2.7 EFFICIENT IRRIGATION

Efficient irrigation follows three steps: sensing and data collection, information and data processing, and actuator control. Data can come from a variety of sources:

- Environmental parameters such as temperature, humidity, air pressure, solar radiation, speed and direction of wind, etc.
- Plant parameters such as leaf wetness, plant height, fruit and stem size and sap flow, etc.

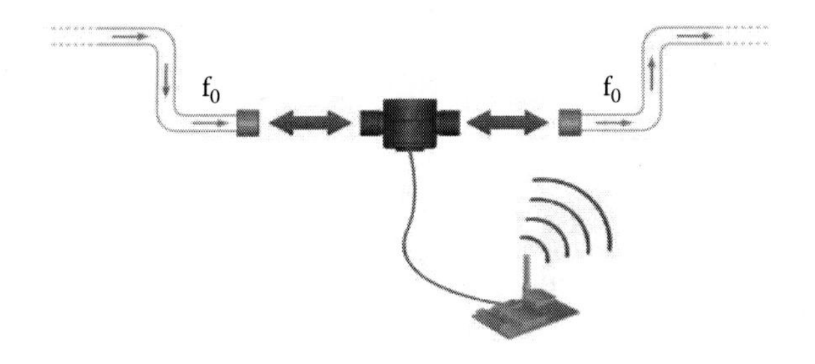

FIGURE 2.9 Sensor/actuator mechanism.

- Soil parameters such as temperature, humidity, and conductivity
- Time of day

Data sensing. Air temperature, humidity, and pressure can be monitored in the location where the gateway node is planted. Leaf wetness, plant height, and fruit size can be collected by installing sensors in specific locations to get precise information. To preserve battery power, sensors are active only during sessions of data collection or data transmission to UAVs. Otherwise, they are in hibernation or sleep mode.

Data acquisition. The gateway nodes accumulate data and relay them to the UAV when acquired. The UAVs then send them to SCADA in the control center for further processing, analysis, and decision making.

Actuator control. The control signal that invokes the actuator (Figure 2.9) to trigger a watering session comes from a sensor or sensors based on the presence of certain conditions derived from the monitored parameters. In effect, it is a closed loop system in which the decision to activate the actuator that triggers the water valve is derived from the sensed parameters.

2.8 CONCLUSIONS AND RECOMMENDATIONS

Saudi Arabia is one of the largest countries in the world and is characterized by a rich and fertile soil. It has no permanent rivers or lakes, and the rain is limited to few bursts per year. The ever-increasing demand for water to satisfy the population of a typical fast-developing country is a challenge. Saudi recognizes and endeavors to meet the challenge with both continuous informed planning and development, and efficient use of water.

With the focus on agriculture, WSNs are used to ensure that crops are provided with the needed amount of water that allows them to grow and yield the intended quality and quantity. Care must be taken to minimize or eliminate an excess of water as it is both harmful to crops and translates into financial loss for farmers. Since farming land in Saudi is mainly circular or rectangular parcels, then data collection from the WSNs planted on the parcels can be challenging as fixed communication infrastructures are either nonexistent or costly and unjustifiable.

WSNs and UAVs are the solution to farming in Saudi. Sensors coupled with actuators as a closed loop system can reduce or even eliminate water waste by ensuring that the water dispensed is the adequate amount for optimal crop growth. UAVs address the tedious communication and data collection problem. Data from the sensor nodes are routed to gateway nodes where they are collected by the UAVs. The role of the unmanned vehicles is to relay the data to the SCADA residing in the control center. Data are further processed in the control center and translated into actionable decisions that enhance farming.

ACKNOWLEDGMENT

The authors gratefully acknowledge the support for work from SNCS Research Center at the University of Tabuk and the Ministry of Higher Education in Saudi Arabia.

REFERENCES

1. K. Chakrabarty and S. S. Iyengar, *Scalable Infrastructure for Distributed Sensor Networks*, Springer, London, U.K., 2005.
2. J. N. Al-Karaki and A. E. Kamal, Routing techniques in wireless sensor networks: A survey, *IEEE Wireless Communications*, 11(6), 6–28, December 2004.
3. S. S. Alwakeel and N. A. Al-Nabhan, A cooperative learning scheme for energy efficient routing in wireless sensor networks, *IEEE 11th International Conference on Machine Learning and Applications (ICMLA 2012)*, Boca Raton, FL, December 12–15, 2012, pp. 463–468.
4. I. F. Akyildiz, W. Su, Y. Sankarasubramaniam, and E. Cayirci, A survey on sensor networks, *IEEE Communications Magazine*, 40(8), 102–116, August 2002.
5. R. Beckwith, D. Teibel, and P. Bowen, Report from the field: Results from an agricultural wireless sensor network, *Proceedings of the 29th Annual IEEE International Conference on Local Computer Networks (LCN'04)*, Tampa, FL, pp. 471–478.
6. T. Wark, P. Corke, P. Sikka, L. Klingbeil, Y. Guo, C. Crossman, P. Valencia, D. Swain, and G. Bishop-Hurley, CSIRO; Transforming agriculture through pervasive wireless sensor networks, *IEEE Pervasive Computing Magazine*, 6(2), 50–57, April–June 2007.
7. Alabama Cooperative Extension Systems, Basics of Crop Sensing, June 2011, ANR-1398.
8. Trading Economics, Agricultural Land (% of land area) in Saudi Arabia, August, 2013. Available: http://www.tradingeconomics.com/saudi-arabia/agricultural-land-percent-of-land-area-wb-data.html, accessed August 4, 2013.
9. Worldstat Info, Saudi Arabia, July, 2013. Available online at: http://en.worldstat.info/Asia/Saudi_Arabia/Land, accessed July 7, 2013.
10. M. Makkawi, Water resources of Saudi Arabia, *National Water Plan*, Vol. I. Prepared by the British Arabian Advisory Company for Ministry of Agriculture and Water Riyadh, KSA, 2010.
11. A. S. Al-Turbak, Water resources supply and demand in Saudi Arabia from National View Point, July, 2013. Available: http://edu.txtshr.com/docs/index-8274.html, accessed July 7, 2013.
12. A. Al-Ibrahim, Water use in Saudi Arabia: Problems and policy implications, *Journal of Water Resources Planning and Management (ASCE)*, 116(3), 375–388, 1990.
13. F. G. Costa, J. Ueyama, T. Braun, G. Pessin, F. S. Osorio, and P. A. Vargas, The use of unmanned aerial vehicles and wireless sensor network in agricultural applications, *2012 IEEE International Geoscience and Remote Sensing Symposium*, Munich, Germany, 2012, pp. 5045–5048.

14. E. P. de Freitas, T. Heimfarth, I. F. Netto, C. E. Lino, C. E. Pereira, A. M. Ferreira, F. R. Wagner, and T. Larsson, UAV relay network to support WSN connectivity, *International Congress on Ultra Modern Telecommunications and Control Systems*, Moscow, Russia, 2010, pp. 309–314.
15. M. A. M. Marinho, E. P. de Freitas, J. P. C. Lustosa da Costa, A. L. F. de Almeida, and R. T. de Sousa, Using cooperative MIMO techniques and UAV relay networks to support connectivity in sparse Wireless Sensor Networks, *2013 International Conference on Computing, Management and Telecommunications (ComManTel)*, Ho Chi Minh City, Vietnam, 2013, pp. 49–54.
16. S. R. Bieniawsk, E. W. Saad, J. L. Vian, and el-H. M. Aggoune, Virtual environment systems and methods, US Patent 08,068,983, 2011. Assigned to the Boeing Company, Seattle, WA.
17. J. Lin, A monitoring system based on wireless sensor network and an SoC platform in precision agriculture, *2008 11th IEEE International Conference on Communication Technology*, Hangzhou, China, November 2008, pp. 101–104.
18. A. Force and R. Academy, The realization of precision agriculture monitoring system based on wireless sensor network, *2010 International Conference on Computer and Communication Technologies in Agriculture Engineering*, 2010, pp. 89–92.

Section II

Environment

3 Scaling Smart Environments

Diane J. Cook

CONTENTS

ABSTRACT

The recent remarkable progress in computing power, sensors and embedded devices, and wireless networking combined with data mining and cloud computing paradigms has enabled researchers and practitioners to create smart environments for useful application. However, existing designs have only been tested in size-limited settings. In this chapter, we discuss recent technological advances in smart environment design and data mining techniques that allow the technologies to scale more easily. We also highlight new analyses that can be performed on smart environment sensor data when such scaling is made possible.

3.1 INTRODUCTION

Since the miniaturization of microprocessors, computing power has been embedded in familiar objects such as home appliances and mobile devices; it is gradually pervading almost every level of society. In the last decade, machine learning and pervasive computing technologies have matured to the point where this power is not only integrated with our lives but it can provide context-aware, automated support in our everyday environments. One physical embodiment of such a system is a smart environment. In these environments, computer software that plays the role of an intelligent agent perceives the state of the physical environment and residents using sensors, reasons about this state using artificial intelligence techniques, and then takes actions to achieve specified goals.

During perception, sensors embedded in the environment generate readings, while residents perform their daily routines. The sensor readings are collected by a computer network and stored in a database that an intelligent agent uses to generate useful knowledge such as patterns, predictions, and trends. On the basis of this information, a smart environment can select and automate actions that meet the goals of the application.

The potential uses of smart environment technology for applications such as health monitoring and energy-efficient automation are viewed by many as *extraordinary* [1]. However, most implementations of this technology to date are somewhat narrow and are performed in controlled laboratory settings. These limitations are due in large part to the difficulty of creating a fully functional smart environment infrastructure. In fact, while realistic smart environment prototypes have been designed [2–5], implementing these smart environments is so cumbersome that meetings have been organized [6] to discuss ways to scale such pervasive computing systems and to share valuable data that have been successfully captured in such settings.

Smart environment technologies have matured to the point where we need to consider how they will scale to real-time responsiveness, complex situations, and smart communities. In order to scale, attention needs to be given to how to design smart environment infrastructures that are lightweight and easy to install, which will allow the number of smart environment deployments to grow dramatically. In addition, smart environment capabilities such as activity discovery, recognition, and tracking need to work out of the box with minimal user training. In this chapter, we describe methods to achieve these goals. By scaling smart environments, we also demonstrate the new types of data collections and analyses that can be performed at a population-wide scale. All of the ideas are evaluated using data collected from the CASAS smart home project at Washington State University.

3.2 SCALING SMART ENVIRONMENT DESIGN

In order to scale the number of environments that employ ambient intelligence technologies, smart environment infrastructures need to be designed that are easy to install and ready to use out of the box. The CASAS *smart home in a box* (SHiB) is designed to do this.

3.2.1 CASAS SHiB Design

The CASAS SHiB software architecture components are shown in Figure 3.1. During perception, control flows up from the physical components through the middleware

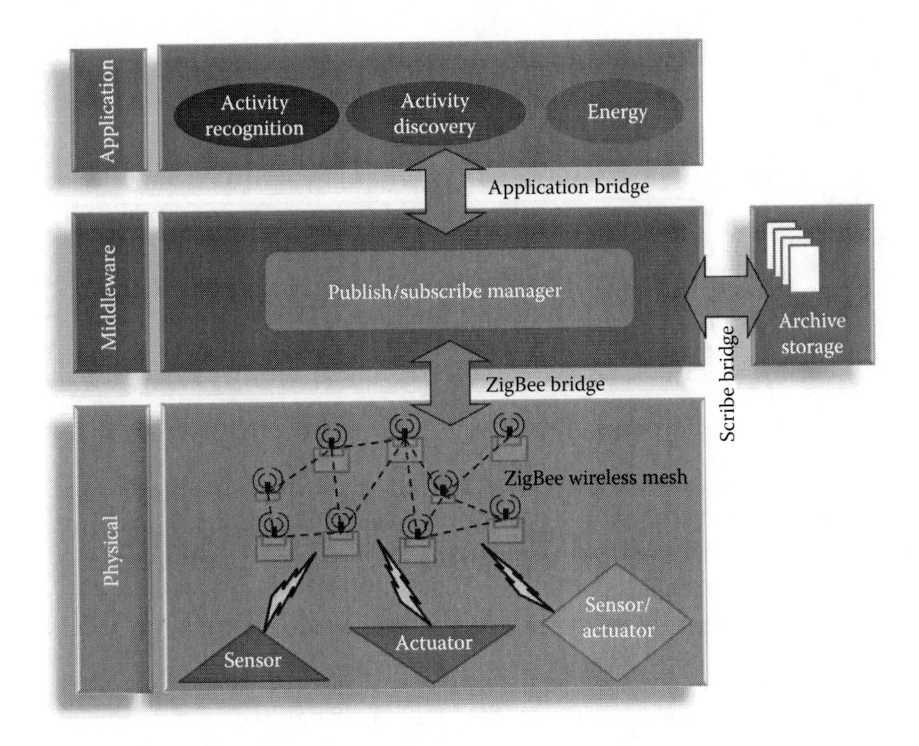

FIGURE 3.1 CASAS smart home components.

to the software applications. When taking an action, control moves down from the application layer to the physical components that automate the action. Our goal is that each of the layers is lightweight, extensible, and ready to use as is, without additional customization or training.

The CASAS physical layer contains hardware components including sensors and actuators. The architecture utilizes a ZigBee wireless mesh that communicates directly with the hardware components. The middleware layer is governed by a publish/subscribe manager. The manager provides named broadcast channels that allow component bridges to publish and receive messages. In addition, the middleware provides valuable services including adding time stamps to events, assigning UUIDs, and maintaining site-wide sensor state. Every component of the CASAS architecture communicates via a customized XMPP bridge to this manager. Examples of such bridges are the ZigBee bridge, the Scribe bridge that archives messages in permanent storage, and bridges for each of the software components in the application layer.

The CASAS architecture is easily maintained, easily extended, and easily scaled. The architecture is easily maintained because the communication bridges use lightweight APIs that support a wide variety of messages in a free-form manner. As a result, the middleware is compact and stable—it has had only one update in 5 years. CASAS is extendable because new bridges can be configured and integrated without changing or even restarting the middleware. We have designed

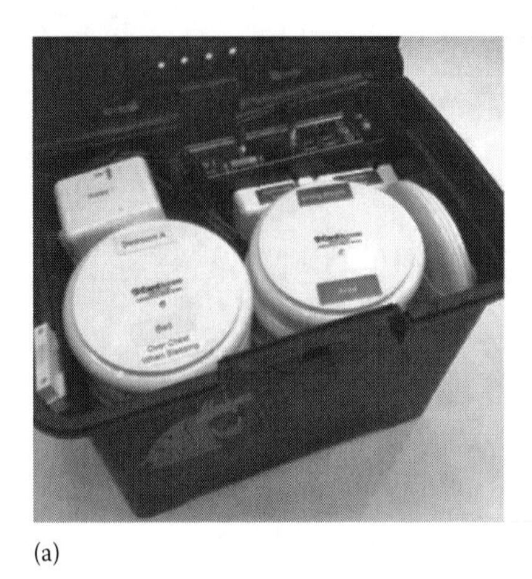

(a)

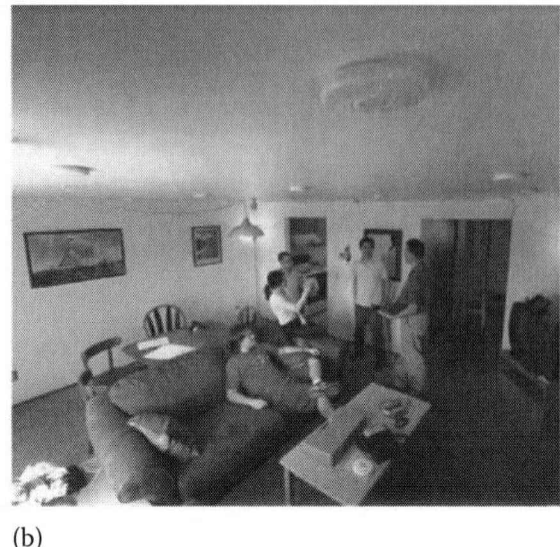

(b)

FIGURE 3.2 CASAS *SHiB* kit (a) and smart home installation site (b).

bridges that link multiple smart homes together, which allows CASAS to scale to communities of smart homes.

All of the CASAS components fit within a single small box, as is shown in Figure 3.2. The current box contains physical components in the form of sensors that are prelabeled with the intended location. Additional sensors and controllers can be included when needed. The middleware, database, and application components reside on a small, low-power computer with an ITX form factor server. While this layout is designed to allow each smart home to run independently and locally, smart homes can also securely upload events to be stored in a relational database or in the cloud.

3.2.2 Smart Home in a Box Usability

The simplicity of the CASAS *SHiB* design has made it possible for our research group to install a large number of smart home testbeds. To date, we have installed 32 smart home testbeds and many of the corresponding datasets are available on the project web page at http://ailab.wsu.edu/casas. A total of 19 datasets represent single-resident sites, 4 represent sites with two residents, and the rest house larger families or residents with pets. With the CASAS streamlined design, our team can install a new smart home in approximately 2 h and can remove the equipment in 30 min, with no changes or damage to the home. The design of the CASAS smart home also keeps installation costs down. Once the home is installed, the resident must consider maintenance of the equipment. The CASAS SHiB includes a software agent that alerts residents if sensor battery levels are getting low or if a sensor suddenly stops reporting events. In practice, this does not happen often, as the batteries typically last over a year.

To test the usability of the CASAS SHiB kit, we conducted a study in an on-campus three-bedroom apartment. We recruited participants to visit the apartment, one at a time, and install a CASAS smart home. The study included 20 participants aged 21–62 years (mean 33 years), 8 males and 12 females, with a variety of background and technological familiarity.

Each of the participants was given a written document explaining the smart home parts and installation process, and each was given a CASAS smart home kit as shown in Figure 3.2a. All of the participants were able to complete the installation without difficulties. The average installation time was just over 1 h. On a scale of 1 (simple) to 10 (impossible), participants rated the difficulty of installation for them as 2.53 ($\sigma = 1.07$). The most difficult issue they faced was trying to determine the optimal placement of sensors.

3.3 SCALING ACTIVITY RECOGNITION

In order to scale smart environment technologies, we need to provide capabilities that work out of the box, with no customization or training. One such key capability is the ability to automatically recognize activities that are being performed by residents in the environments. Here, we describe a method for enabling activities to be recognized from complex smart environment sensor data, without the need for explicit training in a new environment.

Intelligent systems that focus on the needs of a human require information about the activities being performed by the human. At the core of these systems, then, is activity recognition, which is a challenging and well-researched problem. Sensors in a smart home generate events that consist of a date, a time, a sensor identifier, and a sensor message. The generally accepted approach to activity recognition is to design and/or use machine learning techniques to map a sequence of sensor data to a corresponding activity label. Online activity recognition, or recognizing activities in real time from streaming data, introduces challenges that do not occur in the case of offline learning with presegmented data. However, this is an approach to activity recognition that needs to be considered in order to scale the capabilities of smart environments.

The CASAS activity recognition software, called AR, provides real-time activity labeling as sensor events arrive in a stream. To do this, we formulate the learning problem as that of mapping the sequence of the *k* most recent sensor events to a label that indicates the activity corresponding to the last (most recent) event in the sequence. The sensor events preceding the last event define the context for this last event. For example, the sequence of sensor events consisting of

2011-06-15 03:38:23.271939 BedMotionSensor ON
2011-06-15 03:38:28.212060 BedMotionSensor ON
2011-06-15 03:38:29.213955 BedMotionSensor ON

could be mapped to a *Sleep* activity label.

3.3.1 Previous Work

Activity recognition is not an untapped area of research. Because the need for activity recognition algorithms is great when providing context-aware services in a smart environment, researchers have explored a number of approaches to this problem [7]. The approaches can be broadly categorized according to the type of sensor data that is used for classification, the model that is designed to learn activity definitions, and the realism of the environment in which recognition is performed.

Sensor data: Researchers have found that different types of sensor information are effective for classifying different types of activities. When trying to recognize ambulatory movements (e.g., walking, running, sitting, climbing stairs, and falling), data collected from accelerometers positioned on the body have been used [8,9]. More recent research has tapped into the ability of a smartphone to act as a wearable/carryable sensor with accelerometer and gyroscope capabilities. Researchers have used phones to recognize gesture and motion patterns [10,11].

For other activities that are not as easily distinguishable by body movement alone, researchers observe an individual's interaction with key objects in the space such as medicine containers, key, and refrigerators [12–14]. Objects are tagged with shake sensors or RFID tags and are selected based on the activities that will be monitored. Other researchers rely upon environment sensors including motion detectors and door contact sensors to recognize ADL activities that are being performed [15–17].

For recognition of specialized classes of activities, researchers use more specialized sources of information. As an example, Yang et al. [18] collected computer usage information to recognize computer-based activities including multiplayer gaming, movie downloading, and music streaming. In addition, some researchers such as Brdiczka et al. [19] videotape smart home residents and process the video to recognize activities. Because our study participants are uniformly reluctant to allow video data or to wear sensors and because object sensors require frequent charging and are not practical in participant homes, our data collection has consisted solely of passive sensors that could be installed in a smart environment.

Activity models: The number of machine learning models that have been used for activity recognition varies as greatly as the number of sensor data types that have been explored. Naive Bayes (NB) classifiers have been used with promising results for offline learning of activities [19–22] when large amounts of sample data are available. Other researchers [8,16] have employed decision trees to learn logical descriptions of the activities, and still others [23] employ kNNs. Gu et al. [12] take a slightly different approach by looking for emerging frequent sensor sequences that can be associated with activities and can aid with recognition.

An alternative approach that has been explored by a number of research groups is to exploit the representational power of probabilistic graphs. Markov models [17,20,24,25], dynamic Bayes networks [14], and conditional random fields (CRFs) [26,27] have all been successfully used to recognize activities, even in complex environments. Researchers have found that these probabilistic graphs, along with neural network approaches [25,28], are quite effective at mapping presegmented sensor streams to activity labels.

Recognition tasks: A third way to look at earlier work on activity recognition is to consider the range of experimental conditions that have been attempted for activity recognition. The most common type of experiment is to ask subjects to perform a set of scripted activities, one at a time, using the selected sensors [11,14,19,28]. In this case, the sensor sequences are well segmented, which allows the researchers to focus on the task of mapping sequences to activity labels.

Building on this foundation, researchers have begun looking at increasingly realistic and complex activity recognition tasks. These setups include recognizing activities that are performed with embedded errors [20], with interleaved activities [29], and with concurrent activities performed by multiple residents [17,30,31]. The next major step that researchers have pursued is to recognize activities in unscripted settings (e.g., in a smart home while residents perform normal daily routines) [16,25]. These naturalistic tasks have relied on human annotators to segment, analyze, and label the data. However, they do bring the technology even closer to practical everyday usage. The realism of activity recognition has been brought into sharper focus using tools for automated segmentation [12,19] and for automated selection of objects to tag and monitor [13].

3.3.2 Online Activity Recognition with AR

One feature that distinguishes previous work in activity recognition from scalable activity recognition is the need to perform continuous activity recognition from streaming data, even when multiple residents are present and activities may be interleaved. In such realistic situations, the data cannot be segmented into separate sensor streams for different activities. Instead, we adopt the approach of moving a sliding window over the sensor event stream and identifying the activity that corresponds to the most recent event in the window. Formally, the sensor event sequence s_1, s_2, …, s_N is divided into windows of equal number of sensor events s_1, s_2,…, $s_{\Delta s}$, and the ith window is represented by the sequence $[s_{i-\Delta s}, s_i]$. Sliding windows have been used with promising results in other work [29], but not yet for activity recognition in unscripted settings. The approach we describe and evaluate here study considers data collected from environmental sensors such as motion and door sensors, but other types of sensors could be included in these approaches as well.

The approach we describe for online activity recognition can be adapted to many different classifiers. Here, we report results for NB, hidden Markov model (HMM), CRF, and support vector machine (SVM) classifiers. These approaches are considered for this task because they traditionally are robust in the presence of a moderate amount of noise and are designed to handle sequential data. Among these three choices, there is no clear best model to employ—they each utilize methods that offer strengths and weaknesses for the task at hand.

The NB classifier uses relative frequencies of feature values as well as the frequency of activity labels found in sample training data to learn a mapping from activity features, D, to an activity label, a, calculated using the formula arg $\max_{a \epsilon A} P(a|D) = P(D|a)P(a)/P(D)$. In contrast, the HMM is a statistical approach in which the underlying model is a stochastic Markovian process that is not observable (i.e., hidden), which can be observed through other processes that produce the sequence of observed features. In our HMM, we let the hidden nodes represent activities and the observable nodes represent combinations of feature values. The probabilistic relationships between hidden nodes and observable nodes and the probabilistic transitions between hidden nodes are estimated by the relative frequency with which these relationships occur in the sample data.

Like the HMM, the CRF model makes use of transition likelihoods between states as well as emission likelihoods between activity states and observable states to output a label for the current data point. The CRF learns a label sequence that corresponds to the observed sequence of features. Unlike the HMM, weights are applied to each of the transition and emission features. These weights are learned through an expectation maximization process based on the training data.

Our last approach employs SVMs to model activities. SVMs identify class boundaries that maximize the size of the gap between the class boundary and the training data points. We employ a one vs. one SVM paradigm that is computationally efficient when learning multiple classes with possible imbalance in the amount of available training data for each class (Figure 3.3). For the experiments reported in this chapter, we used the libSVM implementation of Chang and Lin [32].

To provide input to the classifiers, we define features that describe a data point i that corresponds to a sequence of sensor events. This fixed dimensional feature vector x_i includes the time of day for the first and last sensor events (discretized into four equal-length bins), the time span of the k-event sensor window, and a count of events for each sensor within the window. Each vector x_i is tagged with the label y_i of the last sensor event in the window. The label y_i corresponds to the activity label associated with the last sensor event in the window. Although a fixed window size k could be identified that works well for a given dataset, this approach requires additional customization from the user. To increase the generalizability of the approach, the window size k is dynamically adjusted by AR based on the most likely activities that are being observed and the activity duration that is typical for those activities.

3.3.3 Scaling Activity Recognition to Complex Spaces

One of the problems associated with windowing based on sensor events that do not have a constant sampling rate is that windows may contain sensor events that are widely spread apart in time. An illustration of this problem is presented in Figure 3.4 (left). This is an example of a sequence of sensor events from our dataset. Notice the time stamp of the last two events in this sequence. There is a gap of nearly one and half hours between these sensor events. All the sensor events that define the context of the last event have occurred in the *distant* past. Thus, in the absence of any weighting scheme, even though the sensor event corresponding to the end of the Personal_Hygiene activity occurred in the past, it has an equal influence on defining the context of the event corresponding to Enter_Home. In order to more greatly consider sensor events that are likely associated with the current activity, activity recognition can use a time-weighting scheme when taking into account the relative temporal distance between the sensors.

Let $\{t_{i-\Delta s}, t_{i-\Delta s+1}, \ldots, t_i\}$ represent the timing of the sensor events in the ith window. We take into account the difference in time for each sensor event with respect to t_i, for computing the feature vector describing the window. In particular, AR uses an exponential function to compute the weights. The contribution of sensor event j within the interval Δs to the feature vector describing the ith window is given as $C(i, j) = \exp(-\chi^{(t_i - t_j)})$, where the value of χ determines the rate of temporal

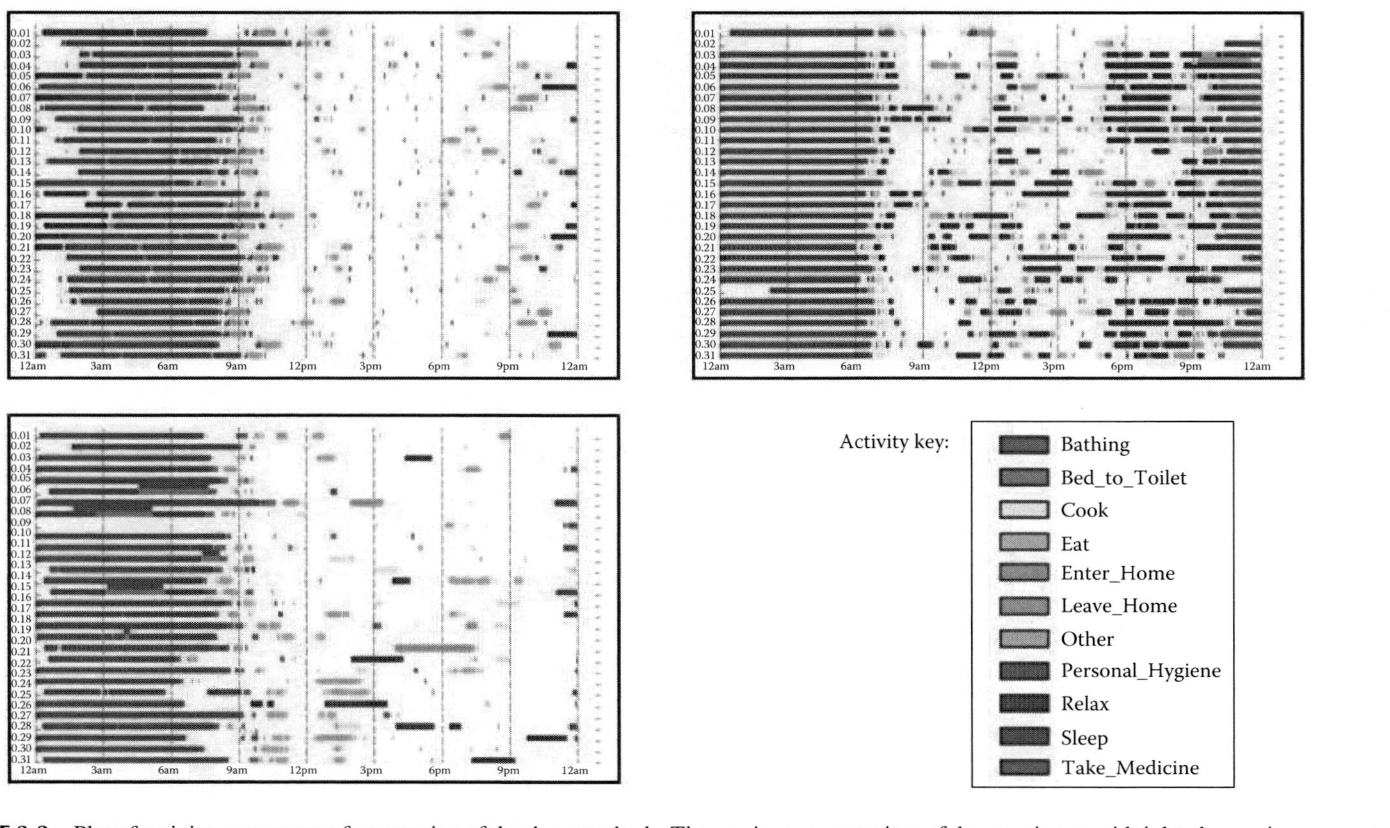

FIGURE 3.3 Plot of activity occurrences for a portion of the three testbeds. The x axis represents time of day starting at midnight, the y axis represents a specific day.

```
2009-07-19 10:18:59.406001 LivingRoom ON
2009-07-19 10:19:00.406001 Bathroom OFF Personal_Hygiene end
2009-07-19 10:19:03.015001 OtherRoom OFF
2009-07-19 10:19:03.703001 LivingRoom OFF
2009-07-19 10:19:07.984001 LivingRoom ON
2009-07-19 10:19:11.921001 LivingRoom OFF
2009-07-19 10:19:13.203001 OtherRoom ON
2009-07-19 10:19:14.609001 Kitchen ON
2009-07-19 10:19:17.890001 OtherRoom OFF
2009-07-19 10:19:18.890001 Kitchen OFF
2009-07-19 10:19:24.781001 FrontDoor On Leave_Home begin
2009-07-19 10:19:28.796001 FrontDoor OFF
2009-07-19 10:19:31.109001 FrontDoor CLOSE Leave_Home end
2009-07-19 12:05:13.296001 FrontDoor OPEN Enter_Home begin
```

```
2009-07-23 19:59:58.093001 Bathroom ON
2009-07-23 20:00:02.390001 Bathroom OFF
2009-07-23 20:00:04.078001 Bathroom ON
2009-07-23 20:00:08.000001 LivingRoom ON
2009-07-23 20:00:08.640001 OtherRoom ON
2009-07-23 20:00:09.343001 Bathroom OFF Personal_Hygiene end
2009-07-23 20:00:12.296001 LivingRoom OFF
2009-07-23 20:00:22.171001 Kitchen ON
2009-07-23 20:00:25.140001 OtherRoom OFF
2009-07-23 20:00:27.187001 FrontDoor ON Leave_Home begin
2009-07-23 20:00:27.437001 Kitchen OFF
2009-07-23 20:00:30.140001 FrontDoor OFF
2009-07-23 20:00:32.046001 FrontDoor ON
2009-07-23 20:00:36.062001 FrontDoor OFF
2009-07-23 20:00:39.343001 Front Door ON
2009-07-23 20:00:43.671001 FrontDoor OFF
2009-07-23 20:00:46.265001 FrontDoor CLOSE Leave_Home end
```

FIGURE 3.4 Illustration of time dependency (left) and sensor dependency (right) for windowing.

influence decay. The simple count of sensor event types within a window is now replaced by a sum of the time-based contributions of each of the sensor events within the window.

A second challenge arises in situations when a sensor event corresponds to the transition between two activities (or different activities performed by multiple residents in parallel) and some of the events occurring in the window are not actually related to the event being labeled. An example of this situation is shown in Figure 3.4 (right). This particular sequence of sensor events from one of the testbeds represents the transition from the *Personal_Hygiene* activity to the *Leave_Home* activity. Notice that all the initial sensor events in the window occur in the bathroom, whereas the second set of sensor events are from an unrelated area of the apartment near the front door. While this certainly defines the context of the activity, since the sensors from a particular activity dominate the window, the chances for a wrong conclusion about the last sensor event of the window are greater. AR overcomes this problem by utilizing a weighting scheme based on mutual information measure between the sensors.

Mutual information is typically defined as a measure of the mutual dependence of two random variables. In AR, each individual sensor is a random variable with two outcomes, namely, ON and OFF. The mutual information, or dependence, between two sensors is then defined as the chance that these two sensors occur successively in the entire sensor stream. The mutual information between two sensors S_i and S_j is defined as $MI(i, j) = (1/N)\sum_{k=1}^{N-1}(s_k, S_i)(s_{k+1}, S_j)$, where $\delta(s_k, S_i)$ is 1 if $s_k = S_i$ and is 0 otherwise.

As a result, sensors that are adjacent to each other and are likely to trigger readings or events in succession will have high mutual information, whereas sensors at opposite ends of the environment will tend to have much lower mutual information. As with the time weighting, each event in the window is weighted with respect to the mutual information between it and the last event in the window (the one being labeled). Instead of computing a count of different sensor event types, AR now computes the sum of contributions of each sensor event based on time weights and mutual information.

TABLE 3.1
Summary of Windowing-Based Classifier Accuracy on Smart Environment Testbeds

Dataset	B1 (%)	B2 (%)	B3 (%)	Average (%)
NB	92.91	90.74	88.81	90.82
HMM	92.07	89.61	90.87	90.85
CRF	85.09	82.66	90.36	86.04
SVM	90.95	89.35	94.26	91.52

We compare the performance of the alternative machine learning models using data collected over 6 months in three separate smart home environments, each housing one resident. Table 3.1 summarizes the recognition accuracy using threefold cross validation. All of the classifiers perform well at recognizing the 10 predefined activities that are listed (not including the *Other* class) and plotted in Figure 3.3. The SVM performs consistently best, however, so we focus on this classifier when evaluating the approach for scalability.

3.3.4 Scaling Activity Recognition to Multiple Spaces

Next, we want to determine the ability of the learned model to generalize over multiple environments and multiple residents. If we can train such a model on a large number of existing spaces, it will be able to recognize and track activities in a new space *out of the box* without excessive initial data gathering and labeling.

One challenge we face in training a model to learn activity models from multiple spaces is defining a common representation for sensors in the space. The experiments described in earlier sections used unique sensor identifiers for each environment. In order to learn a more general model, we map each unique sensor to a functional area in the environment, such as bedroom, bathroom, and dining room, as shown in Figure 3.4. An alternative mechanism would be to use machine learning algorithms to define a mapping between sensor functions in disparate spaces, as has been explored previously [33].

To evaluate the ability of the models to recognize generalizable activities out of the box, we collected sensor data in 18 separate smart apartments, each housing one resident and each utilizing the CASAS SHiB infrastructure. All residents performed their normal daily routines during the data collection. One month of data was annotated manually to provide ground truth activity labels for 10 selected activities. We evaluate performance as the percentage of sensor events that were correctly labeled across all of the apartments using fivefold cross validation, with no additional customized training for each apartment. Table 3.2 shows the confusion matrix that was generated from this experiment.

As the matrix indicates, some activities are easier to recognize than others. This is because some activities, such as cooking, have a fairly unique spatial–temporal signature. Other activities are more challenging because they overlap with other

TABLE 3.2
Activity Recognition Confusion Matrix

	Automatically Generated Activity Label										
Ground Truth Activity Label	**Bed–Toilet Transition**	**Cook**	**Eat**	**Enter_Home**	**Leave_Home**	**Personal_Hygiene**	**Phone**	**Relax**	**Sleep**	**Work**	**Accuracy**
Bed–Toilet Transition	**18,288**	143	261	0	0	22,233	0	3	5,866	38	**0.39**
Cook	3	**370,300**	1,616	11	11	172	4	140	28	1,917	**0.99**
Eat	53	20,528	**9,871**	4	0	41	1	979	118	27,052	**0.17**
Enter_Home	0	195	0	**1,606**	107	3	0	4	57	126	**0.77**
Leave_Home	0	5	0	59	**316**	3	0	0	1	4	**0.81**
Personal_Hygiene	15,769	928	81	3	3	**295,616**	0	77	1,216	921	**0.94**
Phone	0	21	2	0	0	4	**8**	34	73	1,072	**0.01**
Relax	6	1,282	322	13	0	178	8	**2,030**	1,459	2,735	**0.25**
Sleep	33,900	66	33	1	0	279	0	60	**65,189**	306	**0.65**
Work	37	2,875	10,544	66	17	489	20	497	237	**71,684**	**0.83**

The diagonal entries indicate the activities that were correctly categorized. The accuracy for each individual activity is shown in the last column.

activity classes or not enough training data are available to learn the model. The weighted average accuracy is 84%, which indicates that the models are fairly robust even when they are used out of the box in new, distinct home settings.

3.4 SCALING BEHAVIOR MODELING WITH ACTIVITY DISCOVERY

Recognizing activities from streaming data introduces new challenges because data must be processed that do not belong to any of the targeted activity classes. One way to handle unlabeled data is to design an unsupervised learning algorithm to discover activities from unlabeled sensor data. Segmenting unlabeled data into smaller classes improves activity recognition performance because the *Other* class is no longer dominant in terms of size, as what frequently happens in activity recognition datasets. Another important reason to discover activity patterns from unlabeled data is to characterize and analyze as much behavioral data as possible, not just predefined activity classes. Such unlabeled data need to be examined and modeled in order to get a complete view of everyday life.

3.4.1 Previous Work

The algorithm we describe here for activity discovery, called AD, builds on a rich history of discovery research, including methods for mining frequent sequences [12,34], mining frequent patterns using regular expressions [35], constraint-based mining [36], mining frequent temporal relationships [37], and frequent-periodic pattern mining [38].

More recent work extends these early approaches to look for more complex patterns. Ruotsalainen et al. [39] design the GAIS genetic algorithm to detect interleaved patterns in an unsupervised manner. Other approaches have been proposed to mine discontinuous patterns [40–42] in sequential datasets and to allow variations in occurrences of the patterns [43]. Huynh et al. [44] explore the use of topic models and LDAs to discover daily activity patterns in wearable sensor data.

3.4.2 Activity Discovery with AD

Like earlier approaches to sequence mining, our AD activity discovery algorithm searches the space of candidate sensor event sequences ordered by increasing sequence length. Because the space of candidate patterns is exponential in the size of the input data, we employ a greedy search to find the sequence pattern that best compresses the input dataset.

A pattern in AD consists of a sequence definition and all of its occurrences in the data. The initial state of the search algorithm is the set of pattern candidates consisting of all uniquely labeled sensor identifiers. The only operators of the search are the *ExtendSequence* operator and the *EvaluatePattern* operator. The *ExtendSequence* operator extends a pattern definition by growing it to include the sensor event that occurs before or after any of the instances of the pattern.

During discovery, the entire dataset is scanned to create initial patterns of length one. After this initial pass, the whole dataset does not need to be scanned again.

Instead, AD extends the patterns discovered in the previous iteration using the *ExtendSequence* operator and will match the extended pattern against the patterns already discovered in the current iteration to see if it is a variation of a previous pattern or is a new pattern. In addition, AD employs an optional pruning heuristic that removes patterns from consideration if the newly extended child pattern evaluates to a value that is less than the value of its parent pattern.

AD uses a beam search to identify candidate sequence patterns by applying the *ExtendSequence* operator to each pattern that is currently in the open list of candidate patterns. The patterns are stored in a beam-limited open list and are ordered based on their value.

The search terminates upon exhaustion of the search space. Once the search terminates and AD reports the best patterns that were found, the sensor event data can be compressed using the best pattern. The compression procedure replaces all instances of the pattern by single-event descriptors, which represent the pattern definition. AD can then be invoked again on the compressed data. This procedure can be repeated a user-specified number of times. Alternatively, the search and compression process can be set to repeat until no new patterns can be found that compress the data. We use the last mode for experiments in this discussion.

We evaluate candidate patterns based on their ability to minimize the size of the original dataset when it is compressed using the pattern definition, following the minimum description length (MDL) principle [45]. Because each occurrence of a pattern is replaced by a single event labeled with a pattern identifier, the description length of a pattern P given input data D is calculated as $DL(P)+DL(D|P)$, where $DL(P)$ is the description length of the pattern definition and $DL(D|P)$ is the description length of the dataset compressed using the pattern definition. Because human behavior patterns contain a great deal of variation, we employ an edit distance measure to determine if a sensor sequence is sufficiently similar to a pattern to be considered an instance of the pattern. This measure counts the minimum number of add, delete, or transpose operations that are needed to transform a sensor sequence to one that is equivalent to the pattern definition.

To determine the fit of a variation to a pattern definition, we compute the edit distance using the Damerau–Levenshtein measure [46]. This measure counts the minimum number of operations needed to transform one sequence, x, to be equivalent to another, y. In the case of the Damerau–Levenshtein distance, the allowable transformation operators include change of a symbol (in our case, a sensor event), addition/deletion of a symbol, and transposition of two symbols. AD considers a sensor event sequence to be equivalent to another if the edit distance is less than 0.1 times the size of the longer sequence. The edit distance is computed in time $O(|x| \times |y|)$.

As an example, Figure 3.5 shows a dataset where the sensor identifiers are represented by varying patterns. AD discovers four instances of the pattern P in the data that are sufficiently similar to the pattern definition. The resulting compressed dataset is shown as well as the pattern P' that is found in the new compressed dataset.

Figure 3.6 provides a visualization of the three top activity patterns that are discovered when AD is applied to a dataset that combines sensor events from the three testbeds described in Section 3.3.3. The pattern in (a) contains a sequence consisting

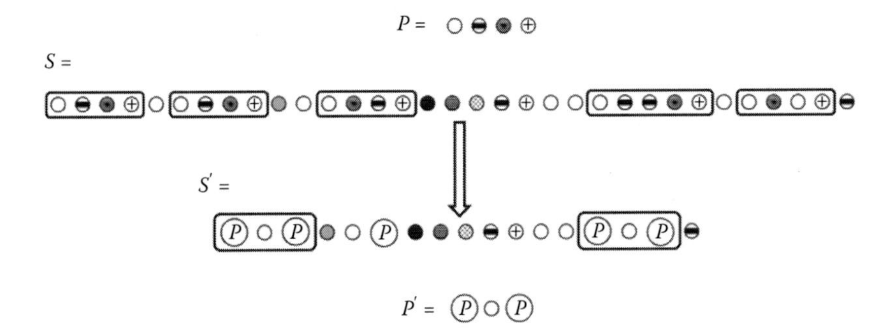

FIGURE 3.5 Example of the AD discovery algorithm. A sequence pattern (P) is identified and used to compress the dataset. A new best pattern (P') is found in the next iteration of the algorithm.

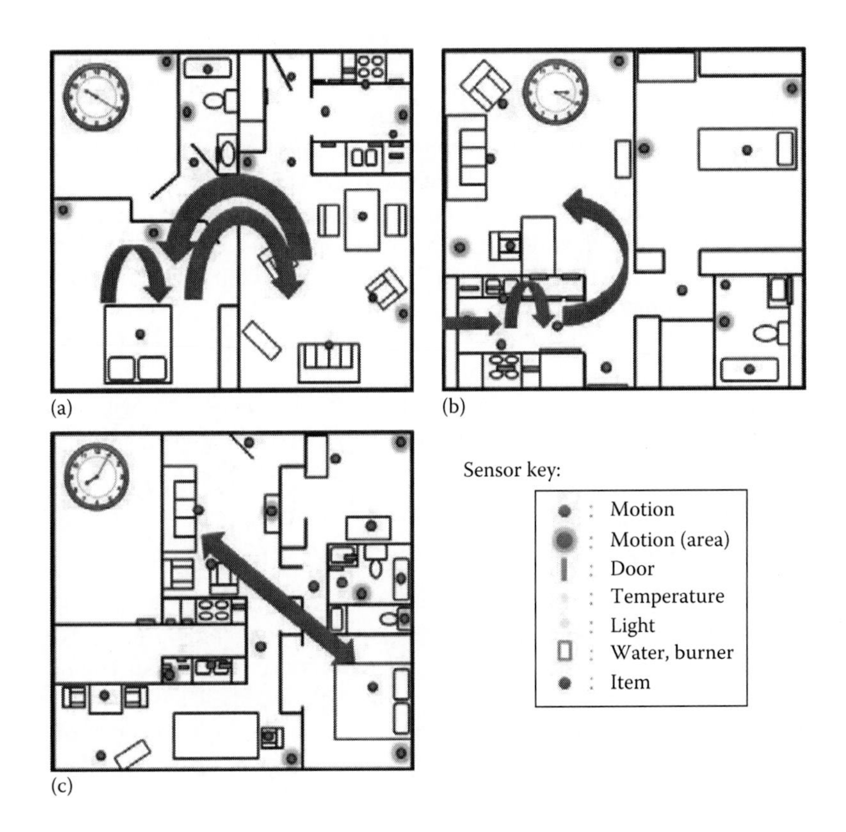

FIGURE 3.6 Visualization of discovered patterns: $P1$ (a), $P2$ (b), and $P3$ (c).

TABLE 3.3
AR Recognition Accuracy for Predefined Activities and Other Category with and without AD-Discovered Patterns

Dataset	B1 (%)	B2 (%)	B3 (%)
No AD patterns + other category	60.55	49.28	74.75
With AD patterns + other category	71.08	59.76	84.89

of motion in the bedroom followed by the living room and back to the bedroom, around 10:20 in the evening. Many of these events occur prior to sleeping and may represent getting ready for bed. The pattern in (b) consists of a front door closing followed by a series of kitchen events and then a living room event, usually in the late morning or midafternoon. This could represent a number of different activities that occur after returning home, such as putting away groceries or getting a drink. The pattern in (c) consists of a sequence of events alternating between the bedroom, a work area, and the living room, after waking up in the morning. This pattern might represent a resident gathering items needed for their daily routine. Other patterns represent transitions between activities or activities that are recognizable but do not appear on the list of predefined activities, such as spending extended time in a secondary bedroom that is used for guests or crafts.

Activity discovery can also be used to improve the performance of activity recognition algorithms. Section 3.3.3 shows that classifiers can identify activities, even in real time, when applied to predefined activities alone. On the other hand, Table 3.3 summarizes the performance of the SVM when the *Other* class is included. The problem is that approximately half of the sensor events are unlabeled, so the *Other* class dominates the dataset.

When we apply AD to this dataset, the *Other* activity category is decomposed into patterns corresponding to the AD-discovered activities. When AR is applied to recognize the predefined activities and the discovered activities, recognition accuracy improves by approximately 10% on average, as shown in Table 3.3.

3.5 VISUALIZATION AND ANALYSIS OF LARGE SMART ENVIRONMENT SENSOR DATASETS

Because the CASAS SHiB is simple to install and maintain, data have been collected in multiple testbeds over multiple years. This type of longitudinal dataset allows researchers to monitor changes in behavior that may indicate a chance in cognitive or physical health. Parameters are monitored, which include overall activity level, sleep quality, and times spent on individual activities of interest. A visualization of these parameters for one resident is shown in Figure 3.7. The visualization in the upper right box plots dots for each dataset of interest (in this case, 20 apartments at an assisted care facility) to allow side-by-side comparison of these parameters between individuals in the target population.

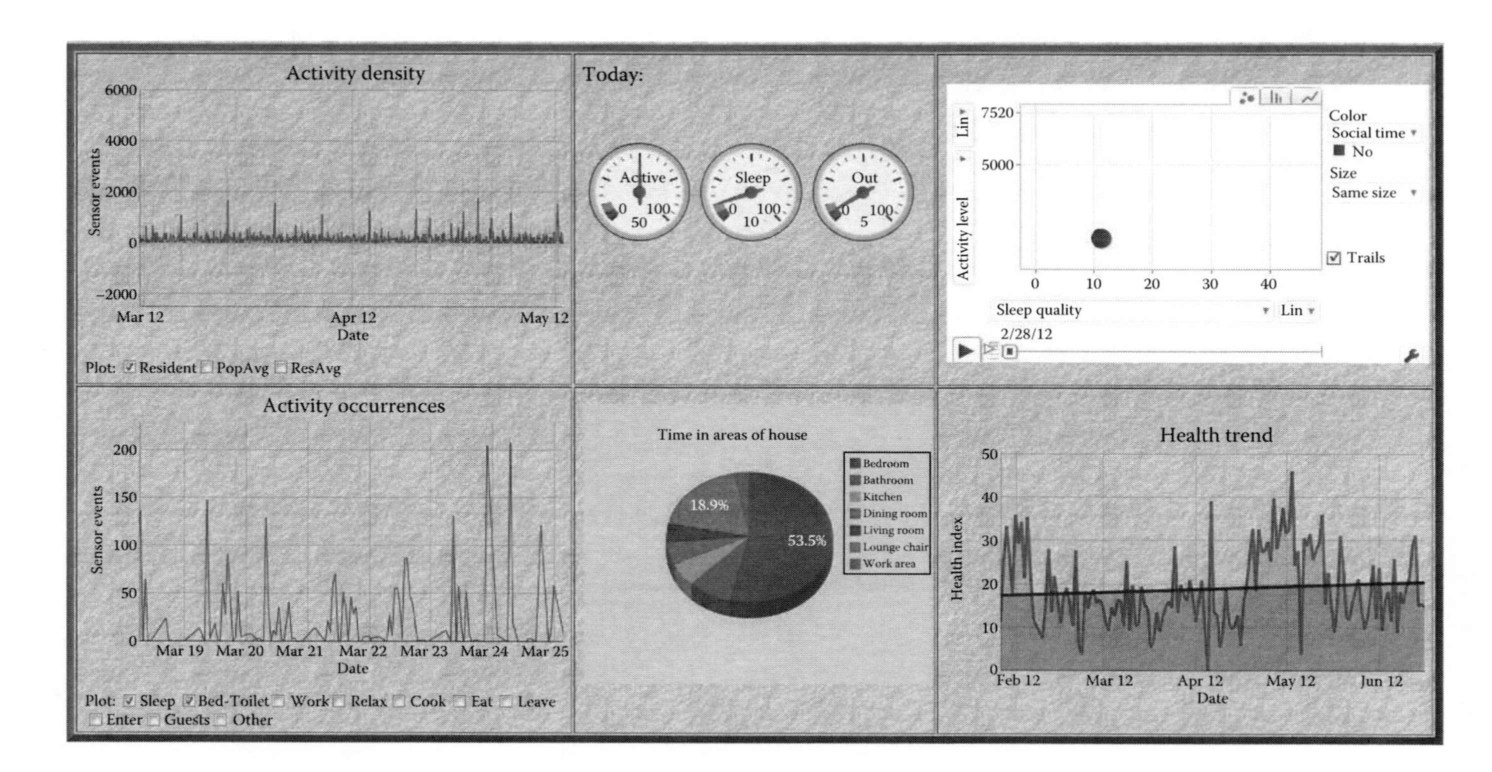

FIGURE 3.7 Activity trends for a smart home resident.

A second application of interest is supporting energy-efficient behavior in the home. By identifying activities that occur in the home and concurrently monitoring whole-home energy usage, we can predict energy that will be consumed for a particular activity. In addition to providing this information to a resident (see Figure 3.8), the smart home can promote energy-efficient behavior [47] and automate control of selected devices to support activities in a more energy-efficient manner.

One type of analysis that is not found in the literature is a population-wide analysis of resident behavior using smart home data. While analyzing behavioral features across a larger demographic would be beneficial for many researchers in psychology, sociology, and technology, gathering data at a significant scale has not yet been a practical goal. Using the *SHiB* design, we are able to start investigating questions that apply to demographic groups, families, and communities.

As a first step, we consider behavioral properties for the CASAS datasets we have collected. In particular, we want to identify how activity levels vary throughout the day for an entire cohort. We also want to determine how individuals spend their home time in terms of individual activities. We also want to determine how consistent the functions are across the group.

Figure 3.9 shows the results of these two analyses for the 18 smart apartments mentioned in Section 3.3.4. As the plots indicate, there is a clear pattern for the entire group in which activity levels are low in the early hours of the day but then increase, peaking at midmorning, midafternoon, and early evening. The exact activity levels vary quite a bit across the population, which may be due to mobility differences and due to sensor granularity within the home. In contrast, the variance across the population for time devoted to various activities is quite a bit smaller. As the graph shows, the most time is dedicated to sleep, while other activities receive less time such as taking medicine (which is typically quick) and cleaning the home (which may not happen as often as other activities). Larger variances exist for the enter activity (which takes into account time spent outside the home) and for bed toilet transitions, which do vary dramatically by age, health, and sleep quality. Being able to run such large-scale analyses will provide a valuable tool for understanding behavior that is central to many research fields including sociology and psychology as well as technology development.

3.6 CONCLUSIONS AND FUTURE WORK

In this chapter, we highlight the capabilities of a smart home system that can be deployed, evaluated, and scaled when the smart environment architecture is made simple and lightweight. As a next step, we would like to evaluate the ease with which additional sensor modalities (e.g., RFID, smartphones) can be incorporated into the architecture and will design applications that more extensively utilize device controllers. We would also like to expand the scope of the data collection to include a greater diversity of resident demographics and to perform longitudinal studies. Finally, we would like to design smart environment automation strategies that provide safe and energy-efficient support of resident daily activities.

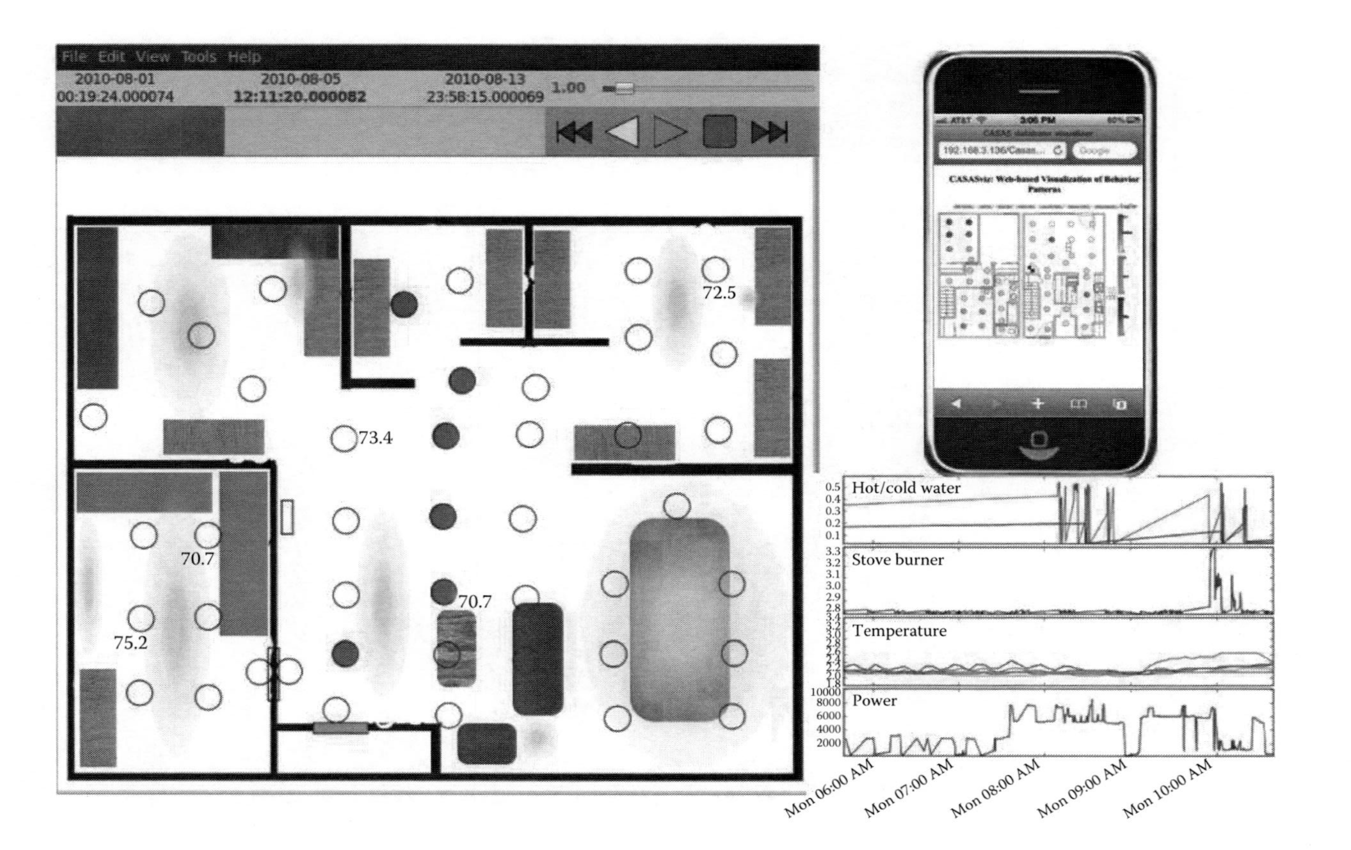

FIGURE 3.8 Snapshot of CASAS activity visualizer. The visualizer renders sensor events on a computer or mobile device while plotting usage of resources such as electricity.

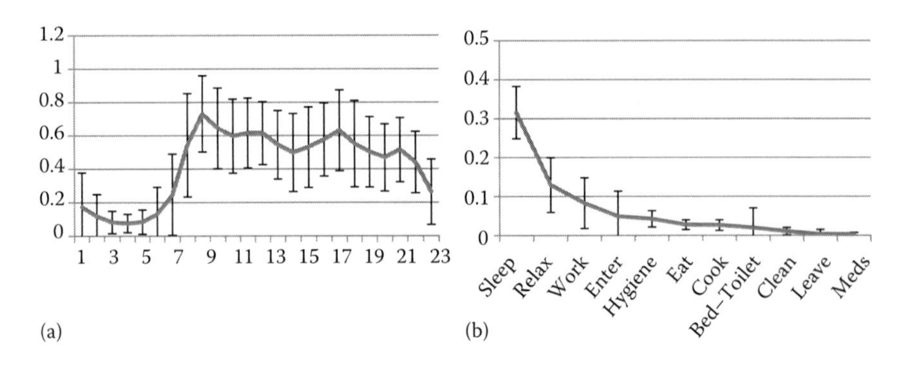

FIGURE 3.9 Plot of relative activity level as a function of the hour of the day (a) and relative activity duration as a function of the activity class (b).

ACKNOWLEDGMENTS

The author would like to thank Jim Kusznir, Aaron Crandall, Narayanan Krishnan, Allan Drassal, Leah Zulas, and all the members of the CASAS team for their contributions to this work. This material is based upon work supported by the National Science Foundation under Grant Number 0852172, by the Life Sciences Discovery Fund, and by NIBIB Grant Number R01EB009675.

REFERENCES

1. Department of Health, Speech by the Rt Hon Patricia Hewitt MP, Secretary of State for Health, in *Long-term Conditions Alliance Annual Conference*, 2007.
2. C. Kidd, R. Orr, G. Abowd, C. Atekson, I. Essa, B. MacIntrye, E. Mynatt, and W. Newstetter, The Aware Home: A living laboratory for ubiquitous computing research, in *Proceedings of the International Workshop on Cooperative Buildings*, 1999, pp. 1–9.
3. H. Hagras, V. Callaghan, M. Colley, G. Clarke, A. Pounds-Cornish, and H. Duman, Creating an ambient-intelligence environment using embedded agents, *IEEE Intelligent Systems*, 19(6), 12–20, 2004.
4. A. Helal, W. Mann, H. Elzabadani, J. King, Y. Kaddourah, and E. Jansen, Gator tech smart house: A programmable pervasive space, *IEEE Computer Magazine*, 38, 64–74, March 2005.
5. M. Mozer, The neural network house: An environment that adapts to its inhabitants, in *Proceedings of the AAAI Spring Symposium on Intelligent Environments*, Menlo Park, CA, 1998, pp. 110–114.
6. D. Cook, A. Campbell, S. Das, and R. Want, *NSF Workshop on Pervasive Computing at Scale*, Seattle, WA, 2011.
7. E. Kim, A. Helal, and D. Cook, Human activity recognition and pattern discovery, *IEEE Pervasive Computing*, 9(1), 48–53, 2010.
8. U. Maurer and A. Smailagic, D. Siewiorek, and M. Deisher, Activity recognition and monitoring using multiple sensors on different body positions, in *Proceedings of the International Workshop on Wearable and Implantable Body Sensor Networks*, Boston, MA, 2006, pp. 113–116.
9. J. Yin, Q. Yang, and J. J. Pan, Sensor-based abnormal human-activity detection, *IEEE Transactions on Knowledge and Data Engineering*, 20(8), 1082–1090, 2008.

10. N. Gyorbiro, A. Fabian, and G. Homanyi, An activity recognition system for mobile phones, *Mobile Networks and Applications*, 14, 82–91, 2008.
11. J. Kwapisz, G. Weiss, and S. Moore, Activity recognition using cell phone accelerometers, in *International Workshop on Knowledge Discovery from Sensor Data*, Washington, DC, 2010, pp. 10–18.
12. T. Gu, S. Chen, X. Tao, and J. Lu, An unsupervised approach to activity recognition and segmentation based on object-use fingerprints, *Data and Knowledge Engineering*, 69(6), 533–544, 2010.
13. P. Palmes, H. K. Pung, T. Gu, W. Xue, and S. Chen, Object relevance weight pattern mining for activity recognition and segmentation, *Pervasive and Mobile Computing*, 6(1), 43–57, 2010.
14. M. Philipose, K. P. Fishkin, M. Perkowitz, D. J. Patterson, D. Hahnel, D. Fox, and H. Kautz, Inferring activities from interactions with objects, *IEEE Pervasive Computing*, 3(4), 50–57, 2004.
15. D. Cook, Learning setting-generalized activity models for smart spaces, *IEEE Intelligent Systems,* 27(1), 32–38, 2012.
16. B. Logan, J. Healey, M. Philipose, E. M. Tapia, and S. Intille, A long-term evaluation of sensing modalities for activity recognition, in *Proceedings of the Ninth International Conference on Ubiquitous Computing*, Innsbruck, Austria, 2007.
17. L. Wang, T. Gu, X. Tao, and J. Lu, Sensor-based human activity recognition in a multi-user scenario, in *Proceedings of the European Conference on Ambient Intelligence*, Salzburg, Austria, 2009, pp. 78–87.
18. J. Yang, B. N. Schilit, and D. W. McDonald, Activity recognition for the digital home, *Computer*, 41(4), 102–104, 2008.
19. O. Brdiczka, J. L. Crowley, and P. Reignier, Learning situation models in a smart home, *IEEE Transactions on Systems, Man, and Cybernetics, Part B*, 39(1), 56–63, 2009.
20. D. J. Cook and M. Schmitter-Edgecombe, Assessing the quality of activities in a smart environment, *Methods of Information in Medicine*, 48(5), 480–485, 2009.
21. E. Munguia-Tapia, S. S. Intille, and K. Larson, Activity recognition in the home using simple and ubiquitous sensors, in *Pervasive*, 2004, pp. 158–175.
22. T. van Kasteren and B. Krose, Bayesian activity recognition in residence for elders, in *Proceedings of the IET International Conference on Intelligent Environments*, Ulm, Germany, 2007, pp. 209–212.
23. C. Lombriser, N. B. Bharatula, D. Roggen, and G. Troster, On-body activity recognition in a dynamic sensor network, in *Proceedings of the International Conference on Body Area Networks*, 2007.
24. I. L. Liao, D. Fox, and H. Kautz, Location-based activity recognition using relational Markov networks, in *International Joint Conference on Artificial Intelligence*, Edinburgh, Scotland, 2005, pp. 773–778.
25. D. Sanchez, M. Tentori, and J. Favela, Activity recognition for the smart hospital, *IEEE Intelligent Systems*, 23(2), 50–57, 2008.
26. D. H. Hu, S. J. Pan, V. W. Zheng, N. N. Liu, and Q. Yang, Real world activity recognition with multiple goals, in *International Conference on Ubiquitous Computing*, Seoul, South Korea, 2008, pp. 30–39.
27. D. Vail, M. Veloso, and J. Lafferty, Conditional random fields for activity recognition, in *Proceedings of the International Conference on Autonomous Agents and Multiagent Systems,* Honolulu, Hawaii, 2007.
28. A. Fleury, N. Noury, and M. Vacher, Supervised classification of activities of daily living in health smart homes using SVM, in *Proceedings of the International Conference of the IEEE Engineering in Medicine and Biology Society*, Minneapolis, MN, 2009, pp. 6099–6102.

29. T. Gu, Z. Wu, X. Tao, H. K. Pung, and J. Lu, epSICAR: An emerging patterns based approach to sequential, interleaved and concurrent activity recognition, in *Proceedings of the IEEE International Conference on Pervasive Computing and Communications*, Galveston, TX, 2009, pp. 1–9.
30. Y.-T. Chiang, K.-C. Hsu, C.-H. Lu, and L.-C. Fu, Interaction models for multiple-resident activity recognition in a smart home, in *Proceedings of the International Conference on Intelligent Robots and Systems*, Taipei, Taiwan, 2010, pp. 3753–3758.
31. C. Phua, K. Sim, and J. Biswas, Multiple people activity recognition using simple sensors, in *Proceedings of the International Conference on Pervasive and Embedded Computing and Communication Systems*, 2011, pp. 224–231.
32. C.-C. Chang and C.-J. Lin, LIBSVM: A library for support vector machines, *ACM Transactions on Intelligent Systems and Technology*, 2(27), 1–27, 2011.
33. P. Rashidi and D. J. Cook, Activity knowledge transfer in smart environments, *Pervasive and Mobile Computing*, 7, 331–343, 2011.
34. R. Agrawal and R. Srikant, Mining sequential patterns, in *Proceedings of the International Conference on Data Engineering*, 1995, pp. 3–14.
35. T. Barger, D. Brown, and M. Alwan, Health status monitoring through analysis of behavioral patterns, *IEEE Transactions on Systems, Man, and Cybernetics, Part A*, 35(1), 22–27, 2005.
36. J. Pei, J. Han, and W. Wang, Constraint-based sequential pattern mining: The pattern-growth methods, *Journal of Intelligent Information Systems*, 28(2), 133–160, 2007.
37. A. Aztiria, J. Augusto, and D. Cook, Discovering frequent user–environment interactions in intelligent environments, *Personal and Ubiquitous Computing*, 16(1), 91–103, 2012.
38. E. O. Heierman and D. J. Cook, Improving home automation by discovering regularly occurring device usage patterns, in *IEEE International Conference on Data Mining*, Melbourne, FL, 2003, pp. 537–540.
39. M. Ruotsalainen, GAIS: A method for detecting interleaved sequential patterns from imperfect data, *IEEE Symposium on Computational Intelligence and Data Mining*, 530–534, 2007.
40. J. Pei, J. Han, M. B. Asl, H. Pinto, Q. Chen, U. Dayal, and M. C. Hsu, PrefixSpan: Mining sequential patterns efficiently by prefix projected pattern growth, in *Proceedings of the International Conference on Data Engineering*, 2001, pp. 215–226.
41. M. J. Zaki, N. Lesh, and M. Ogihara, Planmine: Sequence mining for plan failures, in *Proceedings of the International Conference on Knowledge Discovery and Data Mining*, New York, 1998, pp. 369–373.
42. Y.-I. Chen, S.-S. Chen, and P.-Y. Hsu, Mining hybrid sequential patterns and sequential rules, *Information Systems*, 27(5), 345–362, 2002.
43. P. Rashidi, D. Cook, L. Holder, and M. Schmitter-Edgecombe, Discovering activities to recognize and track in a smart environment, *IEEE Transactions on Knowledge and Data Engineering*, 23(4), 527–539, 2011.
44. T. Huynh, M. Fritz, and B. Schiele, Discovery of activity patterns using topic models, in *International Conference on Ubiquitous Computing*, Seoul, South Korea, 2008, pp. 10–19.
45. J. Rissanen, *Stochastic Complexity in Statistical Inquiry*, World Scientific Publishing Company, River Edge, NJ, 1989.
46. V. I. Levenshtein, Binary codes capable of correcting deletions, insertions, and reversals, *Soviet Physics Doklady*, 10(8), 707–710, 1966.
47. A. Faruqui, S. Sergici, and A. Sharif, The impact of informational feedback on energy consumption—A survey of the experimental evidence, *Energy*, 35, 1598–1608, 2010.

4 Localization of a Wireless Sensor Network for Environment Monitoring Using Maximum Likelihood Estimation with Negative Constraints

Kay-Soon Low and Hao Guo

CONTENTS

ABSTRACT

In many environmental monitoring applications, the location of the sensor node is important information. Due to the large number of sensor nodes to be deployed, it is not practical to equip them with global positioning system (GPS) or manually determine their locations. In this chapter, a smart localization algorithm using maximum likelihood estimation (MLE) with negative constraints (NCs) is proposed. Unlike most of the existing methods that only utilize positive constraint information such as internode distances or connectivity, the proposed algorithm also utilizes NC information to achieve more accurate localization. The distribution of sensor nodes' communication ranges is first studied, and the likelihood function of sensor nodes' positions is derived based on both the positive and negative constraints. To reduce the computational cost, a novel iterative optimization procedure is also proposed to find the MLE. Simulation and experimental works show that the proposed MLE localization algorithm with NC improves the localization accuracy by 20% as compared to the conventional MLE approach.

4.1 INTRODUCTION

One of the key challenges of wireless sensor network (WSN) is to determine the sensor nodes' physical locations. This can be achieved by equipping a global positioning system (GPS) to each sensor node. However, such approach is costly, consumes higher power, and is subject to the availability of GPS signal. To overcome these limitations, a number of GPS-less localization systems have been investigated for WSNs [1–16].

Based on the computational architecture, they can be classified as distributed and collaborative centralized algorithms. In distributed algorithms [1–10], computation is distributed across the network. Each node is responsible for computing its own estimated position using the local information. Collaborative centralized algorithms [11–16] assume there is a central node that collects information across the network, and the estimated positions are computed in the central node and the whole network is localized collaboratively. Based on the information used, the localization algorithms can also be classified as range-free and range-based algorithms. Range-free algorithms [8–12] assume that the distance or angle information is not available for the sensor node. They use the network connectivity to proximate the nodes' locations. Range-based algorithms [2–7,13–16] require distance measurements between neighboring sensor nodes. They usually use multilateration or maximal likelihood estimation techniques to find the locations of unknown nodes.

Besides the network connectivity information or internode distance measurements, some works also use negative constraints (NCs) to improve the localization accuracy. The NCs use the observation that if there is no communication link between two sensor nodes, then the distance between them should be longer than their communication range. In Xiao et al. [10], an anchor node is used to give a repulsive virtual force to repulse the estimated position of an unknown node if it is out of the anchor node's communication range. As it is a distributed algorithm, it only uses the NCs between an unknown and an anchor node within two hops. The NCs between two unknown nodes are not utilized. Consider the centralized algorithm; the NCs can be used more advantageously since the central node has the knowledge of the whole network. However, this issue has not been well addressed, and there is a lack of works on the effect of the NCs on the WSN localization performance.

In this chapter, a maximum likelihood estimation (MLE) WSN localization algorithm that uses the NC and internode distance information from received signal strength indicator (RSSI) measurements is proposed. Different from other works that assume the communication coverage of the sensor node is a perfect circle, this paper studies the communication range distribution of the sensor node based on the log-shadowing model. The likelihood function of the positions of the network is derived with the NC and internode distance information. To find the MLE of the network positions, a least square (LS) optimization problem needs to be solved. The complexity of the optimization problem is largely dependent on the number of NCs being used. To reduce the computational costs, a novel iterative optimization procedure is proposed. In each procedure loop, only the important NCs are being used for the localization. Simulation and experimental works show that the proposed MLE localization algorithm with NCs improves the localization accuracy by 20% as compared to the conventional MLE localization without NCs.

The organization of the chapter is as follows: Section 4.2 presents the error modeling of the NCs, as well as the problem formulation using the proposed MLE localization algorithm with NCs. The optimization procedure to find the MLE of the network positions is presented in Section 4.3. Section 4.4 evaluates the proposed localization approach under various network environments through simulation. Section 4.5 describes the experimental system as well as outdoor experimental results. Section 4.6 concludes the work.

4.2 MLE LOCALIZATION ALGORITHM WITH NEGATIVE CONSTRAINTS

The MLE method is a popular approach in obtaining practical estimators [17]. MLE localization algorithms for WSN without NCs can be readily found in some existing works [13–14].

4.2.1 MLE without Negative Constraints

In almost all existing wireless systems, RSSI can readily be measured after two nodes have established wireless communication connection. Due to its low cost, RSSI has become a widely used ranging technology in WSN localizations. From the log-shadowing model [18] for the path loss, it can be derived that the estimated distance $\tilde{d}_{ij}$ between two nodes i and j is lognormally distributed:

$$\ln \tilde{d}_{ij} \sim \mathrm{N}(\ln d_{ij}, \sigma^2) \tag{4.1}$$

Assume that an N-node sensor network comprises of N_1 unknown nodes and N_2 anchor nodes, that is, $N=N_1+N_2$. Let $\mathbf{X}$ be the vector of the unknown nodes' positions:

$$\mathbf{X} = [x_1 \cdots x_{N_1} \quad y_1 \cdots y_{N_1}]^{\mathrm{T}} \tag{4.2}$$

Let d_{ij} $(\mathbf{X}) = \|\theta_i \theta_j\|$, where $i=1,\ldots,N_1$, $j=1,\ldots,N$, and $\theta_i=[x_i\, y_i]^{\mathrm{T}}$ is the location coordinates of node i. Given (4.1), the likelihood function of $\mathbf{X}$ based on the RSSI estimated distances can be formulated as

$$L(\mathbf{X}) = \prod_{i=1}^{N_1} \prod_{j=i+1}^{N} L_{ij}(\mathbf{X}) \tag{4.3}$$

where

$$L_{ij}(\mathbf{X}) = \begin{cases} \dfrac{1}{d_{ij}(\mathbf{X})\sqrt{2\pi\sigma^2}} \exp\left(-\dfrac{(\ln d_{ij}(\mathbf{X}) - \ln \tilde{d}_{ij})^2}{2\sigma^2} \right), & \text{if } \tilde{d}_{ij} \text{ is observed} \\ 1, & \text{otherwise} \end{cases} \tag{4.4}$$

The position vector $\hat{\mathbf{X}}$ that maximizes (4.4) is the maximum likelihood estimate. It is equivalent to the following optimization problem:

$$\hat{\mathbf{X}} = \arg\min_{\mathrm{X}} \left(\sum_{i=1}^{N_1} \sum_{j=i+1}^{N} w_{ij} (\ln d_{ij}(\mathbf{X}) - \ln \tilde{d}_{ij})^2 \right) \tag{4.5}$$

where

$$w_{ij} = \begin{cases} 1, & \text{if } \tilde{d}_{ij} \text{ is observed} \\ 0, & \text{otherwise} \end{cases} \tag{4.6}$$

4.2.2 Negative Constraints and Modeling

Consider Figure 4.1 with only the distance information available to anchor nodes A_1 and A_2; there are two possible estimated positions for sensor node i, namely, i_1 and i_2. With additional information that node i is out of sensor node j's communication range, i's real position is more likely near to i_1 than i_2. This information is termed negative constraint (NC). On the other hand, an RSSI-measured internode distance is called positive constraint.

Empirical studies on real test beds have shown that the assumption of a perfect circular radio range is not accurate. The communication range of a sensor node usually varies in different directions. When two sensor nodes cannot establish communication, the signal strength between the two nodes i and j is below a certain threshold η. Let R be j's communication range in certain direction, and then from the log-shadowing model [18],

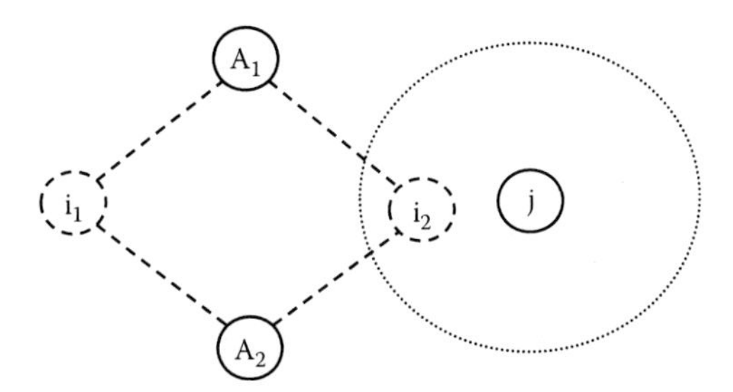

FIGURE 4.1 The NC illustration.

$$\eta = P_0 - 10\alpha \log_{10} R + v_{ij} \tag{4.7}$$

where

P_0 is the RSSI measurement at 1 m distance

α is the path-loss exponent

v_{ij} is a normally distributed random number with zero mean and variance of σ_v^2, that is, $v_{ij} \sim N(0, \sigma_v^2)$

Given η, the communication range R between two nodes follows the lognormal distribution:

$$\ln R \sim N(\ln \bar{R}, \sigma^2) \tag{4.8}$$

where

$\sigma = (\sigma_v \text{ In } 10)/(10\alpha)$

$\bar{R} = 10^{(P_0 - \eta)/(10\alpha)}$ and $\bar{R}$ is called the nominal communication range in this chapter

Let $f_j(R)$ be the probability dense function (pdf) of j's communication range distribution in any direction, from (4.8):

$$f_j(R) = \frac{1}{\sqrt{2\pi}\sigma R} \exp\left(-\frac{(\ln R - \ln \bar{R})^2}{2\sigma^2} \right) \tag{4.9}$$

The likelihood function of node i's position based on this negative constant that i is out of j's communication range can be formulated as

$$L_{ij}(\mathbf{X}) = \int_0^{d_{ij}(\mathbf{X})} f_j(R)\, dR = c_j(d_{ij}(\mathbf{X})) \tag{4.10}$$

where $c_j(R)$ is the cumulative distribution function (cdf) of j's communication range:

$$c_j(R) = \frac{1}{2} + \frac{1}{2} \operatorname{erf}\left(\frac{\ln R - \ln \bar{R}}{\sqrt{2}\sigma} \right) \tag{4.11}$$

4.2.3 MLE with Negative Constraints

Combining the positive constraint and NC information, the overall likelihood function of **X** can be formulated as

$$\bar{L}(\mathbf{X}) = \prod_{i=1}^{N_1} \prod_{j=i+1}^{N} L_{ij}(\mathbf{X}) \tag{4.12}$$

where

$$L_{ij}(\mathbf{X}) = \begin{cases} \dfrac{1}{d_{ij}(\mathbf{X})\sqrt{2\pi\sigma^2}} \exp\left(-\dfrac{(\ln d_{ij}(\mathbf{X}) - \ln \tilde{d}_{ij})^2}{2\sigma^2}\right), & \text{if } \tilde{d}_{ij} \text{ is observed} \\ \dfrac{1}{2} + \dfrac{1}{2}\operatorname{erf}\left(\dfrac{\ln d_{ij}(\mathbf{X}) - \ln \bar{R}}{\sqrt{2}\sigma}\right), & \text{otherwise} \end{cases} \tag{4.13}$$

The maximum likelihood estimate $\hat{\mathbf{X}}$ can be obtained as

$$\hat{\mathbf{X}} = \arg\min_{\mathrm{X}} \left(\sum_{i=1}^{N_1} \sum_{j=i+1}^{N} w_{ij} \left(\frac{(\ln d_{ij}(\mathbf{X}) - \ln \tilde{d}_{ij})^2}{2\sigma^2} \right) + (w_{ij} - 1) \ln \left(\frac{1}{2} + \frac{1}{2}\operatorname{erf}\left(\frac{\ln d_{ij}(\mathbf{X}) - \ln \bar{R}}{\sqrt{2}\sigma} \right) \right) \right) \tag{4.14}$$

4.3 POSITION ESTIMATION

The optimization problems defined in both (4.5) and (4.14) can be viewed as nonlinear LS problems. There are several iterative numerical optimization algorithms that can be used for such nonlinear LS problems. For example, the Gauss–Newton, line-search, and trust-region methods. An initial value is required for such iterative algorithms. In general, a good initial value would result in faster convergent to the global optimum. This can be achieved by estimating the initial positions using low-cost localization algorithms such as DV-hop, DV-distance, or multi-dimensional scaling (MDS).

Compared to the objective function without NCs, the objective function (4.14) is more complicated especially in sparse networks. Let c be the network connectivity, which is defined as the average number of neighboring nodes of each unknown node. There will be $(1/2)cN_i$ summation terms in the optimization problem defined in (4.5). However, if all the NCs are being used, the number of summation terms will be $(1/2)(N_i+2N_a)\,N_i$ as defined in (4.14). For a network with 100 sensor nodes with a connectivity less than 10, direct minimization of (4.14) would require 10 times more computation. As many of the NCs provide little information on the position estimation, not all the NCs are included to the objective function. For example, if two estimated positions are already very far from each other in the initial value, there is almost no difference between the results from MLE with and without the corresponding NC. Thus, an iterative procedure is developed in this study to select the NCs and find the position estimate. The procedure is described in the following:

Step 1: Obtain initial value

Obtain initial value by localizing the network from other low-cost algorithms (DV-distance, MDS, etc.). DV-distance is used in the following discussions.

Step 2: MLE

Estimate the sensor nodes' positions without considering any NCs, through minimizing the objective function (4.5) and using the initial value obtained from Step 1.

Step 3: NC selection

Check if any NCs are violated by the estimated positions. If no new NC is violated, go to Step 6. Otherwise, go to Step 4.

An NC is said to be violated if

1. There is no communication links between its associated two sensor nodes
2. The distance between the two estimated positions is smaller than a predefined distance. Usually, the predefined distance is related to the nominal communication range, that is, if $d_{ij}(\hat{\mathbf{X}}) < k \cdot \bar{R}$, the NC between node i and j is violated, and k is termed as the NC selection factor

The factor k needs to be chosen carefully before the localization. If k is too small, very few NCs will be violated and the localization result will be very similar to the result without NCs. On the other hand, if k is too large, too many NCs will be included and the computational cost for the optimization will be increased significantly. As discussed in Section 4.4.1, the best choice of k would be 1.1 for the case study used in this paper.

Step 4: Objective function update

Update the objective function by including the new violated NCs. The new optimization problem becomes

$$\hat{\mathbf{X}} = \arg\min_{\mathrm{X}} \left(\sum_{i=1}^{N_1} \sum_{j=i+1}^{N} w_{ij} \left(\frac{(\ln d_{ij}(\mathbf{X}) - \ln \tilde{d}_{ij})^2}{2\sigma^2} \right. \right.$$

$$\left. \left. + (w_{ij} - 1) v_{ij} \ln \left(\frac{1}{2} + \frac{1}{2} \operatorname{erf} \left(\frac{\ln d_{ij}(\mathbf{X}) - \ln \bar{R}}{\sqrt{2}\sigma} \right) \right) \right) \right) \tag{4.15}$$

with

$$v_{ij} = \begin{cases} 1, & \text{NC between } i \text{ and } j \text{ is violated} \\ 0, & \text{otherwise} \end{cases}$$

Step 5: NC optimization

Use the current estimated position as the initial value to minimize the new objective function (4.15), and find the new estimated positions. Go to Step 6 if it exceeds the maximum number of loops; otherwise, go to Step 3.

Step 6: End of the optimization procedure

4.4 SYSTEM PERFORMANCE EVALUATION

In this section, the performance of the proposed localization scheme is evaluated through various simulation studies. The first simulated network is an isotropic network that is placed within a 100×100 unit's sensor field. The sensor field is equally partitioned to 10×10 squares and there is a sensor randomly placed in each of the square. Therefore, there are a total of 100 sensor nodes, assuming that 10% of the nodes are anchor nodes and they are randomly chosen. The range measurement is assumed log-normally distributed, and σ ranges from 0.05 to 0.5 in different simulations.

4.4.1 Choice of the Negative Constraints Selection Factor k

The NC selection factor k is an important parameter. If k is too small, very few NCs will be violated and the optimization problem (4.15) is close to the optimization problem defined in (4.5). Consequently, the localization result will be similar to the result without NCs. On the other hand, large k leads to too many NC inclusion. Equation 4.15 will be close to 4.14 and the computational cost increased significantly. The effect of k on localization accuracy and the number of violated NCs are plotted in Figures 4.2 and 4.3, respectively. The data points on the graph are the averages of 50 simulation trials.

From Figure 4.2, it is observed that the localization error decreases as k increases for all the network topologies and range measurements. This is due to more NCs that are included to constraint the position estimation for larger k. However, the reduction in localization error is less significant when k is larger than 1.1. On the other hand, it is also observed from Figure 4.3 that the number of violated NCs is increased significantly when k is larger than 1.1. Therefore, it may be concluded from these results that a value of 1.1 is a reasonable choice for the NC selection factor k. In the later discussion, k is fixed at 1.1.

4.4.2 Performance Comparison of MLE with and without Negative Constraints

To study the effect of the NCs on localization accuracy, Figure 4.4 compares the localization error of MLE without and with NCs (MLE-NC) under different network connectivities and range measurement errors. From Figure 4.4, the localization accuracy is improved consistently by including the NCs, under different network connectivity and range measurement error σ.

Figure 4.5 plots the percentage of the reduced localization error by the NCs under different network connectivity and range error. From Figure 4.5, the localization

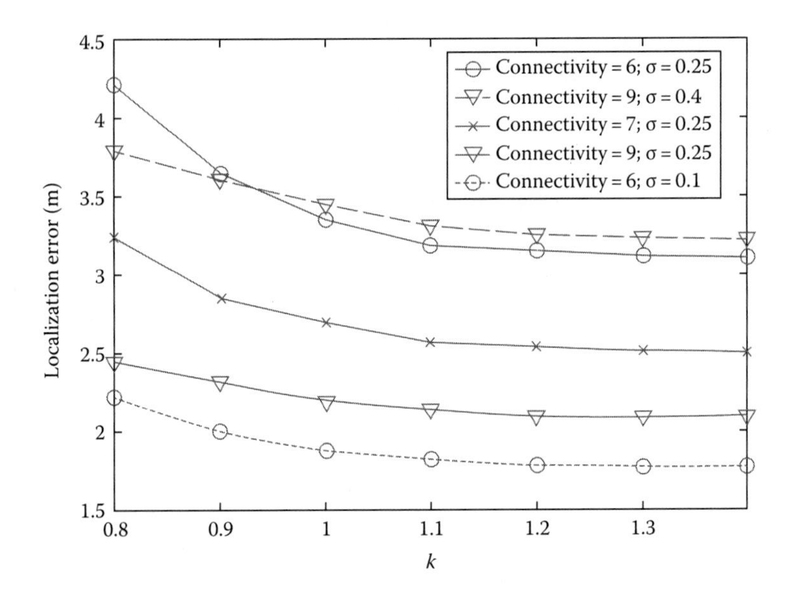

FIGURE 4.2 Effect of k on the localization accuracy.

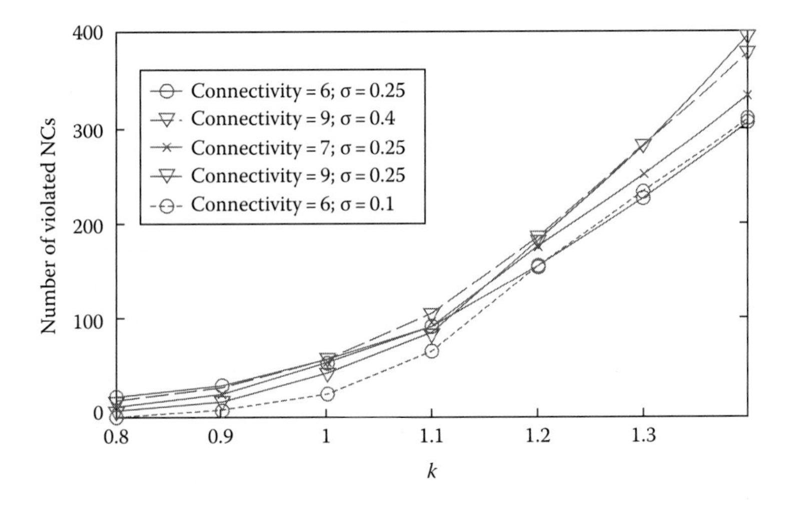

FIGURE 4.3 Effect of k on the number of violated NCs.

accuracy with NC is improved by more than 20% when the network connectivity is less than 8. It is noticed that this accuracy improvement decreases as the connectivity increases. When the connectivity is larger than 9, the localization accuracy improvement from the NCs is less than 15%. This is because higher network connectivity means more positive constraints and less NCs are available. The estimated positions from only the positive constraints are more accurate, and fewer NCs are

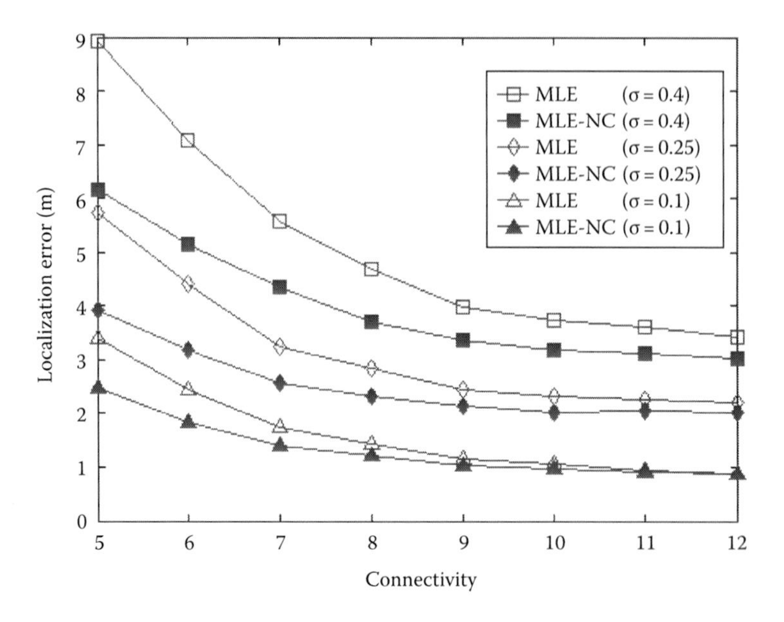

FIGURE 4.4 Comparison of localization accuracy of MLE and MLE-NC.

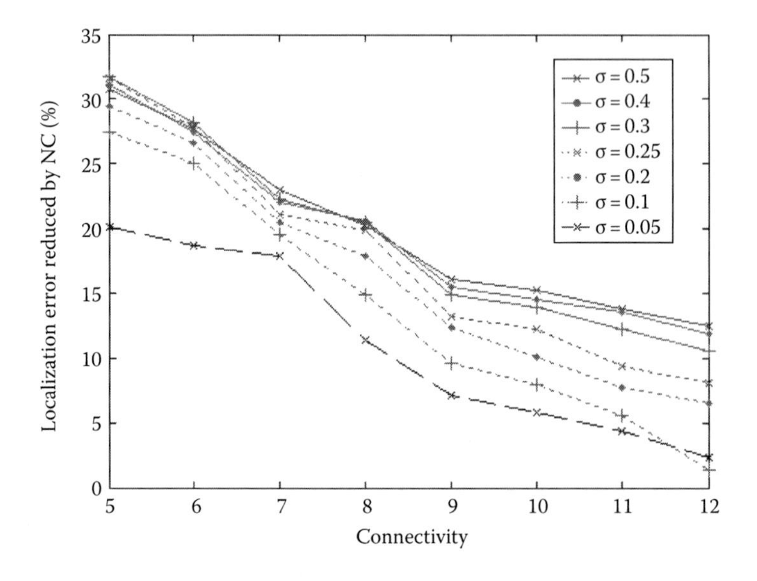

FIGURE 4.5 Localization accuracy improvement by NC under different range measurement errors and connectivities.

violated for higher network connectivity. Therefore, localization accuracy improvement by the NCs is less significant for heavily connected networks.

Figure 4.5 also shows the accuracy improvement by the NCs under a different distance measurement error, that is, σ. The accuracy improvement by the NCs generally increases with the σ, especially when the distance measurement is relatively accurate ($\sigma < 0.25$). This is because less accurate estimated positions from MLE lead to more violated NCs. It is also noticed from (4.8) that a larger σ leads to more irregularity of the communication ranges. This implies that both the NC and positive constraint information become less accurate for a larger σ. Therefore, the improvement from the NCs is relatively consistent when $\sigma > 0.3$.

4.4.3 Performance under Non-Line-of-Sight Environment

In this simulation, we study the performance of NCs under non-line-of-sight (NLOS) environments. Several 10-unit length obstacles are placed within the sensor field in random places and orientations. Any communication link that passes through any obstacles will be blocked, as shown in Figure 4.6. Therefore, some NC information would be misleading since two nodes that are very close to each other would still be out of communication coverage due to the obstacles. Figure 4.7 compares the localization errors of MLE and MLE-NC with different number of obstacles. The average network connectivity is 7 when there is no obstacle and $\sigma = 0.25$ is used in this simulation.

It is noticed that the presence of obstacles causes the network connectivity to be reduced. Consequently, it leads to larger localization error for both methods. Moreover, more obstacles also lower the localization coverage. Some sensor nodes or groups of sensor nodes are isolated from other nodes due to the blocking of their

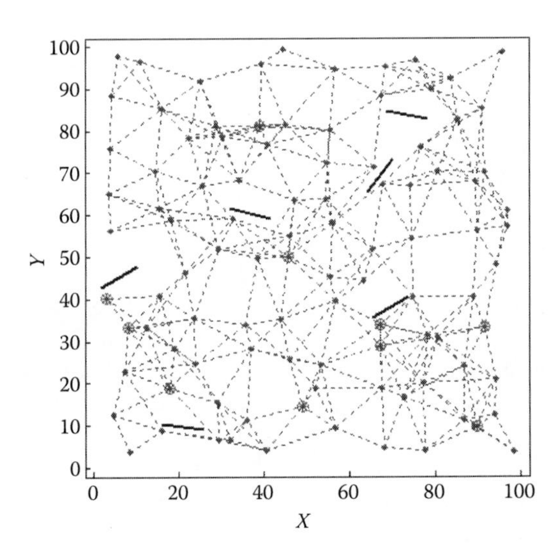

FIGURE 4.6 Network under NLOS environment: '•' unknown nodes; '⊛' anchor nodes; '_ _ _' communication links; and '__' obstacles.

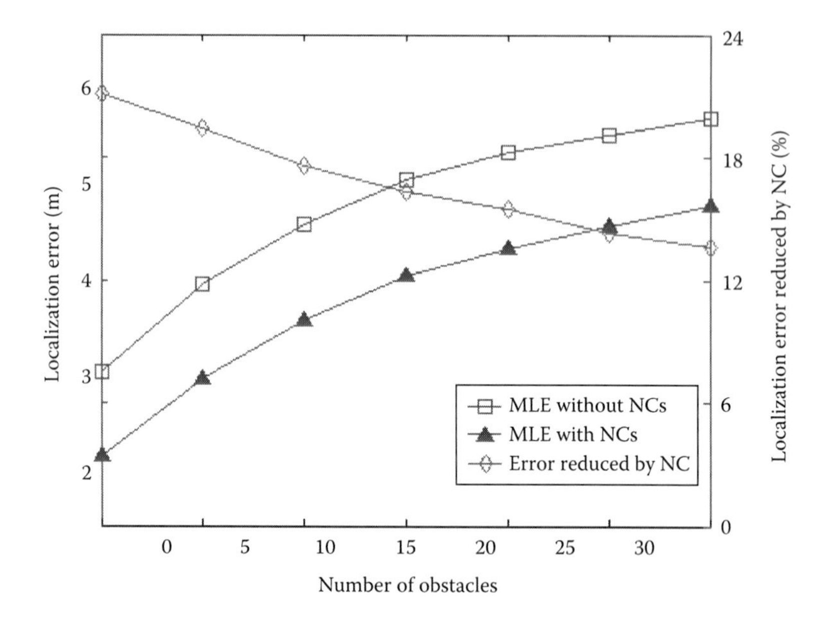

FIGURE 4.7 Localization accuracy under NLOS environments.

communication links, and therefore they are not able to find their estimated positions. The localization error in Figure 4.7 is the average localization error of localized sensor nodes. As shown in Figure 4.7, NCs can still improve the localization accuracy, but the improvement by percentage decreases as the number of obstacles increases.

4.4.4 Localization in Uniform Randomly Distributed Networks

Localization performance for uniform randomly distributed networks is also studied. In this simulation, 100 sensor nodes are randomly placed in the 100×100 unit's sensor field, and 10% of the sensor nodes are anchor nodes. Different from the networks simulated in Sections 4.4.1 and 4.4.2, the networks are generally not isotropic. The sensor node's density varies in different parts of the network, and it is possible that some unknown nodes cannot be localized. An unknown node will not be localized if the sensor node is isolated from the rest of the network or belongs to an isolated group that has less than three noncolinear anchor nodes. Figure 4.8 plots the localization coverage of uniform randomly distributed networks under different communication ranges. The localization coverage is defined as the percentage of the unknown nodes that can be successfully localized. From the figure, nearly all the unknown nodes can be localized when the communication range is larger than 22 m, while the localization coverage drops to 70% with 12 m communication range. The figure also shows the relationship between the network connectivity and communication range. The network connectivity increases almost linearly with the communication range.

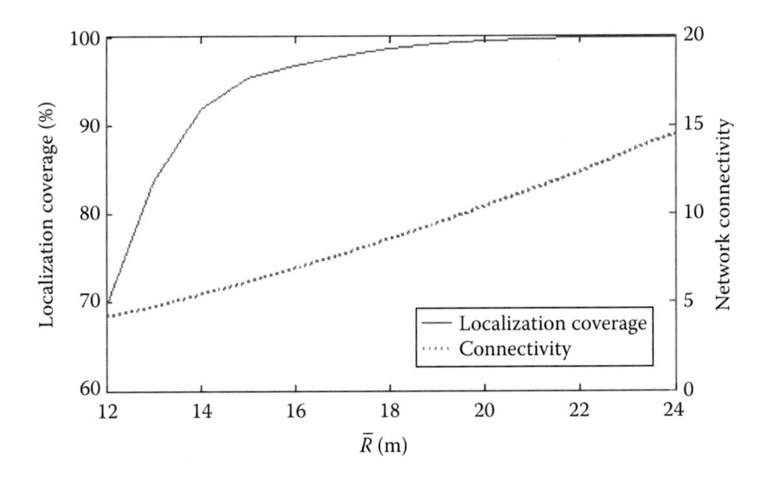

FIGURE 4.8 Localization coverage for uniform randomly distributed networks.

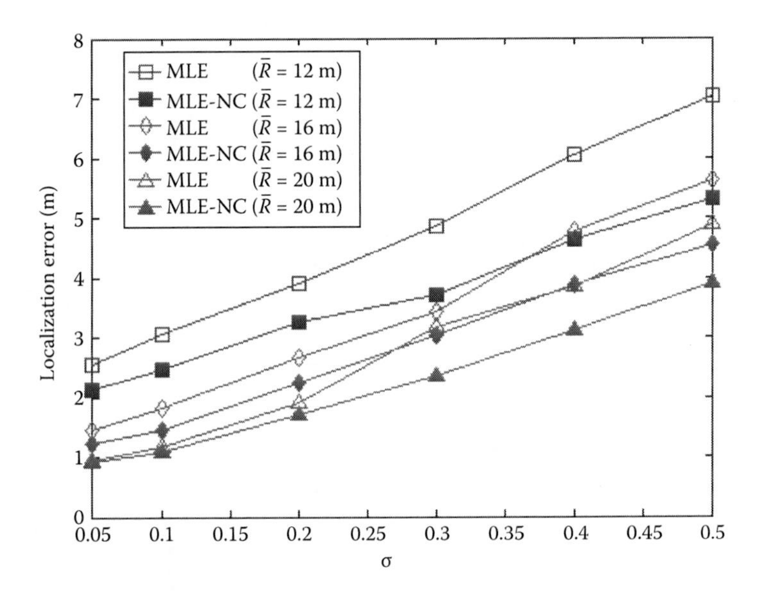

FIGURE 4.9 Localization error for uniform randomly distributed networks.

To study the localization performance over uniform randomly distributed networks, Figure 4.9 compares the localization error of MLE without and with NCs (MLE-NC) under different communication ranges and range measurement error. As shown in Figure 4.9, MLE-NC consistently outperforms MLE in terms of localization accuracy. By including the NCs, the localization error is reduced by 5%–30% depending on the range measurement error and the communication ranges. Similar to

TABLE 4.1
Computational Costs

	MLE Optimization	NC Optimization
CPU time (s)	58.3	27.4
Number of iteration	146.0	57.0
Objective function	24,168.0	9,026.0

Section 4.4.2, the localization accuracy improvement is more significant for larger range measurement error and lower network connectivity.

4.4.5 Computational Cost and Performance Trend

During the position estimation, most computation is spent on the MLE optimization and the NC optimization (Steps 2 and 5 of the optimization procedure). Both two steps are LS optimization problems, and Gauss–Newton algorithm with line-search method is used to perform the optimizations. Table 4.1 lists the average required central processing unit (CPU) time, the number of iterations, and the objective function evaluations before meeting the stopping criterion. The stopping criterion is set as the objective function value changes within 0.1. It is noted that Step 5 may be executed more than once. The values of the NC optimizations in Table 4.1 are the sums of all procedure loops. The average number of procedure loops is about 1.1.

As shown in Table 4.1, the NC optimization converges much faster than the MLE optimization. Even though its objective function is more complicated, NC optimization requires less than half the CPU time of the MLE optimization. Overall, the proposed MLE-NC scheme requires less than 1.5 times of computational cost of conventional MLE scheme.

4.5 EXPERIMENTAL RESULTS

Experimental measurement has been conducted in a 90 m×90 m park located in the university campus. The network consists of 25 sensor nodes. Four of them are anchor nodes and are placed at the corners of the sensor field. The rest of the sensor nodes are unknown nodes and are placed in a pseudorandom manner as shown in Figure 4.10 over Google Earth map. Each sensor node is equipped with an XBee ZNet 2.5 OEM RF module, which is able to measure the RSSI. Before the deployment, calibration has been conducted to determine the parameters used in the path-loss equation by measuring the RSSI with respect to reference distance. From the calibration, the standard deviation of the log range error is determined with $\sigma = 0.22$. The nominal communication range is 40 m, which leads to the average connectivity of 7.62 for the unknown nodes.

To avoid data collision, a multiple access protocol is used in the experiment. The sensor nodes collect their internode RSSI measurements asynchronously. A sensor node polls its neighboring sensor nodes in sequent and requests them to send

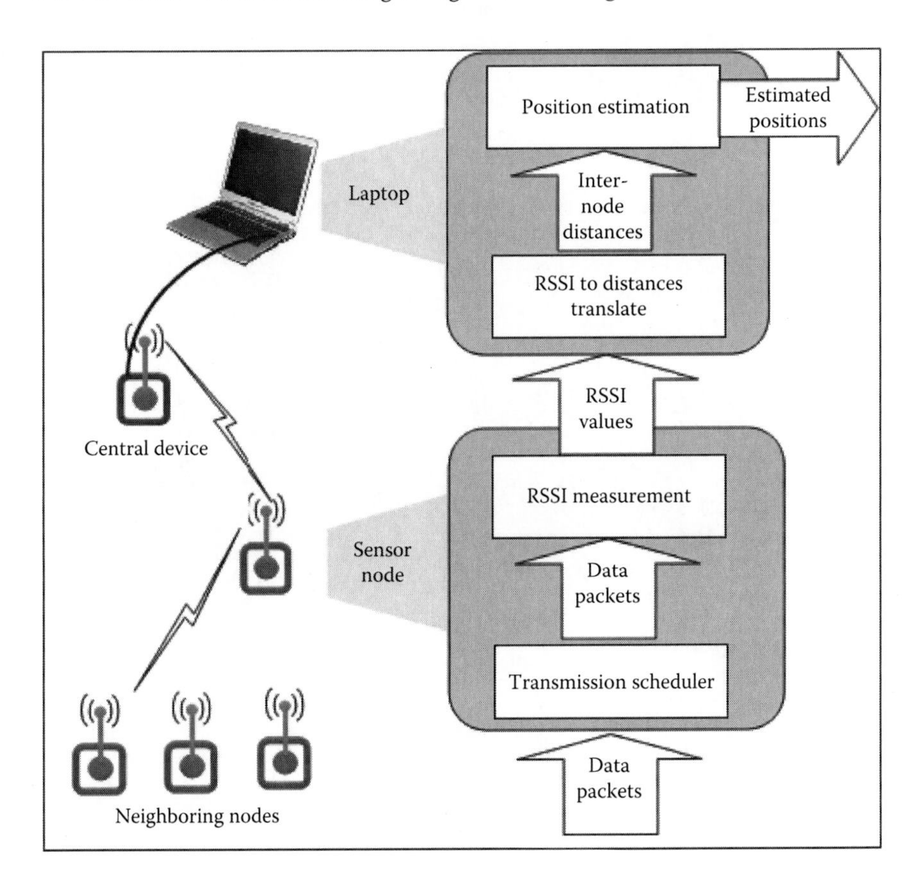

FIGURE 4.10 Experimental system structure.

a data packet. The sensor node measures the RSSI of the received packets and transmits it to a central device, which uploads the data to a laptop. The laptop translates the RSSI measurements to distances and estimates the sensor nodes' positions through the optimization procedure proposed in Section 4.3. Figure 4.11 depicts this system structure.

The experimental results of both the MLE localization without and with the NCs are shown in Figure 4.11. The average localization errors are 5.27 and 4.14 m, respectively. Thus, the NCs reduce the localization error by about 21%.

To study the performance of the proposed localization scheme under different network topology and connectivity, after the localization of the whole network as depicted in Figure 4.10, a few numbers of the unknown nodes are randomly taken out from the network and the rest of the sensors are relocalized by using the remaining positive constraint and NC information. Table 4.2 shows the localization error of the new networks. As shown in Table 4.2, the localization error is reduced by about 20% by including NCs, and generally the improvement is more significant when the network connectivity is low. This is consistent with the finding in Section 4.4.2.

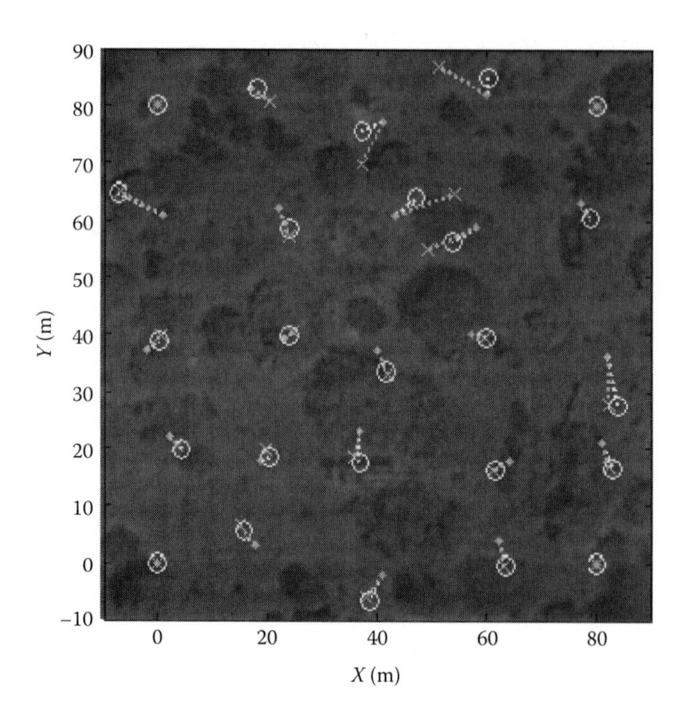

FIGURE 4.11 Experimental network located at Yunan garden in university campus and localization result: '•' unknown nodes' real positions; '❄' anchor nodes; 'x' estimated positions without NCs; '_ _ _' difference between estimated positions without NCs and real positions; 'O' estimated positions with NCs; and '_ _ _' difference between estimated positions with NCs and real positions.

TABLE 4.2
Summary of Experimental Results

Number of Unknown Nodes	Connectivity	Localization Error without NCs (m)	Localization Error with NCs (m)	Improvement by the NCs (%)
21	7.62	5.27	4.17	20.9
18	6.56	6.28	5.21	17.1
15	5.62	7.53	5.84	22.4
12	4.5	9.63	7.21	25.2

4.6 CONCLUSIONS

In this chapter, the likelihood function of sensor nodes' positions with NC information is derived and a new localization algorithm with NCs for WSNs is proposed. The effect of the NCs on the localization accuracy of the MLE localization algorithm has been investigated. Simulation and experimental works show that the proposed approach improves the localization accuracy by 20%.

REFERENCES

1. Langendoen, K. and Reijers, N., Distributed localization is wireless sensor networks: A quantitative comparison, *Computer Networks*, 43, 499–518, 2003.
2. Guo, H., Low, K. S., and Nguyen, H. A., Optimizing the localization of a wireless sensor network in real time based on a low cost microcontroller, *IEEE Transactions on Industrial Electronics*, 58(3), 741–749, March 2011.
3. Nguyen, H. A., Guo, H., and Low, K. S., Real time estimation of sensor node's position using particle swarm optimization with log barrier constraint, *IEEE Transactions on Instrumentation & Measurements*, 60(11), 3619–3628, November 2011.
4. Yu, K., Guo, Y. J., and Hedley, M., TOA-based distributed localisation with unknown internal delays and clock frequency offsets in wireless sensor networks, *Signal Processing, IET*, 3(2), 106–118, March 2009.
5. Liu, J., Zhang, Y., and Zhao, F., Robust distributed node localization with error management, *Proceedings of the Seventh ACM International Symposium on Mobile Ad Hoc Networking and Computing*, Florence, Italy, May 22–25, 2006.
6. Lim, H. and Hou, J. C., Distributed localization for anisotropic sensor networks, *ACM Transactions on Sensor Networks*, 5(2), Article 11, March 2009.
7. Wei, Q., Han, J., Zhong, D., and Liu, R., An improved multihop distance estimation for DV-hop localization algorithm in wireless sensor networks, *Proceedings of Vehicular Technology Conference*, Quebec City, Quebec, Canada, 2012, pp. 1–5.
8. Huircan, J. I. and Munoz, C., Zigbee-based wireless sensor network localization for cattle monitoring in grazing field, *Computers and Electronics in Agriculture*, 74(2), 258–264, November 2010.
9. Ou, C. H. and He, W. L., Path planning algorithm for mobile anchor-based localization in wireless sensor networks, *IEEE Sensor Journal*, 13(2), 466–475, February 2013.
10. Xiao, B., Chen, H. K., and Zhou, S., Distributed localization using a moving beacon in wireless sensor networks, *IEEE Transactions on Parallel and Distributed Systems*, 19(5), 587–600, May (2008).
11. Shang, Y., Ruml, W., Zhang, Y., and Fromherz, M. P. J., Localization from connectivity in sensor networks, *IEEE Transactions on Parallel and Distributed Systems*, 15(11), 961–974, November 2004.
12. Li, L. and Kunz, T., Cooperative node localization using nonlinear data projection, *ACM Transactions on Sensor Networks*, 5(1), 1–26, February 2009.
13. Neal, P., Alfred, O. H. et al., Relative location estimation in wireless sensor networks, *IEEE Transactions on Signal Processing*, 51(8), 2137–2148, August 2003.
14. Li, X., Collaborative localization with received-signal strength in wireless sensor networks, *IEEE Transactions on Vehicular Technology*, 56(4), 3807–3817, November 2007.
15. Chan, F., So, H. C., and Ma, W. K., A novel subspace approach for cooperative localization in wireless sensor networks using range measurements, *IEEE Transactions on Signal Processing*, 57(1), 260–269, January 2009.
16. Lui, K. W. K., Ma, W.-K., So, H. C., and Chan, F. K. W., Semi-definite programming algorithms for sensor network node localization with uncertainties in anchor positions and/or propagation speed, *IEEE Transactions on Signal Processing*, 57(2), 752–763, February 2009.
17. Kay, S. M., *Fundamentals of Statistical Signal Processing: Estimation Theory*, Englewood Cliffs, NJ: Prentice-Hall, 1993.
18. Rappaport, T. S., *Wireless Communication: Principles and Practice*, Englewood Cliffs, NJ: Prentice-Hall, 1996.

5 Reconfigurable Intelligent Space and the Mobile Module for Flexible Smart Space

JongSeung Park and Joo-Ho Lee

CONTENTS

ABSTRACT

In this chapter, Reconfigurable Intelligent Space (R+iSpace) is introduced. R+iSpace was proposed to overcome the inadequacies of conventional smart space. The devices in R+iSpace can change their position according to the current requirement of the space. By changing their position, the performance of the entire system can be improved. The Mobile Module, which is called MoMo, is a wall/ceiling surface robot to suit the requirements of R+iSpace. In this chapter, the structure of the prototype MoMo is also described.

5.1 INTELLIGENT SPACE

5.1.1 Previous Research of the Intelligent Space

From the mid-1990s, a lot of research has been carried out on the intelligent environment. This research is based on the fact that lots of sensors can be located around a room, which will constitute an extended environmental system. Examples of such research projects are the Oxygen Project of MIT, EasyLiving Project of Microsoft, DreamSpace of IBM, and Intelligent Space of the University of Tokyo [1–4] (Figure 5.1).

Intelligent Space (iSpace) was first proposed in 1996 by the Hashimoto Laboratory at the University of Tokyo. iSpace is a system that provides appropriate services to users in the space by using various devices and agent robots. Figure 5.2 is a conceptual diagram of iSpace. As shown in the figure, lots of DINDs (Distributed Intelligent Networked Devices) can be seen installed on the ceiling and walls of iSpace. The DIND is a device that includes a processor for information data handling, a network

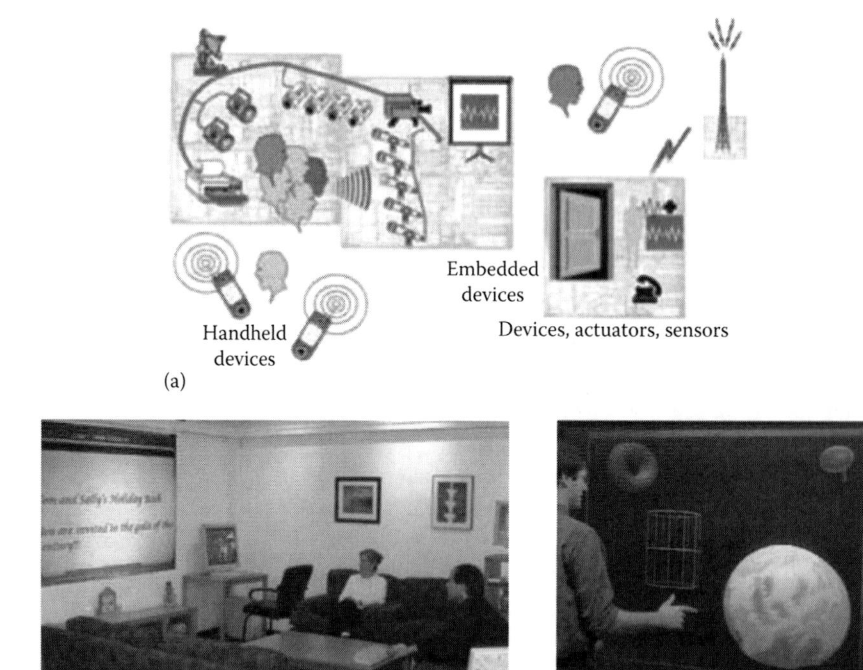

FIGURE 5.1 Examples of intelligent environment [1–3]: (a) Oxygen project, (b) EasyLiving, and (c) DreamSpace.

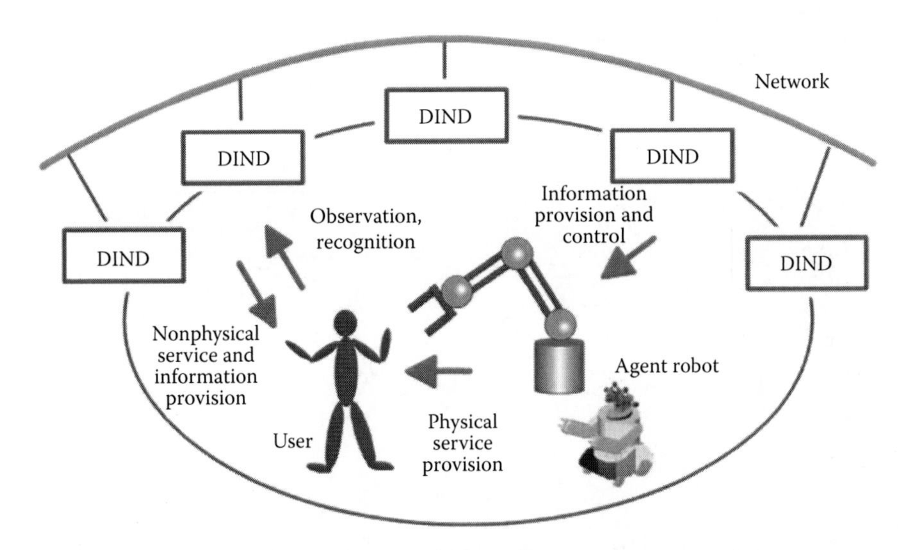

FIGURE 5.2 The conceptual diagram of Intelligent Space.

communication device, and sensors. By using DINDs, iSpace is able to recognize the user's demands. When a user requires a nonphysical service or an information provision, iSpace offers the appropriate service by using devices such as projectors and speakers. When a user requires physical service, iSpace offers the appropriate service by using agent robots. Currently, research based on this concept is ongoing.

5.1.2 Problems in Previous Researches

Nowadays, a lot of research is being carried out to achieve a smart space that will include iSpace in a real environment. However, there are lots of difficulties in achieving this. The main difficulty is the environmental difference. The target environment has many differences in its shape, size, purpose, etc. There are many obstacles, which might be in the form of a large object such as furniture that can block the recognition process of a sensor device. Furthermore, sometimes, its location can be overlapped with that of the installed device. In addition, there is a layout problem with sensors. Since the number of sensors is finite, it is critical to know where to install the sensors for a better performance. However, the situation of a room changes according to the time and people in that room. Thus an optimal layout for one situation can be nonoptimal in the other situations and a versatile layout of sensors cannot be obtained for all possible situations. Therefore, lots of sensors are required to create an intelligent environment with typical buildings, and it is difficult to find an optimal position for the sensors. Figure 5.3 shows an example of this problem.

Figure 5.3 shows images captured by camera devices installed in iSpace. In this figure, a user is pointing at an auto door and agent robots are trying to control them. In the first situation, camera 1 device could recognize the user's demand but camera 2 could not. However, in the second situation, the result is contrary. In the third situation, both cameras could not recognize the user's demand. It means that the

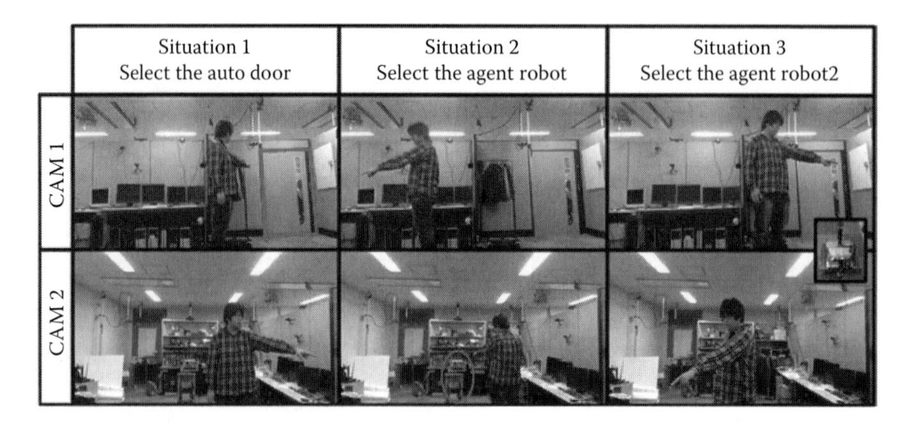

FIGURE 5.3 An example of a problem.

FIGURE 5.4 The ubiquitous display.

optimal position of the sensors changes according to the situation. If the devices can rearrange their position automatically, it will be a simple solution to overcome this problem. The *u*biquitous *d*isplay (UD), shown in Figure 5.4, is a projector device that can rearrange its position to the optimal position according to the situations [5].

As shown in Figure 5.4, the UD has an image projector and a pan-tilt mechanism mounted on a mobile robot. When the UD recognizes a user's position, the UD provides visual information to the user from the optimal position by projecting CG images on the position that the user is looking at. The UD is also able to guarantee the quality of service as it rearranges its position according to the current situation. This can be a solution for the device relocation problem mentioned earlier. However, if all devices are mounted on mobile robots like the UD to achieve a new system that is able to cope with varying situations, new problems might occur. The mobile robot requires many sensors for the localization problem. To suppress the measuring error of mounted sensors to a minimum, the mobile robot should be equipped with a high performance localization system. But in doing this, the insufficient positioning problem will accompany the cost problem. Besides, the sensor device mounted mobile robots have another problem—the problem of coexisting with people in the same place. People often feel uncomfortable living with artificial robots since those robots are not as intelligent as humans and pose the risk of harming them. This is a common open problem for researchers working on developing service robots.

On the basis of these problems, the necessary conditions that a new system must have can be summarized as follows:

- The new system should be reconfigurable by rearranging its devices.
- The rearrangement of devices should be easy and the positioning accuracy should be high.
- The system must not interfere with people's movement.

5.2 NEW SYSTEM

5.2.1 Concept of R+iSpace

R+iSpace is an extended system of iSpace that can overcome the problems mentioned earlier. R+iSpace stands for *Reconfigurable Intelligent Space.* R+iSpace can rearrange the position of devices according to the current situation. To provide mobility for its devices, the R+iSpace adopts the originally designed wall/ceiling moving Mobile Module (MoMo). The devices for iSpace are mounted on the MoMo. The MoMo satisfies all the necessary conditions. The architecture of MoMo is explained in Section 5.3.

As shown in Figure 5.5, not only special functional devices but also general devices, for example, electronics, illuminations, embellishments, and furniture, can be mounted on the MoMo. Since the MoMo moves on the wall and ceiling, the MoMo does not interfere with people in the same space. The R+iSpace works as follows. When a user comes into the R+iSpace, the sensor devices such as cameras measure the user's direction and position. Other devices share this information, and they compute their optimal positions to rearrange their position. Thus, the R+iSpace is able to reconfigure the character of the environment.

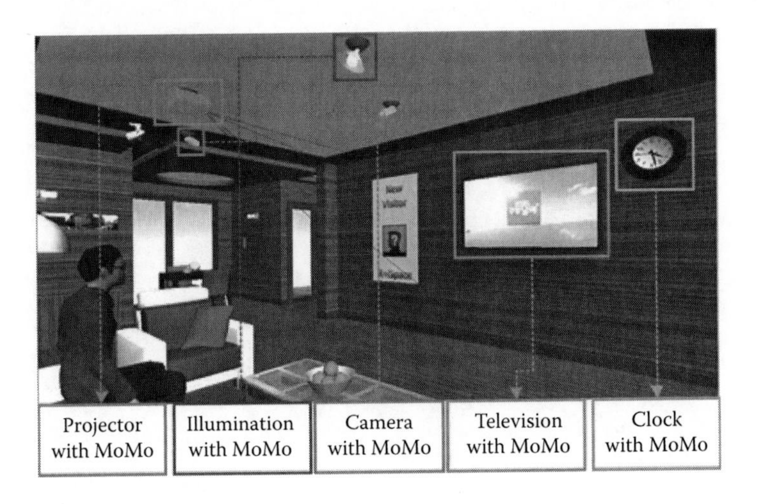

FIGURE 5.5 An example of R+iSpace.

5.2.2 Sticking and Moving Mechanism of the MoMo

As mentioned in the previous section, the MoMo is a type of robot that moves on the wall and ceiling. Previous researches have discussed several methods by which robots can stick to the wall or ceiling. Figure 5.6 shows the conventional method of movement on the wall and ceiling.

The first method is to use vacuum power. The *NINJA-I* in Figure 5.6a is a typical robot using vacuum power [6]. These kinds of robots are designed to climb mainly glass-like surfaces. However, ordinary spaces are not made of glasses. Besides, this method requires a vacuum pump to work continuously to create the vacuum state

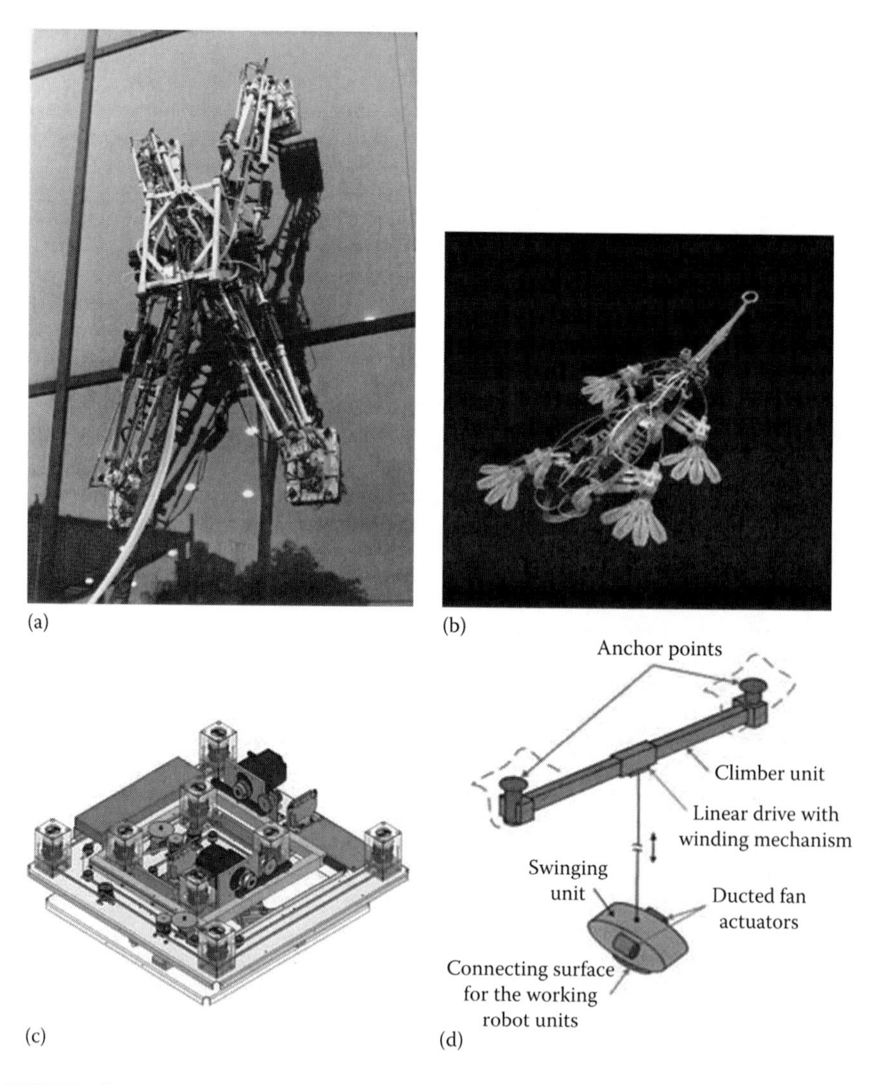

FIGURE 5.6 Climbing robots in previous researches: (a) NINJA-I, (b) StickyBot, (c) HangBot, and (d) ACROBOTER.

and to sustain it. Since it is noisy and energy consuming, it is not appropriate to use this for MoMo as more than one will be used simultaneously in iSpace.

The second method is to make the robot's leg with sticky materials [7]. The StickyBot in Figure 5.6b is a very famous robot in this field. It can move on ordinary walls and ceilings without any modification to the walls or ceilings. However, these robots have other problems. The sticky force is reduced with use over time and hence the position of the robot cannot be guaranteed.

The last method is to use mechanical equipments [8,9]. The HangBot and ACROBOTER in Figure 5.6c and d are designed to use this method. These robots are hung by using mechanical equipment, attached to the wall. For example, the HangBot requires a specially designed ceiling to move on or hang from the ceiling. The ACROBOTER needs to have *Anchor point* installed on the wall and ceiling.

Usually the MoMo is used inside a building and the MoMo can stick to the wall for a long time. Therefore, the third method is most appropriate for the R+iSpace system. The MoMo uses nut/screw mechanism for sticking to the wall because this is practical and easy. The MoMo has screws and the environment, where the R+iSpace is actualized, has nut-boxes fixed at specific positions of the wall and the ceiling. The MoMo is fixed to the wall by screws fastened to the nut on the wall and the ceiling. The position error of MoMo is therefore limited to a certain range; it is almost the same as the error in the installation of nut-boxes. Therefore, the position of the rear-ranged device is obtained easily and its accuracy is high, reliable, and robust. There is an additional merit in using this method. Generally, a wall-climbing robot, which uses the other method, requires energy to maintain its current position. However, the nut/screw mechanism does not require additional energy to maintain the current position, as the screws are fixed to the nut holes and cannot be unscrewed except manually. Figure 5.7 shows the structure of an R+iSpace.

The nut-box, which is inserted to the wall and the ceiling, consists of a nut and a nut-box. The structure of a nut-box is shown in Figure 5.8. As shown in Figure 5.8, the nut-box is made up of a front part and a rear part and the nut is placed inside the nut-box. The front part of the nut-box has a 50° inclination near the opening hole.

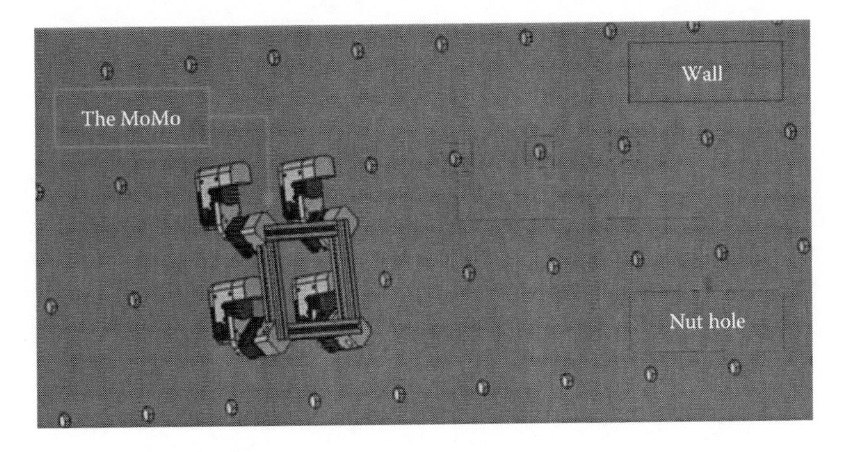

FIGURE 5.7 The structure of R+iSpace.

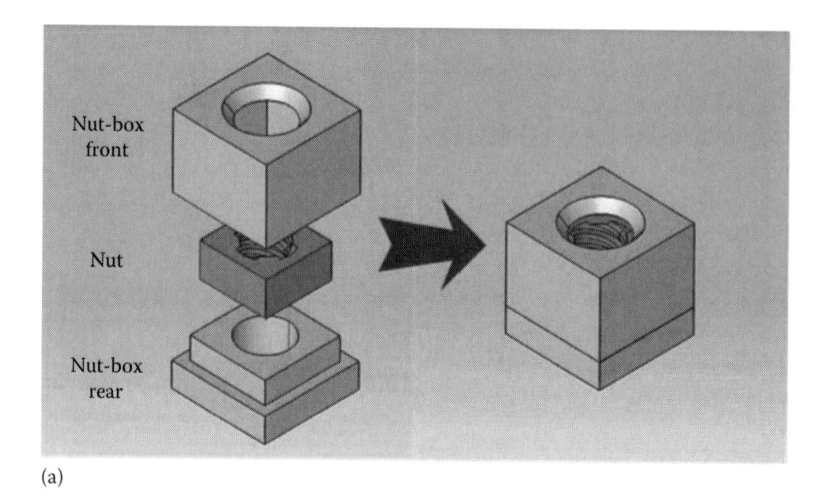

(a)

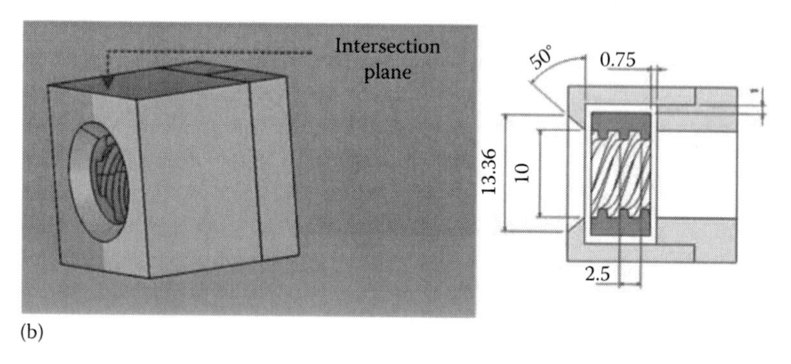

(b)

FIGURE 5.8 The structure of the Nut-Box: (a) the components of the Nut-Box and the assembled Nut-Box and (b) the cross-sectional view of the Nut-Box.

This slope ensures that a little error between the axis of the screw and the axis of the nut hole can be neglected. Furthermore, as shown in the cross-section view in Figure 5.8, there are some spaces between the nut and the case. These spaces also help to ignore the error.

5.2.3 Electrical Structure of R+iSpace

Basically, the devices of R+iSpace have the same structure as DINDs in the iSpace. The DIND is composed of a processor part, a sensor (or an output device) part, and a communication part [4]. The actuator part is added to the DIND for mobility in R+iSpace. The devices mounted on the MoMo have to rearrange their position according to the situation. Some devices use DC power source but others require AC power source. Therefore, wireless communication, which does not limit the area, and stable power supply are needed. The electrical structure of R+iSpace and the MoMo are shown in Figure 5.9. The MoMo can be divided into two types. The first type uses only DC power source. Therefore, it has a battery and a wireless

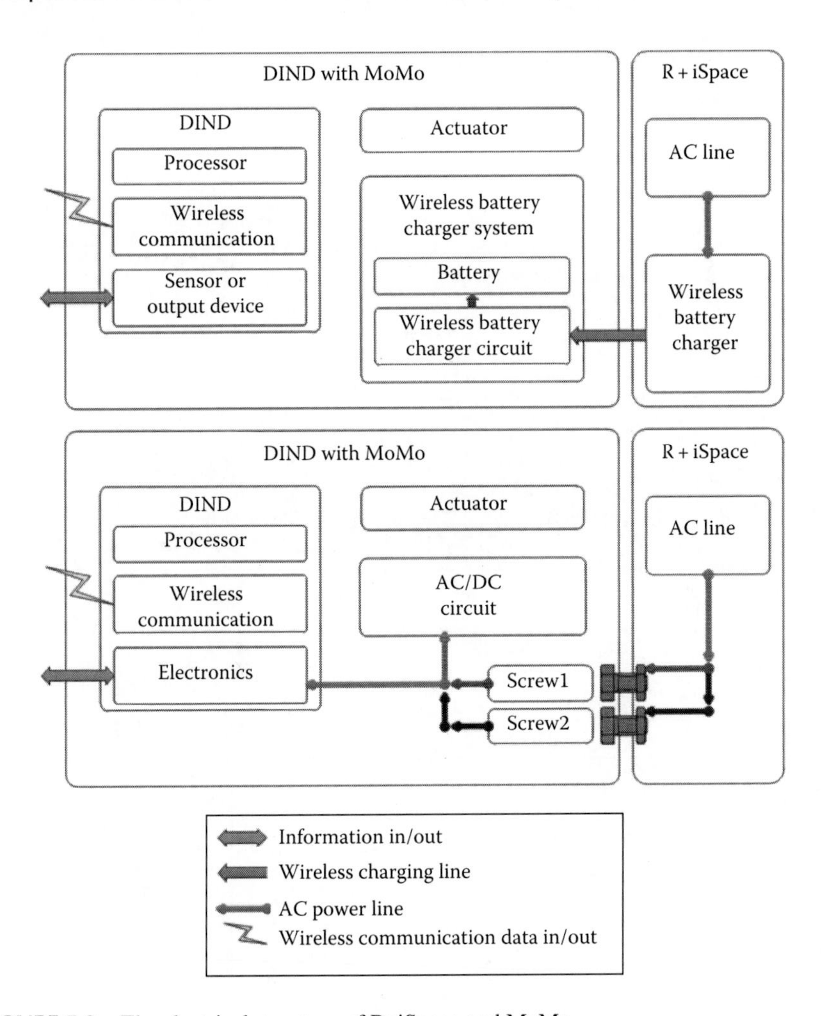

FIGURE 5.9 The electrical structure of R+iSpace and MoMo.

battery charging system. For this type of MoMo, the wireless battery charger has to be installed in some area of R+iSpace. The second type of MoMo uses AC power source. This MoMo has to be connected to the AC line of the space through electrically conductive screws and nuts. The AC power connects directly to the electronics and the AC/DC converter which supply DC power to MoMo.

5.2.4 Software Algorithm of R+iSpace

As mentioned earlier, sensor devices measure the user's absolute position and direction for generating optimal positions for the MoMos. Each device calculates the duration of movement until the optimal position is reached and shares this information with other devices. The devices, which have the same application, execute the application taking into consideration the moving time of the MoMo and the importance of the device, for example, if there were two camera devices in

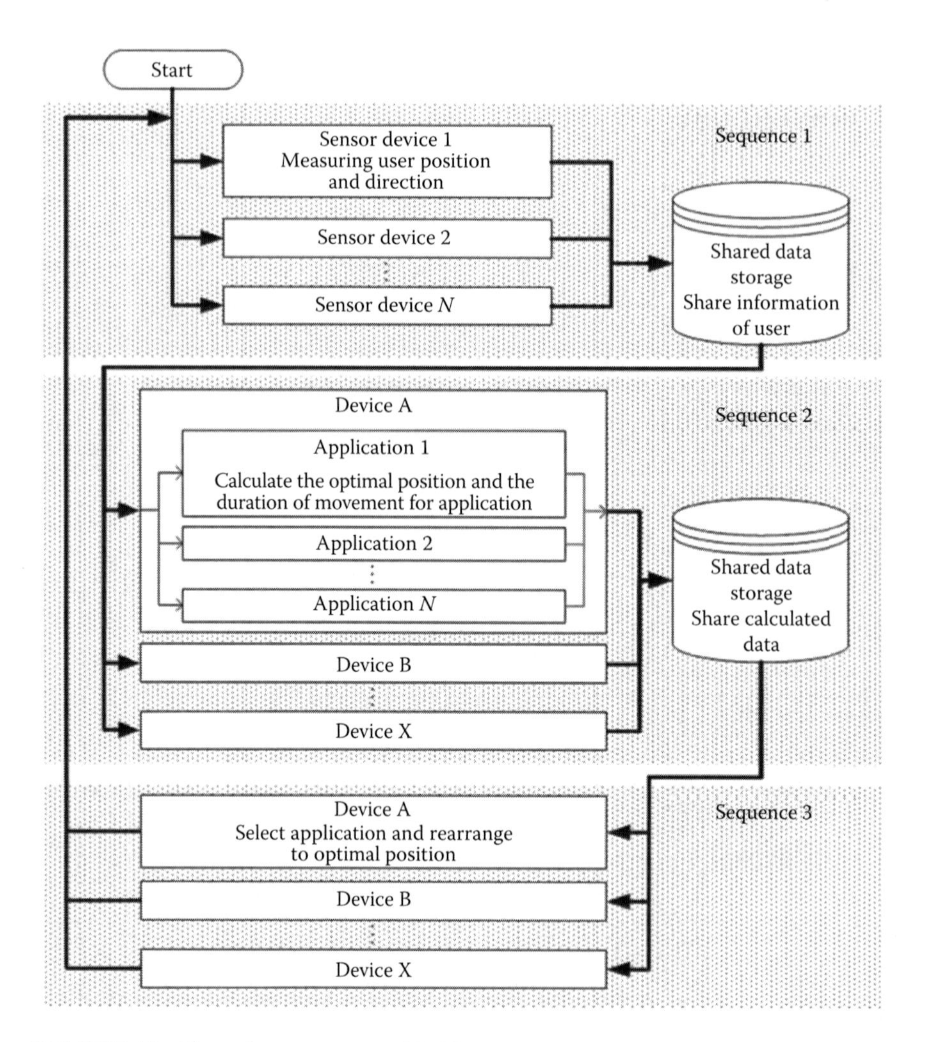

FIGURE 5.10 The software process of R+iSpace.

the R+iSpace and they had three applications to perform, e.g., gesture recognition application, facial expression recognition application, and recording application for overall space. Above all, according to the priority of the application, R+iSpace selects the applications that have to be executed on priority. Then, the R+iSpace distributes each application to a device depending on the shortness of the duration time of movement to reach the target position. Figure 5.10 shows the entire software sequence of the R+iSpace. All devices perform Sequence 1, Sequence 2, and Sequence 3 in consecutive order.

5.2.5 Cooperation Control Mode

Fundamentally, every MoMo in the R+iSpace is of the same size and has the same specification. It has a nut hole, which is inserted into the wall, consistently and

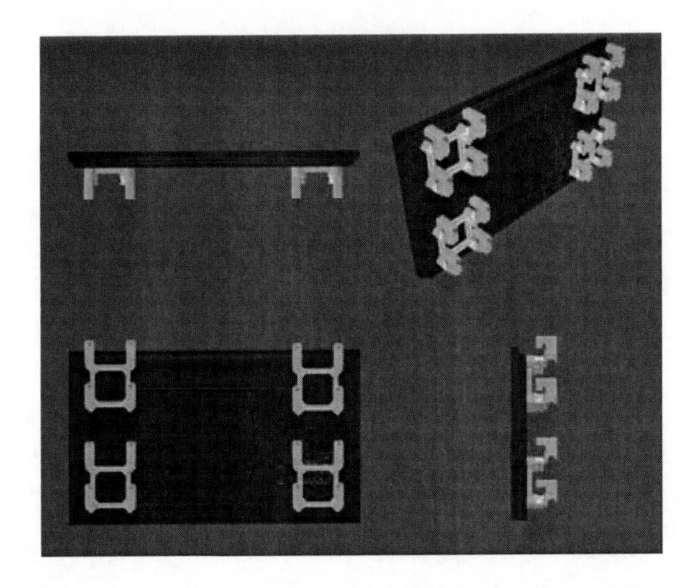

FIGURE 5.11 An example of cooperation control mode for television.

simply. However, some devices cannot be mounted on one MoMo because they are too large or too heavy. For these devices, the MoMo provides a cooperation control mode. An example of the cooperation control mode is shown in Figure 5.11. Usually, a television is too large and heavy for one MoMo to move. Therefore, more than one MoMo is needed to handle the television. The cooperation control mode is used to handle a large device by using multiple MoMos. Multiple MoMos are combined with the large device and their movement is synchronized. With this method, theoretically, the MoMo can handle any big and heavy device. There are plans for the future to move even people using MoMos.

5.3 PROTOTYPE MOMO

The prototype MoMo is composed of pinning parts and the panning part. The role of the pinning part is to fix the MoMo to the wall or the ceiling. The panning part, which plays the role of legs of the MoMo, controls the pinning part that is to be located on the next nut hole. The panning part is connected to the pinning parts through four panning actuators that rotate the pinning parts, and each pinning part has one actuator for fixing to a nut hole. The panning part rotates the MoMo's body too by rotating all the panning actuators in the same direction simultaneously. The CAD image of a prototype MoMo is shown in Figure 5.12.

5.3.1 Pinning Part of the Prototype MoMo

The pinning part consists of an actuator, an actuator gear, a screw body, a bridge part, a screw gear and a sponge. The structure of the pinning part is shown in Figure 5.13.

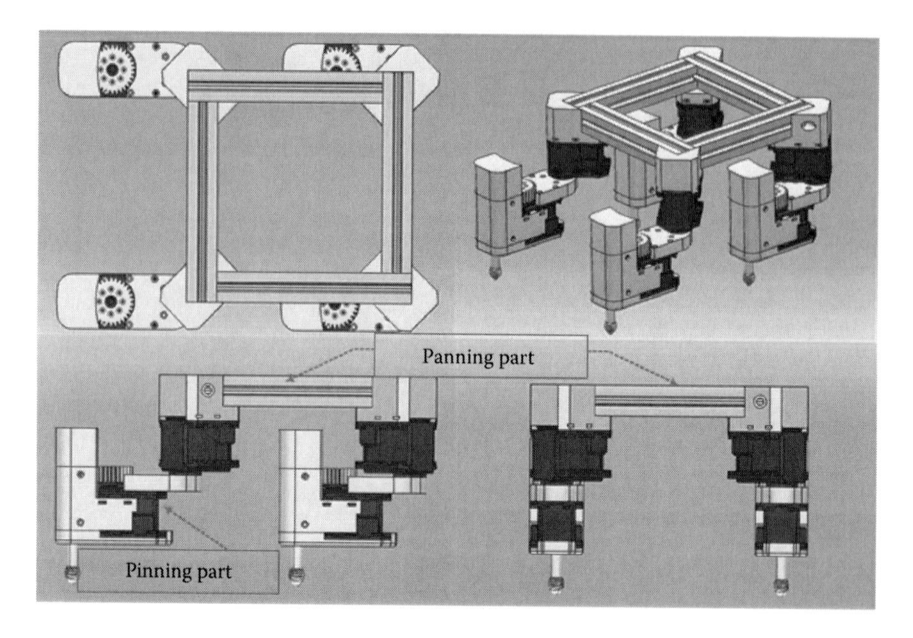

FIGURE 5.12 Whole-view CAD image of prototype MoMo.

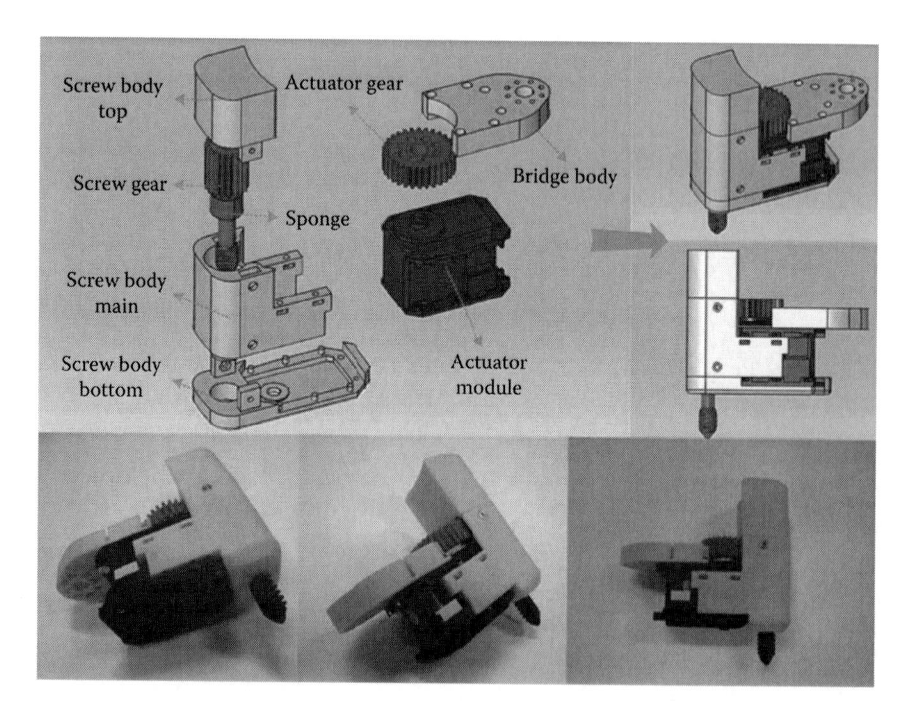

FIGURE 5.13 The components of pinning part and assembled pinning part.

TABLE 5.1
Primary Specification of MX-28T

MCU	ST Cortex-M3
Position sensor	Contactless absolute encoder
Resolution	0.088°
Stall torque	2.5 N m (at 12 V, 1.4 A)
No load speed	55 rpm (at 12 V)
Command signal	Half duplex asynchronous serial communication
Protocol type	(8 bit, 1 stop, no parity), baud rate (8000 bps–4.5 Mbps)
Supported control mode	Joint mode
	Wheel mode
Control algorithm	PID controller (position, torque, angular velocity)

The actuator module of the pinning part is a Dynamixel MX-28T motor, which is made by ROBOTIS Inc. The primary specification of MX-28T is shown in Table 5.1. The MX-28T in the pinning part is controlled in the wheel mode by the angular velocity PID controller. The actuator gear is attached to the output horn in the actuator module. The module value of this gear is 1. The outer diameter of the gear is 32 mm and the pitch circle diameter is 30 mm. The screw body consists of a top part, a bottom part, and the main part. The top part restricts the translational movement of the screw gear in the upward direction. The bottom part is in contact with the wall. The role of the main part is to connect the screw gear and the actuator module. It has four threads in a 10 mm pitch. These threads facilitate the translational motion of the screw gear apart from the rotational motion. The bridge body connects the pinning part to an actuator module in the panning part. The detailed view of the parts of screw gear is shown in Figure 5.14.

As shown in the figure, the screw gear consists of the screw thread part, the gear part, and the cylinder part. The screw thread part allows the screw to move out to the nut hole and tightens and loosens with the nut. The tip of the screw is sharp and is at a slope of 80°. This helps the screw to ignore the error between the axis of screw and the axis of nut hole like the incline of the nut-Box. The gear part is connected to the actuator gear. This part is acted on by the torque from the actuator gear. The module value of this gear also is 1. The outer diameter is 20 mm and the pitch circle diameter is 18 mm. The ratio between the actuator gear and the screw gear is 0.6. The cylinder part is designed to bridge the gap between the MoMo and the wall. When the MoMo moves on the wall or the ceiling, there may be gaps bigger than the normal ones. It may cause a problem in the MoMo's movement. The cylinder part helps reduce the gap. Figure 5.15 shows the screw tightening sequence when there is a gap. As shown in Figure 5.15, after the screw gear passes through a nut region, the thread of the screw body cannot affect the motion of screw. If the screw gear enters deeper into the nut's thread, the gap will be reduced. The tension power of the sponge, which is located inside the pinning part, assists in the interlocking of the threads, when the screw is being loosened.

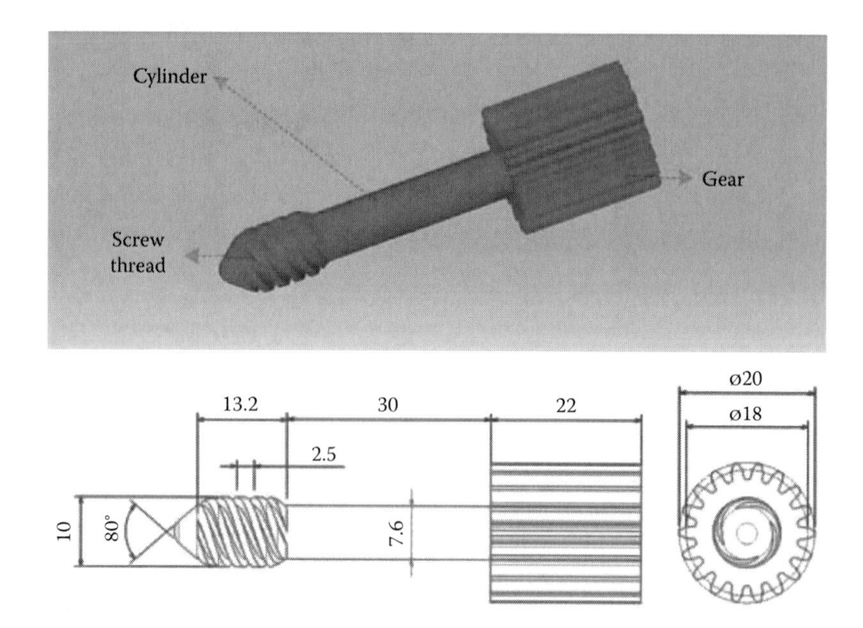

FIGURE 5.14 The detailed view of the screw gear.

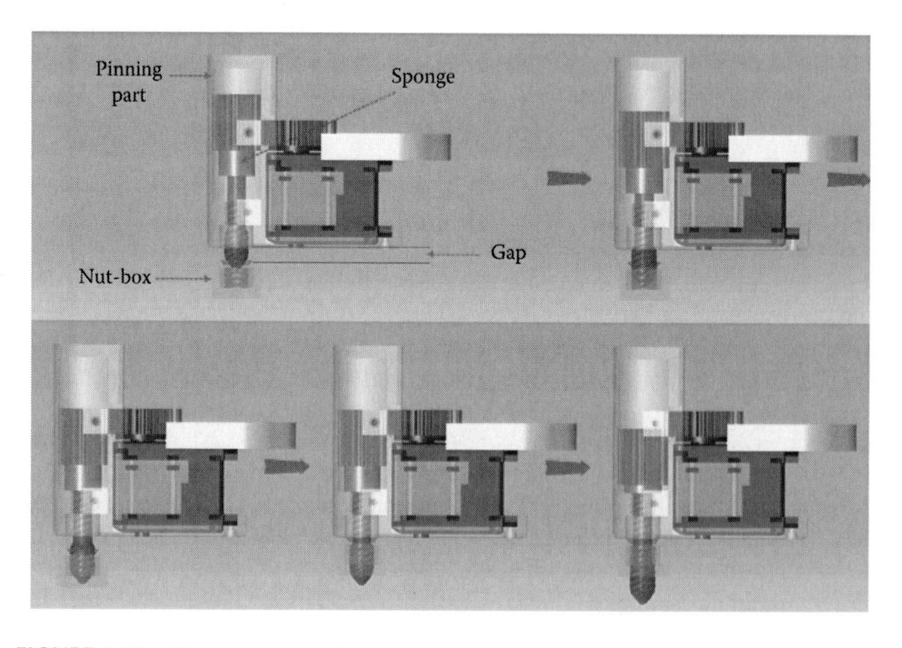

FIGURE 5.15 The screw tightening sequence with a gap.

5.3.2 Panning Part of the Prototype MoMo

The MoMo has one panning part. Figure 5.16 shows the structure of the panning part with the camera projector and its pan-tilt actuators. The four actuators of the panning part are similar to the one in the pinning part. These actuator modules are operated in the joint mode. The role of the panning part is to rotate the pinning parts and to move the MoMo. When all the screws of pinning parts are tightened, the panning part moves the MoMo to the desired position. When a screw of a pinning part is loosened, the actuator module, which is connected to the screw, rotates the pinning part to the position where the next nut hole is located. Table 5.2 shows the overall specification of the prototype MoMo.

5.3.3 Gait Sequence of the Prototype MoMo

The gait sequence of the MoMo is designed with emphasis on stability. According to the proposed gait sequence, at least three screws are fixed to the nuts on the wall or the ceiling during the MoMo's movement. The gait sequence is shown in Figure 5.17. It illustrates the MoMo's movement to the next position to its right. The zero-energy consuming state means that the MoMo is fixed to the wall and no energy is required to keep it in its position. The zero-energy consuming state is shown in Figure 5.17(1) and (15). This state occurs at specific postures of the MoMo when it is moving on the wall. This posture does not need additional energy to maintain current posture

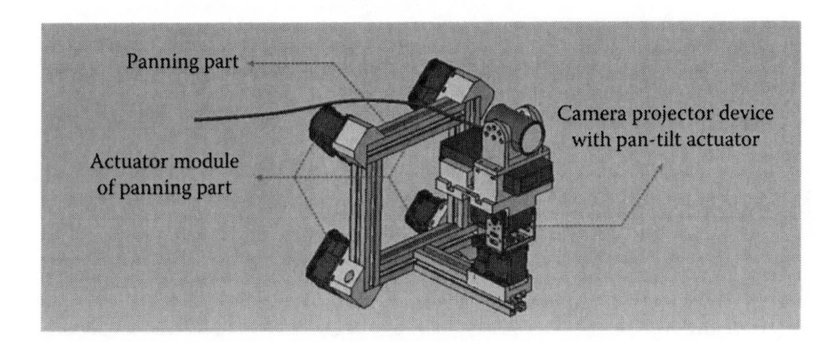

FIGURE 5.16 The structure of the panning part.

TABLE 5.2
Specification of the Prototype MoMo

Weight (Only Mechanical Parts)	Approx. 1.6 kg
Size (mm)	235 (W) × 190 (H) × 110 (D)
Number of actuators	8EA
Locomotion direction	Two directions (X and Y)
Locomotion distance for one step	150 mm
Maximum payload	Approx. 12 kg

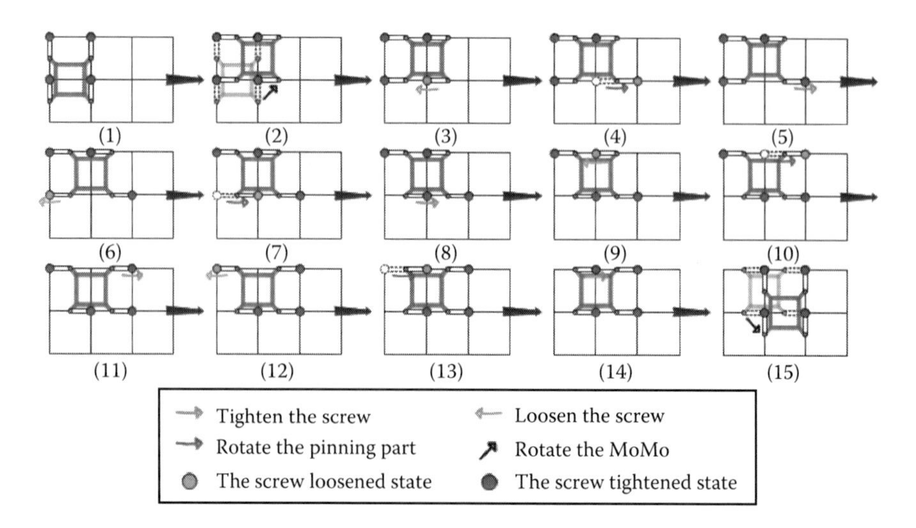

FIGURE 5.17 The specification of the prototype MoMo.

because all the joints' torques are zero as each actuator of the panning part is in singular state against gravity. Moreover, it is a stable posture. Therefore, when the gait of the MoMo ends, the MoMo moves to the zero-energy consuming state. When the MoMo is moving on the ceiling, it is always in the zero-energy consuming state.

5.4 EXPERIMENT

5.4.1 Purpose of the Experiment

The purpose of the following experiments is to verify that R+iSpace satisfies the necessary conditions mentioned in Section 5.1.2. Basically, the MoMo does not use additional sensors for localization. The MoMo moves using only encoders of the actuator modules since the nut holes are placed at regular intervals. The first experiment is performed to verify the gait of the prototype MoMo. Through this experiment, the rearrangement ability of the mounted device in the R+iSpace is confirmed. Moreover, it helps to bring to light any mechanical modification that may be necessary. In the second and third experiments, the third necessary condition is verified. Through these experiments, the usability of MoMo for the research objectives can be verified. The prototype MoMo and the prototype wall for the experiment are shown in Figure 5.18.

5.4.2 Gait Sequence Experiment

As shown in Figure 5.18, a camera device was mounted on the MoMo. In this experiment, the MoMo's move to next position and the duration for one step movement is measured. In addition, any problems in its movement on the wall, are noted. The result of the gait sequence experiment is shown in Figure 5.19. The duration for one step movement is approximately 46 s. This speed is not fast enough for the MoMo

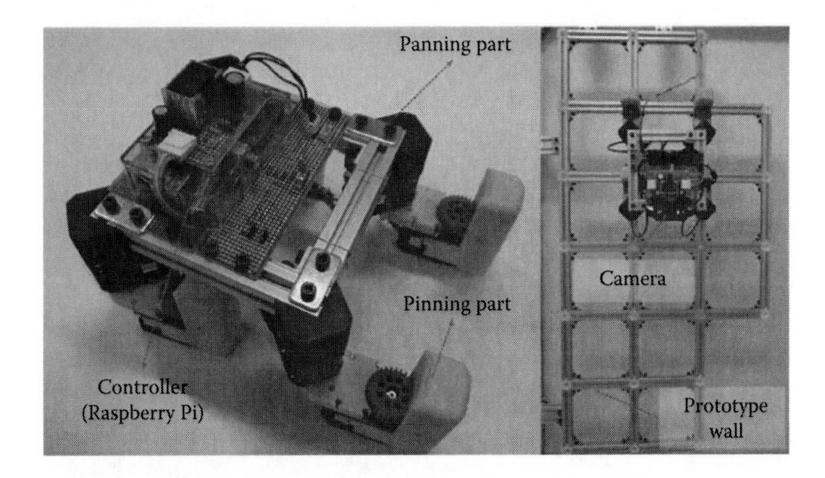

FIGURE 5.18 The prototype MoMo and the prototype wall.

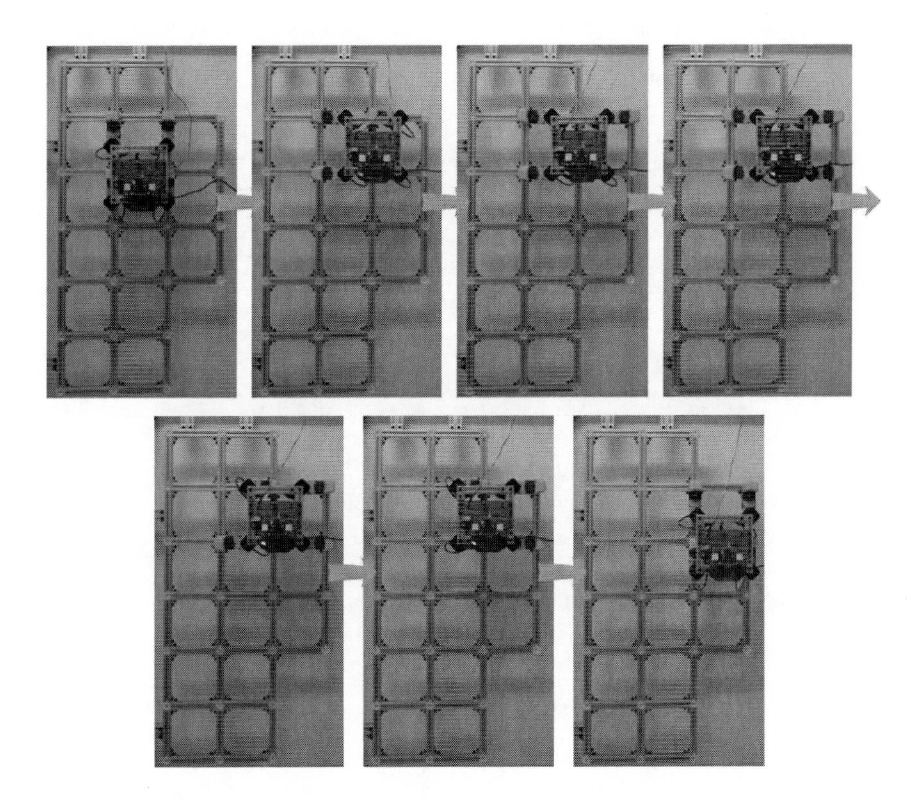

FIGURE 5.19 The result of gait sequence experiment.

FIGURE 5.20 The environment of repeatability error experiment.

to be used in R+iSpace. The main reason for the slow speed is that too much time is spent in the loosening and tightening of screws. However, the prototype MoMo could rearrange its position for stability. The MoMo could find the next nut hole without any additional sensors and the error between the axis of screw and the axis of nut hole is a negligible quantity.

5.4.3 Repeatability Error Experiment

The experiment to test the accuracy of the estimated position was performed in two ways. The first one is known as the repeatability error experiment. This experiment verifies the error in the distance of the same position during iterative movements. This experiment was performed in the following order. The MoMo was located at position *C* as shown in Figure 5.20 and estimated the current position through the well-calibrated camera mounted on the MoMo with a checkerboard. The MoMo moved to the position *R* and returned to the position *C* and then estimated the current position again. The experiments were repeated in other directions too. Through this experiment, the error distance, which occurs as a result of the mechanical tolerance, the control performance, etc., was verified. The result of this experiment is shown in Figure 5.21. The sphere in Figure 5.21 is for comparison. The center of the sphere is located at the position of first estimated data and the diameter of sphere is 1.6 mm. All the estimated positions after returning to position *C* are located inside the sphere. This means that the position error in the rearrangement is not greater than 0.8 mm. This error is negligible.

5.4.4 Position Error Experiment

The purpose of this experiment is to confirm the expected positions. The MoMo moved on a predetermined path and measured its position at every step. The path of the MoMo is shown in Figure 5.22. The MoMo moved from position 1 to position 10

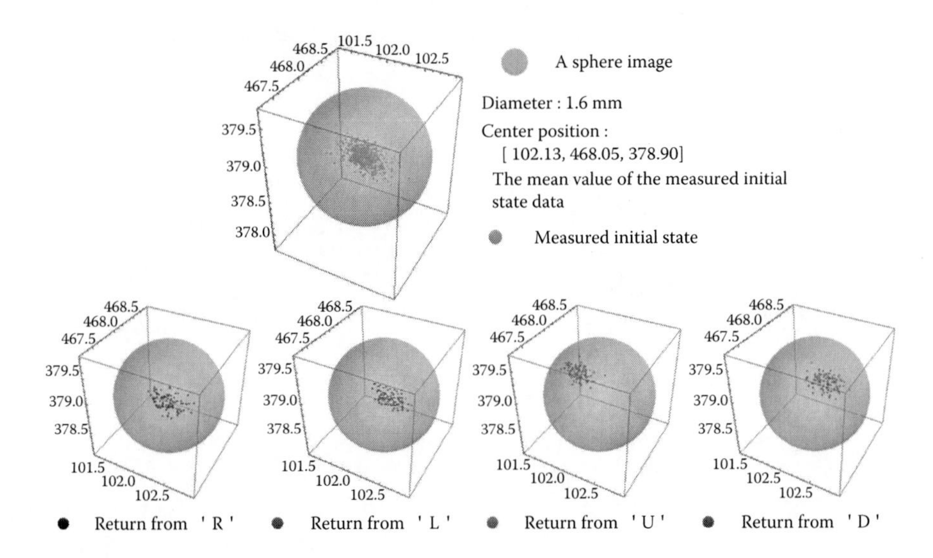

FIGURE 5.21 The result of repeatability error experiment.

FIGURE 5.22 The environment of position error experiment and the captured image.

and at each step its current position was measured by using the mounted camera. The result of this experiment is shown in Table 5.3. The measured positions are given by the position of the camera's optical center based on a global coordinate system where the origin is on the checkerboard. The expected positions are calculated by adding 15 cm distance to the expected position of previous state as the interval between the successive nut holes is 15 cm. As shown in Table 5.3, the maximum distance error is approximately 8.9 mm. These errors are within acceptable limits for application in iSpace. Table 5.3 shows that the distance error of the MoMo's position is not cumulative, which is an important result. These results show that the MoMo can be placed reliably, robustly, and easily at the appropriate location.

TABLE 5.3
Result of Position Error Experiment

	Measured Position	Expected Position	Distance Error (cm)
1	(−81.47, 16.47, −113.58)	(−81.47, 16.47, −113.58)	0.0000
2	(−81.25, 17.06, −98.46)	(−81.35, 16.68, −98.58)	0.4111
3	(−81.05, 17.52, −83.63)	(−81.23, 16.90, −83.58)	0.6541
4	(−80.76, 17.60, −68.76)	(−81.10, 17.11, −68.59)	0.6238
5	(−80.91, 17.71, −53.92)	(−80.98, 17.32, −53.59)	0.5166
6	(−80.73, 2.68, −53.36)	(−80.62, 2.36, −52.62)	0.8202
7	(−80.49, 2.85, −68.08)	(−80.74, 2.15, −67.61)	0.8840
8	(−80.59, 2.76, −82.58)	(−80.87, 1.93, −82.61)	0.8740
9	(−80.92, 2.53, −97.24)	(−80.99, 1.72, −97.61)	0.8896
10	(−80.98, 2.25, −112.34)	(−81.12, 1.51, −112.60)	0.7970

5.5 CONCLUSION AND FUTURE WORK

In this chapter, the R+iSpace, which can overcome the inadequacies of previous intelligent environments, was introduced. The R+iSpace can prevent the failure of applications in almost all situations by rearranging devices. For rearranging devices, the MoMo was introduced and three experiments were performed to verify the performance of the MoMo.

As shown in the results of experiments, the MoMo is a very useful system to rearrange the device in iSpace. However, the MoMo is also found to have several problems. The first problem is that the movement of the prototype MoMo requires a lot of time. In particular, the motion of loosening or tightening screws needs approximately 4 s. The prototype MoMo checks the torque load data from the actuator for estimating its state. However, this data has some errors and delays, which is responsible for the slow speed. The second problem is that the trajectory of the device is too big while the MoMo moves. It is seen that the mounted device cannot be used while it is moving because of the big moving motion. Thus, a new mechanism, which can overcome these problems, is required.

The concept of R+iSpace and MoMo is valid not only for iSpace but also for other sensor network–based environmental systems, and future research will aim at improving the performance.

REFERENCES

1. http://www.research.ibm.com/natural/dreamspace/, accessed May 9, 2014.
2. B. Brumitt et al., EasyLiving: Technologies for intelligent environments, *Proceedings of the Second International Symposium on Handheld and Ubiquitous Computing, HUC 2000*, Bristol, U.K., Springer-Verlag, New York, pp. 12–29, 2000.
3. http://lucente.us/past/career/natural/dreamspace/, accessed May 9, 2014.
4. J.-H. Lee and H. Hashimoto, Intelligent space—Its concept and contents, *Advanced Robotics Journal*, 16(4), 265–280, 2002.

5. J.-H. Lee, Human centered ubiquitous display in intelligent space, *The 33rd Annual Conference of the IEEE Industrial Electronics Society, 2007, IECON 2007*, Taipei, Taiwan, pp. 22–27, 2007.
6. S. Hirose and A. Nagakubo, Walking and running of the quadruped wall-climbing robot, *Proceedings of the IEEE International Conference on Robotics and Automation*, San Diego, CA, pp. 1005–1012, 1994.
7. S. Kim, M. Spenko, S. Trujillo, B. Heyneman, V. Mattoli, and M.R. Cutkosky, Whole body adhesion: Hierarchical, directional and distributed control of adhesive forces for a climbing robot, *IEEE International Conference on Robotics and Automation*, 10–14 April 2007, Roma, Italy, pp. 1268–1273, 2007.
8. R. Fukui et al., Hangbot: A ceiling mobile robot with robust locomotion under a large payload (key mechanisms integration and performance experiments), *Proceedings of IEEE International Conference on Robotics and Automation (ICRA)*, Shanghai, China, pp. 4601–4607, 2011.
9. G. Stepan et al., ACROBOTER: A ceiling based crawling, hoisting and swinging service robot platform, *Beyond Gray Droids: Domestic Robot Design for the 21st Century Workshop at HCI 2009*, Cambridge, U.K., 2009.

Section III

Energy

6 Sensor Networks for Energy Sustainability in Buildings

Alessandra De Paola, Marco Ortolani, Giuseppe Lo Re, Giuseppe Anastasi, and Sajal K. Das

CONTENTS

ABSTRACT

The topic of energy saving in buildings is increasingly raising the interest of researchers for its practical outcomes in terms of economic advantages and long-term environmental sustainability. Many sensory devices are currently available that allow precise monitoring of every physical quantity; in particular, it is possible to obtain estimates of energy consumption, which can be used to enact proper energy-saving strategies. Such devices may be considered part of a complex sensor infrastructure permeating the whole site of interest, which may be characterized by the adopted protocols and architectural models.

This work provides a comprehensive review of the current literature about sensory devices for energy consumption measurement and global architectures for implementing energy saving in buildings.

6.1 INTRODUCTION

The issue of energy efficiency is nowadays raising the interest of researchers and developers all over the world, due to the ever-increasing awareness about the economic and environmental costs of a misuse of available resources. The attention devoted to the development of models for sustainable global energy consumption stimulates the adoption of suitable policies for cutting unnecessary energy consumption; however, in order to effectively enforce such policies, it is necessary to properly characterize energy consumptions so as to identify the main causes of wastes. Specialized studies [1] show that a relevant fraction of worldwide energy consumption is tightly related to indoor systems for residential, commercial, public, and industrial premises.

In literature, many systems have been proposed, with varying degrees of complexity, for building energy management; they often rely on the assumption that the environment is permeated by a large set of sensory and actuator devices, remotely controllable according to some defined policy, in order to bring the environmental conditions closer to the user's desires while also taking into account some globally defined constraints (see Figure 6.1).

This assumption is supported by the off-the-shelf availability of cheap and unintrusive sensors that may be easily distributed in the environment in order to sense relevant measurements. Wireless sensor networks (WSNs) [2], for instance, are one of the most interesting and investigated approaches for reliable remote sensing, and wireless sensor and actuator networks (WSANs) extend their functionalities by adding control devices, that is, actuators. Such networks do not only passively monitor the environment, but represent the tool by means of which the system interacts with the surrounding world and modifies the environment according to the observed data in order to meet high-level goals (e.g., energy efficiency).

This work will chiefly focus on the sensory infrastructure for the specific purpose of energy consumption monitoring in buildings, without neglecting its potential use

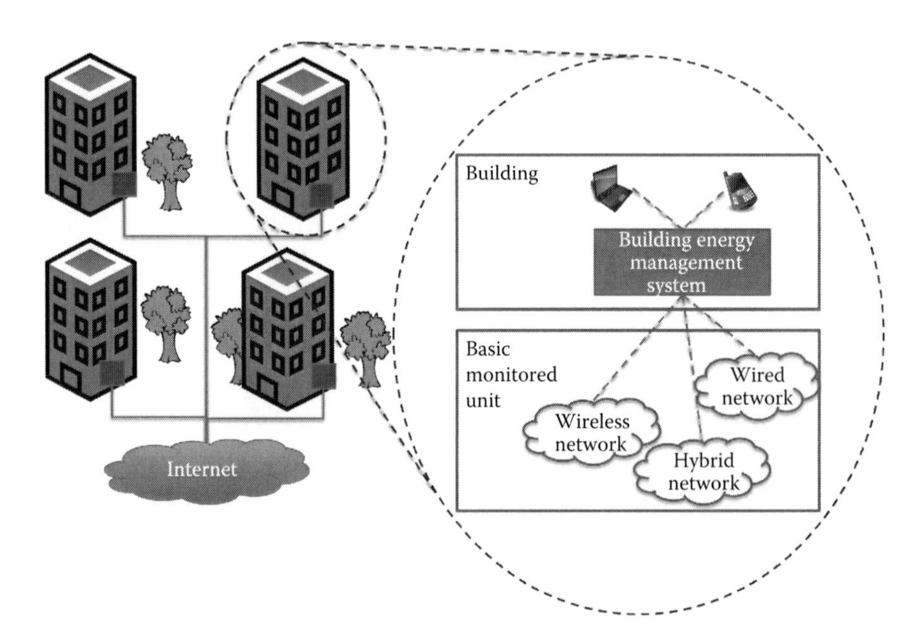

FIGURE 6.1 Exemplary infrastructure for a building energy management system.

in the context of an overall, complex system. This entails taking into account not only specialized technologies used for the construction of efficient buildings but also the overall information and communication technology (ICT) control architecture. A tight correlation exists among the chosen technology, the architectural paradigm, and the energy-saving policies. Indeed, the choice over the available technologies and the architectural paradigm represents a constraint over the viable energy-saving policies; on the other hand, the complexity of such policies is dictated by the corresponding complexity of the sensory and actuator infrastructure.

In the following, we will provide an overview on the solutions reported in literature about the basic devices composing the sensory infrastructure for energy monitoring. The possible approaches for the realization of global monitoring systems will also be discussed, as well as their integration into comprehensive architectures. Finally, we will conclude by providing some insights about higher-level intelligent data analysis techniques that may leverage the use of the discussed architectures.

6.2 TOWARDS ECO-SUSTAINABILITY IN BUILDINGS

Four main general approaches have been identified in the literature for optimizing, or at least reducing, the electrical energy consumptions in buildings [3], namely, user's awareness about energy consumptions, reduction of standby consumptions, scheduling of flexible tasks, and adaptive control of electrical equipments.

The simplest approach to energy efficiency consists in providing appropriate feedbacks about energy consumptions to users so as to increase their awareness and encourage eco-friendly behaviors. User awareness has been leveraged in many commercial and prototype systems such as Google PowerMeter, Microsoft Hohm, Berkeley Energy Dashboard, AlertMe, and Cambridge Sensor Kit (CSK) for energy. Providing simple feedbacks can valuably influence the user's behavior [4]. However, to reduce costs, these systems typically provide only aggregate measures of energy consumption. Hence, they do not allow to identify the specific device or behavior causing the highest energy waste.

More detailed information is provided by monitoring systems that can measure the energy consumption of individual appliances, such as the device-level energy monitoring system proposed in [5], which allows users to monitor and compare the energy consumption of each appliance in a building, through a web-based interface. A hybrid solution, consisting in a monitoring system providing real-time information about the energy consumption of heating, ventilation, and air conditioning (HVAC) and artificial lighting, is proposed in [6]. Instead of relying on a fine-grained energy monitoring, it exploits a simple rule-based approach for correlating energy consumptions with contextual information.

Although user awareness is the basic approach to energy efficiency, its effectiveness is quite limited. Experimental studies carried out in a real building have shown that the sole provision of feedbacks is not sufficient to ensure significant energy savings in the long term. Specifically, the authors of [7] measured the energy consumption in the building before and after the installation of a monitoring system. While a reduction of more than 30% was observed in the week immediately following the installation, the energy savings became negligible just 1 month after the beginning of the experiment. Moreover, the authors of [8] claim that feedbacks can actually stimulate a virtuous behavior in home users, while they are not so effective in a work environment.

Another viable approach consists in eliminating, or drastically reducing, energy wastes due to electrical appliances left in standby mode. Despite its apparent simplicity, such an approach can produce significant energy savings. It has been estimated that most consumer electronics (such as TVs, set-top boxes, hi-fi equipments) and office devices (e.g., printers, IP phones) consume more energy in standby mode than in active mode, as they remain in standby for very long times [9]. The standby mode can be detected by monitoring the energy consumption of the specific device. This requires a metering infrastructure, which, of course, should have a very low energy consumption by itself [10]. Once the standby mode has been detected, the device can be switched off. To this end, different strategies can be used to trade off energy saving for user satisfaction. The easiest way is to let the user decide about the time to switch off a device that entered the standby mode [3]. A more sophisticated approach consists in taking into account information related to the user's presence, or in learning their behavior. Such a strategy is used, for instance, in [11]. The authors propose a control system for generic electronic devices that leverages the hourly habits of users and turns off appliances whose relevance for user is currently negligible, while keeping the other ones in standby mode. The energy management device (EMD) [12,13] is able to autonomously detect devices in standby mode and to switch them off. The automatic detection can be performed by exploiting basic knowledge about the time zone and the user's presence and the appliance energy profile and, whenever possible, by catching signals generated by innovative appliances just before entering the standby mode.

The widespread adoption of smart technology in many electrical appliances enables the scheduling of their activity plans for energy optimization. In case of constraints on the energy peak demand, or in the presence of time-dependent fares, ad hoc strategies can be implemented for determining the optimal scheduling of energy-hungry tasks that do not require user interaction (e.g., washing machine, dishwasher). The system proposed in [3] allows the user to specify the exact time (or time period) when a certain task is to be executed by a specific appliance (e.g., dishwasher). Such a policy makes sense only when energy fares vary over time, but their variations are known a priori. In [3], the identification of the most convenient time intervals is left to the user. However, it is foreseeable that energy providers will be able to supply information about the current contractual offer, scheduled shortages, and low-fare hours [14] so that the overall system can autonomously plan for an optimal scheduling of flexible tasks. A similar approach is also used in the AIM project [15]. Here, users are allowed to specify a set of time requirements that will be regarded as constraints, when computing the optimal scheduling, based on information received from the energy provider.

A final approach is based on the consideration that both in residential and commercial buildings, a significant fraction of energy is wasted due to electrical appliances that are left unnecessarily on (e.g., when no user is present). Hence, the most effective approach to energy savings consists in enforcing a more intelligent utilization of such appliances so as to avoid energy wastes. As previously mentioned, HVAC and artificial lighting systems account for the major fraction of energy consumption, both in residential and commercial buildings. Thus, adaptive control on such systems is essential for effective energy management in buildings. On the other side, adaptive control strategies should consider user wellness in addition to energy savings. The enacted policies should not negatively affect the comfort perceived by the user; otherwise, the reaction would be an immediate rejection of any automatic control, thus discarding the possibility of energy saving. For instance, daylight-harvesting systems currently available on the market have a limited penetration because of their side effects on the user's comfort. Half of these systems, whose control is only based on daylight, with no additional intelligence, exhibit an intolerable latency in adapting to external light conditions, and typically, they are soon deactivated by users. The use of intelligent techniques for user-presence detection and/or prediction is advised to adaptively tune the activation time of electrical equipments, especially those whose latency in bringing the environment into the desired conditions is non-negligible (e.g., HVAC systems).

In the next section, we will start by discussing the available devices for the creation of the monitoring infrastructure.

6.3 SENSOR NETWORKS FOR ENERGY MONITORING

Regardless of the particular choice adopted towards sustainability, researchers have specifically considered the basic requirement for enforcing any energy-saving strategy, namely, a precise and timely knowledge of energy consumptions. A relevant issue to be addressed during the design phase of a complex monitoring system is in fact how to select the technology to be used for creating the sensory infrastructure, specifying the required precision and granularity while preserving the users' privacy.

A wide selection of sensory technologies for energy sensing is today available off-the-shelf, and the choice of a given technology directly affects the complexity of the ICT architecture supporting the monitoring system.

When a coarse-grained monitoring is sufficient, sensory devices may be installed at the root of the power distribution network. This represents an extremely simple and inexpensive solution, but even though many devices have been designed to implement such *single-point-of-sensing* approach, it is still advisable to carefully assess the impact of the technology on preexisting premises, as well as its ease of deployment. The authors of [16] propose recognition of electrical appliances and profiling in real time (RECAP), a system for the identification of energy fingerprints of appliances, which relies on a single wireless energy monitoring sensor clipped to the main electrical unit. Using wireless communication links avoids the need for the deployment of a communication infrastructure from scratch. In the described supervised approach, the user is guided through the phase of profile creation for various devices, thus allowing the construction of a fingerprint database. Such data is then used in real time by a neural network aiming at recognizing all appliances operating at a given moment.

At a finer grain, devices for monitoring energy consumption differ mainly with respect to the density of deployment. The monitoring system proposed in [7] exploits a sensory system consisting in a network of heterogeneous wireless sensors, made of AC meters and light sensors. The energy sensing nodes allow to collect active, reactive, and apparent power measurements [10]; each node implements the IPv6/6LoWPAN stack, and the entire WSNs is connected to other TCP/IP networks via a router.

Other solutions for AC power metering through a sensor network have been proposed. In [17], the *plug* network is described, which is composed of nodes fulfilling all the functional requirements of a normal power strip and equipped with an antenna and a CPU. Additionally, each node, thanks to some onboard sensors, is able to provide measurements of temperature, light exposure, and noise. Such a device allows to realize a comprehensive sensor network for the observation of environmental conditions and energy consumptions. However, the conspicuous size and weight of sensor nodes prevented their use for an unobtrusive sensory infrastructure, which is instead a basic requirement for any pervasive system.

The design of a pervasive sensor network for energy monitoring might benefit from a special focus on the more energy-hungry actuators, such as those for offices HVAC, or for domestic appliances. A possible solution consists in using integrated sensor/actuator platforms for energy consumption monitoring, through commercially available devices such as Plogg and WiSensys, as suggested by the authors of [3]. Currently, such devices are still expensive, so it might be convenient to allow for coarser granularity of monitoring, by coupling a single energy sensor to a group of devices.

Monitoring and efficiently managing the energy consumption of the sensing infrastructure itself would deserve a separate discussion. This issue is extremely important in case of sensor nodes powered by batteries with a limited energy supply, as in typical WSNs for environmental and context monitoring; however, this topic is beyond the scope of our present discussion, and we refer the reader to [18] for a detailed overview on power management in WSNs.

Whenever any of the previously mentioned devices is used as part of a complex system, the main issues that need to be addressed with regard to the architectural choices are the integration of various different technologies and the scalability with respect to the diversity of devices and to the number of monitored areas. In particular, the first two points arise from the availability of very different technologies for the same functionality, while quite often one of them is not sufficient to perform all sensory/actuator duties by itself.

As regards the choice of protocols, a possibility consists in the adoption of one of the standards used in the field of home automation, which are typically based on wired communications, or alternatively to use wireless technologies, allowing to create a wireless home automation network (WHAN). The former class of home automation protocols includes, for instance, the Modbus protocol, used for serial connection of electronic devices; it also comprises the standard for communications over different physical media, such as the KNX standard, the LonWorks platform, and the BACnet protocol; other protocols exploit the preexisting power lines in order to connect different devices: those include the HomePlug protocol [19], LonWorks again, or the X10 standard. A brief overview on this class of solutions is reported in [20].

TABLE 6.1
Home Automation Technologies and Protocols

Technology	References
BACnet	http://www.bacnet.org
HomePlug	https://www.homeplug.org/home
Insteon	http://www.insteon.net/pdf/insteondetails.pdf
KNX	http://www.knx.org/knx-standard/knx-specifications
LonWorks	http://www.echelon.com/tecknology/lonworks/lonworks-protocol.htm
Modbus	http://www.modbus.org/specs.php
Plogg	http://www.plogginternational.com
Wavenis	http://www.wavenis-osa.org
WiSensys	http://www.wisensys.com
X10	http://www.x10.com
ZigBee	http://zigbee.org
Z-Wave	http://www.z-wavealliance.org

The latter class of protocols, based on wireless, includes instead the protocols of the ZigBee family, Z-Wave [21], Insteon (which also allows communication over power line and X10 compatibility), and Wavenis technology. The most significant difference with respect to wired system is that wireless technology is more suitable for pervasive and nonintrusive deployment, thanks to the possibility of installing devices virtually everywhere and with negligible impact on the environment even in the presence of high-density deployment. However, such kind of technology makes it harder to design and implement the systems, due to the typical issues of wireless communication, such as the possibility of interference, the presence of reflective surfaces, or the need for multi-hop communication. For a detailed overview on WHAN technologies and the related design issues, the reader may refer to [22]. Table 6.1 provides a reference for the mentioned technologies and protocols.

Regardless of the wired or wireless nature of the communication link, one technology is generally not sufficient on its own to cover all the necessary functionalities of a complex system for managing entire buildings; for instance, it may be required to integrate specialized technology, such as radio frequency identification (RFId) for user tracking, or sensors for energy monitoring, or even all kinds of actuators. It is thus evident that those approaches aimed at the cooperative connection of heterogeneous devices are to be preferred.

6.4 SYSTEMS FOR ENERGY SUSTAINABILITY

A precise and timely knowledge of energy consumptions is an essential requirement for enforcing any saving strategy, and it must be the basic functionality of any energy-aware system. In this section, we will review some of the approaches to monitoring recently proposed in the literature, as well as the corresponding architectures and protocols.

6.4.1 Approaches to Energy Monitoring

Systems for energy monitoring can be classified according to different criteria, for example, the type of sensors they use or the spatial granularity used for collecting data. With respect to sensors, it is possible to distinguish between *direct*, *indirect*, and *hybrid* monitoring systems. Direct monitoring systems use electricity sensors for directly measuring energy consumptions, while indirect systems infer energy consumptions by measuring other quantities such as temperature and/or noise. Finally, hybrid systems rely on both approaches. Direct monitoring systems can be further classified into *fine-grained*, *medium-grained*, and *coarse-grained* systems, depending on the level of spatial granularity they use in collecting data about electrical energy consumptions. The whole taxonomy is graphically summarized in Figure 6.2.

Following this general taxonomy, indirect monitoring systems are so called because they do not use electricity sensors for measuring the energy consumption of appliances. Instead, they indirectly infer information about energy consumptions by measuring other physical quantities that are somewhat related with energy consumptions. This approach leverages the fact that appliances typically affect other observable environmental variables, such as temperature, ambient noise, vibrations, or electromagnetic field. Specifically, data provided by sensors are combined with a consumption model of the appliance in order to obtain an estimate of its energy consumption. An indirect monitoring system is proposed in [23], where a WSN is used to measure physical quantities such as noise, temperature, and vibrations. Each appliance is identified by a specific pattern of its sensory measurements. For instance, switching on a kettle is associated to temperature rising and to a variation in vibration and ambient noise. However, this chapter does not specify how the system is provided with the association between sensory patterns and the specific operating appliance.

Whenever a model for appliance energy consumption is available, any system able of automatically detecting appliances could be used for performing indirect energy monitoring. Those systems include the approach proposed in [24], which exploits information coming from the energy distribution network, other than explicit energy consumption. The proposed approach performs the analysis of the high-frequency electromagnetic interferences (EMIs) generated by the electronic devices powered

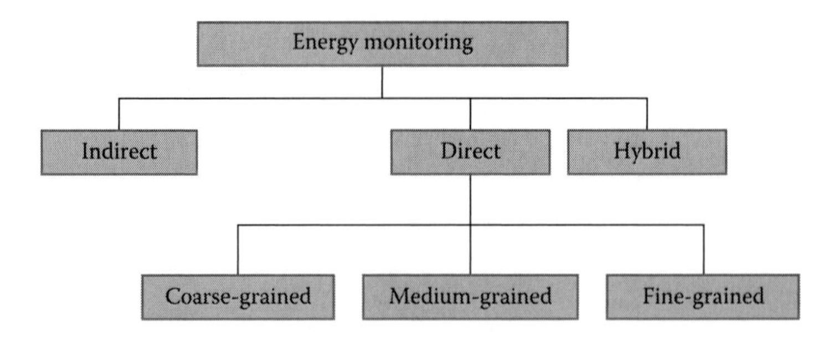

FIGURE 6.2 Taxonomy of energy monitoring systems.

through a switch mode power supply (SMPS) (used in fluorescent lighting and in many electronic devices). Due to the limited applicability to a specific class of actuators, such technology should be just regarded as complementary to the energy monitoring system.

Unlike indirect systems, direct monitoring system measures energy consumptions through ad hoc electricity sensors, typically referred to as power meters. The granularity used for direct energy monitoring spans from a single point of metering to the monitoring of individual appliances. The rationale for using only a single power meter is keeping intrusiveness at a very low level. Accordingly, these coarse-grained systems are referred to as *nonintrusive load monitoring* (*NILM*) systems or *nonintrusive appliance load monitoring* (*NALM*) systems if the focus is on individual appliances. On the opposite side, fine-grained systems allow to monitor individual appliances with a high precision but require the deployment of a large number of power meters. Obviously, the granularity of monitoring affects the approach to the artificial reasoning carried on the collected sensory data and, indirectly, also the possible energy-saving policies that can be used.

The NALM approach has been initially introduced by Hart [25], who proposed a system for measuring current and voltage at the root of the energy distribution network, which is typically organized as a distribution tree. Variations in collected measurements, after preprocessing, are compared to consumption profiles for the various appliances in order to infer their activation or deactivation. Hart's work has been seminal for a number of subsequent works in the field of energy monitoring. The work presented in [26] follows the NALM approach in order to disaggregate the overall consumption into those components that can be attributed to the more relevant end users in a home environment. As in the original NALM approach, this work requires the consumption pattern for each appliance to be manually extracted by a human operator and explicitly coded into the disaggregation algorithm. The NILM/NALM approach provides only a high-level view of the energy flows [7]; it is straightforward to use such information to provide feedback to the users in order to trigger virtuous behavior, which is indeed the aim of some projects, such as the previously mentioned Google PowerMeter and Microsoft Hohm.

Several approaches proposed in literature are based on the processing of measurements collected by a single point of measurement [27] and on the use of complex algorithms, such as genetic algorithms [28] or support vector machines [29] in order to decompose the measurement into its components. However, some authors question the effectiveness of such disaggregation techniques in environments like office rooms, where many loads are based on switched power supplies [7]. A survey of disaggregation techniques for sensing energy consumption is presented in [30].

The alternative approach to a single point of sensing consists in monitoring energy consumption at a finer grain. Brought to its ideal extreme, this approach would require a detailed knowledge of every branch of the power distribution network, which, of course, is not feasible in practice. Works presented in literature only attempt to come close to this ideal goal. The authors of [7] explore several practical techniques for approximately disaggregating the load tree using a relatively sparse set of power meters. They also propose some techniques for modeling and estimating the energy consumption along three directions, namely, *functional*, by identifying which functionality requires a specific slot of energy; *spatial*, by identifying the area where the energy slot is consumed; and *personal*, by identifying the end user of that energy slot.

Within the broad spectrum of granularity, there exists an intermediate position between NILM systems and systems targeting each device individually. The authors of [31] propose to measure energy consumption only for those branches of the energy distribution tree where some particular devices are connected. With respect to a fine-grained approach, this proposal requires the installation of fewer monitoring devices, while in comparison to a NILM system, it allows to monitor the behavior of low-consumption devices, whose fingerprints would otherwise be overshadowed by high-powered devices. In particular, this can be obtained by powering the latter class of devices on a separated circuit. Within one specific branch, it is however necessary to use data analysis algorithms allowing for a disaggregation of partial data. One of the approaches proposed in [31] uses a probabilistic level-based disaggregation algorithm. Samples about active and reactive power are collected at various devices; after normalization, samples are clustered in order to extract representatives for each operating status. Extracted clusters represent the consumption models for each device, starting from which a classifier is built that takes into account all the possible combinations of activated devices. A similar approach is adopted in [32] for the monitoring of the energy consumption of buildings in a university campus. In particular, the proposed monitoring system consists of a power meter at the root of the distribution network of each building, whereas a finer-grain monitoring system is deployed in one of the buildings, by partitioning the network supplying that building into 15 separately monitored circuits. Such partition allows to isolate plug loads, lighting, and the machine/server room for each floor. The authors just report a visual representation of the consumption of the different devices, without proposing any algorithm for further data disaggregation.

Finally, a hybrid approach to monitoring—including both a direct and an indirect part—involves using both specific sensors for energy measurement (typically consisting in a single power meter at the root of the distribution tree) and indirect sensors for recognizing the operating status of appliances. An example of such a complex approach may be found in [33], where the authors propose a monitoring system based on WSNs with magnetic light and noise sensors and including a power meter for monitoring the overall energy consumption. The authors propose an automated calibration method for learning the combination of appliances that best fits the collected sensory data and the global consumption. The calibration method integrates two types of models. Specifically, a model of the influence of magnetic field, depending on two a priori unknown calibration parameters, is used for more complex appliances with many operating modes, whereas appliances with fewer operating modes only require models associating the relative interaction with the user consumption to each specific mode, which is estimated via the noise and light sensors. The main disadvantage of this work is that the calibration is to be performed in situ and cannot be carried out before the deployment since many unpredictable external factors may influence the measured environmental variables. It is worth pointing out that hybrid systems are typically characterized by a coarse-grained direct monitoring of energy, with a single sensor at the root of the energy distribution tree. This is usually coupled with a fine-grained indirect monitoring.

6.4.2 Architectures and Protocols

To be effective, the previously described approaches need to be implemented as part of an overall automated system capable of enforcing effective use of electrical appliances so as to reduce electrical energy consumptions in the building without negatively affecting the user's comfort. The logical layers, which are likely present in such architectures, are depicted in Figure 6.3.

A number of architectural solutions have been proposed in literature, which can be analyzed and compared from different viewpoints, such as *architectural model* (i.e., centralized or distributed), *internal organization* (i.e., single-layer or multilayer organization), *networking protocols*, and ability to support *heterogeneity* in sensing technologies.

One of the proposed solutions is a monitoring system based on web-enabled power outlets [5]. Each appliance is connected through a (Plogg) power outlet, that is, a power meter that measures the energy consumption of the appliance and sends the acquired information to a gateway using a standard communication protocol (e.g., Bluetooth or ZigBee).

A further evolution consists in a direct integration of power meters, and possibly any other smart device, by exploiting the *web of things* (WoT) paradigm. The latter is the extension of the well-known *internet of things* (IoT) paradigm to the web [34]. Following the WoT approach, any smart object (e.g., power meter, sensor/actuator device) hosts a tiny web server. Hence, it can be fully integrated in the web by reusing and adapting technologies and patterns commonly used for traditional web content.

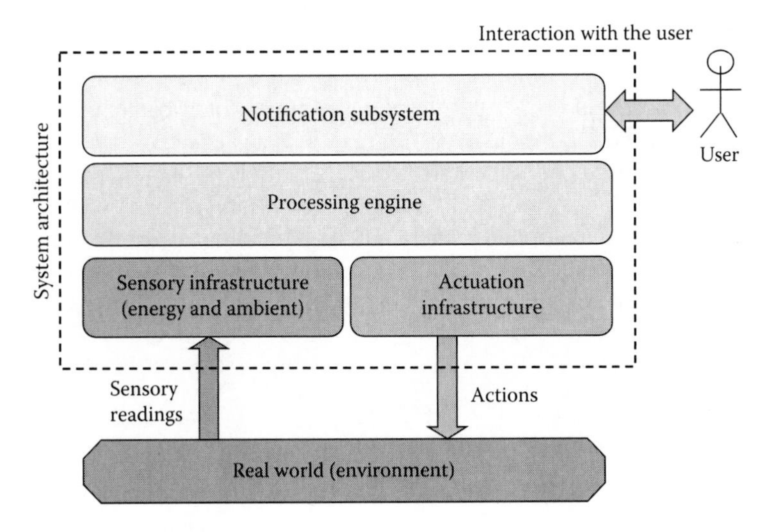

FIGURE 6.3 Logical layers of an architecture for building energy management.

A centralized architecture is also leveraged by the *iPower* system [35]; here, a central server interacts with heterogeneous sensory and actuator devices. Specifically, a WSN is used to monitor environmental conditions and to measure energy consumptions, while actuation is performed by X10 devices connected to the server via power line communication (PLC). Since wireless sensors have a limited transmission range, they may not be able to communicate directly with the server. Hence, to extend the system coverage, sensing devices send their data to a local base station. Base stations are then connected to the server through an Ethernet high-speed local area network. To manage heterogeneity with a sufficient degree of abstraction, *iPower* relies on a multilayer architecture.

A more complex architecture, capable of providing advanced support to heterogeneous sensory and actuator infrastructures, is used in the *Sensor9k* system [14], which is internally organized according to a three-tier model. The *physical* layer includes all the sensory and actuation devices, the *middleware* layer is composed of a set of building blocks for implementing basic services, and, finally, the *application* layer hosts the control logic and consists of various ambient intelligence (AmI) applications. The inclusion of a *physical abstraction interface* ensures support against the heterogeneity of physical devices, as it takes care of exporting higher-level abstractions identifying the basic monitored units. Furthermore, it deals with basic connectivity issues among devices and groups together all the functionalities related to message relaying, monitoring and control of the infrastructure health, and reconfiguration due to changes in the underlying physical infrastructure. *Sensor9k* aims to address scalability with the number of monitored areas, which is typically the major limitation of centralized solutions.

The idea of a hierarchical architecture with gateways interconnecting different technologies is also proposed in [11], in the context of the AIM project [15]. The main goal is the construction of a bridge between a smart home and the smart power grid in order to control the energy consumption of appliances.

Unlike the previously mentioned architecture, the sensor–actuator network proposed in [36] is a fully distributed architecture that does not fall into the same general scheme, since the control logic is widely embedded into the sensor and actuator infrastructure. The resulting wireless network consists of three components, namely, a WSN for environmental monitoring, a set of actuator devices, and a number of control nodes that interconnect sensors and actuators. The system is specifically targeted to lighting system control, and because of the tight binding between the sensor and actuator systems, the introduction of additional functionalities does not appear straightforward. Moreover, the scarcity of computing and storage resources available in wireless nodes makes it impractical to accomplish complex functionalities, such as database maintenance.

The main features of the described architectures are summarized in Table 6.2.

6.5 FINAL CONSIDERATIONS

The significance of the adoption of energy-saving strategies in building management has now been fully recognized both from industry and academy. In this work, we aimed at providing a detailed overview of the available sensory devices for energy

TABLE 6.2
Comparison among Different Architectures

	Web-Enabled iPower Outlets	iPower	Sensor9k	AIM Architecture	Sensor–Actuator Networks
Ambient sensor technologies	None	WSN	WSN, RFId, user action sensors	WSN, RFId	WSN
Energy sensor technologies	Power meters	Wireless power meters, power actuators (X10)	Root power meter, wireless power meters	EMDs	None
Architecture model	One-tier	Multitier	Multitier	Multitier	Zero-tier
Abstraction for heterogeneity	None	OSGi	Open GIS-based	OSGi	None
Control logic deployment	None	Centralized	Centralized	Distributed	Distributed

consumption measuring and of the different energy monitoring systems, analyzing the trade-off between costs and precision.

Even though devising a sensory infrastructure for energy monitoring is a fundamental step in the creation of an energy-aware system for environment management, this is not sufficient to obtain a relevant impact with respect to energy saving.

In order to improve the overall outcome, it would be necessary to devise complex systems made up of multiple specialized modules, that is, a sensing component, a data processing engine, a user interface, and an actuation component. Maintaining the emphasis on the sensory module, which has been the focus of this work, different solutions have been examined; however, efficient energy monitoring requires a broader view, which necessarily includes reasoning about other quantities of interest, ranging from physical quantities, such as temperature or lighting, to higher-level ones, such as user activities. Technologies for environment and context monitoring have been in fact widely discussed in several specific surveys [20,22,37].

In order to obtain a coherent vision of the sensory infrastructure, particular relevance derives from the adoption of a global architecture able to integrate all the different technologies pervading the buildings. A few approaches presented in literature have been described here, targeting energy awareness; some works have also been published on the topic of choosing the ICT architecture for developing smart spaces [38,39]. To the best of our knowledge, however, none of them focuses on the suitability with respect to the specific goal of energy saving.

Furthermore, in our opinion, the global sensory infrastructure must be driven by a data processing engine implementing the full system logic, starting from data preprocessing up to user/appliance profiling, prediction of the quantities of interest, and the overall planning. Such engine might require the use of advanced techniques of artificial intelligence (AI) in order to have the system comprehensively show an *intelligent behavior.* Many works presented in literature focus on the design of individual intelligent functionalities, such as user profiling, predicting the occupancy status of the monitored premises, or detecting the activity patterns of users, even though different approaches may also be found where comprehensive modular architectures, either agent-based or service-oriented, allow for embedding several intelligent modules as previously mentioned in the present work [14,40–42].

There is still however a yet unattended demand for a comprehensive analysis of the issues related to the design of a complete system for energy saving in buildings, where the topics of the present work would fit as a basic building block.

REFERENCES

1. US Energy Information Administration. International Energy Outlook 2010–Highlights. Report DOE/EIA-0484(2010), 2010.
2. D. Estrin, L. Girod, G. Pottie, and M. Srivastava. Instrumenting the world with wireless sensor networks. In *Proceedings of the International Conference on Acoustics, Speech, and Signal Processing (ICASSP 2001)*, Salt Lake City, UT, pp. 517–520, May 2001.
3. F. Corucci, G. Anastasi, and F. Marcelloni. A WSN-based testbed for energy efficiency in buildings. In *Proceedings of the 16th IEEE Symposium on Computers and Communications (ISCC'11)*, Kerkyra (Corfu), Greece, pp. 990–993. IEEE, 2011.
4. S. Darby. The effectiveness of feedback on energy consumption. Technical Report, Environmental Change Institute, University of Oxford, Oxford, U.K., 2006.
5. M. Weiss and D. Guinard. Increasing energy awareness through web-enabled power outlets. In *Proceedings of the Ninth International Conference on Mobile and Ubiquitous Multimedia*, Limassol, Cyprus, p. 20. ACM, 2010.
6. K. Kobayashi, M. Tsukahara, A. Tokumasu, K. Okuyama, K. Saitou, and Y. Nakauchi. Ambient intelligence for energy conservation. In *Proceedings of the 2011 IEEE/SICE International Symposium on System Integration (SII)*, Kyoto, Japan, pp. 375–380. IEEE, 2011.
7. X. Jiang, M. Van Ly, J. Taneja, P. Dutta, and D. Culler. Experiences with a high-fidelity wireless building energy auditing network. In *Proceedings of the Seventh ACM Conference on Embedded Networked Sensor Systems*, Berkeley, CA, pp. 113–126. ACM, 2009.
8. S. Taherian, M. Pias, G. Coulouris, and J. Crowcroft. Profiling energy use in households and office spaces. In *Proceedings of the First International Conference on Energy-Efficient Computing and Networking*, Passau, Germany, pp. 21–30. ACM, 2010.
9. International Energy Agency. *Cool Appliance—Policy Strategies for Energy Efficient Homes.* IEA Publications, Paris, France, 2003.
10. X. Jiang, S. Dawson-Haggerty, P. Dutta, and D. Culler. Design and implementation of a high-fidelity ac metering network. In *Proceedings of the 2009 International Conference on Information Processing in Sensor Networks*, San Francisco, CA, pp. 253–264. IEEE Computer Society, 2009.
11. A. Capone, M. Barros, H. Hrasnica, and S. Tompros. A new architecture for reduction of energy consumption of home appliances. In *Proceedings of the European Conference of the Czech Presidency of the Council of the EU "Towards eEnvironment"*, Prague, Czech Republic, pp. 1–8, 2009.

12. S. Tompros, N. Mouratidis, M. Caragiozidis, H. Hrasnica, and A. Gavras. A pervasive network architecture featuring intelligent energy management of households. In *Proceedings of the First International Conference on PErvasive Technologies Related to Assistive Environments*, Athens, Greece, pp. 1–6. ACM, 2008.
13. S. Tompros, N. Mouratidis, M. Draaijer, A. Foglar, and H. Hrasnica. Enabling applicability of energy saving applications on the appliances of the home environment. *IEEE Network*, 23(6): 8–15, 2009.
14. A. De Paola, S. Gaglio, G. Lo Re, and M. Ortolani. Sensor9k: A testbed for designing and experimenting with WSN-based ambient intelligence applications. *Pervasive and Mobile Computing*, 8(3): 448–466, 2012.
15. The AIM Consortium. AIM—A novel architecture for modelling, virtualising and managing the energy consumption of household appliances, 2008.
16. A. G. Ruzzelli, C. Nicolas, A. Schoofs, and G. M. P. O'Hare. Real-time recognition and profiling of appliances through a single electricity sensor. In *Proceedings of 2010 Seventh Annual IEEE Communications Society Conference on Sensor Mesh and Ad Hoc Communications and Networks (SECON)*, Boston, MA, pp. 1–9. IEEE, 2010.
17. J. Lifton, M. Feldmeier, Y. Ono, C. Lewis, and J. A. Paradiso. A platform for ubiquitous sensor deployment in occupational and domestic environments. In *Proceedings of the Sixth International Conference on Information Processing in Sensor Networks*, Cambridge, MA, pp. 119–127. ACM, 2007.
18. G. Anastasi, M. Conti, M. Di Francesco, and A. Passarella. Energy conservation in wireless sensor networks: A survey. *Ad Hoc Networks*, 7(3): 537–568, 2009.
19. M. K. Lee, R. E. Newman, H. A. Latchman, S. Katar, and L. Yonge. HomePlug 1.0 powerline communication LANs-protocol description and performance results. *International Journal of Communication Systems*, 16(5): 447–473, 2003.
20. V. Ricquebourg, D. Menga, D. Durand, B. Marhic, L. Delahoche, and C. Loge. The smart home concept: Our immediate future. In *Proceedings of the 2006 First IEEE International Conference on E-Learning in Industrial Electronics*, Hammamet, Tunisia, pp. 23–28. IEEE, 2006.
21. M. T. Galeev. Catching the z-wave. *Embedded Systems Design*, 19(10): 28, 2006.
22. C. Gomez and J. Paradells. Wireless home automation networks: A survey of architectures and technologies. *Communications Magazine, IEEE*, 48(6): 92–101, 2010.
23. A. Schoofs, A. G. Ruzzelli, and G. M. P. O'Hare. Appliance activity monitoring using wireless sensors. In *Proceedings of the Ninth ACM/IEEE International Conference on Information Processing in Sensor Networks*, Stockholm, Sweden, pp. 434–435. ACM, 2010.
24. S. Gupta, M. S. Reynolds, and S. N. Patel. ElectriSense: Single-point sensing using EMI for electrical event detection and classification in the home. In *Proceedings of the 12th ACM International Conference on Ubiquitous Computing*, Copenhagen, Denmark, pp. 139–148. ACM, 2010.
25. G. W. Hart. Nonintrusive appliance load monitoring. *Proceedings of the IEEE*, 80(12): 1870–1891, 1992.
26. M. L. Marceau and R. Zmeureanu. Nonintrusive load disaggregation computer program to estimate the energy consumption of major end uses in residential buildings. *Energy Conversion and Management*, 41(13): 1389–1403, 2000.
27. C. Laughman, K. Lee, R. Cox, S. Shaw, S. Leeb, L. Norford, and P. Armstrong. Power signature analysis. *Power and Energy Magazine, IEEE*, 1(2): 56–63, 2003.
28. M. Baranski and J. Voss. Genetic algorithm for pattern detection in NIALM systems. In *Proceedings of the 2004 IEEE International Conference on Systems, Man and Cybernetics*, The Hague, the Netherlands, Vol. 4, pp. 3462–3468. IEEE, 2004.

29. S. N. Patel, T. Robertson, J. A. Kientz, M. S. Reynolds, and G. D. Abowd. At the flick of a switch: Detecting and classifying unique electrical events on the residential power line. In *Proceedings of the Ninth International Conference on Ubiquitous Computing*, Innsbruck, Austria, pp. 271–288. Springer-Verlag, 2007.
30. J. Froehlich, E. Larson, S. Gupta, G. Cohn, M. Reynolds, and S. Patel. Disaggregated end-use energy sensing for the smart grid. *Pervasive Computing, IEEE*, 10(1): 28–39, 2011.
31. A. Marchiori, D. Hakkarinen, Q. Han, and L. Earle. Circuit-level load monitoring for household energy management. *IEEE Pervasive Computing*, 10(1): 40–48, 2010.
32. Y. Agarwal, T. Weng, and R. K. Gupta. The energy dashboard: Improving the visibility of energy consumption at a campus-wide scale. In *Proceedings of the First ACM Workshop on Embedded Sensing Systems for Energy-Efficiency in Buildings*, Berkeley, CA, pp. 55–60. ACM, 2009.
33. Y. Kim, T. Schmid, Z. M. Charbiwala, and M. B. Srivastava. Viridiscope: Design and implementation of a fine grained power monitoring system for homes. In *Proceedings of the 11th International Conference on Ubiquitous Computing*, Orlando, FL, pp. 245–254. ACM, 2009.
34. D. Guinard, V. Trifa, F. Mattern, and E. Wilde. From the Internet of things to the Web of things: Resource-oriented architecture and best practices. In *Architecting the Internet of Things,* U. Dieter, H. Mark, and M. Florian (eds.), pp. 97–129. Springer Berlin Heidelberg, 2011.
35. L. W. Yeh, Y. C. Wang, and Y. C. Tseng. iPower: An energy conservation system for intelligent buildings by wireless sensor networks. *International Journal of Sensor Networks*, 5(1): 1–10, 2009.
36. M. Gauger, D. Minder, P. J. Marron, A. Wacker, and A. Lachenmann. Prototyping sensor-actuator networks for home automation. In *Proceedings of the Workshop on Real-World Wireless Sensor Networks (REALWSN'08)*, New York, pp. 56–60. ACM, 2008.
37. L. Benini, E. Farella, and C. Guiducci. Wireless sensor networks: Enabling technology for ambient intelligence. *Microelectronics Journal*, 37(12): 1639–1649, 2006.
38. D. J. Cook and M. Schmitter-Edgecombe. Assessing the quality of activities in a smart environment. *Methods of Information in Medicine*, 48(5): 480, 2009.
39. D. J. Cook and S. K. Das. How smart are our environments? An updated look at the state of the art. *Pervasive and Mobile Computing*, 3(2): 53–73, 2007.
40. D. J. Cook, M. Youngblood, E. O. Heierman III, K. Gopalratnam, S. Rao, A. Litvin, and F. Khawaja. MavHome: An agent-based smart home. In *Proceedings of the First IEEE International Conference on Pervasive Computing and Communications (PerCom)*, Fort Worth, TX, pp. 521–524, 2003.
41. A. Holmes, H. Duman, and A. Pounds-Cornish. The iDorm: Gateway to heterogeneous networking environments. In *Proceedings of the International ITEA Workshop on Virtual Home Environments*, Paderborn, Germany, pp. 1–8, 2002.
42. S. Helal, W. Mann, H. El-Zabadani, J. King, Y. Kaddoura, and E. Jansen. The Gator Tech Smart House: A programmable pervasive space. *Computer*, 38(3): 50–60, 2005.

7 Wireless Sensor and Actor Networks for Monitoring and Controlling Energy Use in the Smart Grid

Melike Erol-Kantarci and Hussein T. Mouftah

CONTENTS

ABSTRACT

Wireless sensor and actor networks (WSANs) have promising applications in a large number of fields, recently including the smart grid among many others. Smart grid is the future electricity grid that adopts two-way communications and energy flow between the operators and the consumers for the betterment of electrical services. Efficient energy use is among the fundamental goals of the smart grid.

Hence, utilizing WSANs for smart energy use has been a significant research topic recently. In this chapter, we will summarize the research on WSAN-based energy management in the smart grid. In the smart grid, WSANs can be used to monitor and control power consumption of the consumers. WSAN-based residential energy management schemes can schedule appliances such that the use of electricity from the grid during peak hours is reduced, which consequently reduces the need for the power from the peaker plants and reduces the carbon footprint of the household. Moreover, it is possible to use locally generated power according to time of day and sell the excess energy to the grid. For instance, consumers may monitor their appliances while they are away from home; in addition, they can select from which resource to supply electricity to their appliances. Thus, consumers need energy and demand management tools to make decisions to reduce their costs and at the same time to reduce their contribution to peak load. In this chapter, we will first give an overview of the application areas of WSANs in the power grid including generation, transmission and distribution, and consumer segments. We will then discuss the incentives for smart energy use, followed by available communication technologies for WSANs such as Zigbee, low-power Wi-Fi, Z-wave, and Wavenis, and compare their applicability in smart energy use applications. We will further summarize the WSAN-based smart energy use tools in the literature. Finally, we will conclude our chapter by discussing the challenges and outlining the open issues.

7.1 SMART ENERGY USE AND SMART GRID

In proportion to growing population and industry, nations are becoming more and more dependent on energy. Electrical power is one of the most fundamental functions in a modern society. The lack of power, that is, outages, has caused major financial losses in the past years. On the other hand, the generation of electrical power may threaten the sustainability of earth resources since the energy sector is one of the major contributors to greenhouse gas (GHG) emissions, hence global warming. GHG refers to several different gases including carbon dioxide (CO_2), methane (CH_4), nitrous oxide (N_2O), sulfur hexafluoride (SF_6), hydrofluorocarbons (HFCs), and perfluorocarbons (PFCs). It has been reported that electricity generation is the largest source of GHG emissions in the United States, causing 40% of the total emissions across the United States only [1].

Recently, sustainability has become a major concern for most governments, and they have been pursuing low carbon economy goals in concert with the so-called Kyoto protocol. Smart grid, electric transportation, and Future Internet are few examples of the technologies that are aligned with the low carbon economy goals. Smart grid integrates information and communication technologies (ICTs) to power system operations and benefits from two-way communications to increase reliability, security, and efficiency of the electrical services while reducing the GHG emissions of the electricity production process [2,3]. There is a wide range of communication technologies that can be employed to support two-way communications, such as WiMAX, LTE, Wi-Fi, and cognitive radio [4–9].

Monitoring the power grid for potential losses and monitoring the energy use of consumers are among the several features provided by the smart grid. In this chapter, we will focus on the smart energy use aspect since it is important to reach the goal of low carbon economies. In the traditional power grid, coarse-grained energy use monitoring and demand control is available. In fact, analog metering serves as an energy monitoring tool while demand response programs implement demand control to some extent. However, analog metering provides accumulated consumption data, which is only useful for billing. On the other hand, demand response is only available for industrial consumers or commercial buildings, which is handled by basically calling consumers to reduce their consumption at peak hours. For instance, in commercial buildings, when the building operator receives a request from the utility, they cycle off the heating, ventilation, and air conditioning (HVAC) to avoid penalties or to receive credits. In industrial plants, generally, equipment cooling is cycled off in response to utility demand [10]. In the traditional grid, demand management does not reach to residential consumers due to scalability issues. Without an efficient communication infrastructure, it is not feasible for the utility to reach out to millions of customers.

In the smart grid, two-way communications, smart meters, and the advanced metering infrastructure (AMI) provide the means for demand management for residential consumers. Two-way communications refer to communication between the customer and the utility. Customer consumption data are forwarded to the utility every 10–15 min intervals via smart meters, and the utility control signals, if available, are transferred to the customers again via smart meters. Smart meters are digital meters that house communication modules. They can provide hourly consumption data and display hourly electricity price and some smart meters may control appliances. A network of smart meters is called the AMI. With the help of AMI, remote meter readings have become possible. Currently, most premises in North America are equipped with smart meters. In Figure 7.1, an AMI network is presented.

Another complementary technology for smart energy use is the wireless sensor and actor networks (WSANs) [10–12]. WSANs consist of actors and sensors that are small, low-cost microelectromechanical systems (MEMS) that collect measurements from their surroundings, process and store these measurements, and transmit them to a sink node or a control center [13]. WSANs are deployed in a wide range of environments, and they are employed in various applications including target tracking, surveillance, health monitoring, disaster relief, seismic monitoring, wild life monitoring, structural integrity verification, and hazardous environment exploration [14]. Despite a wide range of WSAN applications, the use of WSANs in the power grid has been recently explored, particularly after the invention of the smart grid. WSANs can be used in generation, transmission and distribution (T&D), and customer segments of the smart grid.

Bulk power is either generated at fossil fuel–based plants or nuclear power plants or hydro plants. In addition, wind, ocean, or solar power can be used to generate electricity, which are called renewable energy. Those are intermittent resources; therefore, they generally have small contribution to the energy mix. Smart grid aims to increase the penetration of renewable energy generation by adopting intelligent techniques that allow utilization of wind and solar power more effectively. WSANs play a significant role in condition monitoring of power plants and remote renewable

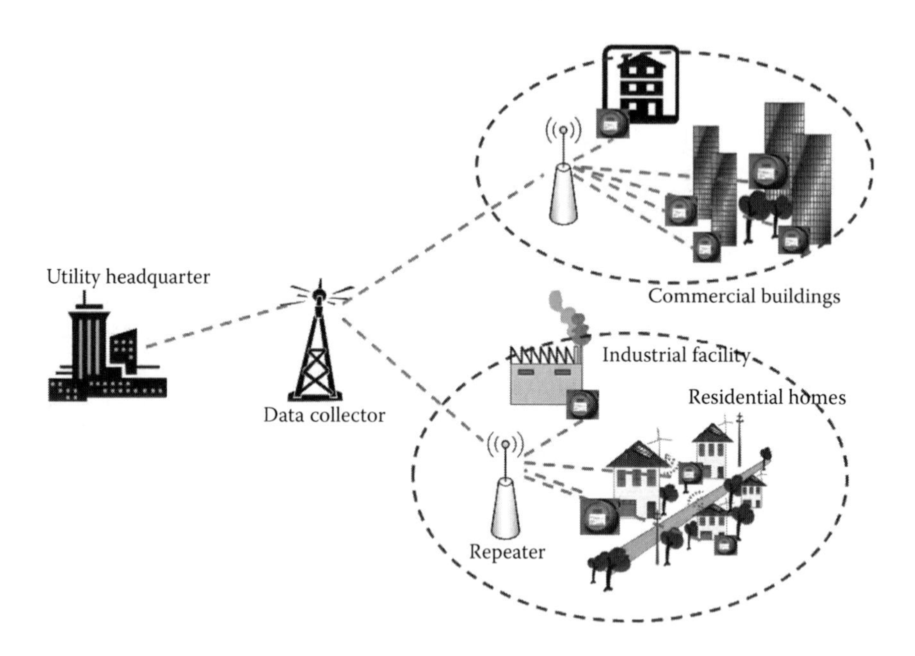

FIGURE 7.1 AMI network.

energy generation sites. They provide low-cost and flexible monitoring capabilities. WSANs are also useful in monitoring T&D equipment such as substations, overhead power lines, underground power lines and controlling switches, and relays. In the traditional power grid, substations and the transformers inside the substations are monitored with remote terminal units (RTUs) of the supervisory control and data acquisition (SCADA) system while power lines are monitored by multiple sensors collecting data on sag, conductor strength, temperature, heating, icing, wind speed, and contact with vegetation and animals. Those sensors are not interconnected, while some advanced sensors are capable of communicating through the cellular networks. Besides, generation and T&D WSANs will play a critical role on the consumer side. Consumers can be roughly classified into three groups based on their power needs and usage patterns, which are industrial, commercial, and residential consumers. In this chapter, we will primarily focus on smart energy use of residential consumers and small offices. Smart grid will enable to communicate, monitor, and possibly control the power consumption of the consumers pervasively without disturbing their business or comfort. In this context, neighborhood area networks (NAN) and home area networks (HAN) as well as the AMI are the enablers of smart energy use. In the following sections, we will discuss the communication technologies for WSANs to implement NANs and HANs.

A HAN includes appliances, pool pump, electrical vehicle, and lights. Some of these energy-consuming devices need to be always on, or they have certain schedules that cannot be modified. These kinds of appliances are called nonshiftable loads. For instance, a light that needs to be on is nonshiftable. On the other hand, some appliance loads can be shifted to a further time or its load can be stretched over a longer time. For instance, a washing machine can start its cycle later than the time it is turned on, or the electric vehicle load can be stretched over the overnight instead of charging within a couple of hours during the peak hours. WSANs are ideal tools to provide pervasive smart energy use techniques to consumers. Meanwhile, consumers need to be provided with some incentives. In the smart grid, this is possible by dynamic pricing schemes. In the following sections, we will review dynamic pricing schemes as well as the studies that consider WSANs for smart energy use.

The rest of the chapter is organized as follows. In Section 7.2, we introduce the dynamic pricing schemes. In Section 7.3, we present the communication standards for WSANs. In Section 7.4, we discuss the WSAN-based smart energy use applications. We summarize the chapter and discuss the open issues in Section 7.8.

7.2 DYNAMIC PRICING AND SMART ENERGY USE

In dynamic pricing policies, the price of electricity varies with time, as opposed to widely used flat rates. There are several policies for time-varying pricing, which are time of use (TOU), critical peak pricing (CPP), peak time rebate (PTR), and real-time pricing. All of these pricing policies require communications between the utility and the customers. Previous work has shown that dynamic pricing can reduce peak load [15]; therefore, a smart grid employing dynamic pricing provides advantages for utilities and consumers. For instance, smart outlets or smart appliances (or devices) may determine their schedule based on price signals from the utility or by following the electricity market prices using web services.

7.2.1 Time of Use

The market price of bulk electricity varies during a daily cycle in deregulated markets. Demand and supply determine the final market price, which varies throughout the day. In Figure 7.2, we present the demand for 4 days in April using the database of the Australian Energy Market Operator (AEMO) [16]. AEMO stores the historical database for load and price for five Australian jurisdictions. The figure presents the demand peaks starting at 12 a.m. for New South Wales (NSW) jurisdiction. Morning and evening peaks are clearly observed in the plot.

During peak hours, the market price of the electricity increases. This is depicted in Figure 7.3. Increasing demand, as well as using expensive fuels in supplying these demands, increases the price of the electricity. Hydro plants or nuclear power plants generate electricity to meet the base load. During peak hours, load exceeds the base load; therefore, additional resources are needed. Power plants with low response times usually are brought online to supply the peak load. In general, they use expensive fossil fuels. As a result, the price of electricity increases during peak hours. In the traditional flat rate pricing policy, an average rate is selected.

According to statistical data collected over the years, in most parts of the world, there are two peak demand periods during the months of winter and one peak period during summer. In the winter, peaks occur during mornings and evenings, while

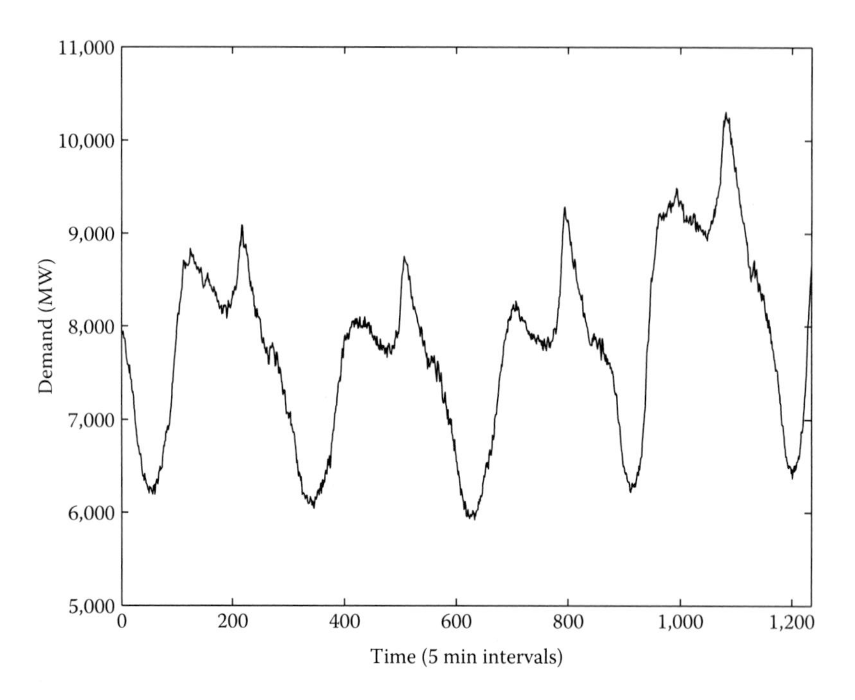

FIGURE 7.2 Electricity demand in NSW on 4 days in April. (From Erol-Kantarci, M. and Mouftah, H.T., Prediction-based charging of PHEVs from the smart grid with dynamic pricing, *Proceedings of First Workshop on Smart Grid Networking Infrastructure in LCN 2010*, Denver, CO, October 2010.)

in the summer, the peak occurs in the afternoon due to air conditioning. In TOU pricing policy, electricity is more expensive during peak hours than the nonpeak hours. Some utilities have mid-peak pricing as well, where the price of electricity is between peak and nonpeak hours.

7.2.2 Critical Peak Pricing

CPP has been implemented in the traditional grid to reduce the load of industrial or commercial consumers on several days in order to prevent grid failure. When the load is expected to exceed the supply, then large-scale consumers are charged more for their electricity usage. For instance, a forecast of a very hot day may trigger CPP. CPP may be applied to residential customers, as well. However, a large volume of customers need to respond to make the required reduction in peak, and advanced communication technologies need to be implemented. It is not practical to call residential customers and ask them to reduce their consumption. Only with the smart grid, the implementation of CPP for residential customers may become feasible. For instance, air conditioning appliances may be cycled off during critical peaks after customers are notified and they agree to collaborate in peak reduction. Two-way communication is essential for all those types of actions, which is one of the fundamental novelties brought by the smart grid concept.

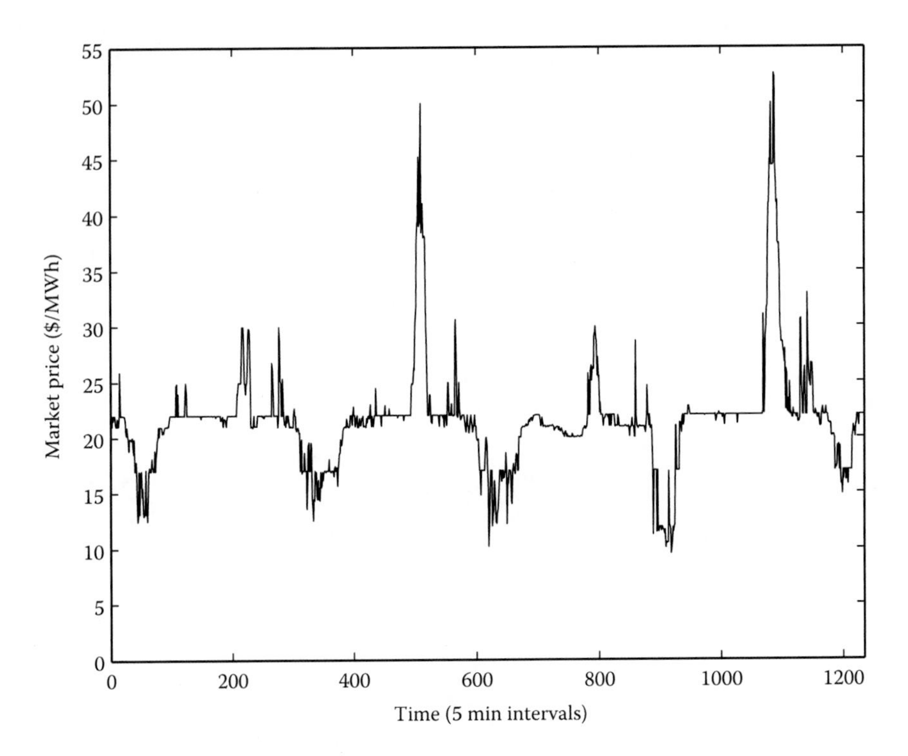

FIGURE 7.3 Electricity price in NSW on 4 days in April. (From Erol-Kantarci, M. and Mouftah, H.T., Prediction-based charging of PHEVs from the smart grid with dynamic pricing, *Proceedings of First Workshop on Smart Grid Networking Infrastructure in LCN 2010*, Denver, CO, October 2010.)

7.2.3 Peak Time Rebate

In CPP, customers are charged more for electricity usage during peak hours, whereas in PTR pricing policy, utilities reward customers with credits for their corporation on critical peak periods.

7.2.4 Real-Time Pricing

Real-time pricing, also sometimes called as dynamic pricing, reflects the actual price of the electricity in the market to the customer bills. The cost of electricity generation varies with the source of electricity, as explained in the previous section. Bulk power generators (suppliers) sell electricity, and the final price is determined after importer's bids and a settlement is reached. Generally, the final price is determined by the regional independent system operator (ISO). The ISO arranges a settlement for the electricity price of the next day or next hour. Since it is not possible to know the day-ahead or hour-ahead demand, load forecasting techniques are applied that can be either short-term forecasts, which are usually from 1 h to 1 week, or medium forecasts, which are usually from a week to a year, and long-term forecasts, which are longer than a year [18]. Forecasts consider various factors such as the historical load and weather data, the number of customers in different categories, the appliances in the area, and their characteristics including age.

7.3 WIRELESS COMMUNICATION TECHNOLOGIES FOR WSANs

WSANs can be implemented via several communication technologies, which are Zigbee, low-power Wi-Fi, Z-wave, or Wavenis. Zigbee and Wi-Fi are based on Institute of Electrical and Electronics Engineers (IEEE) standards, namely, IEEE 802.15.4 and IEEE 802.11, while Z-wave and Wavenis are proprietary technologies.

IEEE 802.15.4 standard defines the physical and medium access control (MAC) layer access of sensor nodes. Routing and applications are defined in the Zigbee protocol stack [19]. Zigbee operates in several different industrial scientific and medical (ISM) bands in various countries. In North America, it operates in the 915 MHz band and uses 13 channels. In Europe, it operates in the 868 MHz band using one channel. Meanwhile, 16 channels in the 2.4 GHz band can be used worldwide and these are the widely used channels by Zigbee.

Zigbee has data rates of 250, 100, 40, and 20 kbps, which are low when compared to Wi-Fi. In addition, it has an approximate range of 30 m indoors, which is also lower than Wi-Fi. Zigbee targets low-data rate and short-range application domains. The range of the WSAN is extended usually by multihop communications. Zigbee sensor network can be organized in a star topology, cluster-tree topology, or mesh topology as seen in Figure 7.4. Note that WSANs and wireless sensor networks (WSNs) can be organized in the same topology; therefore, Figure 7.4 only shows the WSN. From hereafter, we will use WSANs and WSNs interchangeably. Zigbee-based WSNs are ideal for HAN deployments.

One of the unique advantages of Zigbee is energy efficiency, thanks to its duty-cycling mechanism. Zigbee has two modes for channel access, namely, beacon-enabled and beaconless modes. In the beacon-enabled mode, personal area network (PAN) coordinator synchronizes the nodes in the network via beacons. A beacon duration is divided into two periods, active and inactive periods. Nodes communicate only in the active period, and they sleep in the inactive period; thus, they implement duty cycling. Initially, WSNs have been considered for local data collection and

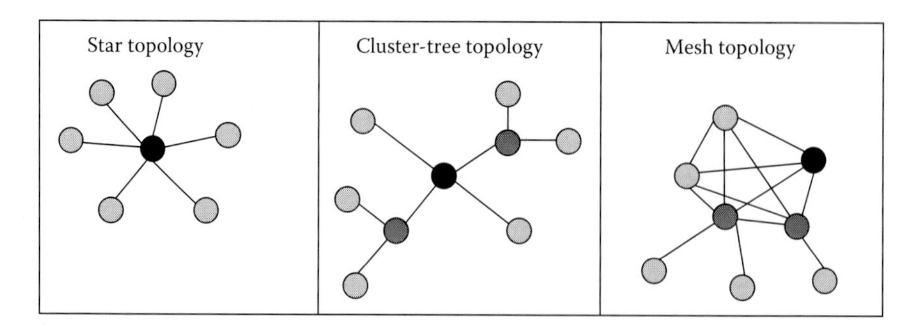

FIGURE 7.4 WSAN topologies.

their connectivity to the web has been omitted. Recently, with the emerging IPv56 and the IETF RFC 4944 on IPv6 over low-power wireless personal area networks (6LoWPAN), IP addressing has become possible for WSNs [20].

Wi-Fi is being widely used for wireless access in many residential and commercial premises [21]. The use of Wi-Fi in WSNs has recently become possible with the advances in ultra-low-power Wi-Fi technology. Ultra-low-power Wi-Fi is based on the IEEE 802.11b/g standard [22]. The traditional Wi-Fi has high power consumption; therefore, it has not been convenient for the limited battery sensor nodes. The ultra-low-power Wi-Fi adopts a duty-cycling mechanism similar to Zigbee. It offers data rates around 1–2 Mbps and has a range of 10–70 m indoors [23,24]. Wi-Fi has the advantage of being interoperable since low-power Wi-Fi can be deployed at the HAN, and the traditional Wi-Fi can serve for the AMI and the NAN. Traditional Wi-Fi is based on IEEE 802.11 standard family, and the data rates range from 1 to 100 Mbps, while the range can go up to 500 m. Wi-Fi operates in the 2.4 GHz ISM band. One of the main challenges regarding Zigbee and Wi-Fi is coexistence. As they may operate in the same 2.4 GHz ISM band, they experience interference issues.

Z-wave and Wavenis are other alternative communication technologies for WSNs. Z-wave is a proprietary wireless communication protocol initially developed for home automation by ZenSys (currently owned by Sigma Designs) [25]. Wavenis is a wireless protocol stack developed by Coronis Systems [26]. Different from Zigbee and Wi-Fi, Z-wave operates in the 908 MHz ISM band while Wavenis operates in the 433, 868, and 915 MHz bands in Asia, Europe, and the United States, respectively. Z-wave has data rate of 40 kbps, whereas Wavenis has a data rate of 100 kbps.

A recent amendment to IEEE 802.15.4, namely, IEEE 802.15.4a standard, also called as ultra-wide band (UWB), offers increased data rates for WSNs that can reach up to 27.24 Mbps [27]. UWB utilizes 16 channels between 250 and 750, 3,244 and 4,742, and 5,944 and 10,234 MHz. UWB is ideal for short-range, high data rate applications in HANs and NANs.

In [28], the authors have evaluated several wireless technologies for indoor environments. A typical small flat in Germany and a multidwelling building have been considered for comparisons. In the experimental setup, wireless links have been formed between the gas furnace, water meter, indoor temperature sensor, display, washing machine, and an outdoor temperature sensor. IEEE 802.15.4 has shown to perform better than the other technologies, that is, Bluetooth and Konnex (KNX). KNX is a Europe-based communication technology for home automation, which has been standardized in the 1990s. IEEE 802.15.4 operating at the 868 MHz has the highest reliable indoor coverage among all the considered technologies. However, it has the disadvantage of operating in a single channel.

7.4 WSAN-BASED SMART ENERGY USE

Smart energy use refers to managing the electricity consumption of home appliances such as air conditioner, dishwasher, dryer, washing machine, oven, refrigerator, as well as the home charging of electric in favor of reduced expenses and load [29–31]. WSAN-based smart energy use schemes employ sensors, actors, and smart meters.

Sensors provide information on energy consumption, while actors may turn off an appliance or change its settings when needed, while smart meters may provide information on the price of electricity or receive certain control signals from the utility. In the following section, we will summarize the WSAN-based smart energy use schemes in the literature.

7.4.1 WSN-Based In-Home Energy Management

In-home energy management (iHEM) applications aim to reduce the cost of electricity usage at home and reduce peak load, while causing the least comfort degradation for the consumers. A WSN-based iHEM approach has been recently introduced in [29]. From hereafter, we will call this scheme shortly as iHEM. iHEM coordinates appliance schedules by taking into consideration both smart grid signals and consumer preferences. iHEM applies to shiftable appliances only. Consumers may turn on their appliances at any time, regardless of peak time concern, while iHEM computes a more convenient schedule, if available, and suggests that to the customer. If the customer agrees, the load of the appliance is shifted to a later off-peak hour. iHEM employs a central controller, which is in charge of communication with the smart meter, as well as communicating with the appliances when they are turned on. A simple optimization model, which is called optimization-based residential energy management (OREM), has been proposed in order to compare the performance of iHEM. We will first introduce iHEM followed by OREM.

iHEM employs appliances with communication modules, a WSN with several sensor nodes, a central controller, and a smart meter. Appliances act as a reduced function device (RFD) within the WSN. They only generate packets and do not forward packets. Sensor nodes act as full function devices (FFD); hence, they can become relay nodes for the appliance packets. At the same time, they collect temperature, humidity, and presence data for smart home-related applications. Central controller collects the packets of the appliances, communicates with the renewable energy resources such as rooftop solar panels or backyard wind turbines, and also communicates with the smart meter to receive price signals. Smart meter communicates with the utility side and receives price information and handles outage notification. The controller avoids scheduling the appliances during peak hours unless renewable energy is available. After the appliances are scheduled according to TOU prices, the user is notified of the new schedule. If the user agrees to use the appliance according to this schedule, then he or she sends a confirmation message to the controller. The user has the freedom to use the appliance right away in order not to compromise comfort. This process is repeated every time a controllable appliance is turned on.

The performance of iHEM is compared with a benchmark optimization scheme, that is, OREM. OREM aims to minimize the total energy expenses by scheduling the appliances in the appropriate timeslots. The objective function of OREM is given in Equation 7.1:

$$\text{Minimize} \sum_{i=1}^{I}\sum_{j=1}^{J}\sum_{t=1}^{T}\sum_{k=1}^{K} E_i D_i U_t S_t^{ijk} \tag{7.1}$$

The constraints of OREM are as follows:

$$\sum_{k=1}^{K}\sum_{i=1}^{I} D_i S_t^{ijk} \le \Delta_t, \quad \forall t \in T, \forall j \in J \tag{7.2}$$

$$\sum_{t=1}^{T} D_i S_t^{ijk} = D_i, \quad \forall i \in I, \forall j \in J, \forall k \in K \tag{7.3}$$

$$\sum_{t=1}^{m-1} S_t^{ijk} + \sum_{t=m+2}^{T} S_t^{ijk} = 0, \quad \forall i \in I, \forall j \in J, \forall k \in K, m = a_{ijk} \tag{7.4}$$

$$\sum_{t=m}^{m+1} S_t^{ijk} = 1, \quad \forall i \in I, \forall j \in J, \forall k \in K, m = a_{ijk} \tag{7.5}$$

where
E_i is the power consumption of appliance i
D_i is the length of the cycle of appliance i
U_t is the unit price for slot t
a_{ijk} is the timeslot of the arrival of request k of appliance i on day j
S_t^{ijk} is the ratio of timeslot occupied by request k of appliance i on day j
Δ_t is the length of timeslot t
$D_{\max}$ is the maximum allowable delay
d_i is the delay of appliance i

Inequality (7.2) ensures that the total duration of the cycles of the scheduled appliances does not exceed the length of the timeslot that is assigned for them. Equation 7.3 ensures that an appliance cycle is fully accommodated without experiencing any interruptions. Equations 7.4 and 7.5 ensure that the maximum delay is limited by an upper bound as the request is accommodated either in the present or in the next timeslot.

Performance of iHEM has been evaluated by a discrete event simulator, and OREM has been solved in ILOG CPLEX optimization suite. The interarrival times between two requests are set as negative exponentially distributed with a mean of 12 h. During morning peak periods and evening peak periods, the interarrival time is negative exponentially distributed with a mean of 2 h. Several shiftable appliances have been considered in simulations, that is, a washer, dryer, dishwasher, and coffee maker whose energy consumption is 0.89, 2.46, 1.19, and 0.4 kWh and the duration of cycle is 30, 60, 90, and 10 min, respectively. TOU rates have been selected as follows: the on-peak, mid-peak, and off-peak prices are taken as 9.3, 8.0, and 4.4 cent/kWh, respectively. For winter, on-peak hours are considered to be between 6 a.m. and 12 p.m. and 6 p.m. and 12 a.m. Mid-peak hours are from 12 to 6 p.m. Weekdays between 12 and 6 a.m., and the weekends are off-peak periods. We compare the performance of iHEM with OREM and when there are appliances with priorities. These appliances are not shifted; they are turned on regardless of peak hours. This may correspond to a case where users have either preconfigured

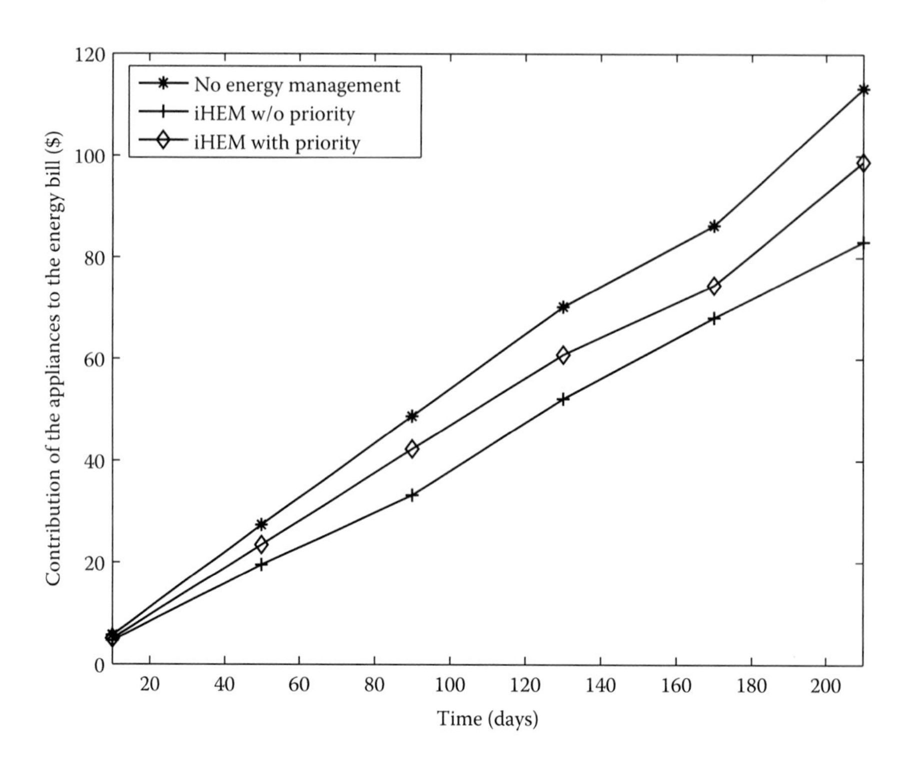

FIGURE 7.5 Total contribution of appliances to the energy bill with priority-based scheduling.

a subset of their appliances as high priority appliances or several appliances are not able to communicate with the energy management unit (EMU). In Figure 7.5, iHEM without priority denotes when all the appliances are treated equally. As seen in the figure, savings of the consumer reduce when appliances have priorities because this reduces the flexibility of scheduling the appliances in the off-peak hours. Similarly, the peak load increases for *iHEM with priority* as seen from Figure 7.6. Nevertheless, iHEM succeeds to reduce the expenses and the load in both scenarios.

7.4.2 Appliance Usage Forecasting

A household appliance usage forecasting mechanism has been developed within the course of the bright energy equipment (BEE) project by the researchers at the Politecnico di Milano. BEE project implements smart energy use through two-phase analysis. In the first phase, wireless power meter sensor network (WPSN) collects data from the appliances. The power consumption of the appliances, their time of usage, and their usage duration are measured over a certain time, that is, N days. After N days, a prediction algorithm is used to predict the future usage of the appliances. This initial period is called as the training period. By predicting the future demand, some settings of the demand management tool are refined without the need of the user to input his or her preferences.

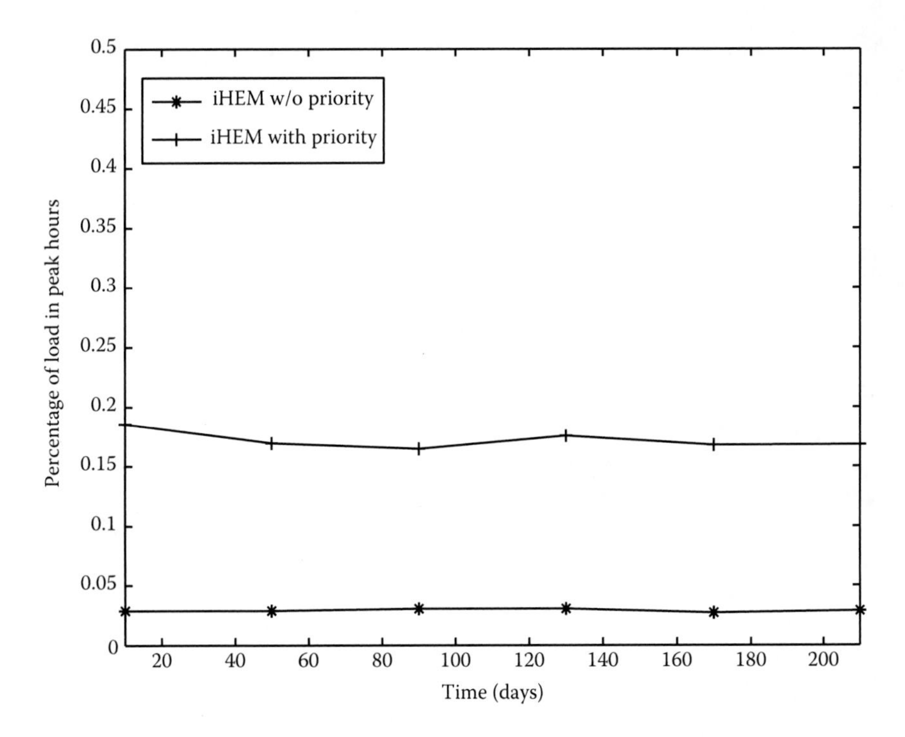

FIGURE 7.6 Contribution of the appliances to the total load on peak hours with priority-based scheduling.

BEE projects use a WPSN with star topology where appliances are the RFDs of the WSN. An appliance is assumed to be *on* if its power consumption exceeds a certain threshold. The authors have shown that at the end of 60 days of training, the *on–off* status of air conditioner, dishwasher, and oven has been predicted with over 90% reliability. Similar findings have been confirmed from a testbed implemented within a 45 m^2 residential unit, occupied by one person.

7.4.3 iPower

Intelligent and personalized energy conservation system by WSNs (iPower) is an energy conservation tool for residential premises as well as small offices [32]. It focuses on energy conservation regardless of the features that become available with the smart grid. Based on user presence and ambient conditions, lights or air conditioning appliances or computers are turned off.

iPower employs a WSN, a control server, power-line control devices, and user identification devices. WSN monitors the rooms with light, sound, and temperature sensors. When there is a change in the monitored room, sensor nodes send notification to the gateway, which then communicates with the intelligent control server of the house or the building. Intelligent control server may turn off an appliance or adjust its settings according to the profiles of the users who are present in the room. Users are recognized by their identification tags. The control commands are delivered via power-line communications. iPower can also work in the interactive mode where the server sends an alarm signal to notify the people in the room that the appliances, lights, and HVAC will be turned off. If the room is occupied, then the occupant may signal their presence by making noise or moving a sensor-attached furniture, and the server does not turn off the devices.

7.4.4 Energy Information System for Flats

An in-home energy information system has been presented in [33]. Real-time data acquisition, visualization, analysis, and control of home appliances can be performed with the proposed tool. Energy and cost savings are displayed to the users to provide information on the output of smart energy use policies.

Zigbee-based sensors and actors have been used to monitor the consumption of appliances as well as control their status. Smart energy controller can be implemented as an embedded system running either Linux or embedded WinXP where the software has been implemented using Java. The controller software has been arranged in four sections:

1. *Active system management*: This module manages and controls the multithreading system.
2. *Energy storage archive*: This module provides rapid data access as well as historical data. Historical data are stored after aggregation for 1 year. These data are stored for future load forecasting applications.
3. *Energy device unit*: Data access and acquisition is handled by this module. It acts as a gateway between different data formats and different interfaces.
4. *Energy service unit*: This module has multiple functions. First, it allows data access for browser or desktop use. Control of appliances depending on the utility tariff is also executed by this module.

The communication with the Zigbee sensors and the controller is presented in Figure 7.7. Controller receives the status of the device via *GET-DevInfo* and *DevInfo* messages. If the tariff is on-peak, then *SET-POW = OFF* message is sent to turn the appliance off. Otherwise, *SET-POW = ON* message is sent. In addition to those control messages, heartbeat messages are exchanged periodically between the controller and the sensors.

To display the energy use, a thin client browser application has been designed. The display shows the actual power consumption, which is refreshed every 2 s. It is also possible to visualize a histogram of energy usage for the past 24 h via this browser application.

7.5 DISTRIBUTED POWER CONSUMPTION MEASUREMENT SYSTEM

Distributed power consumption measurement system employs point of load power meters to collect detailed consumption information [34]. The measurements collected by the point of power meters are fine grained, for instance; it is possible

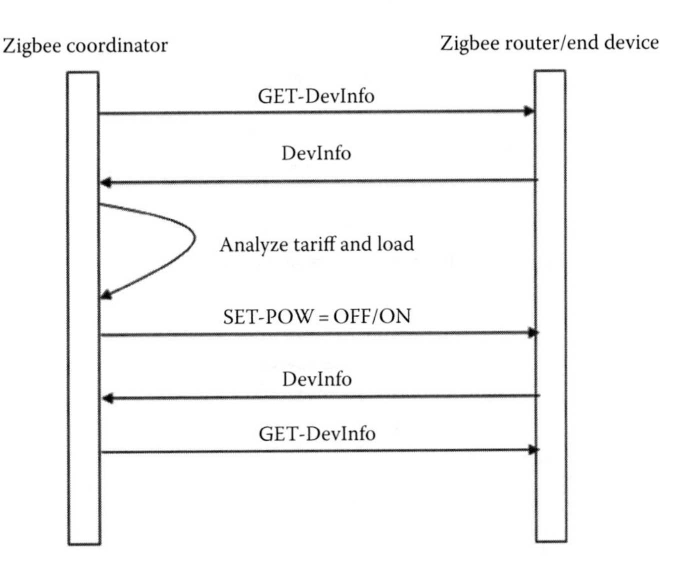

FIGURE 7.7 Communication diagram for the energy information system.

to infer that the door of the fridge has been opened given that sampling interval is adequate to capture the instance. Off-the-shelf point of load power meters have been used in this study. Those meters can measure the electricity consumption on one-phase, real, and reactive power consumption. They also store the measurements in nonvolatile memory. They can remotely turn on or off a device where the communication between those meters is established via Zigbee.

In Figure 7.8, we present the basic components of the measurement system. It consists of multiple Zigbee D-Bus interfaces, analyses components, data collector components

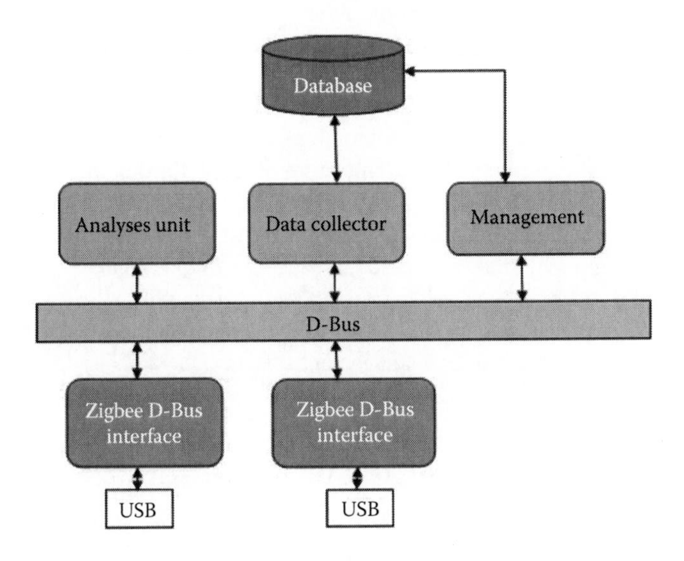

FIGURE 7.8 System model.

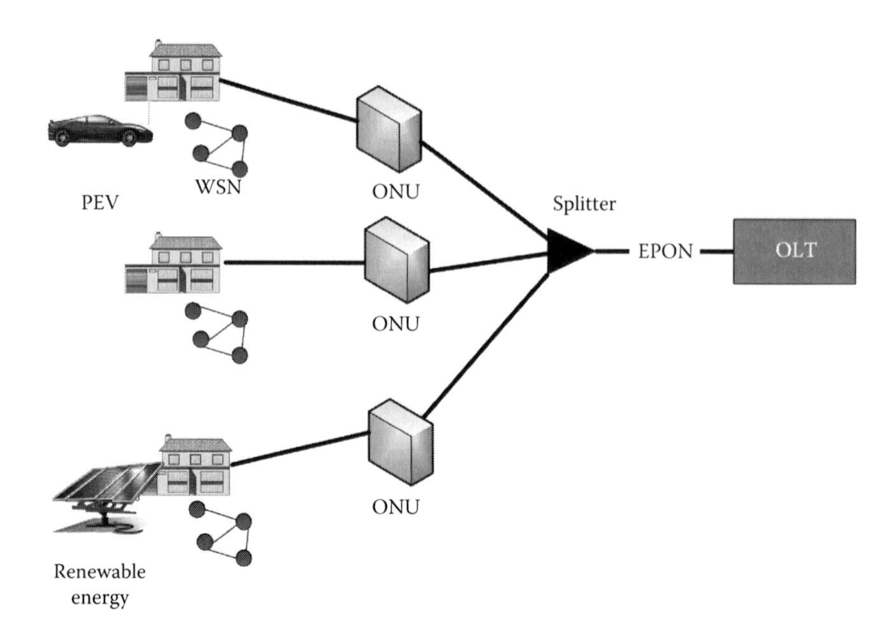

FIGURE 7.9 Fiber-wireless network architecture.

saving collected data and the results of the analyses to databases, and a management component, which configures the components of the system for autonomous operation.

7.6 FIBER-WIRELESS SENSOR NETWORKS

In [35], the authors have proposed an intrahome and interhome scheduling for smart energy use. In intrahome scheduling, appliances in individual homes are delayed up to 60 min to correspond to mid-peak or off-peak usage. In the centralized interhome scheduling algorithm, requests of several homes are collected at the optical line terminal (OLT), which limits the maximum total power consumption of appliances. This scheme relies on the communication architecture presented in Figure 7.9. OLTs function as the control unit of an Ethernet passive optical network (EPON). Optical network units (ONUs) are installed at customer premises while two or more customers may share one ONU. In addition to this broadband access network, fiber optic sensors and WSN are employed to monitor the status of power distribution network components such as breakers, switches, voltage regulators, and transformers. The fiber optic sensors communicate with the distribution management system (DMS) using dedicated wavelength channels, while WSN uses Zigbee.

7.7 WSN-BASED PERSONAL LIGHTING MANAGEMENT

Simple lighting control systems that rely on occupancy have been used for many years, and their energy saving properties have been known for long. With the advent of smart grid, more advanced control systems can offer further savings. In [36],

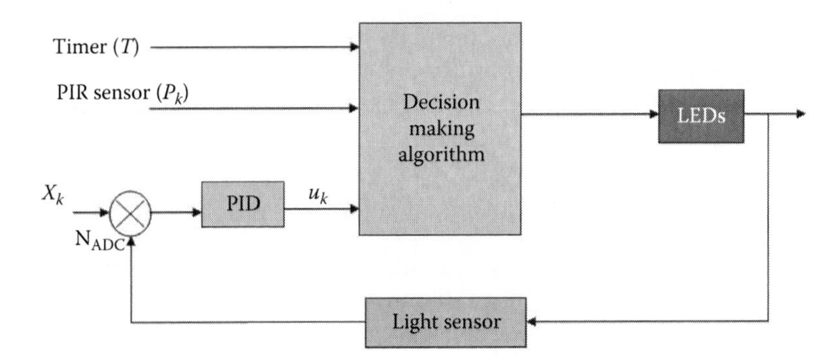

FIGURE 7.10 Closed-loop personal lighting control scheme.

the authors have proposed using wireless sensors for personal control of the dc grid-powered networked light emitting diode (LED) lighting. The sensor network is distributed into the workspace within a building, and the sensors collect ambient data for indoor environment quality monitoring. The sensor network is also responsible for real-time control of the human–lighting interaction as well as optimizing the energy usage of the building lights. The proposed personal lighting management system reduces energy waste while improving the comfort of the users.

WSN-based personal lighting system consists of nine sensors that report to a sink node using wireless links, a personal computer (PC), and a digital addressable lighting interface (DALI) controller to control 14 sets LED lighting arrays. The sink and the DALI controller are connected via RS232 serial communication protocol. The light intensity and occupancy measurements collected by the WSN are used to control the *on–off* status of LED lights as well as configure their luminance, such that the lighting condition is maintained 300,500 lux. The closed-loop control scheme is presented in Figure 7.10. Here, X_k denotes the set point of sensor k, and N_{ADC} is the measured lumen value. They are fed into a proportional integral derivative controller (PID controller) whose output u is used by the decision making algorithm. The other parameters of the decision making algorithm are the timer, T, that is used to check the time of day condition, while the output of the pyroelectric infrared (PIR) sensor is a binary value, which is 1 if the space is occupied. The output of the decision making algorithm controls the light intensity.

The authors have implemented the WSN-based personal lighting system in a 70 m^2 office using ultra-low-power MSP430F2274 microcontrollers and a CC2500 2.4 GHz wireless radio transceivers arranged in a star topology. They have shown that WSN-based personal lighting system provides enhanced energy conservation and improved user experience.

7.8 SUMMARY AND OPEN ISSUES

Electricity generation process is one of the fundamental contributors to the GHG emissions and consequently threatens earth's sustainability. Smart grid aims to reduce the GHG emissions by integrating ICTs to power system operations and benefiting from two-way communications between the utility and the customers. In the smart grid, electricity consumption of the customers is managed by using the fine-grained consumption data made available by the smart meters, WSNs, and intelligent mechanisms.

Smart energy use involves management of home appliances as well as office lights and PCs. In general, intelligent mechanisms aim to shift the shiftable loads to off-peak hours as well as avoiding unnecessary energy consumption. Energy saving while maintaining user comfort is one of the key challenges of smart energy use. For this reason, consumers are generally provided with some incentives through dynamic pricing schemes. In this chapter, we have reviewed the dynamic pricing schemes. We have explained TOU pricing, CPP, PTR, and real-time pricing. In addition, user comfort can be maintained by ambient intelligence that is achieved via WSANs. WSANs can use various communication technologies such as Zigbee, Wi-Fi, Z-wave, and Wavenis. We have provided a brief overview of the state-of-the-art technologies.

Finally, we have included WSAN-based smart energy use applications that are available in the literature. A WSN-based residential energy management scheme is introduced, which is called in-home energy management. iHEM coordinates shiftable appliance schedules such that customers are provided by cost savings and the utility benefits from reduced peak load. The second scheme forecasts appliance usage after collecting samples for N days. A prediction algorithm is used to predict the future usage of the appliances. The third scheme, intelligent and personalized energy conservation system by WSNs (iPower), is an energy conservation tool for residential premises as well as small offices. The fourth scheme is a real-time data acquisition, visualization, analysis, and control tool for home appliances. The fifth scheme is a distributed power consumption measurement system that employs point of load power meters to collect detailed consumption information. The sixth scheme describes the fiber-wireless sensor network (Fi-WSN) architecture for smart grid energy management tools. The final scheme proposes a personal lighting management system that reduces energy waste while improving the comfort of the users in small offices.

There are several unaddressed issues in the literature regarding smart energy use. Security is one of the fundamental challenges in this context. Smart grid security is a broad topic covering the security of communication protocols, security of devices, and security of smart grid assets [37–44]. WSAN-based smart energy management schemes may become vulnerable to attacks if security mechanisms are not implemented as an integral part of those systems. For instance, the modification of smart meter data by malicious attackers may result in incorrect billing, inaccurate load statistics, false forecasting, and false pricing decisions. In [45], the authors have presented an Internet-based load alteration attack, which is implemented by compromising the load control command signals of a utility. It has been shown that these types of attacks risk the stability of the grid once they compromise a large number of smart meters. On the other hand, misconfiguration of smart grid-integrated consumer devices may also increase the vulnerability of the grid either by sending modified user data to the utility or by sending modified utility signals to the users [46]. These attacks can have serious consequences since the devices on consumer premises are relatively easy to be compromised, and the utility has little or no control over those devices. Attackers may also access the memory of consumer devices, extract the keys used for authentication, and insert malicious software, which could spread to other devices in the AMI [47].

Privacy needs to be carefully handled for WSAN-based smart energy use schemes as well. Availability of high-resolution electricity consumption data may violate the privacy of the consumer as it is possible to obtain a detailed information on the activities in a residential premise such as absence or presence of a person, sleep cycles, meal times, and shower times by accessing fine-grained electricity consumption data [48].

Secure and privacy-preserving communication for smart energy use is still an open issue. In WSNs, security mechanisms usually increase the cost of the sensor. Limited processing and memory resources pose a challenge for computationally intensive public key cryptography mechanisms; meanwhile, symmetric keys are also challenging to implement due to their communication overhead. Therefore, there is a need for lightweight and cost-effective security mechanisms for WSNs.

Another aspect of smart energy use is electricity theft prevention [49]. In developed countries, electricity theft is usually related to marijuana-growing operations, which require high amount of energy to power up air conditioners and lights. In this context, detailed consumption information collected by the smart meters and AMI offers sophisticated methods to determine electricity theft. Smart grid forensics is a newly emerging research field that can play a significant role in investigation of electricity theft and prevent electricity wastage.

REFERENCES

1. US Environmental Protection Agency Report on Inventory of US GHG Emissions and Sinks: 1990–2008, Available online at: http://www.epa.gov/climatechange/emissions/usinventoryreport.html. Last accessed on February 4, 2014.
2. E. Santacana, G. Rackliffe, T. Le, X. Feng, Getting smart, *IEEE Power and Energy Magazine*, 8(2), 41–48, March–April 2010.
3. S. M. Amin, B. F. Wollenberg, Toward a smart grid: Power delivery for the 21st century, *IEEE Power and Energy Magazine*, 3(5), 34–41, 2005.
4. J. Gao, Y. Xiao, J. Liu, W. Liang, P. Chen, A survey of communication/networking in smart grids, *Future Generation Computer Systems (Elsevier)*, 28(2), 391–404, 2012.
5. V. C. Gungor, F. C. Lambert, A survey on communication networks for electric system automation, *Computer Networks Journal (Elsevier)*, 50, 877–897, May 2006.
6. V. C. Gungor, D. Sahin, T. Kocak, S. Ergut, C. Buccella, C. Cecati, G. P. Hancke, Smart grid technologies: Communication technologies and standards, *IEEE Transactions on Industrial Informatics*, 7(4), 529–539, 2011.
7. C. Lo, N. Ansari, The progressive smart grid system from both power and communications aspects, *IEEE Communications Surveys and Tutorials*, 14(3), 799–821, 2012.
8. H. T. Mouftah, M. Erol-Kantarci, Smart grid communications: Opportunities and challenges, *Handbook of Green Information and Communication Systems*, M. S. Obaidat, A. Anpalagan, I. Woungang (eds.), pp. 631–663. Elsevier, 2012.
9. P. P. Parikh, M. G. Kanabar, T. S. Sidhu, Opportunities and challenges of wireless communication technologies for smart grid applications, *Power and Energy Society General Meeting, 2010 IEEE*, Minneapolis, MN, July 25–29, 2010.
10. M. Erol-Kantarci, H. T. Mouftah, Wireless multimedia sensor and actor networks for the next-generation power grid, *Elsevier Ad Hoc Networks Journal*, 9(4), 542–511, 2011.

11. V. C. Gungor, B. Lu, G. P. Hancke, Opportunities and challenges of wireless sensor networks in smart grid, *IEEE Transactions on Industrial Electronics*, 57(10), 3557–3564, October 2010.
12. S. Ullo, A. Vaccaro, G. Velotto, The role of pervasive and cooperative Sensor Networks in Smart Grids communication, *15th IEEE Mediterranean Electrotechnical Conference (MELECON)*, pp. 443–447, IEEE, 2010.
13. I. F. Akyildiz, W. Su, Y. Sankarasubramaniam, E. Cayirci, Wireless sensor networks: A survey, *Computer Networks Journal (Elsevier)*, 38(4), 393–422, March 2002.
14. J. Yick, B. Mukherjee, D. Ghosal, Wireless sensor network survey, *Computer Networks Journal (Elsevier)*, 52, 2292–2330, 2008.
15. G. Barbose, R. Bharvirkar, C. Goldman, N. Hopper, B. Neenan, Killing two birds with one stone: Can real-time pricing support retail competition and demand response? *Proceedings of the 2006 ACEEE Summer Study on Energy Efficiency in Buildings*, Pacific Grove, CA, August 2006.
16. Australian Energy Market Operator (AEMO). Market price of electricity. http://www.aemo.com.au/data/price_demand.html. Last accessed on April 2012.
17. M. Erol-Kantarci, H. T. Mouftah, Prediction-based charging of PHEVs from the smart grid with dynamic pricing, *Proceedings of First Workshop on Smart Grid Networking Infrastructure in LCN 2010*, Denver, CO, October 2010.
18. E. A. Feinberg, D. Genethliou, Load forecasting, *Applied Mathematics for Restructured Electric Power Systems, Power Electronics and Power Systems*, J. H. Chow, F. F. Wu, J. A. Momoh (eds.). Springer, 2005, pp. 269–285.
19. IEEE 802.15.4 standard. Available online at: http://standards.ieee.org/about/get/802/802.15.html. Last accessed on November 2012.
20. RFC4919: IPv6 over Low-Power Wireless Personal Area Networks (6LoWPANs). Available online at: http://tools.ietf.org/html/rfc4919. Last accessed on September 2012.
21. IEEE 802.11 standard. Available online at: http://standards.ieee.org/about/get/802/802.11.html. Last accessed on April 2012.
22. S. Tozlu, Feasibility of Wi-Fi enabled sensors for Internet of Things, *Seventh International Wireless Communications and Mobile Computing Conference (IWCMC)*, Istanbul, Turkey, July 4–8, 2011, pp. 291–296.
23. Ultra-low power WiFi chips of Gainspan Inc. Available online at: http://www.gainspan.com/products/gs2000. Last accessed on October 2012.
24. Ultra-low power WiFi chips of Redpine Signals Inc. Available online at: http://www.redpinesignals.com/Renesas/index.html. Last accessed on September 2012.
25. M. T. Galeev, Catching the Z-wave, *EE Times Design*, February 2006. Available online at: http://www.eetimes.com/design/embedded/4025721/Catching-the-Z-Wave. Last accessed on September 2012.
26. C. Gomez, J. Paradells, Wireless home automation networks: A survey of architectures and technologies, *IEEE Communications Magazine*, 48(6), 92–101, June 2010.
27. E. Karapistoli, F. N. Pavlidou, I. Gragopoulos, I. Tsetsinas, An overview of the IEEE 802.15.4a standard, *IEEE Communications Magazine*, 48(1), 47–53, January 2010.
28. N. Langhammer, R. Kays, Performance evaluation of wireless home automation networks in indoor scenarios, *IEEE Transactions on Smart Grid*, 2012, in press.
29. M. Erol-Kantarci, H. T. Mouftah, Wireless sensor networks for cost-efficient residential energy management in the smart grid, *IEEE Transactions on Smart Grid*, 2(2), 314–325, June 2011.
30. M. Erol-Kantarci, H. T. Mouftah, Using wireless sensor networks for energy-aware homes in smart grids, *IEEE Symposium on Computers and Communications (ISCC)*, Riccione, Italy, June 22–25, 2010.

31. M. Erol-Kantarci, H. T. Mouftah, Management of PHEV batteries in the smart grid: Towards a cyber-physical power infrastructure, *Proceedings of Workshop on Design, Modeling and Evaluation of Cyber Physical Systems (IWCMC11)*, Istanbul, Turkey, July 5–8, 2011.
32. L. Yeh, Y. Wang, Y. Tseng, iPower: An energy conservation system for intelligent buildings by wireless sensor networks, *International Journal of Sensor Networks*, 5(1), 1–10, 2009.
33. I. Kunold, M. Kuller, J. Bauer, N. Karaoglan, A system concept of an energy information system in flats using wireless technologies and smart metering devices, *IEEE Sixth International Conference on Intelligent Data Acquisition and Advanced Computing Systems*, Prague, Czech Republic, September 15–17, 2011, pp. 812–816.
34. T. Kovacshazy, G. Fodor, C. B. Seres, A distributed power consumption measurement system and its applications, *12th International Carpathian Control Conference (ICCC)*, Velke Karlovice, Wellness Hotel Horal, Czech Repubic, May 25–28, 2011, pp. 224–229.
35. M. Maier, Fiber-wireless sensor networks (Fi-WSNs) for smart grids, *13th International Conference on Transparent Optical Networks (ICTON)*, Stockholm, Sweden, June 26–30, 2011, pp. 1–4.
36. Y. K. Tan, T. P. Huynh, Z. Z. Wang, Smart personal sensor network control for energy saving in DC grid powered LED lighting system, *IEEE Transactions on Smart Grid*, 2013.
37. M. Amin, Securing the electricity grid, *The Bridge, US National Academy of Engineering*, 40(1), 13–20, Spring 2010. Last accessed on October 2012.
38. M. Amin, Toward a more secure, strong and smart electric power grid, *IEEE Smart Grid Newsletter*. Available online at http://smartgrid.ieee.org/january-2011/67-toward-a-more-secure-strong-and-smart-electric-power-grid. Last accessed February 4, 2014.
39. P. McDaniel, S. McLaughlin, Security and privacy challenges in the smart grid, *IEEE Security & Privacy*, 7(3), 75–77, May–June 2009.
40. The Smart Grid Interoperability Panel, Cyber Security Working Group, Guidelines for smart grid cyber security: Vol. 1, Smart grid cyber security strategy, architecture, and high-level requirements, August 2010. Available online at: http://csrc.nist.gov/publications/nistir/ir7628/nistir-7628_vol1.pdf. Last accessed on October 2011.
41. The Smart Grid Interoperability Panel, Cyber Security Working Group, Guidelines for smart grid cyber security: Vol. 2, Privacy and the smart grid, August 2010. Available online at: http://csrc.nist.gov/publications/nistir/ir7628/nistir-7628_vol2.pdf. Last accessed on October 2011.
42. The Smart Grid Interoperability Panel, Cyber Security Working Group, Guidelines for smart grid cyber security: Vol. 3, Supportive analyses and references, August 2010. Available online at: http://csrc.nist.gov/publications/nistir/ir7628/nistir-7628_vol3.pdf. Last accessed on October 2011.
43. US Department of Energy Office of Electricity Delivery and Energy Reliability, Study of security attributes of smart grid systems—Current cyber security issues, April 2009. Available online at: http://www.inl.gov/scada/publications/d/securing_the_smart_grid_current_issues.pdf. Last accessed on October 2011.
44. US Congress, Office of Technology Assessment, Physical vulnerability of electric system to natural disasters and sabotage, *OTA-E-453*. Washington, DC: US Government Printing Office, June 1990. Available online at: http://www.fas.org/ota/reports/9034.pdf. Last accessed on October 2012.
45. H. Mohsenian-Rad, A. Leon-Garcia, Distributed Internet-based load altering attacks against smart power grids, *IEEE Transactions on Smart Grid*, 2011.

46. Y. Simmhan, A. G. Kumbhare, B. Cao, V. Prasanna, An analysis of security and privacy issues in smart grid software architectures on clouds, *IEEE International Conference on Cloud Computing (CLOUD)*, Washington, DC, July 4–9, 2011, pp. 582–589.
47. T. Goodspeed, D. R. Highfill, B. A. Singletary, Low-level design vulnerabilities in wireless control systems hardware, *Proceedings of the SCADA Security Scientific Symposium (S4)*, Orlando, FL, January 21–22, 2009, pp. 3-13–3-26.
48. M. A. Lisovich, D. K. Mulligan, S. B. Wicker, Inferring personal information from demand-response systems, *IEEE Security & Privacy*, 8(1), 11–20, January–February 2010.
49. M. Erol-Kantarci, H. T. Mouftah, Smart grid forensic science: Applications, challenges and open issues, *IEEE Communications Magazine*, 51(1), 68–74, 2013.

8 Mobile Monitoring Application to Support Sustainable Behavioral Change toward Healthy Lifestyle

Valerie M. Jones, Rene A. de Wijk, Jantine Duit, Ing A. Widya, Richard G.A. Bults, Ricardo J.M. Batista, Hermie Hermens, and Lucas P.J.J. Noldus

CONTENTS

ABSTRACT

We describe the development of body area networks (BANs) incorporating sensors and other devices to provide intelligent mobile services in healthcare and well-being. The first BAN applications were designed to simply transmit biosignals and display them remotely. Further developments include analysis and interpretation of biosignals in the light of context data. By including feedback loops, BAN telemonitoring was also augmented with teletreatment services. Recent developments include incorporation of clinical decision support by applying techniques from artificial intelligence. These developments represent a movement toward smart healthcare, making health BAN applications more intelligent by incorporating feedback, context awareness, personalization, and decision support.

The element of decision support was first introduced into the BAN health and well-being applications in the Food Valley Eating Advisor (FOVEA) project. Obesity and overweight represent a growing threat to health and well-being in modern society. Physical inactivity has been shown to contribute significantly to morbidity and mortality rates, and this is now a global trend bringing huge costs in terms of human suffering and reduction in life expectancy as well as uncontrolled growth in demand on healthcare services. Part of the solution is to foster healthier lifestyle. A major challenge however is that exercise and dietary programs may work for the individual in the short term, but adherence in the medium and long term is difficult to sustain, making weight management a continuing struggle for individuals and a growing problem for society, governments, and health services. Using information and communication technologies (ICT) to support sustainable behavioral change in relation to healthy exercise and diet is the goal of the FOVEA monitoring and feedback application. We strive to design and develop intelligent BAN-based applications that support motivation and adherence in the long term. We present this healthy lifestyle application and report results of an evaluation conducted by surveying professionals in related disciplines.

8.1 INTRODUCTION

This chapter outlines work on design and development of mobile monitoring systems based on body area networks (BANs) and their application in the healthcare and well-being domains. We focus on a particular mobile application designed to support sustainable lifestyle change toward healthier living. Obesity and overweight represent a growing threat to health and well-being in modern society. Physical inactivity has been shown to contribute significantly to morbidity and mortality rates, and this trend affects not only in the west but now is apparent worldwide, bringing huge costs in terms of human suffering and loss of life expectancy as well as uncontrolled growth in demand on healthcare services. Part of the solution is to foster healthier lifestyle in the population in order to avoid or at least reduce the health consequences of physical inactivity and unhealthy diet for individuals and society at large. However, a drawback of exercise and dietary programs is that they may work for the individual in the short term, but adherence in the medium and long term is difficult to sustain, making weight management a continuing struggle for individuals and a growing problem for society, governments, and health services.

The focus of this chapter is a mobile application designed to support users in following and maintaining healthy dietary and exercise programs. The program is framed in consultation with professionals such as dieticians, nutritionists, and physical therapists and is personalized to the individual's needs and preferences. The intention is to aid maintenance of the intervention and promote continuing healthy living habits over time.

The first version of the application was developed during the Food Valley Eating Advisor (FOVEA) project and work continues under the European Institute of Innovation and Technology (EIT) ICT Labs Healthy Consumption program. The work of the FOVEA and EIT Healthy Consumption projects builds on work on remote monitoring and treatment teleservices BANs conducted by the Telemedicine Group at the University of Twente since 2001 [1]. Later in the chapter, we present the results of a market survey that was designed to elicit professional attitudes to the FOVEA mobile weight management application.

Section 8.2 describes the design, implementation, and main results of the FOVEA market survey. Section 8.3 summarizes the findings and Section 8.4 presents discussion and conclusions based on analysis of the results. First, we explain the context in terms of successive developments in BANs for smart healthcare.

8.1.1 Body Area Networks for Smart Healthcare

The mobile weight management application represents an extension of previous work on health BANs, used for monitoring of chronic and acute health conditions, in the direction of health and well-being applications, namely, in the direction of primary prevention. Here, we describe the development of health BANs for telemonitoring/teletreatment at the University of Twente and their evolution toward provision of more advanced intelligent services and applications.

We define a health BAN as a network of communicating devices worn on, around, or in the body that provides mobile health services to the user. In our generic architecture, a BAN consists of a Mobile Base Unit (MBU; handling communication, storage, and local processing) and a set of devices, which may include a number of sensors, actuators, and other devices. So the concept is broader than a sensor network since sensors and possibly also other devices are integrated together with a processing platform/communications gateway (the MBU). The MBU has been implemented on a range of mobile platforms (PDAs and smartphones) and a variety of mobile operating systems.

The BANs are supported by a BAN server (back-end system) that provides persistent storage and other BAN support services. Figure 8.1 shows the generic architecture of the BAN developed during the MobiHealth project and extended subsequently in other projects.

Sensor data are collected by the BAN, processed, and transmitted to a remote (healthcare) location via the MBU and the BAN back end. The generic architecture, a first prototype health BAN, and a number of variants of the BAN for different clinical

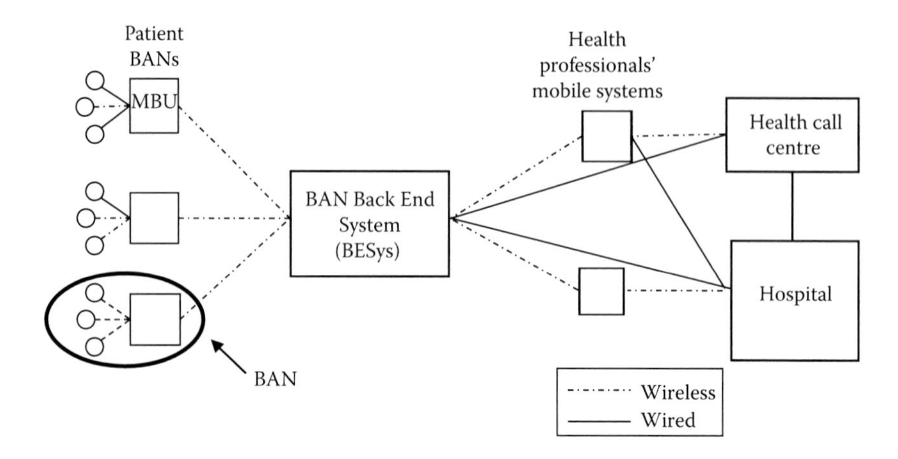

FIGURE 8.1 Generic architecture of the MobiHealth BAN system.

applications were prototyped and trialled during Information Society Technologies (IST) MobiHealth [2,3], which demonstrated that physiological measurements could be collected by sensors connected to a PDA or mobile phone and transmitted over general packet radio service (GPRS) and universal mobile telecommunications service (UMTS) to a remote location (e.g., a hospital or health call center). Trials were conducted on nine patient groups between 2002 and 2004; trials included telemonitoring of patients with cardiac arrhythmias, COPD patients, pregnant women, and casualties in trauma care.

Later variants of the BAN for different clinical applications including monitoring of epilepsy and chronic pain were developed in the Dutch Freeband Awareness project [4,5], the European eTen project HealthService24 [6], and the European eTen project MYOTEL [7,8]. In the Dutch project FOVEA, we extended the scope to include primary prevention in the well-being domain by addressing healthy lifestyle issues in the population at large.

Each clinical application requires a specific set of sensors as well as development of application-specific software and user interfaces. Sensors that have been integrated into the BAN to date include electrodes for measuring electrocardiography (ECG) and electromyography (EMG), pulse oximeter, motion sensors (step counters, 3D accelerometers), temperature, and respiration sensors. Apart from sensors, other devices that have been incorporated into different variants of the BAN include positioning devices, alarm buttons, and a multimodal biofeedback device.

Figure 8.2 shows the physical components of the epilepsy monitoring BAN developed during the Awareness project. The epilepsy BAN incorporates an Xsens MT9-B inertial sensor sensing 3D acceleration, three electrodes (Ag/AgCl contact electrodes) to measure ECG, and the Mobi8-MT9 sensor system. Three-dimensional accelerometer data are sampled at 128 Hz and the ECG signal is sampled at 1024 Hz.

In this case, the MBU is implemented on a PDA (HTC P3600). Simple rule-based decisions are made on the basis of analysis of the biosignal data and data fusion. If measured activity level is low and heart rate increase reaches a predefined

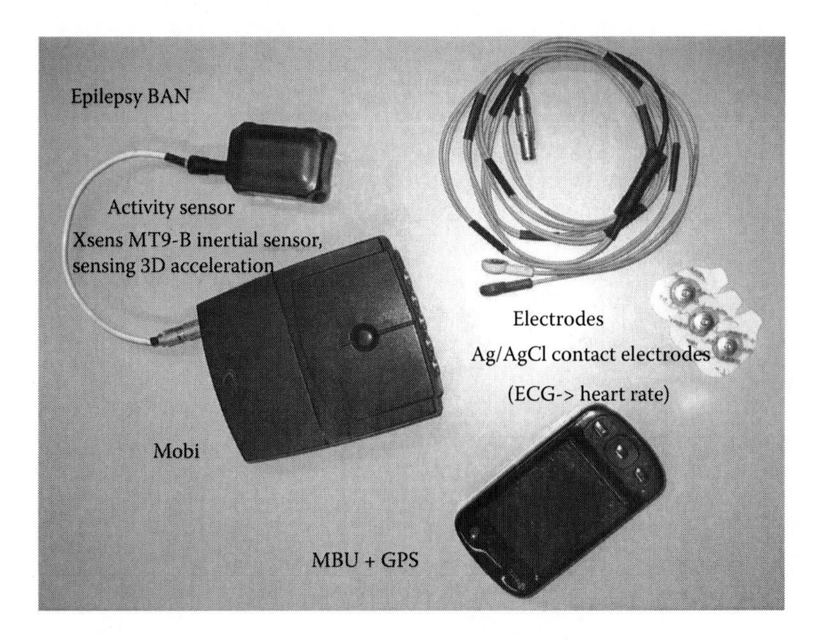

FIGURE 8.2 An early BAN: the epilepsy BAN incorporated electrodes and an activity sensor.

threshold, the event is labeled as a possible seizure. If posture is lying, or changes to lying, the probability that the patient is having a seizure is revised upwards. Heart rate is derived from the ECG signal by RR interval analysis. The beat to beat heart rate is converted to heart rate change by calculating the difference between mean heart rate in two moving time windows of 10 and 120 s. Activity level and posture (lying or not, detected by reference to the earth's gravitational field) are calculated every 10 s. The internal GPS device of the PDA, together with cell-ID information of the mobile network service provider, is used for location determination so that appropriate assistance can be dispatched to the patient if a seizure is detected. The specialist is notified in case of a detected seizure and can view the patient's biosignals and location on a health professional portal. This BAN uses a combination of external sensors and sensors onboard the PDA.

The first BAN applications in the MobiHealth project were designed to simply transmit biosignals and display them remotely, in order to investigate the feasibility of the use of 3G networks to support useful and usable applications and services related to telemonitoring. In later projects, we introduced analysis and interpretation of biosignals in the light of context data, and by including feedback loops, BAN telemonitoring services were augmented with teletreatment services that, for example, provide information, advice, coaching, and/or (bio)feedback to patients. In FOVEA, we introduced the element of decision support for the BAN user. FOVEA is discussed in more detail in the following section.

In the European FP7 project MobiGuide, we are developing the smart health BAN further, focusing on implementing mobile and distributed clinical decision support using knowledge-based systems (KBS) techniques from the field of artificial intelligence, including inferencing techniques applied to declarative and procedural knowledge bases. The work of the University of Twente in MobiGuide is focused on design, development, and implementation of evidence-based decision support on the MobiGuide BANs, with formalized clinical guidelines forming the core of the knowledge bases. Focus applications of the project are patient guidance services for patients with atrial fibrillation, gestational diabetes, and gestational hypertension; hence, specific variants of the patient BAN including (clinical) application-specific sensors and software components are under development. However, the MobiGuide system is intended to be generic and extensible to serve other chronic conditions.

In MobiGuide, the patients use variants of the patient BAN with the appropriate sensor sets to monitor biosignals (e.g., heart rate, blood pressure, blood glucose levels). Measurements are transmitted to a smartphone running Android and from there to a powerful fixed back-end system that includes the MobiHealth back-end system together with a set of specialized servers hosting, for example, declarative and procedural knowledge bases, guideline libraries, inferencing engines, electronic medical records (EMRs), and personal health records (PHRs). The MobiGuide decision support components, which have access to the patient's historical clinical data, including hospital records, analyze the data; alert the patient about actions that should be taken; ask the patient questions, in the case that additional information is needed; and make recommendations regarding lifestyle changes or advise the patient to contact a care provider. The knowledge bases are evidence-based, incorporating computer-interpretable versions of the relevant clinical guidelines, and are distributed between the smartphone and the back end. Inferencing as well as knowledge is also distributed between the BAN and the back end such that the BAN can perform autonomous reasoning if connection with the back end is lost; otherwise, the mobile and fixed parts of the distributed clinical decision support system collaborate together to give best advice to the patient. Figure 8.3 shows the major health BAN projects of the Telemedicine Group at the University of Twente and how successive generations of projects introduced smarter processing and interpretation of sensor data in order to provide more intelligent teleservices in healthcare.

Over the years, these developments represent an evolution toward smart healthcare, making health BAN applications more intelligent by incorporating feedback, context awareness, personalization, and clinical decision support. As well as improving the intelligence of mobile services to patients, the use of (distributed and real time) clinical decision support that accepts real-time streaming biosignals as input can also provide a basis for machine learning, enabling adaptive clinical decision support by personalizing and adapting strategies to individuals and their changing needs, aiding compliance and potentially supporting sustainable behavioral change.

8.1.2 Mobile Application for Weight Management

The objective of the FOVEA project is to change consumer behavior in the direction of a healthier lifestyle by applying behavioral theory with support from ICT, including ambulatory monitoring technology based on our experience of developing health BANs. The aim is to support sustainable behavioral change with respect to food and drink consumption and exercise in order to improve health and well-being

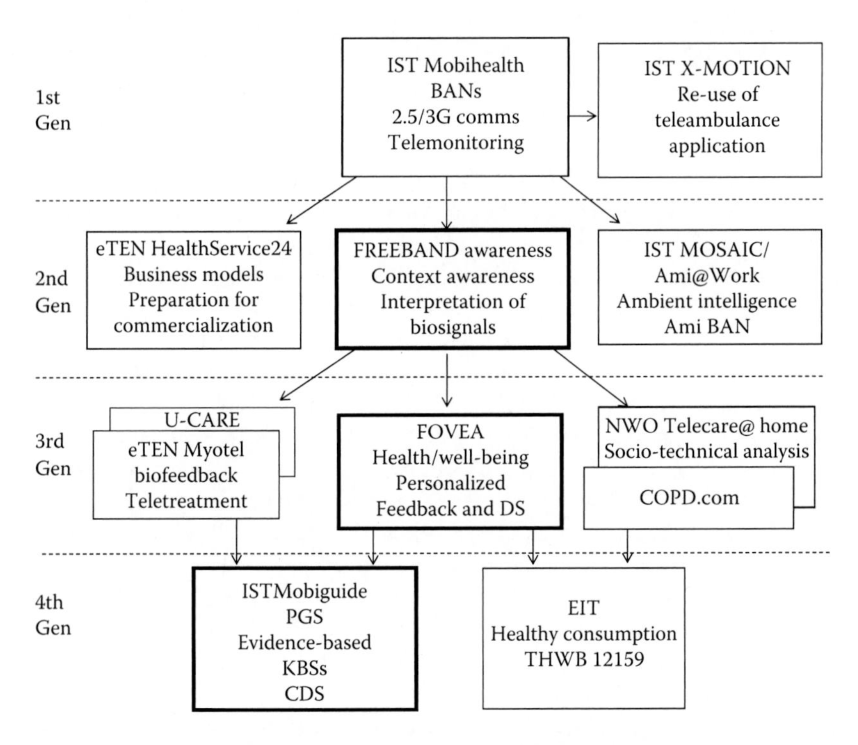

FIGURE 8.3 Major health BAN projects of the Telemedicine Group at the University of Twente. Current research activities are related to the fourth generation: DS, decision support; PGS, patient guidance services; KBSs, knowledge-based systems; CDS, clinical decision support.

and prevent chronic illness. The approach is to provide real-time personalized feedback and advice at the point of decision making. In the weight management application, the advice is tailored to the individual's weight management goals and stage of change. We target the *inclined abstainer* who is an *external eater* in the *action* stage of the *stages of change* model [9,10]. The Restaurant of the Future (RoF) [11] in Wageningen, the Netherlands, provides an instrumented environment that was used in FOVEA and later in EIT Health Consumptions as a test bed for interactive research in a real-life setting. The RoF infrastructure includes video cameras for behavioral observation, weight scales at the checkouts, and automatic registration of individuals' food and drink consumption at the point of sale terminals. It also offers possibilities for altering the ambient environment in order to investigate effects on physiology and behavior of subtle changes in environmental factors.

A prototype of the entire FOVEA system, integrating components from the RoF infrastructure with the food database of the restaurant supplier, was developed, and a study was conducted in 2011 at the RoF involving 30 users selected from regular visitors to the RoF who have BMI 25.00–29.99 (World Health Organization [WHO] classification *overweight, pre-obese* [12]). A dietician formulated a plan with each user, including up to five lunch compositions (each with up to five food items) to match the user's personal goals and preferences. These trial users were surveyed after the experiment in the ROF and the results of that survey are reported elsewhere.

The mobile part, the FOVEA BAN, is a small part of a larger distributed FOVEA system. The mobile system designed is to give real-time monitoring and personalized feedback. The FOVEA mobile system conforms to our BAN architecture; in this case, the BAN consists of a smartphone and a single sensor: the smartphone's onboard accelerometer. The FOVEA BAN was implemented as a smartphone application on a Samsung Galaxy S running Android. Food and drink consumption is registered on the smartphone, physical activity is monitored, and feedback is given in real time. The mobile application detects the different self-service buffets in the RoF. By means of indoor positioning, a map of the layout of the RoF, connection with the food database, and knowledge of the individual's targets, the mobile application is able to guide the user away from less healthy options (using a cue avoidance strategy) and toward healthier options and balanced meal compositions.

The FOVEA BAN stores the user's profile and keeps track of their energy balance throughout the day. In the RoF, the BAN can be used to display the food and beverage selections on offer, allowing the user to check an item before making a selection. *Good* and *bad* selections are highlighted to help the user make healthy consumption decisions. Energy values of food and beverage items on offer are stored in a food database and are used to track total kilocalories of items selected. Registration of food and beverage consumption on the smartphone enables real-time estimation of energy intake and helps the user to manage their daily energy budget.

Once the user enters the restaurant, the BAN begins to discover the buffets, which are identified by Bluetooth beacons. The floor plan of the restaurant is displayed on the smartphone screen together with a description of the buffets where the items recommended by the nutritionist can be found. When the user selects one of the discovered buffets, information about those food items is retrieved from the restaurant's database and presented to the user. When a food selection (and confirmation) is made, it is added to the smartphone's database and its calories are added to the user's daily energy intake.

In addition to calculating energy intake, real-time measurement of physical activity (using the smartphone's onboard tri-axial accelerometer) is used to estimate energy expenditure (EE) in real time. EE estimation is based on a step counting application from MobiHealth B.V. The algorithm has been validated against the OMRON Walking Style Pro. The step count is input to an EE algorithm that was developed during the FOVEA project. The algorithm uses the Harris–Benedict method of calculating basal metabolic rate (BMR).

Figure 8.4 shows four of the screen displays: the restaurant floor plan, the list of buffets, the energy screen, and the impact of consuming one item (in this case a can of *Coca Cola*) on energy balance on the right. The last screen shows that consumption of a certain soft drink is not compliant with the current lunch composition (selected by the user on the basis of advice from the dietician adapted to user preferences) and would send their temporary energy balance negative. Thus, the user can do a what-if analysis before finalizing his selection.

In the following section, we describe a survey of 95 potential stakeholders/users of FOVEA. This survey is designed primarily to elicit professional attitudes to and feedback on the weight management application and on the wider applicability of the FOVEA system in general.

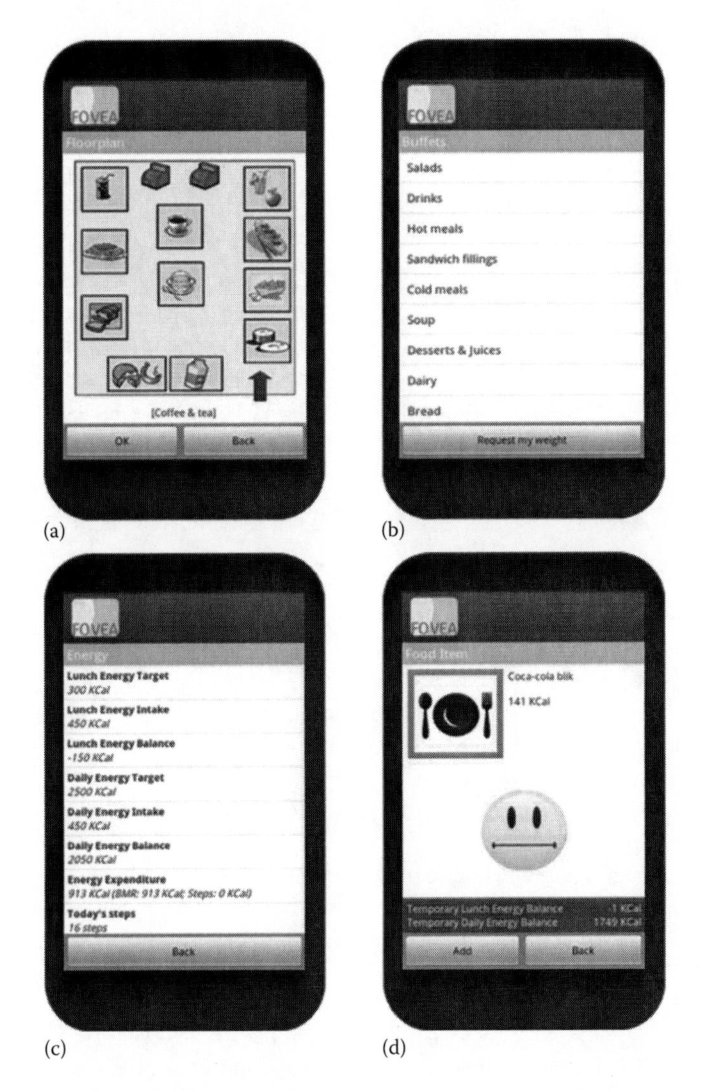

FIGURE 8.4 Four of the FOVEA screenshots: (a) floor plan, (b) buffets, (c) energy screen, and (d) impact of one food item on energy balance.

8.2 SURVEY

The objective of the survey is to research the business case for FOVEA by surveying a range of stakeholders including users, developers, industry, and researchers in relevant fields in order to establish stakeholder requirements needed to arrive at a FOVEA service that can be deployed full scale. Stakeholders are offered different usage scenarios and revenue models. Section 8.2.1 describes the methodology used. Section 8.2.2 characterizes the sample and Section 8.2.3 presents the results. The results have many implications for the evolving design, including issues relating to long-term adherence and hence sustainability.

8.2.1 Methodology

The online survey was aimed at professionals from fields relevant to the FOVEA system in general and to weight management application in particular. The online questionnaire was developed by Val Jones with help from members of the FOVEA project and was administered by Noldus Information Technology B.V. using FormSite. An invitation to participate was sent by e-mail. The following is an extract from the invitation:

> As part of the FOVEA system for monitoring, reasoning and feedback we have developed a personalised application to support health and wellbeing. One application of the FOVEA system is designed to achieve a healthy balance between eating behaviour and exercise.... The aim is to support sustainable behavioural change with respect to healthy living.

Respondents were presented with a description of the FOVEA system and the weight management application. The questionnaire elicits the stakeholders' attitudes and feedback as follows:

- General attitudes toward healthy lifestyle
- Specific feedback on the FOVEA system and weight management application
- Perceived economic value (using willingness to pay [WTP] under different revenue models and different usage scenarios)
- Privacy and data ownership
- Potential extension to other (lifestyle or heath) applications

In addition, the survey invites these professionals, if interested, to answer some user-oriented questions relating to their interest or otherwise as potential *end users* of the weight management application. The survey includes multiple choice questions and free text responses. A small pilot of 13 test subjects was conducted during June 2011 and the comments and feedback of the pilot subjects were used to improve the survey. The full survey was launched on June 28. The survey was closed on August 1, 2011. The results were analyzed by Jantine Duit of Noldus Information Technology B.V. using the FormSite analysis software, which produces the colored histograms seen in the following section.

8.2.2 Sample

The invitation to participate was sent to 554 individuals drawn from a selection of 523 contacts from the database of Noldus Information Technology B.V., selected on grounds of relevant profession/research area, plus the FOVEA Stakeholders Group (12); selected participants at the Artificial Intelligence in Medicine (AIME) 2011 conference (15); and colleagues of a related project (4). Ninety four e-mail invitations bounced with a failed delivery message, 22 produced an out-of-office response, and hence 116 invitations that we know of were not delivered or not delivered in time. When the survey closed, there were 95 responses. Assuming that the remaining 438 were delivered, this gives a response rate of 95/438 × 100 = 22%, quite high for this type of survey, especially considering the time of year (summer vacation).

Responses were received from 23 countries; the majority of respondents are currently living in Europe (with the Netherlands and Germany most numerous), followed by the United States. The remainder are living in United Arab Emirates, South Africa, the Russian Federation, China, Canada, Australia, and Argentina.

Responses were evenly split between male and female (46:47) indicating that (at least in this self-selected subsample of the sample) males are equally interested in lifestyle and weight management as females. This may simply reflect professional interest rather than personal interest since the survey was targeted at relevant professional groups.

Age bands 21–30 to 61–70 were represented, with the majority falling into age range 31–50. As can be expected, given the target population, the respondents were almost exclusively highly educated (university/college degree or higher degree) and/or professionally qualified.

In Section 8.2.3, a sample of questions and responses are presented.

8.2.3 Survey Results

In this section, we present the results of the survey. For reasons of space, we omit some questions and many of the histograms, showing the result summary only.

8.2.3.1 General Attitudes toward Healthy Lifestyle in Terms of Importance of Diet and Exercise

Q1–4, relating to attitude toward healthy lifestyle in terms of importance of diet and exercise, show that a majority (83/95) are interested in healthy lifestyle and/or weight management. The majority of those were interested in healthy lifestyle and/or weight management as private individuals (54/83) as well as from a professional viewpoint (49/83). Almost all (93/95) agreed that a healthy diet is important or very important and that regular exercise is important or very important (94/95). The conclusion, not surprisingly, is that these respondents have a positive belief in the importance of healthy lifestyle.

8.2.3.2 Reactions to the FOVEA Weight Management Application in Particular

Q5–14 sought reactions to the FOVEA weight management application in particular. Before these questions were presented, a series of informative screens were displayed that explain the application in some detail illustrated by some of the smartphone screenshots from the weight management app. Note: * indicates a mandatory question.

Q5. How do you rate the FOVEA smartphone application as a helpful diet and exercise aid?

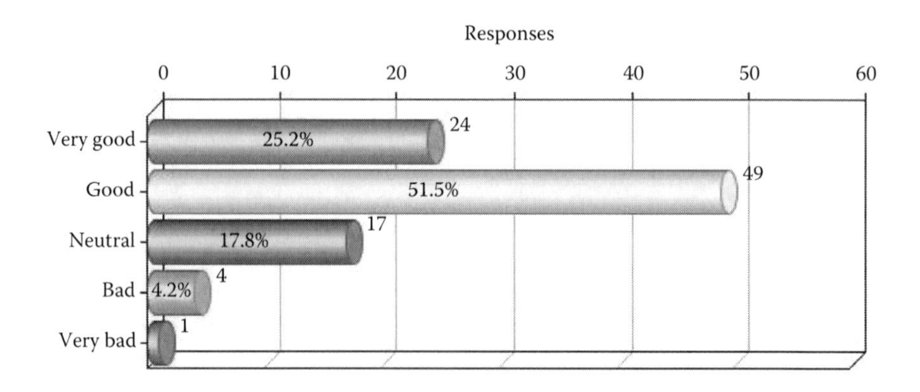

N = 95
73/95 judged the FOVEA smartphone application as very good or good as a helpful diet and exercise aid. Seventeen replied neutral, and five judged it as bad or very bad.

Q6. How do you see the usefulness of the FOVEA application for supporting healthy eating choices?
A similar pattern was observed for the usefulness of the FOVEA application for supporting healthy eating choices.

Q7. Would you be interested in using the FOVEA application? *
The respondents were more evenly split on the question of whether they were interested in using the FOVEA application: yes, 44/95; no, 34; and I don't know, 17. The 44 who answered Q7 positively were asked (Q8) on whose behalf they were interested in using the FOVEA application. Multiple responses were allowed (and used). 24/44 were interested in the application themselves as a professional, 3 for their patients or clients; 30 were interested themselves as private individuals; and 10 were interested on behalf of family or friends. So to summarize, more than half were interested in their capacity as private individuals as well as in a professional capacity.

Q9. What kinds of feedback on diet and exercise would you like to see? [free text response]
Out of N = 95, 35 responded. Several responses were negative; the criticisms included the opinion that the system was too complicated or overambitious. Another criticized the imbalance between diet and exercise and highlighted some shortcomings of accelerometry for EE estimation. Another suggests that GPS monitors would be a better choice for exercise monitoring. However, current GPS solutions would miss out all indoor exercise. Indeed accurate estimation of total EE (indoors as well as outdoors) does remain a challenge.

Further, respondents wanted positive feedback when they make good choices and warnings when they make poor choices, with warnings possibly including likely consequences of poor choices on body weight; tracking of progress in diet and exercise including longitudinal information; better balance between diet and exercise information and relating the two; more feedback relating to exercise; menu or food suggestions; special diet information; nutritional profile and value of food options and more visuals.

Q10. What positive aspects of the FOVEA application do you see? [free text response]

R = 74/95

Most responses were positive. The features mentioned as positive aspects fall into the following categories: awareness raising at point of purchase and consumption; tracking (of physical activity, of food intake); linking of eating behavior with physical activity and health implications; daily and real-time checking (of food intake and activity); user friendliness, attractiveness, ease of use, convenience, and simplicity; easy for consumers to understand; ability to check out food choices and meal options; good real-time feedback that can have immediate effects on behavior; objective information and advice; implications of choices are visible; personalization of feedback; support of planning, goal setting, and record keeping; information enables informed decision making to achieve balance; automated calorie counting and warnings; use of positive and negative reinforcement; user has a chance to override the system, teaching self-control; usable at all times; and targeting behavioral change.

Some respondents were skeptical: one replied to the effect that if the system worked, it would provide very good feedback but was doubtful whether the system would meet its claims. Other respondents saw no positive aspects whatever.

Q11. What negative aspects of the FOVEA application do you see? [free text response]

R = 74/95

6/74 respondents replied that they saw no negative aspects to the FOVEA application. Others saw a number of negative aspects. One felt cheated, seeing the survey as a marketing exercise. The other responses are summarized in the following: limited environment of use (*of the current prototype—one restaurant only*); cost and effort of building and maintaining food databases in all relevant settings (and doubts about likelihood of honesty and accuracy for some food suppliers); lack of inclusion (not everyone has [can afford, wants to use] a smartphone, may exclude older people or those with lower literacy); doubts about compliance, privacy, intrusiveness, and data security; too complicated and too cumbersome to use; taking enjoyment away from eating; locus of control (warnings too strong, too controlling, big brother effect, perceived limitation of freedom could lead to reactance effects in some individuals); doubts about the utility and added value of the information/app (motivated people are usually already well-informed about calorific values and EE and nonmotivated people won't use it, will ignore advice, or will cheat); doubts about accuracy of energy intake and expenditure estimation, especially in circumstances such as high-intensity workouts or training with weights.

Q12. Are there any additional features you would like to have? [free text response]

R = 49/95

Out of 49/95 responses, 13 answered with variants of no (e.g., *no, none, is enough, can't think of any*).

One respondent pointed out that purchase doesn't imply consumption (this is a constant challenge in this kind of research).

Other additional features suggested were the following: more nutritional information, including sugar (for diabetics), lactose, gluten for intolerances; function for scanning products in the supermarket or at home for calorific value; planning function (for meals, shopping, exercise, etc.); more physiological measures (absorption of, e.g., sugar or fat aliments, pulse rate, heart rate variability, % body fat, hydration levels based on consumption); suggestions for more healthy meal options and sustainable options, recipe creation; better activity measurement, for example, based on metabolic equivalents (MET) tables or global positioning system (GPS); feedback from an exercise professional preferably someone with American College of Sports Medicine (ACSM) certification; alarms; sleep monitoring; suggestion of extra exercise to offset over consumption of calories; social reinforcement—buddy system, reward system for good choices, and competition among friends.

Q13. Where do you think that FOVEA would be most useful? #

Reponses to this multiple response question show that roughly a third of the respondents recognized the usefulness of FOVEA in each of the following: restaurants, supermarkets, and at home and slightly fewer selected at work (21/95) and at the gym (16/95). 27/95 selected everywhere. Seven selected other and made further comments; responses under *Other* included views that the application needs to be ubiquitously available to be effective and that it should be on a personal device. Note: # indicates multiple responses are allowed.

Q14. Who do you think would benefit most from FOVEA? #

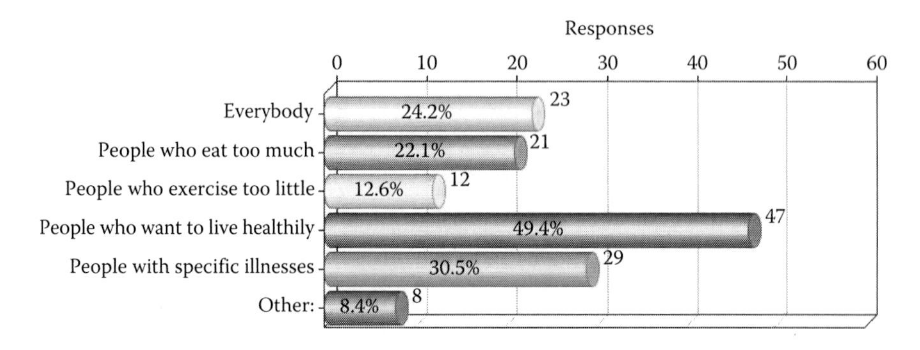

N = 95

23/95 thought everybody would benefit, people who eat too much (21) or exercise too little (12), 47/95 mentioned people who want to live healthily, and 29 mentioned people with specific illnesses.

8.2.3.3 Willingness to Pay

Q15–16 relate to WTP [13] using two business models: one-off payment for a smartphone application and monthly subscription to a smartphone service, respectively. Any currency is accepted; responses were converted to euros before the graphs were produced.

Q15. If you could buy the FOVEA application for your phone, how much would you be willing to pay for it? Please indicate which currency you are using.

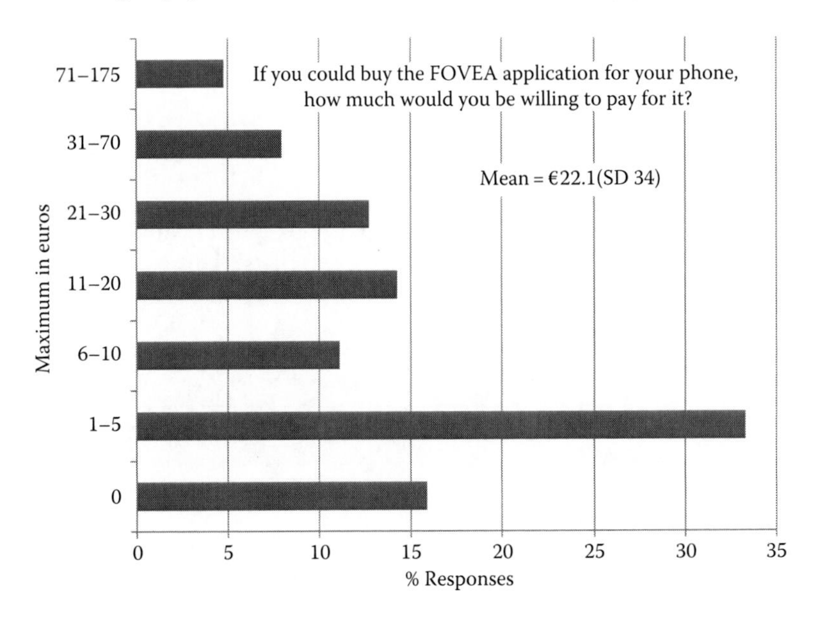

N = 63

Respondents' WTP as a one-off payment for a phone app, based on the description given, ranged from €0 to €150, with mean €22.1, SD 34.

Q16. If the FOVEA application was available by monthly subscription, how much would you be willing to pay for the service per month? Please indicate which currency you are using.

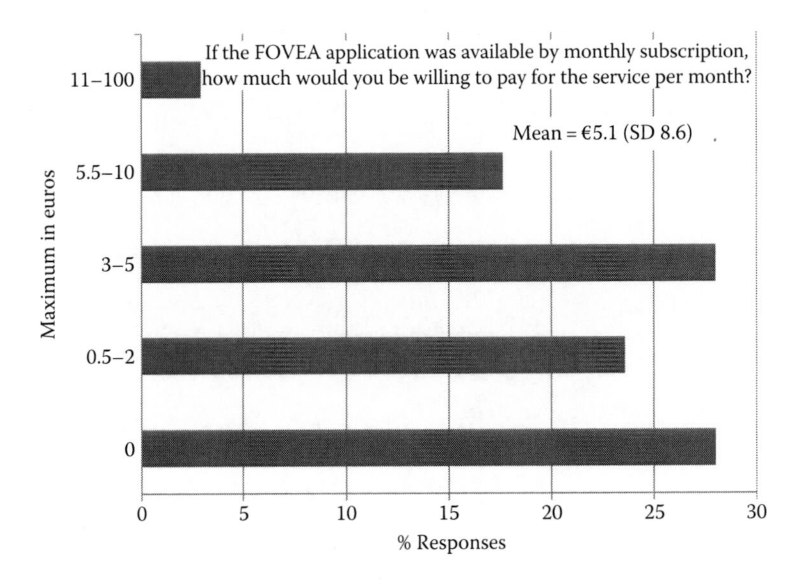

N = 68
Respondents' WTP as a monthly subscription for a phone app, based on the description given, ranged from €0 to €50, with mean €5.1, SD 8.6.

The respondents are then presented with the possibility of extending the current FOVEA service beyond the limited setting of the experimental prototype (one restaurant) as described in the informative screens, to (potentially) a scenario where the service is available at *all times and places where food and drink purchasing decisions are made.*

Q17. Would you like to see the FOVEA application extended to link to the product databases in other restaurants/food shops/supermarkets? This way it could assist food purchasing decisions in other places and at other times of the day. *

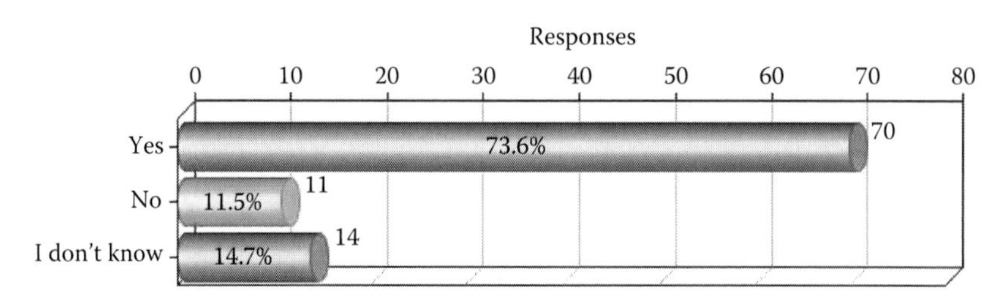

N = 95
73% (70/95) said they would like to see this extended service.

Q18. Would this increase your desire to buy the application? *

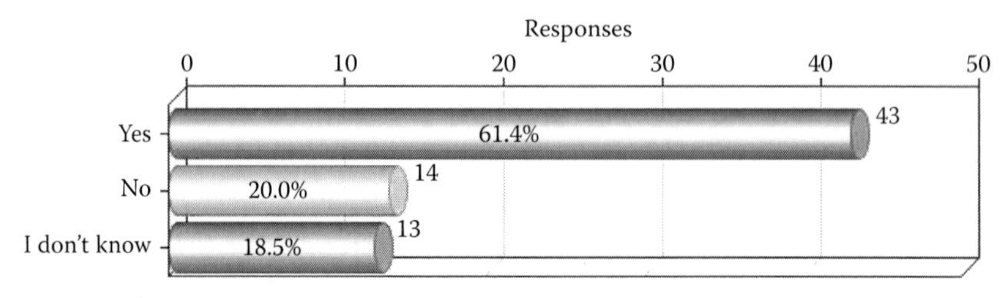

N = 70
Sixty-one percent of the respondents to Q18 (43/70) said this would increase their desire to buy the application.

The WTP questions are then repeated in relation to this new scenario in Q19–20.

Q19. If the FOVEA application could link to all the shops and restaurants where you shop and eat, how much would you be willing to pay for it as a one-off payment? Please indicate which currency you are using.
N = 32
Respondents' WTP as a one-off payment for a phone app, based on the extended scenario, ranged from €1.5 to €175, with mean €35, SD 37.

The extended scenario saw a rise in mean WTP from €22.1 to €35 for one-off payment.

Q20. If the FOVEA application could link to all the shops and restaurants where you shop and eat, how much would you be willing to pay for it per month? Please indicate which currency you are using.

N = 33

Respondents' WTP as a monthly subscription for a phone app, based on the extended scenario, ranged from €2 to €20, with mean €7.6, SD 17.

The extended scenario saw a rise in mean WTP from €5.1 to €7.6 for monthly subscription.

8.2.3.4 Privacy Issues

Q21–23 elicit opinions on privacy issues.

Q21. Do you see the FOVEA application as intrusive or not?

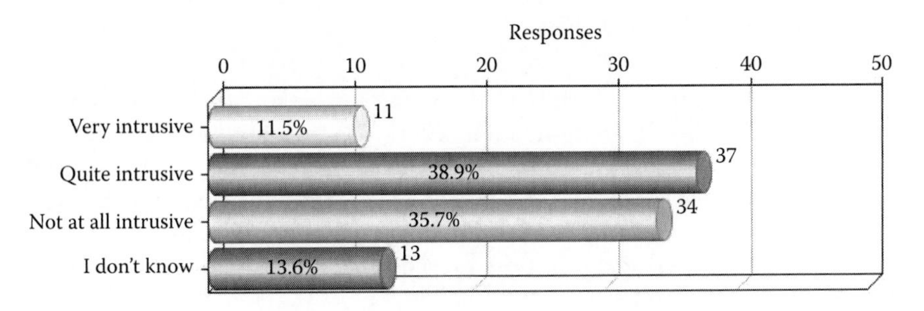

N = 95

Just over half of the respondents (48/95, 50.4%) saw the application as quite intrusive or very intrusive.

Q22. Who do you think should be responsible for holding the personal information of the user?

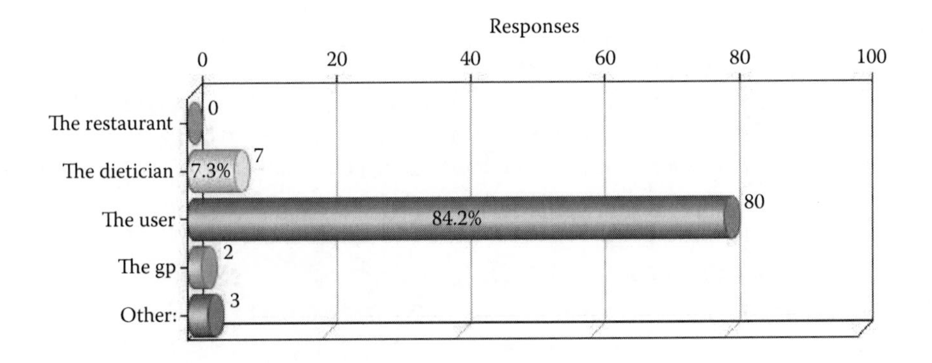

N = 95

The majority of respondents considered that personal information should be held by the user (80/95, 84.2%).

Under *Other*, one response was blank, the other two were

Other: the user (so we revise the aforementioned result to 81/95)
Other: the smartphone/computer system

Q23. Please comment on your answer, if you wish:
Free text responses showed broad agreement that the data belonged to the user and should only be seen by the user, except where the user grants access to certain professionals to classes of data relevant to their care function. Several responses related motivation to sustainability.

The FOVEA system is a general real-time measurement, reasoning, and feedback system for lifestyle applications and hence is much more generic than the single application described here (weight management). Q24 explains this context and elicits ideas about other potential applications.

Q24. The system can in principle be augmented with any body-worn sensors that have a (wireless) communication interface. In addition, new software applications can be implemented to provide all kinds of services involving real-time monitoring, processing of biosignals and context data, and user feedback and/or e-coaching. Assuming availability of the necessary sensors, what other applications of the FOVEA real-time measurement, reasoning, and feedback system might be of interest to you?
Of the 47 respondents who filled in this question, 12 responded some variant of *no* or *don't know*. Further five simply answered with a variant of Y, without specifying an application. The suggestions of alternative applications were research, monitoring sleepiness vs. alertness, stress monitoring, healthcare/well-being applications (chronic disease management [diabetes, high cholesterol, hypertension], children's health and exercise, rehabilitation), e-coaching, monitoring athletes during training, and control household devices.

The remainder did not suggest alternative applications as such but rather suggested extensions to the current application (e.g., monitoring blood glucose and hormones in relation to nutrient intake, measure time, and distance covered during exercise) or additional individual parameters to be monitored (e.g., heart rate, temperature, BP, lung capacity, body fat percentage, biomarkers of health status, fluids), additional sensors (e.g., step counter, movement recorder) or extra features (e.g., graphing, transmission of data, ability to combine group data).

8.2.3.5 Other Comments

Q25. If you have any other comments on FOVEA, please add them here.
Responses to Q25 were mainly positive, mentioning, for example, a good application concept, usefulness, and convenience. Some questioned whether motivation would be sustained and whether sufficient feedback was given. Another issue was the complexity of providing adequate coverage of the huge range of food items available. One responder felt the product was not viable.

8.2.3.6 Professionals as Potential End Users

Q27–34 are directed at end users of the weight management application. Before Q26, a screen invites the respondent, if interested, to adopt the role of potential user and answer the user-oriented questions. They are given the option to skip this section. 28/95 (29.4%) of respondents exercised the option to answer the user-oriented section. Therefore, for Q27–34, max. N is 28.

Q27. Are you happy with your weight at present?

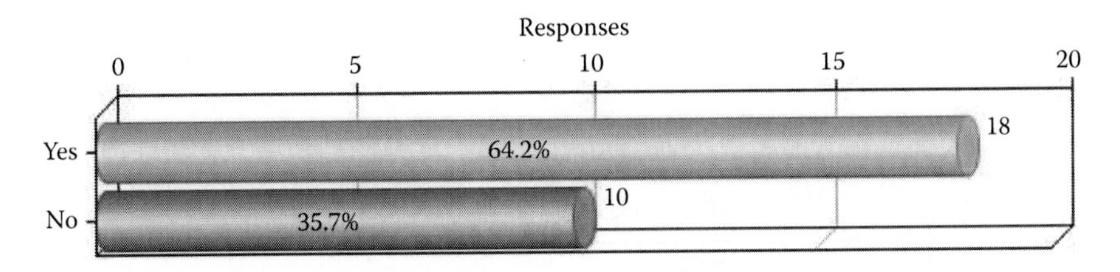

N = 28

Of all the respondents, 64.2% (18/28) claimed to be happy with their present weight (although responses to the following question [Q28] show that nearly half considered themselves to be overweight); the remainder consider themselves to be the right weight. None considered themselves underweight, very underweight, or very overweight.

Q28. Do you consider yourself to be:

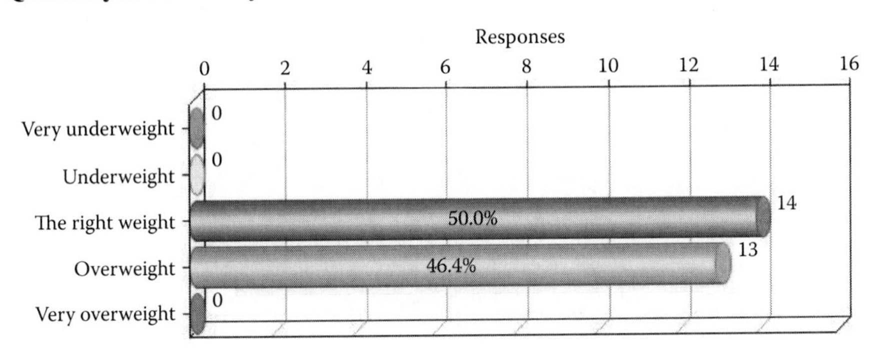

N = 28

Q29. Do you want to make changes to your diet?

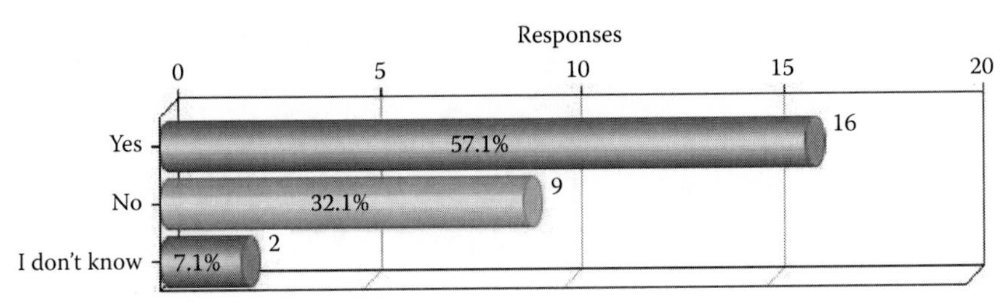

N = 28
Q29 shows that 57.1% (16/28) stated that they wanted to make changes to their diet, and Q30 shows that 64.2% (18/28) stated that they wanted to make changes to their exercise habits. Of all the respondents, 67.8% (19/28) say they have tried to change their weight by means of diet and/or exercise in the past (Q31) reporting this was very successful (26.3%: 5/19), quite successful (63.1%: 12/19), and not at all successful (10.5%: 2/19) (Q32).

Q30. Do you want to make changes to your exercise habits?

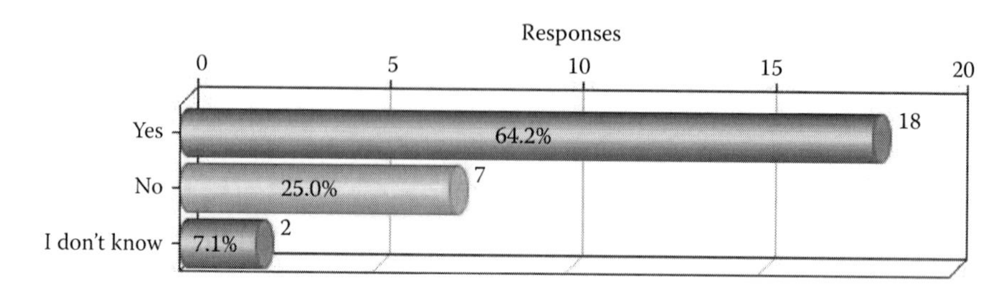

N = 28
18/28 said they wanted to change their exercise habits, 7 do not, and 2 don't know.

Q31. Have you tried in the past to change your diet and/or exercise pattern in order to lose weight or increase in weight? *

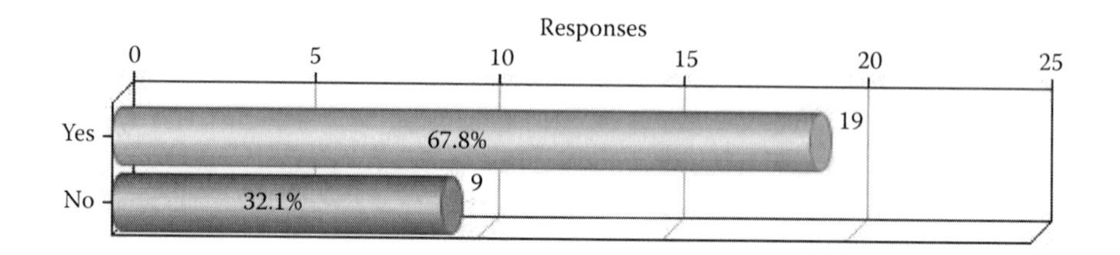

N = 28
19/28 have tried in the past to change their diet and/or exercise pattern in order to lose weight or increase weight.

Q32. How successful were you in losing weight or increasing in weight? *

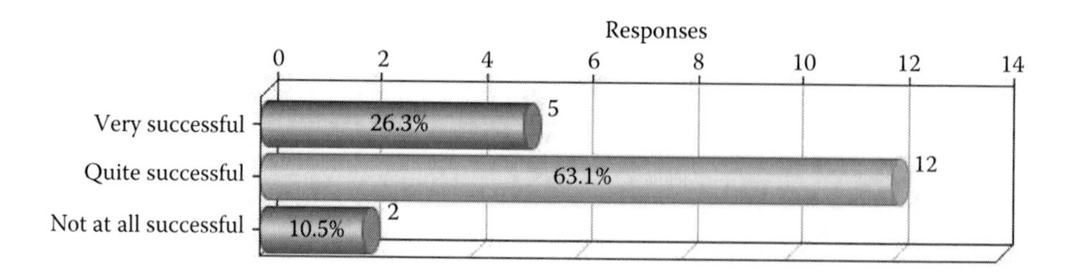

N = 19
Q32 was only presented to respondents who answered *yes* to the previous question, hence N = 19, despite this being a mandatory question.

Q33. How long did you maintain your new weight?

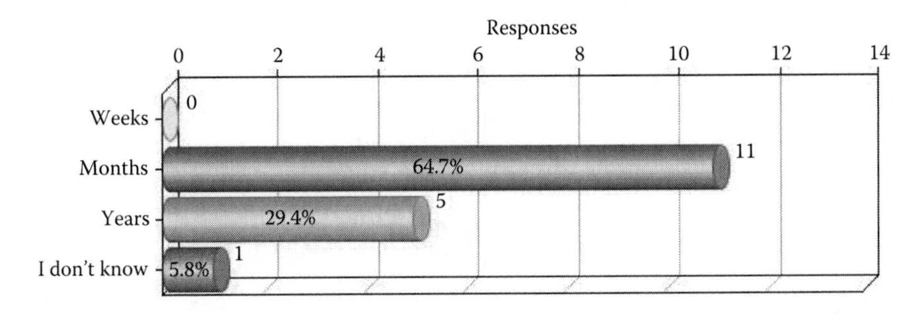

N = 17
Of all the respondents, 64.7% (11/17) reported success in maintaining their new weight over a period of months and 29.4% (5/17) over years.

Q34. Has a health professional ever told you that you were: #

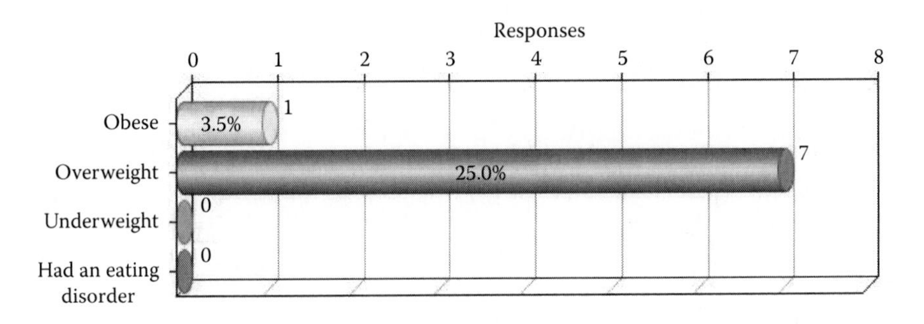

N = 28
7/28 reported being told they were overweight, and 1/28 obese, by a health professional (compare with 13/28 self-reported overweight [Q28]).

The remaining questions are related to demographics and are previously summarized in the description of the sample.

Q40. Please describe the kind of organization that you work for as precisely as possible.
The responses describing the organizations respondents work for (sometimes multiple organizations per respondent) are classified in the following:

- Research institute/university (41) with some specifying further as medicine, nutrition, food and drink research organizations, exercise science, rehabilitation, and emotion research

- Food industry (10): fragrance and flavor supplier, manufacturer of healthy foods, dairy company, food company, project development in food and health, research lab of dairy company, research center in large food company, Wageningen University RoF, spice and seasoning manufacturers, and food manufacturer in consumer packaged goods (CPG) (frozen foods, condiments, sauces) and Foodservice (condiments, sauces)
- Healthcare (7): child rehabilitation, healthcare company, hospital, research association, and obesity clinic/care (3)
- Other industries (6): SAP AG, IT hardware, software and services provider, cosmetic company, consumer product goods, fast-moving consumer goods industry, and telecommunications
- Government (2): public government and government research agency
- R&D (2): R&D and Fortune 100 company in corporate R&D
- Press and media (1)
- Sport (1)
- Miscellaneous: agency, consultant on various funded grants, user-centered design, top support and innovation lab, independent, and market research agency

A small minority of the professionals surveyed belonged to consumer groups (6/94), patient groups (4/94), or other relevant groups (11/94).

Most questions were optional and therefore could be skipped. 94/95 respondents stayed with the survey and answered all up to the final question.

8.3 SUMMARY OF FINDINGS

This section summarizes the findings from the market survey, according to the different sections of the questionnaire.

8.3.1 General Attitudes toward Healthy Lifestyle

Results relating to attitude toward healthy lifestyle in terms of importance of diet and exercise (Q1–4) show that a majority (83/95) are interested in healthy lifestyle and/or weight management. The majority of those were interested in healthy lifestyle and/or weight management as private individuals (54/83) as well as from a professional viewpoint (49/83). Almost all (93/95) agreed that a healthy diet is important or very important and that regular exercise is important or very important (94/95).

8.3.2 Specific Feedback on the FOVEA System and Weight Management Application

Q5–14 invite opinions on the FOVEA weight management application. 73/95 respondents judged the FOVEA smartphone application as a helpful diet and exercise aid as very good or good. Seventeen replies were neutral, and five judged it as bad or very bad. A similar pattern was observed for the usefulness of the FOVEA application for supporting healthy eating choices. Respondents were more or less evenly split on the question of whether they were interested in using the FOVEA application.

Of those who were interested, 24/44 said they were interested in the application themselves as a professional, 3 for their patients or clients; 30 were interested themselves as private individuals; and 10 were interested on behalf of family or friends. So, more than half were interested in their capacity as private individuals as well as the (overlapping) interest in a professional capacity.

Q9 (*What kinds of feedback on diet and exercise would you like to see*) produced many positive suggestions and also some criticisms. Critical points included are the following: an imbalance in the application between diet and exercise (focus on diet, not enough focus on exercise), the shortcomings of accelerometry for accurate measurement of exercise intensity, and a plea for inclusion of expertise from exercise professionals.

Many constructive suggestions were made, such as inclusion of *positive feedback for good choices and warnings for poor choices that include likely consequences for effect of poor choices on body weight; progress over time, weight tracking, and correlation to progress; plan vs. reality = balance; and guidelines and suggestions for exercise.*

Other ideas were *suggestions for food alternatives* and *suggestions for healthy food in the restaurant* and *assistance on healthy choices.* Several respondents asked for more or more specialized nutritional and dietary information, for example, *vegetarian, lactose or gluten information,* and nutritional profile and value of food options.

Main positive aspects of the FOVEA application reported were information/raised awareness at point of purchase and consumption, real-time personalized feedback, tracking (of physical activity, of food intake) enabling informed decision making to help achieve a balance; automation (of calorie counting, of warnings), targeting of behavioral change, and use of positive and negative reinforcement. Respondents also mentioned user friendliness, attractiveness, ease of use, convenience, and simplicity as positive features.

When asked about negative aspects of the FOVEA application, notable points are related to the following: the cost and effort of building and maintaining food databases in all relevant settings (and also doubt was expressed on the likelihood of honesty and accuracy in case of some food suppliers); lack of inclusion (not everyone has [or can afford, or wants to use] a smartphone), so the application may exclude older people or those with lower (computer) literacy; likelihood of compliance; and issues relating to privacy, intrusiveness, data security, and locus of control (*too controlling, big brother effect*). Further doubts were expressed about the utility and added value of the information/app (balance is too much toward diet and not enough toward physical activity; doubts about accuracy of energy intake and expenditure estimation, especially in circumstances such as high-intensity workouts or training with weights).

When asked were there any desired additional features, one respondent pointed out that purchase doesn't imply consumption. This is a constant challenge in this kind of research. Perhaps we should adopt the principle of *intention to eat*, as in *intention to treat* that is commonly used in Health Services Research.

Other desired additional features mentioned include the following: more nutritional information, including sugar (for diabetics), lactose, and gluten for people with intolerances; a function for scanning products in the supermarket or at home (e.g., for calorific value); planning functions (e.g., for meals, shopping, exercise); and additional physiological measures (e.g., absorption of sugar or fat aliments, pulse rate, heart rate variability, % body fat, and hydration levels based on consumption). Several respondents would appreciate suggestions from the app for more healthy meal options, or sustainable food options, also recipe creation/ideas. Again the issue of accuracy of activity measurement was raised; better activity measurement, for example, based on MET tables or GPS was suggested by one respondent; another wanted feedback from an exercise professional *preferably someone with ACSM certification.* Other suggested additions were alarms, sleep monitoring, and suggestions for extra physical exercises to compensate for overconsumption of calories. Several suggested the use of social reinforcement: *buddy system, reward system for good choices,* and *competition among friends.*

When asked *Where do you think that FOVEA would be most useful*? roughly a third of respondents recognized the usefulness of FOVEA in restaurants, supermarkets, and at home and slightly fewer selected at work (21/95) and the gym (16/95). 27/95 selected everywhere.

When asked about who would benefit most from FOVEA, 23/95 thought that everyone would benefit, 21 ticked *people who eat too much*, 12 ticked *people who* exercise *too little*, 47 mentioned people who want to live healthily, and 29 mentioned people with specific illnesses.

8.3.3 Perceived Economic Value

Q15–16 relate to WTP [13] using two payment models: one-off payment for a (smart) phone application and monthly subscription to a (smart)phone service, respectively. Respondents' WTP as a one-off payment for a phone app, based on the description given in the questionnaire, ranged from €0 to €150, with mean €22.1, SD 34. Respondents' WTP as a monthly subscription for a phone app, based on the description given, ranged from €0 to €50, with mean €5.1, SD 8.6.

The respondents were then presented with the possibility of extending the current FOVEA service beyond the limited setting of the experimental prototype (one restaurant) as described in the informative screens, to (potentially) a scenario where the service is available at all times and places where food and drink purchasing decisions are made. The majority (70/95) said they would like to see the FOVEA application extended to link to the product databases in other restaurants/food shops/supermarkets to assist food purchasing decisions in other places and at other times of the day. 43/70 said this would increase their desire to buy the application. The WTP questions were then repeated with the proviso that *the FOVEA application was linked to all the shops and restaurants where you shop and eat.* Respondents' WTP as a one-off payment for a phone app, based on this extended scenario, ranged from €1.5 to €175, with mean €35, SD 37. The extended scenario saw a rise in mean WTP from €22.1 to €35 for one-off payment. Respondents' WTP as a monthly subscription for a phone app, based on this extended scenario, ranged from €2 to €20, with mean €7.6, SD 17. The extended scenario saw a rise in mean WTP from €5.1 to €7.6 for monthly subscription.

8.3.4 Privacy and Data Ownership

Q21–22 elicit opinions on privacy issues. 48/95 respondents (50.4%) saw the application as quite intrusive or very intrusive. The majority considered that personal information should be held by the user (80/95, 84.2%).

In summary, we can say that there was broad agreement that the data belonged to the user and should only be seen by the user, except when the user grants access to certain professionals to classes of data relevant to their (care) function.

8.3.5 Potential Extension to Other (Lifestyle or Heath) Applications

Q24 (*...Assuming availability of the necessary sensors, what other applications of the FOVEA real-time measurement, reasoning and feedback system might be of interest to you?*) produced a number of suggestions of alternative applications as follows: *research*, monitoring sleepiness vs. alertness, stress monitoring, healthcare/well-being applications (chronic disease management [diabetes, high cholesterol, hypertension], children's health and exercise, rehabilitation), *e-coaching, monitoring athletes during training*, and *control household devices.*

8.4 USER-ORIENTED QUESTIONS

Q27–34 are directed at end users of the weight management application. Before Q26, a screen invites the respondent, if they are interested, to adopt the role of potential user and answer the user-oriented questions, but also giving the option to skip this section. 28/95 (29.4%) of respondents exercised the option to answer the user-oriented section.

A total of 18 of the 28 respondents to the user-oriented section claimed to be happy with their present weight (although responses to the following question [Q28] show that nearly half considered themselves to be overweight); the remainder consider themselves to be the right weight. None considered themselves underweight, very underweight, or very overweight.

A total of 16/28 stated that they wanted to make changes to their diet and 18 stated that they wanted to make changes to their exercise habits. 19/28 say they have tried to change their weight by means of diet and/or exercise in the past; 5/19 reported this was very successful, 12/19 quite successful, and 2/19 not at all successful. 18/28 want to change their exercise habits, 7 do not want to change their exercise habits, and 2 answered *don't know.*

19/28 have tried in the past to change their diet and/or exercise pattern in order to lose weight or increase weight. Of these 19, a total of 5 reported that this was very successful, 12 quite successful, and 2 not at all successful. 11/17 reported success in maintaining their new weight over a period of months and 5/17 over a period of years.

7/28 reported being told they were overweight, and 1/28 obese, by a health professional (compare with 13/28 self-reporting as overweight).

8.5 DISCUSSION AND CONCLUSIONS

Results relating *to general attitude toward healthy lifestyle* in terms of importance of diet and exercise show that a large majority of the sample reported that they were interested in healthy lifestyle and/or weight management. The majority of those were interested as private individuals as well as from a professional viewpoint. All but two (N = 95) agreed that a healthy diet is important or very important and all but one that regular exercise is important or very important.

Regarding *feedback on the FOVEA system and weight management application*, 73/95 respondents judged the FOVEA smartphone application as a helpful diet and exercise aid. Respondents were evenly split on the question of whether they were interested in using the FOVEA application. Critical points included focus on diet, not enough emphasis on exercise, and the shortcomings of accelerometry for accurate measurement of exercise intensity.

Main *positive aspects of the FOVEA application* reported were information/raised awareness at point of purchase and consumption, real-time personalized feedback, tracking (of physical activity, of food intake) enabling informed decision making to help achieve a balance; automation (of calorie counting, of warnings), targeting of behavioral change, and use of positive and negative reinforcement. Respondents also mentioned user friendliness, attractiveness, ease of use, convenience, and simplicity as positive features.

Negative aspects of the FOVEA application raised related to the cost and effort of building and maintaining food databases in all relevant settings; lack of inclusion (not everyone has, or can afford, or wants to use) a smartphone, so the application may exclude older people or those with lower (computer) literacy; likelihood of compliance; and issues relating to privacy, intrusiveness, data security, and locus of control and big brother effect. Some doubts were expressed about the utility and added value of the information/app (e.g., doubts about accuracy of energy intake and expenditure estimation, especially in circumstances such as high-intensity workouts or training with weights).

Roughly a third of respondents recognized the *utility of the FOVEA* weight management application in the following settings: restaurants, supermarkets, and at home.

Regarding *perceived economic value*, questions relating to WTP used two payment models: one-off payment for a (smart)phone application and monthly subscription to a (smart)phone service, respectively. Respondents' WTP as a one-off payment for a phone app, based on the description given in the questionnaire, with the limited usage scenario of the RoF only, ranged from €0 to €150, with mean €22.1, SD 34. Respondents' WTP as a monthly subscription for a phone app, based on the description given in the questionnaire, with the limited usage scenario of the RoF only, ranged from €0 to €50, with mean €5.1, SD 8.6.

The majority said they would like to see *extension of the FOVEA application* to link to the product databases in other restaurants/food shops/supermarkets to assist food purchasing decisions in other places and at other times of the day. 43/70 said this extended scenario would increase their desire to buy the application. Respondents' WTP as a one-off payment for a phone app, based on this extended scenario, ranged from €1.5 to €175, with mean €35, SD 37. The extended scenario saw a rise in mean WTP from €22.1 to €35 for one-off payment. Respondents' WTP as a monthly subscription for a phone app, based on this extended scenario, ranged from €2 to €20, with mean €7.6, SD 17. The extended scenario saw a rise in mean WTP from €5.1 to €7.6 for monthly subscription.

Regarding *privacy and data ownership*, there was broad agreement that the data belonged to the user and should only be seen by the user, except when the user grants access to certain professionals to classes of data relevant to their function.

Regarding *potential extension to other (lifestyle or heath) applications*, suggested alternative applications were made as follows: research, monitoring sleepiness vs. alertness, stress monitoring, healthcare/well-being applications (chronic disease management [diabetes, high cholesterol, hypertension], children's health and exercise, rehabilitation), e-coaching, monitoring athletes during training, and control household devices.

Regarding *attitudes as potential users themselves*, a total of 18/28 of these (professional) respondents claimed to be happy with their present weight (although responses to another question show that nearly half considered themselves to be overweight); the remainder consider themselves to be the right weight. None considered themselves underweight, very underweight, or very overweight. A total of 16/28 stated that they wanted to make changes to their diet and 18 stated that they wanted to make changes to their exercise habits. 19/28 say they have tried to change their weight by means of diet and/or exercise in the past; 5/19 reported this was very successful, 12/19 quite successful, and 2/19 not at all successful. 18/28 want to change their exercise habits. 19/28 have tried in the past to change their diet and/or exercise pattern in order to lose weight or increase weight. Of these 19, a total of 5 reported that this was very successful, 12 quite successful, and 2 not at all successful. 11/17 reported success in maintaining their new weight over a period of months and 5/17 over a period of years. 7/28 reported being told they were overweight, and 1/28 obese, by a health professional (compare with 13/28 self-reporting as overweight).

In December 2012, a further experiment was conducted at the ROF using the new version of the system developed during EIT Healthy Consumption. Results of that experiment will be reported elsewhere. The results of the FOVEA project including the survey described in this chapter and other FOVEA surveys and experiments feed into the EIT Healthy Consumption project.

According to the WHO, more than 1.4 billion adults were overweight worldwide in 2008. Of these adults, over 200 million men and nearly 300 million women were obese [14]. Lee et al. [15] published results in *The Lancet* showing that physical inactivity caused 9% of premature mortality worldwide in 2008.

Even among this well-informed and well-motivated group of subjects, attempts to maintain healthy weight were shown to be poor in the long term; less than a third managed to sustain their new weight over a period of years (Q33). In order to support healthy lifestyle in way that sustains motivation and brings long-term results, we need to continue multidisciplinary study in order to evaluate and improve ICT support such as the FOVEA BAN application in order to contribute positively to this burgeoning health problem. If we can be successful, the health benefits can be great. Lee et al. estimate that if inactivity were decreased by only 10%, more than half a million deaths could be averted every year worldwide. Furthermore, if physical inactivity were eliminated completely, this would increase the life expectancy of the world's population by 0.68 years [15].

ACKNOWLEDGMENTS

FOVEA (http://www.wageningenur.nl/nl/show/FOVEA-Food-Valley-Eating-Advisor-1.htm) was funded by the Dutch Ministry of Economic Affairs and the Province of Gelderland under the program Pilot Gebiedsbericht Innovatiebeleid. EIT Healthy Consumption project (THWB—Health & Well-being 12159) is partially funded by the EIT ICT Labs. The MobiGuide project is partially funded by the European Commission under the 7th Framework Program, grant #287811.

REFERENCES

1. V.M. Jones, R.G.A. Bults, D.M. Konstantas, and P.A.M. Vierhout, Healthcare PANs: Personal Area Networks for trauma care and home care, *Presented at Fourth International Symposium on Wireless Personal Multimedia Communications (WPMC)*, Aalborg, Denmark, 2001.
2. MobiHealth project: http://www.mobihealth.org/, accessed February 7, 2014.
3. V. Jones, A. van Halteren, I. Widya, N. Dokovsky, G. Koprinkov, R. Bults, D. Konstantas, and R. Herzog, MobiHealth: Mobile Health Services based on body area networks, *M-Health: Emerging Mobile Health Systems*, R.H. Istepanian, S. Laxminarayan, and C.S. Pattichis (eds.), Springer, New York, pp. 219–236 (624pp.), 2006.
4. FREEBAND AWARENESS project: http://www.utwente.nl/ctit/research/projects/concluded/bsik/freeband/projects/awareness/, accessed February 7, 2014.
5. V.M. Jones, H. Mei, T. Broens, I. Widya, and J. Peuscher, Context aware body area networks for telemedicine, *Proceedings Pacific-Rim Conference on Multimedia (PCM) 2007*, City University of Hong Kong, Hong Kong, China, December 11–14, 2007.
6. eTen HealthService 24 project: http://www.healthservice24.com, accessed February 7, 2014.
7. MYOTEL project: http://www.myotel.eu/, accessed February 7, 2014.
8. V.M. Jones, R. Huis in't Veld, T. Tonis, R.B. Bults, B. van Beijnum, I. Widya, M. Vollenbroek-Hutten, and H. Hermens, Biosignal and context monitoring: Distributed multimedia applications of body area networks in healthcare, *Proceedings of the 2008 IEEE 10th Workshop on Multimedia Signal Processing (MMSP2008)*, Cairns, Queensland, Australia, D. Feng, T. Sikora, W.C. Siu, J. Zhang, L. Guan, J.-L. Dugelay, Q. Wu, and W. Li (eds.), pp. 820–825, October 8–10, 2008. IEEE Catalog Number CFP08MSP-CDR. (c) IEEE 2008.
9. W.F. Velicer, J.O. Prochaska, J.L. Fava, G.J. Norman, and C.A. Redding, Smoking cessation and stress management: Applications of the transtheoretical model of behavior change. *Homeostasis* 38: 216–233, 1998.
10. I. Widya, R.G.A. Bults, R. de Wijk, B. Loke, N. Koenderink, R.J. Mendes Batista, V.M. Jones, and H.J. Hermens, Requirements for a Nutrition Education Demonstrator, *17th International Working Conference on Requirements Engineering: Foundation for Software Quality (RefsQ2011)*, Essen, Germany, D. Berry and X. Franch (eds.), *LNCS 6606*, Springer-Verlag, Berlin, Germany, pp. 48–53, March 28–30, 2011.
11. Restaurant of the Future, Wageningen University and Research centre, Wageningen, the Netherlands. http://www.restaurantvandetoekomst.wur.nl/UK/, accessed February 7, 2014.

12. World Health Organisation (WHO), BMI classification. http://apps.who.int/bmi/index.jsp?introPage=intro_3.html, accessed February 7, 2014.
13. J.A. Olsen and R.D. Smith, Theory versus practice: A review of 'willingness-to-pay' in health and health care. *Health Economics* 10: 39–52, 2001.
14. World Health Organization, Obesity and overweight, 2013. http://www.who.int/mediacentre/factsheets/fs311/en/, accessed February 7, 2014.
15. I. Lee, E.J. Shiroma, F. Lobelo, P. Puska, S.N. Blair, and P.T. Katzmarzyk, Effect of physical inactivity on major non-communicable diseases worldwide: An analysis of burden of disease and life expectancy. *The Lancet* 380(9838): 219–229, July 21, 2012, doi: 10.1016/S0140-6736(12)61031-9.

9 Cooperative and Self-Organizing Sensor Networks
The Enabler for Smarter Grids

Alfredo Vaccaro, Antonio Iacoviello, and Marjan Popov

CONTENTS

ABSTRACT

The complexity of future smart grids and the expected high level of uncertainties in these networks might radically affect the required security and reliability of power systems operation. Wide-area measurement systems (WAMSs) involve the use of system-wide information and the communication of selected local information to a remote location to counteract the propagation of large disturbances and reduce the probability of potential catastrophic blackouts. It is expected that WAMS will in the future generally improve the reliability and security of energy production, transmission, and distribution, particularly in power networks with a high level of operational uncertainties. To realize these benefits, researchers and designers of high-performance WAMS are revisiting numerous design issues and assumptions pertaining to scale, reliability, heterogeneity, manageability, and system evolution over time.

Following these research directions, in this chapter, we outline the important role played by cooperative and self-organizing smart sensor networks. In particular, we explore the possibility of decentralizing the WAMS processing and synchronization functions on a network of interactive smart units equipped by distributed consensus protocols. Detailed simulation studies aimed at assessing the effectiveness of the proposed computing and synchronization paradigm are presented and discussed.

9.1 INTRODUCTION

For many years, the electrical power system was organized according to a *static model* where electricity was generated by large power generation plants, transmitted to the loads by transmission lines, and finally distributed to the consumers by the distribution systems. Until a few years ago, the technological investments were almost exclusively devoted to the transmission system improvements, mainly as a result of the progressive concentration of power plants in cost-effective locations. In details, substantial resources were dedicated for the construction of new transmission lines aimed at increasing the network interconnection level and supporting the transfer of large amounts of electricity over long distances with minimal losses. At the same time, the distribution networks have been considered as the less important element of the electrical energy chain and characterized by a completely passive role. As a consequence, the criteria and the principles adopted for distribution systems planning and operation have not been enhanced over the years.

Today, electrical power systems are subjected to a host of challenges: the need for siting new generation near load centers, transmission expansion to meet growing demand, distributed resources, congestion management, grid ownership versus system operation and reliability coordination, etc. Some of the major challenges facing the electrical industry today include balancing between resource adequacy, reliability, economics, environmental, and other public purpose objectives to optimize transmission and distribution resources to meet the growing demand [1]. Besides, the operational environment of future electrical grids will become increasingly rigorous due to continually evolving functions of a power system from operation jurisdiction to control responsibly—coupled with the rising demand and expectation for reliability [2]. In addressing these challenges, power system operators must face the following critical issues:

- The rising level of network interconnections, which makes power networks more vulnerable with respect to dynamic perturbations.
- More components will operate at or near their limits, and the systems are enduring much more frequent changes of operating conditions.
- The increasing number of smaller geographically dispersed generators that could sensibly raise the number of power transactions.
- The difficulties arising in prediction and modeling the market operator's behavior, governed mainly by unpredictable economic dynamics, which introduce considerable uncertainty in short-term system operation.

- The need of more detailed security studies by assessing the impact of multiple contingencies on the power network (i.e., $N - 2$ security criteria*).
- The high penetration of generating units powered by renewable energy sources, which induces a number of side effects (e.g., highly variable power injections and voltage profile perturbation) introducing additional complexity in power system operation.

To attempt and address these issues, smart grid paradigm has been recognized as the most promising enabling technology. Large-scale deployment of this emerging paradigm is expected to enhance the power systems' efficiency and the use of cleaner energy resources by improving the security and reliability of the electrical infrastructures [3,4]. Besides, it could support the evolution of electrical power systems toward active, flexible, and self-healing web energy networks composed of distributed and cooperative energy resources.

Smart grid technologies include advanced sensing systems, two-way high-speed communications, monitoring and enterprise analysis software, and related services to get location-specific and real-time actionable data in order to provide enhanced services for both system operators (i.e., distribution automation, asset management, advanced metering infrastructure) and end users (i.e., demand side management, demand response) [5,6].

The cornerstone of these technologies is the ability of multiple entities (e.g., devices, software processes) to manage accurate and heterogeneous information. It follows that the development of reliable and flexible distributed measurement systems represents a crucial issue in both structuring and operating smart networks.

To address this complex issue, the deployment of WAMSs could play a strategic role. WAMS involves the use of system-wide information to avoid large disturbances and reduce the probability of catastrophic events by supporting the application of adaptive protection and control strategies aimed at increasing network capacity and minimizing wide-area disturbances.

The main wide-area applications that may be able to benefit from the WAMS are [2]

- Advances in system integrity protection schemes
- Distribution circuit network management
- Dynamical loading of power equipments
- System restoration and smart restoration tools
- Advance warning systems of impending trouble

WAMS requires accurate phasor and frequency information from multiple synchronized devices installed at various power system locations. Time-synchronized measurements are key to implement WAMS systems [7,8]. Presently, phasor measurement units (PMUs) provide accurate synchronized information about voltage and current phasors, frequency, and rate of change of frequency using a high common accuracy time reference [9].

* N–2 security criteria refer to the analysis of electrical network security when two contingencies occur.

The adoption of PMUs allows WAMS to monitor power flows in interconnected areas, and/or heavily loaded lines offer the opportunity to reliably operate the power system closer to its stability limits. Additionally, WAMS based on PMUs can monitor the dynamic behavior of the power system and identify interarea oscillations in real time. The ability to detect and reduce interarea oscillations could allow the system operator to exploit transmission and generation capacity more efficiently. As a result, renewable power generators can be used more effectively, and the marginal cost of power generation can be reduced [10].

Anyway, to realize the benefits of synchronized WAMS in future smart grids, several open problems need to be addressed.

In particular, these systems have been traditionally deployed according to client/server-based architectures whose main components are PMUs, phasor data concentrators, application software, and its supporting communication networks. The phasor information jointly with other electrical variables are collected by the PMUs and forwarded over the communication network to the phasor data concentrators that forward the selected information to the central monitoring center.

Many research works conjectured that this hierarchical monitoring paradigm could be not affordable in addressing the increasing network complexity and the massive data exchanging characterizing modern smart grids [1,11]. Unaffordable complexity, hardware redundancy, network bandwidth, and data storage resources are the main barriers imposed by technology and costs [12,13].

As a consequence, researchers and designers of high-performance WAMS are revisiting numerous design issues and assumptions pertaining to scale, reliability, heterogeneity, manageability, and system evolution over time [14,15].

Following these research directions, in this chapter, we outline the important role played by cooperative and self-organizing smart sensor networks [16,17]. In particular, we explore the possibility of decentralizing the WAMS processing and synchronization functions on a network of interactive PMUs equipped by distributed consensus protocols [18,19].

The application of consensus protocols allows distributed PMUs to reach an agreement on key pieces of information or on a common that enable them to cooperate in a coordinate fashion. This decentralized computing and synchronization paradigm doesn't require neither explicit point-to-point message passing nor routing protocols. It spreads information across the communication network by updating each PMU state by a weighted average of its neighbor states. At each step, every PMU computes a local weighted least-squares estimate, which converges to the global maximum-likelihood solution.

Thanks to this feature the PMUs can both synchronize their local acquisitions and compute, in a completely decentralized way, the many important variables characterizing the global power system operation (i.e., mean grid voltage magnitude, power losses, regulation costs) without the need for a central fusion center acquiring and processing all the PMU measurements.

The resulting monitoring architecture exhibits several advantages over traditional client/server-based paradigms as far as less network bandwidth, less computation time, and ease to extend and reconfigure are concerned. These features make the overall WAMS architecture highly scalable, self-organizing and distributed, and thus an ideal candidate for addressing wide-area monitoring in smart grids.

Simulation results obtained on the 118 bus IEEE test network are presented and discussed in order to prove the effectiveness of the proposed methodology.

9.2 ROLE OF WIDE-AREA MEASUREMENTS IN FUTURE SMART GRIDS

Recent advances in Information and Communications Technology are leading to significant developments in the use of smart systems for power network monitoring, protection, and control. A broad spectrum of advanced technology solutions and novel protection and control techniques will deliver new modes of integrated power system monitoring and control. In this domain, the large-scale deployment of WAMSs has been recognized as one of the most promising enabling technologies.

9.2.1 Elements of WAMS

WAMSs aim at support power system operators in managing the increasing complexities characterizing modern power networks by an integrated implementation of pervasive measurement technologies, advanced applicative tools, and large-scale communication infrastructures. They have been typically considered as a stand-alone infrastructure aimed at enhancing the conventional functions of existing supervisory control and data acquisition (SCADA) systems by enabling real-time wide-area situational awareness [20]. This has been obtained by generating critical information on many important aspects of power grid operation that provide decision support to power system operation and planning and to the control/protective systems design. These information may also be obtained by processing and integrating historical data into an adaptive knowledge base aimed at classifying the current power system state and detecting incipient faults [4].

WAMSs require reliable and accurate phasor and frequency information from multiple synchronized devices installed at various locations of the power network. This can be obtained by deploying a network of time-synchronized PMUs that provide accurate information about voltage and current phasors, frequency, and rate of change of frequency using a high common accuracy time reference [21].

Correct PMU operation requires a common and accurate timing reference determining the instant at which the samples of the buses' voltages and currents are acquired. This may be obtained by synchronizing these samples to a common timing reference furnished by a synchronizing source that may be internal or external to the PMUs. The timing signals should be (1) referenced to Coordinated Universal Time, (2) available without interruption at all measurement locations throughout the interconnected power system, and (3) characterized by a degree of availability, reliability, and accuracy suitable with the power system requirements [21].

To address these needs, the employment of GPS-based timing signals has been widely adopted for PMU synchronization. The main benefit arising by the application of the GPS technology is that it does not require the deployment of primary time and time dissemination systems assuring, at the same time, a set of intrinsic advantages as far as wide-area coverage, easy access to remote sites, and adaptability to changing network patterns are concerned.

WAMSs have been traditionally deployed according to hierarchical architectures based on distributed sensors organized on several network layers and a wide variety of data acquisition, transmission, concentration, and processing technologies. A typical WAMS architecture is depicted in Figure 9.1 [10]. It is based on the following main components: PMUs, phasor data concentrators (whose number depends by the system requirements), application software, and communication networks. Analyzing this figure, it is worth observing as phasor information and other variables are collected from the power network by PMUs and forwarded over different communication media to the phasor data concentrators. The latter forward selected information to the *Super PDC*, which is connected to the monitoring center. The WAMS applications can be run directly at this level, along with the data archiving functionalities.

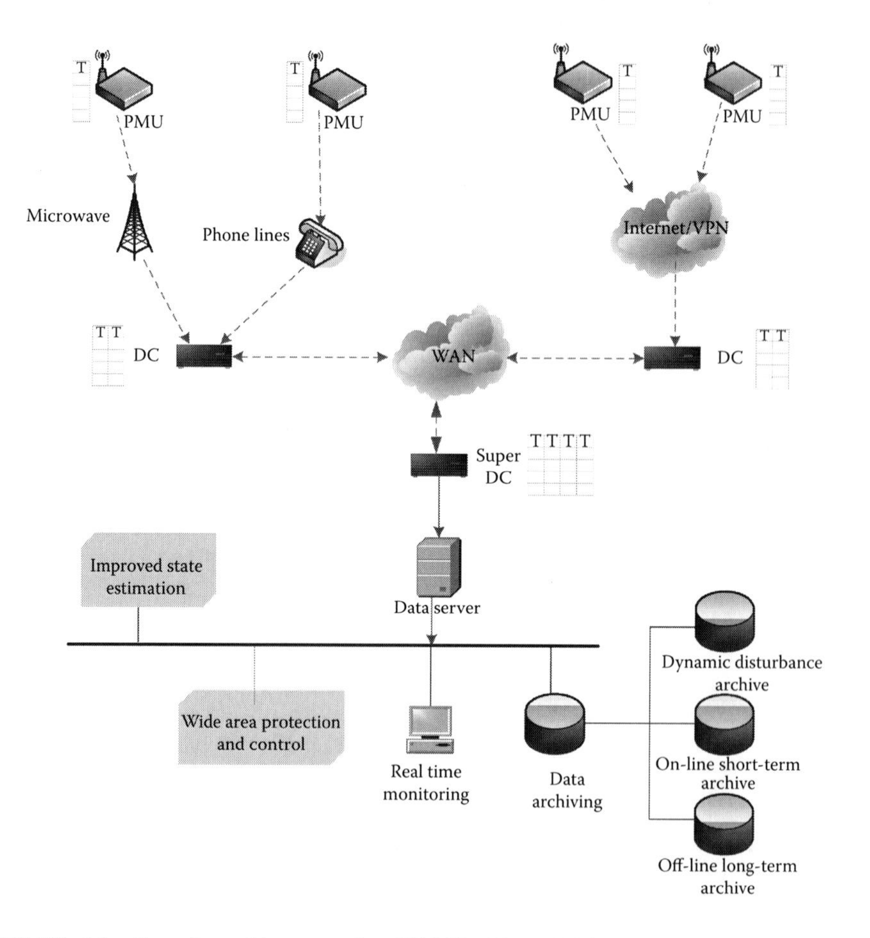

FIGURE 9.1 Generic architecture of a WAMS. (From Dickerson, B., Substation time synchronization, *Protection, Automation and Control World*, Summer 2007, pp. 39–45.)

These applications depend on the number and locations of the available PMUs and in particular on the following:

- If a limited number of PMUs are available, WAMS data processing can only partially describe the power system operation state. In this case, the typical application include
  - Voltage stability monitoring for transmission corridors
  - FACTS control using feedback from remote PMU measurements

  Obviously, each application is characterized by specific requirements in terms of minimum number and location of synchronized measurements.
- On the contrary, if a sensible number of PMUs are available into the power system, then real-time information on the actual operation state of the entire grid could be assessed. In this case, more advanced applications based on a detailed network model view can be implemented including [22]
  - Loadability calculation based on OPF studies
  - Topology detection and state estimation
  - Distribution circuit network management
  - System restoration and smart restoration tools
  - Advance warning systems of impending trouble

  Also in this case, each application requires that a sufficient number of PMUs have been strategically located into the power system. These applications allow the WAMSs to evolve toward the so-called wide-area measurement protective and control systems (WAMPAC).

Field experiences have shown that the application of WAMSs on interconnected power networks allows system operators to avoid large disturbances and reduce the probability of catastrophic blackouts. Besides, it can support the implementation of adaptive protection and control strategies aimed at increasing transmission network capacity and minimizing wide-area disturbances.

Anyway, to realize the benefits of synchronized WAMS in future smart grids, several open problems need to be addressed as outlined in the next section.

9.2.2 Emerging Needs in Smart Grid Monitoring

While WAMSs have been commercially available since the early 1990s, the development of its applications in future smart grids is still in its infancy. A particular challenge here is to deploy such a system according to pervasive and highly scalable architectures, which will be immune and robust to different kinds of contingencies that might affect its operation and consequently be reflected to the reliable operation of the smart grid.

In details, as previously outlined, WAMSs have been traditionally deployed according to client/server-based architectures where the PMU time synchronization is obtained by processing satellite signals.

As outlined by many papers [12,13,15], this hierarchical monitoring architecture exhibits some intrinsic disadvantages that could hinder its application in future smart grids where the constant growth of the electrical network complexity and the need for pervasive monitoring ask for more scalable, more flexible monitoring paradigms.

In details, it is expected that smart grid will increase the current data acquisition in about four orders of magnitude leading centralized WAMS architectures to becoming rapidly saturated. Consequently, the streams of data acquired by the distributed PMUs could not provide system operators with the necessary information to act on in the time frames necessary to minimize the impact of a disturbance. Even in the presence of fast models aimed at converting the data into information, the smart grid operator must face with the following challenges:

- Communication bottlenecks.
- Intractable control and optimization problems.
- Some problems can't be solved even with supercomputers.
- Events can occur due to limitations of controlling large-scale renewable energy.
- EMS, DMS system complexity continues to grow.
- Centralized infrastructure can be a security target.

Besides, the satellite-based synchronization relies on information transfer over air communications media. This wireless nature of satellite communications links and the weak power levels of satellite signals make them vulnerable to radiofrequency interference. Any electromagnetic radiation source can act as an interference source if it can emit potential radio signals in the satellite signal frequency bands [21]. Consequently, the research for reliable and effective synchronization techniques aimed at providing redundant PMU timing signals is of paramount importance for a correct and safe WAMS operation. These *backup systems* come into effect in the case of a GPS unavailability providing a more reliable timing source. Furthermore, if one signal is degraded or unavailable, the PMU should still operate within overall system requirements.

In trying and addressing these issues, researchers and designers of WAMSs are revisiting numerous design issues and assumptions pertaining to scale, reliability, heterogeneity, manageability, and system evolution over time [1,11].

These papers suggest a shift toward a more distributed architecture for WAMSs. In addressing these needs, the most promising enabling technologies include

- The conceptualization of advanced monitoring paradigms aimed at distributing the intelligence at PMU level
- The deployment of pervasive communication networks aimed at allowing PMUs to communicate with remote centers, with other PMUs, and with all the systems at substation level

Anyway, the identification for the more effective WAMS architecture in a smart grid context is still embryonic and needs both theoretical and practical improvements.

In our opinion, this issue should be solved by determining a suitable trade-off between a cost-effective solution with better management and higher performance and reliability.

In particular, it is well noted that a centralized WAMS manages the intelligence from a single central location. This makes the monitoring process easy to maintain and provides greater control to the systems operators. The drawback to a fully centralized monitoring paradigm is that every data must be routed back to a single application function (which represents a potential single point of failure for the network). Besides, this could lead to slower communications in the presence of large volumes of raw data.

On the other hand, decentralized WAMSs allow local PMUs to make decisions on their own, without communicating with a *master* device. This could improve response times eliminating the single point of failure. However, smart PMUs capable of making their own decisions ask for more processing power and greater upkeep. Consequently, the cost of a fully decentralized WAMS architecture could be greater than that of a fully centralized network.

We strongly believe that the more effective WAMS architecture should be based on the synergic fusion of these two monitoring paradigms.

Armed with such a vision in this chapter, we outline the potential role of cooperative and self-organizing smart sensor networks equipped by distributed consensus protocols in WAMSs.

In particular, we explore the possibility of decentralizing the WAMS processing and synchronization functions on a network of interactive PMUs equipped by distributed consensus protocols [18,19].

The application of consensus protocols allows PMUs to reach an agreement on key pieces of information or on a common that enable them to cooperate in a coordinate fashion. This decentralized paradigm spreads information across the communication network by updating each PMU state by a weighted average of its neighbor states. At each step, every PMU computes a local weighted least-squares estimate, which converges to the global maximum-likelihood solution.

In our opinion, this contribution can be viewed as an intermediate step toward a comprehensive understanding of how to coordinate the PMU acquisition of a large-scale system into a coherent whole.

9.3 SELF-ORGANIZING WAMS ARCHITECTURE

In this section, we propose a WAMS architecture based on the *think locally, act globally* approach. The insight is to start from the theory of information spreading for coordinating a network of cooperative PMUs [23–26]. Each PMU can (1) acquire local bus measurements by querying a set of sensors and (2) communicate only with a limited number of neighbor PMUs.

The WAMS processing and synchronization functions are computed by the PMUs by exchanging and processing local information according to a distributed consensus algorithm.

This is obtained by equipping each PMU by a dynamical system (oscillator) initialized by local information. The oscillators of nearby PMUs are mutually coupled by proper local coupling strategies derived from the mathematics of populations of mutually coupled oscillators [19,27,28]. This bioinspired paradigm allows all the PMUs to reach consensus on general functions of the variables sensed by all the PMUs.

Thanks to this feature, the local PMU acquisitions can be time synchronized and each PMU can compute the most important variables characterizing the global power system operation without the need for a fusion center. Consequently, all the basic WAMS functions could be implemented by the PMUs according to a totally decentralized/nonhierarchical paradigm. This makes the proposed architecture self-organizing, highly scalable, cooperative, and distributed and thus an ideal candidate for smart grid monitoring.

9.3.1 Theoretical Foundations

We consider a set of N PMUs $\Gamma = v_1, \ldots, v_N$, each one monitoring a specific power system bus. The PMUs interact over a communication network described by a graph $G(\Gamma, A)$, where $A = \{a_{ij}\}$ is the adjacency matrix. Let's indicate with $N_i = \{v_j \in \Gamma : a_{ij} \neq 0\}$ the set of neighbors of PMU v_i. It follows that PMU v_j can exchange information to agent v_i if $v_j \in N_i$.

Let's indicate by $x_i \in \mathfrak{R}$ the state of the PMU v_i evolving according to the following differential equation:

$$\dot{x}_i = f(x_i, u_i) \quad i \in [1, N] \tag{9.1}$$

Consequently, the PMU network state evolves according to the *network dynamics* $\dot{x} = F(x,u)$ where $x = (x_1, \ldots, x_N)$ and $F(x,u)$ is the concatenation of the elements $f(x_i, u_i)$ $i \in [1, N]$.

We say that the PMU network reaches a *consensus* if $x_i = x_j \; \forall i, j \in [1, N]$.

Let $\gamma: \mathfrak{R}^N \to \mathfrak{R}$ be a real valued function and $a = x(0)$ is the initial state of the PMU network.

The γ-*consensus problem* is a distributed way to compute $\gamma(a)$ by processing inputs u_i that only depend by the states of PMU v_i and its neighbors [12].

We say that $u_i = k_i(v_{j1}, \ldots, v_{jni})$ is a distributed protocol if the cluster $J_i = (v_{j1}, \ldots, v_{jni})$ of PMUs with indices $(j_1, \ldots, j_{ni})$ satisfies the property $J_i \subset \{v_i\} \cup N_i$ and $|J_i| < N$.

A distributed protocol asymptotically solves the γ-consensus problem if there exists an asymptotically stable equilibrium x^* of $\dot{x} = F(x, k(x))$ satisfying $x_i^* = \gamma(x(0)) \; \forall i \in [1, N]$.

In this chapter, we are mainly interested in computing $\gamma(x) = (1/N)\sum_{i=1}^{N} x_i$, $\gamma(x) = \max_i x_i$, $\gamma(x) = \min_i x_i$ denoted as *average consensus*, *max-consensus*, and *min-consensus*, respectively.

In solving the average consensus problem, we adopted the following linear coupling protocol:

$$u_i = \frac{K}{c_i} \sum_{j \in N_i} h(x_j - x_i) \tag{9.2}$$

where

- $h(\cdot)$ is a monotonically increasing, nonlinear, odd function
- K is the control loop gain
- c_i describes the attitude of the ith PMU to adapt itself to the state variations of the coupled PMUs

It is possible to demonstrate that if a proper choice of the control loop gain K is made, this protocol allows the PMUs to asymptotically reach a consensus with a convergence rate governed by the eigenvalues of the network topology [23,24].

Besides, it is also simple to show that, when the PMUs reach a consensus, they synchronize to the following value:

$$x^* = \frac{\sum_{i=1}^{N} c_i x_i(0)}{\sum_{i=1}^{N} c_i} \tag{9.3}$$

This allows each PMU to compute the weighted average of the sensed variables from all the PMUs in the network without the need for a fusion center.

As far as the max-consensus is concerned, it can be solved by adopting the following distributed protocol:

$$\begin{cases} x_i(k+1) = \max(x_i(k), u_i(k)) \\ u_i(k) = \max\limits_{j \in N_i} x_j(k) \end{cases} \tag{9.4}$$

This protocol allows all the PMUs to iteratively converge to the state of the max leader (namely, the PMU characterized by the highest state) after a maximum of $N-1$ iterations [29]. A similar distributed approach could be used to solve the min-consensus problem.

9.3.2 Synchronization Functions

PMU time synchronization is an important foundation of WAMSs. It relies upon distributed clocks to allow correct analysis of collected PMU data.

The information propagation based on mutual coupling of dynamic systems described in the previous section prevents the application of traditional centralized, hierarchical clock synchronization strategies. In fact, the PMUs can reach time synchronization by adapting their clocks according to the local coupling strategy described in (9.2) without the need for any cluster header since every PMU is a header to emit singles to other PMUs. When this network of coupled oscillator clocks synchronize, then each clock will show the same value, without changing the value once reached. Thanks to this feature, the built-in PMU oscillators are able to lock to a common phase, despite the differences in the frequencies of the individual oscillators.

The effectiveness of this decentralized synchronization paradigm has been assessed in several research works ranging from the biology to the technological area [30].

9.3.3 Monitoring Functions

The employment of the cooperative paradigm described in Section 9.3.1 allows each PMU to execute the WAMS monitoring functions in a fully decentralized way. This is obtained by allowing the PMUs to reach a consensus on the main variables characterizing the global smart grid operation.

In particular, if the PMUs sense the bus voltage magnitude, the following vector of observations could be adopted to initialize the dynamical systems:

$$x_i = (V_i, |V_i - V_i^*|) \tag{9.5}$$

where V_i and V_i^* are the current and nameplate voltage magnitudes at bus i, respectively.

In this case, it is easy to show that the solution of the average consensus problem allows the dynamical systems to synchronize to the mean grid voltage magnitude and the average voltage magnitude deviation:

$$x^* = \left(\frac{\sum_{i=1}^{n} V}{n}, \frac{\sum_{i=1}^{n} |V_i - V_i^*|}{n} \right) \tag{9.6}$$

where n is the number of buses.

Other variables of interest can be easily assessed by a proper selection of the observations' vector.

In particular, if the PMUs sense the active and reactive bus power, the following vector of observations could be adopted to initialize the dynamical systems:

$$x_i = \left(n \cdot (P_{Gi} - P_{Li}), n \cdot c_{pi}(P_{Gi}) \cdot P_{Gi}, n \cdot c_{qi}(Q_{Gi}) \cdot Q_{Gi}\right) \tag{9.7}$$

where

P_{Gi} and P_{Li} are the active power generated and absorbed at the ith bus
Q_{Gi} is the reactive power generated at the ith bus
$c_{pi}(P_{Gi})$ and $c_{qi}(Q_{Gi})$ are the costs of the active and reactive powers generated at the ith bus

In this case, the solution of the average consensus problem allows the dynamical systems to synchronize to the active system losses and to the total cost of active and reactive power:

$$x^* = \left(\sum_{i=1}^{n} (P_{Gi} - P_{Li}), \sum_{i=1}^{n} \left(c_{pi}(P_{Gi}) \cdot P_{Gi}\right), \sum_{i=1}^{n} \left(c_{qi}(Q_{Gi}) \cdot Q_{Gi}\right) \right) \tag{9.8}$$

Other variables of interest (i.e., power quality indexes) could be easily assessed by a proper selection of the vector of observations.

Because of the employment of this biologically inspired paradigm, each PMU can know both the variables characterizing the monitored bus (sensed by inbuilt sensors) and the global variables describing the actual performance of the entire smart grid (assessed by checking the state of the dynamical system). Thus, a comparison between local and global quantities can be made at any time, for any bus, and subsequent actions can be taken in the case that the bus parameters strongly deviate from the actual grid performances.

9.4 SIMULATION STUDIES

This section discusses the application of the proposed self-organizing WAMS architecture in the task of synchronized grid monitoring for the IEEE 118-bus test system (Figure 9.2).

This power network is composed of 118 buses, 186 branches, 91 load buses, and 54 generators [31].

A sensor network composed of 118 cooperative PMUs has been deployed along the power network (one for each bus). The coupling coefficients a_{ij} describing the PMU logical connection have been defined according to the adjacency matrix of the electrical network.

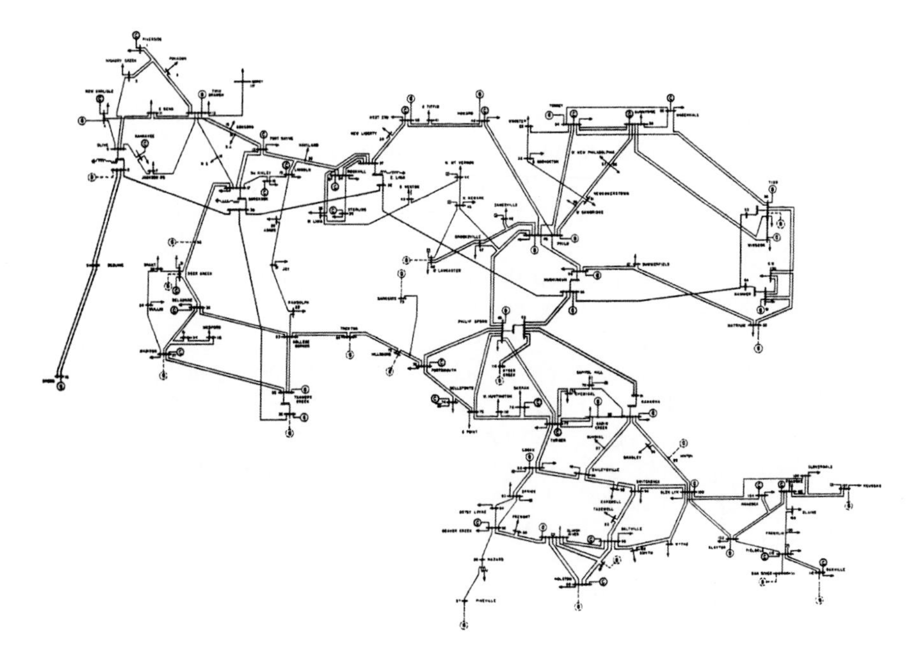

FIGURE 9.2 The IEEE 118-bus test system.

9.4.1 Monitoring Functions

The employment of the proposed monitoring paradigm allows each PMU to compute both the performances of the monitored bus (i.e., the bus index), computed by acquiring local information, and the global grid performances, computed by local exchanges of information with its neighbor PMUs. Consequently, the system operator can assess the main variables characterizing the actual grid operation by inquiring any PMU without the need of a central fusion center acquiring and processing all the PMU acquisitions.

In details, we assumed that each PMU senses the following bus variables: voltage magnitude and generated and demanded active power. The corresponding vector of local observations is organized as follows:

$$x_i = \left(x_1, x_2, x_3\right) = \left(V_i, n \cdot (P_{Gi} - P_{Li}), (V_i - V_i^*)^2\right) \tag{9.9}$$

The solution of the average consensus problem formalized in (9.2) allows the built-in dynamic systems to synchronize to the mean grid voltage magnitude, the active power losses, and the average voltage magnitude deviation:

$$x^* = \left(\sum_{i=1}^{n} \frac{V_i}{n}, \sum_{i=1}^{n} \left(P_{Gi} - P_{Li}\right), \sum_{i=1}^{n} \frac{(V_i - V_i^*)^2}{n}\right) \tag{9.10}$$

In solving this problem, a linear coupling strategy characterized by a constant gain equal to 0.2 has been adopted. The corresponding dynamical system trajectories for a fixed grid operating state are depicted in Figure 9.3.

Moreover, the solution of the maximum/minimum consensus problems formalized in (9.4) allows the PMUs to compute the max/min value for each grid variable. The effectiveness of this feature can be assessed by analyzing Figure 9.4, where the evolution of the PMU states in the task of computing the minimum and maximum grid voltage magnitude is depicted.

As expected, the application of the distributed consensus protocols allows all the PMUs (whose initial states are chosen randomly) to converge to the real values of the variables of interest. This is obtained in about 300 iterations for the active power losses computing and in only 11 iterations for the maximum and minimum bus voltage magnitude. The larger number of iterations required for computing the active power losses is justified by the higher complexity characterizing the average consensus problem solution. In this connection, it is important to note that the adoption of a nonlinear coupling protocol is expected to dramatically reduce the convergence iterations. This topic is currently under investigation by the authors.

As far as the synchronization time is concerned, it is expected to be less than few seconds for wired communications, and it appears suitable with the time constraints characterizing the smart grid monitoring constraints.

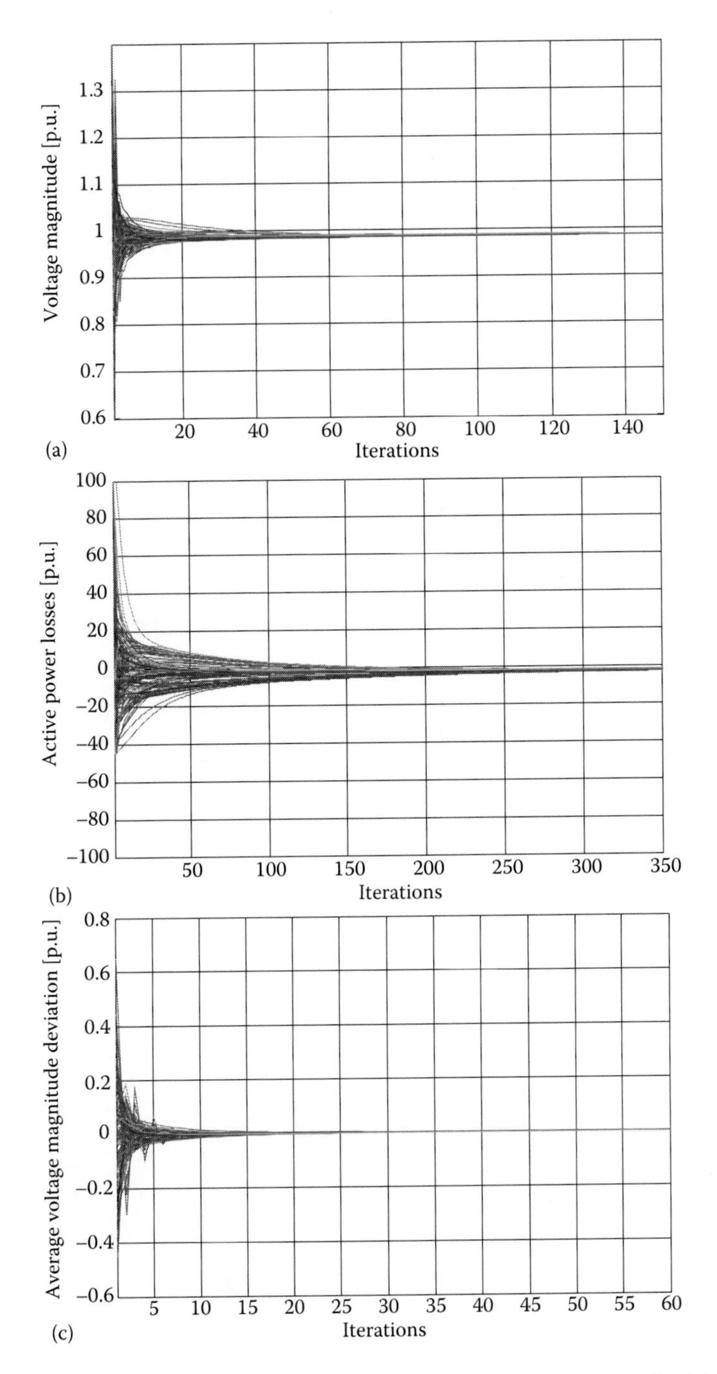

FIGURE 9.3 Evolution of the PMU states in the task of computing global grid variables by solving average consensus problems: (a) mean grid voltage magnitude, (b) active power losses, and (c) average voltage magnitude deviation.

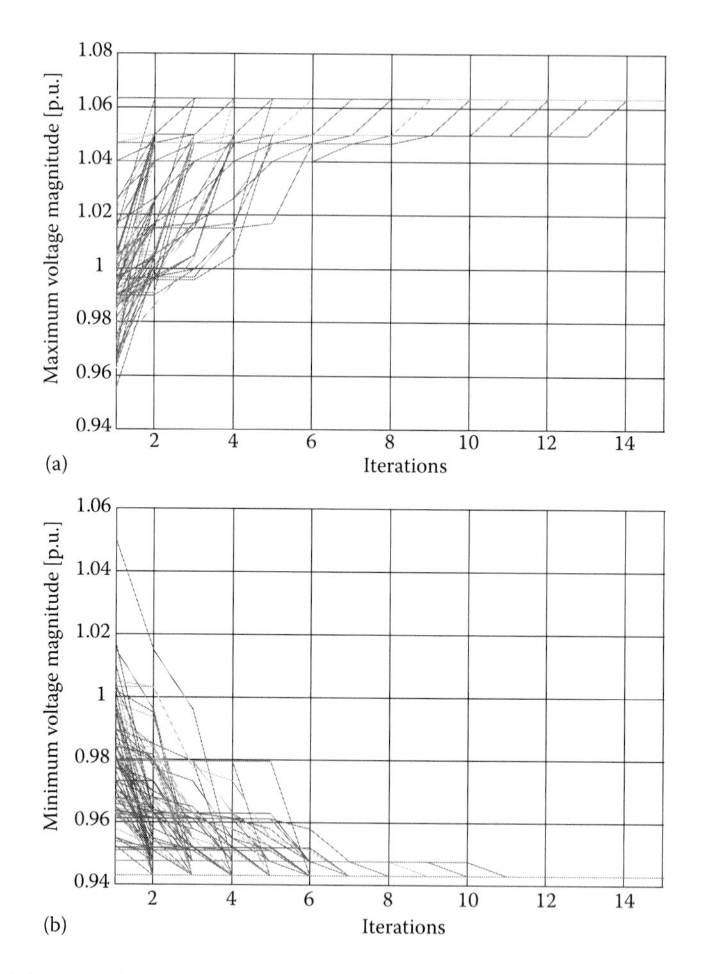

FIGURE 9.4 Evolution of the PMU states in the task of computing global grid variables by solving max-/min-consensus problems: (a) maximum voltage magnitude and (b) minimum voltage magnitude.

9.4.2 Synchronizing Functions

In order to prove the effectiveness of the cooperative paradigm in PMU time synchronization, we tested both linear and nonlinear coupling strategies. For this aim, we considered the following functions:

$$h(\cdot) = \begin{cases} x & \rightarrow \text{Linear} \\ \dfrac{e^{Lx}-1}{e^{Lx}+1} & \rightarrow \text{Exp} \\ \sin\left(2\pi \dfrac{x}{L}\right) & \rightarrow \text{Sin} \end{cases} \tag{9.11}$$

TABLE 9.1
Synchronization Time

	k	L	Iterations
Linear	0.1	—	283
Exp	0.1	3	274
Sin	1.2	50	192

In assessing the synchronization performance, we considered the number of iterations needed to obtain a synchronization error lower than 10^{-4} s. The obtained results are summarized in Table 9.1.

Analyzing these data, it is worth observing as the application of a nonlinear coupling strategy allowed us to sensibly reduce the convergence times. This feature could be also appreciated by analyzing the dynamic evolution of the PMU state reported in Figure 9.5.

These results confirmed the effectiveness of the proposed paradigm in synchronizing the PMU data acquisition. This feature could be considered as a redundant synchronization source that could come into effect in the presence of a GPS unavailability/degradation. A rigorous experimental analysis aimed at characterizing the real performances of this synchronization paradigm is currently under investigation by the authors.

9.5 CONCLUSIONS

Modern trends in smart grids are oriented toward the deployment of monitoring architectures that move away from the older centralized paradigms to systems distributed in the field and characterized by an increasing pervasion of smart and cooperative entities.

In supporting this complex task, this chapter conceptualizes a distributed and self-organizing wide-area monitoring architecture based on cooperative sensor networks.

The distributed PMUs employ traditional sensors to acquire local bus variables and distributed consensus protocols to assess the main variables that characterize the global smart grid operation. These variables could be then amalgamated by applicative software in order to identify proper control actions aimed at improving the grid performances.

The results obtained on a test power system show as this monitoring paradigm allows the PMUs to detect power system anomalies since they know both the performances of the monitored buses, computed by acquiring local information, and the global performances of the entire grid, computed by local exchanges of information with its neighbor PMUs. The convergence of this process is expected to correspond well with the time constraints characterizing the monitoring process in smart grids. This is obtained without the need of a central fusion center acquiring and processing all the PMU acquisitions. This makes the overall monitoring architecture highly scalable, self-organizing, and distributed.

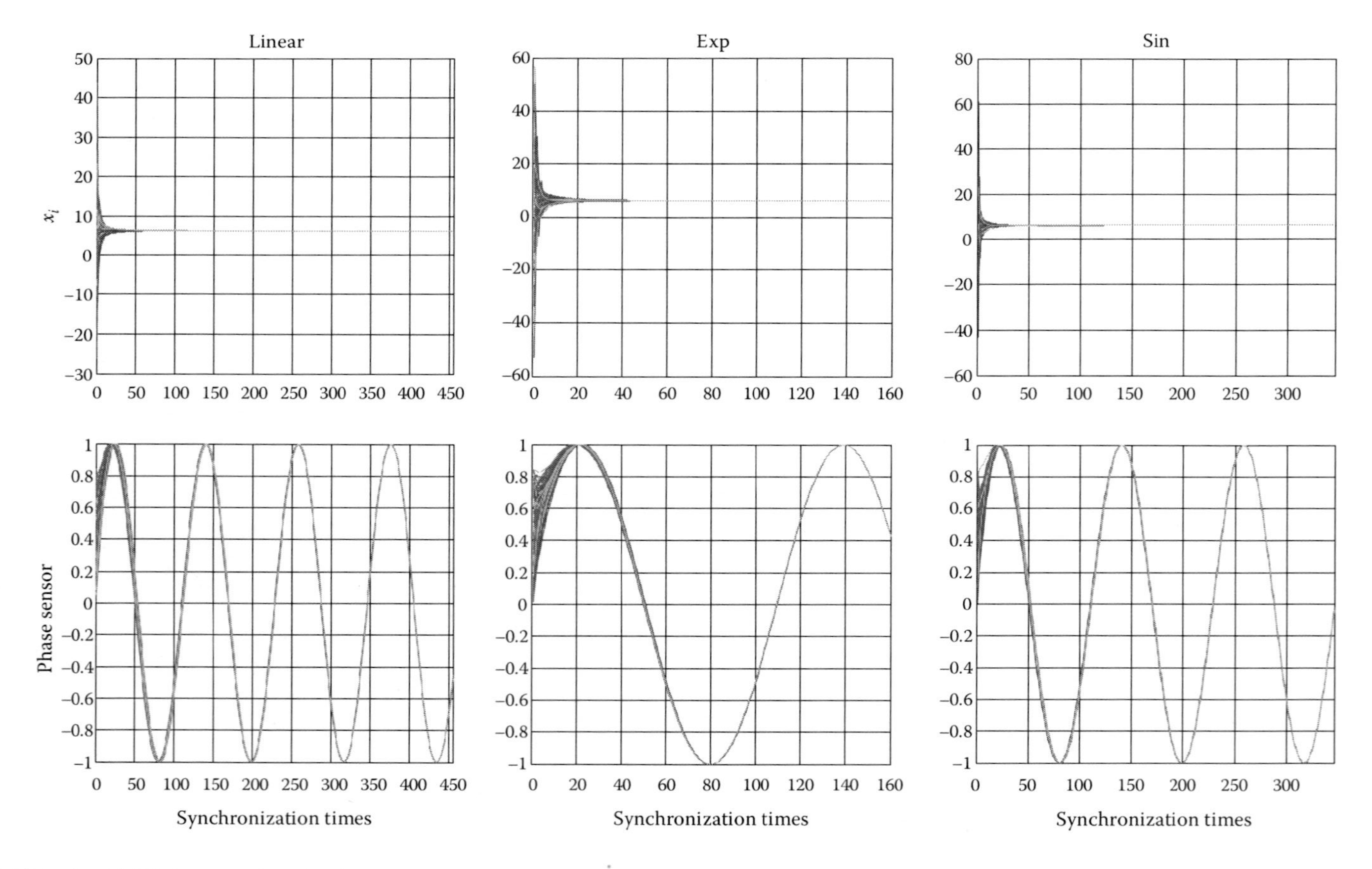

FIGURE 9.5 PMU time synchronization.

REFERENCES

1. V. Madani and R.L. King, Strategies to meet grid challenges for safety and reliability, *International Journal of Reliability and Safety*, 2(1–2), 146–165, 2008.
2. V. Madani, A. Vaccaro, D. Villacci, and R.L. King, Satellite based communication network for large scale power system, *2007 iREP Symposium—Bulk Power System Dynamics and Control—VII, Revitalizing Operational Reliability*, August 19–24, 2007, pp. 1–7, Charleston, SC, 2007.
3. S. Ullo, A. Vaccaro, and G. Velotto, The role of pervasive and cooperative sensor networks in smart grids communication, *Proceedings of the 15th IEEE Mediterranean Electrotechnical Conference (MELECON 2010)*, April 25–28, 2010, pp. 443–447, Valletta, Malta, 2010.
4. V. Terzija, D. Cai, G. Valverde, P. Regulski, A. Vaccaro, M. Osborne, and J. Fitch, Flexible wide area monitoring, protection and control applications in future power networks, *The 10th Institution of Engineering and Technology Conference on Developments in Power System Protection DPSP 2010*, March 29–April 1, 2010, pp. 1–5, The Hilton Deansgate, Manchester, U.K., 2010.
5. A. Vaccaro, M. Popov, D. Villacci, and V. Terzija, An integrated framework for microgrids modeling, control, communication and verification, Invited Paper—*IEEE Proceedings*, 99(1), 119–132, January 2011.
6. S.M. Amin and B.F. Wollenberg, Toward a smart grid, *IEEE Power and Energy Magazine*, 3(5), 34–38, September/October 2005.
7. S. Chakrabarti, E. Kyriakides, T. Bi, D. Cai, and V. Terzija, Measurements get together, *IEEE Power and Energy Magazine*, 7(1), 41–49, 2009.
8. Y. Hu, V. Madani, R. Morales, and D. Novosel, Requirements of large-scale wide area monitoring, protection and control systems, *Proceedings of 10th Annual Fault Disturbance Analysis Conference,* Atlanta, GA, pp. 1–9, 2007.
9. B. Dickerson, Substation time synchronization, *Protection, Automation and Control World*—Summer 2007, 3, 39–45, 2007.
10. V. Terzija, G. Valverde, D. Cai, P. Regulski, V. Madani, J. Fitch, S. Skok, M.M. Begovic, and A. Phadke, Wide area monitoring, protection and control of future electric power networks, *Proceedings of the IEEE*, 99(1), 80–93, January 2011.
11. J. Giri, M. Parashar, J. Trehern, and V. Madani, The situation room: Control center analytics for enhanced situational awareness, *IEEE Power and Energy Magazine*, 10(5), 24–39, September 2012.
12. V.C. Gungor, D. Sahin, T. Kocak, S. Ergut, C. Buccella, C. Cecati, and G.P. Hancke, Smart grid technologies: Communication technologies and standards, *IEEE Transactions on Industrial Informatics*, 7(4), 529–539, November 2011.
13. Q. Yang, J.A. Barria, and T.C. Green, Communication infrastructures for distributed control of power distribution networks, *IEEE Transactions on Industrial Informatics*, 7(2), 316–327, May 2011.
14. M.O. de Mues, A. Alvarez, A. Espinoza, and J. Garbajosa, Towards a distributed intelligent ICT architecture for the smart grid, *Ninth IEEE International Conference on Industrial Informatics (INDIN 2011)*, July 26–29, 2011, Lisbon, Portugal, pp. 745–749, 2011.
15. W. Qi, J. Liu, and P.D. Christofides, A distributed control framework for smart grid development: Energy/water system optimal operation and electric grid integration, *Journal of Process Control*, 21(10), 1504–1516, December 2011.
16. V.C. Gungor, B. Lu, and G.P. Hancke, Opportunities and challenges of wireless sensor networks in smart grid, *IEEE Transactions on Industrial Electronics*, 57(10), 3557–3564, October 2010.

17. P. Palensky and D. Dietrich, Demand side management: Demand response, intelligent energy systems, and smart loads, *IEEE Transactions on Industrial Informatics*, 7(3), 381–388, August 2011.
18. A. Vaccaro, G. Velotto, and A.F. Zobaa, A decentralized and cooperative architecture for optimal voltage regulation in smart grids, *IEEE Transactions on Industrial Electronics*, 58(10), 4593–4602, October 2011.
19. M. di Bisceglie, S. Ullo, and A. Vaccaro, The role of cooperative information spreading paradigms for smart grids monitoring, *Proceedings of the 16th IEEE Mediterranean Electrotechnical Conference (MELECON)*, March 25–28, 2012, Hammamet, Tunisia, pp. 814–817, 2012.
20. J. Hauer and J. DeSteese, *Descriptive Model of a Generic WAMS*, Pacific Northwest National Laboratory, Richland, WA, June 2007.
21. North American SynchroPhasor Initiative (NASPI)–Performance & Standards Task Team (PSTT), Guidelines for Synchronization Techniques Accuracy and Availability, 2009. Available online at https://www.naspi.org/File.aspx?fileID=546.
22. M. Larsson, R. Gardner, and C. Rehtanz, Interactive simulation and visualization of wide-area monitoring and control applications, *Proceedings of the Power Systems Computation Conference*, Liège, Belgium, pp. 1–6, 2005.
23. T.S. Lee and S. Ghosh, The concept of "stability" in asynchronous distributed decision-making systems, *IEEE Transactions on Systems, Man, and Cybernetics, Part B: Cybernetics*, 30(4), 549–561, August 2000.
24. J. Cortés, Distributed algorithms for reaching consensus on general functions, *Automatica*, 44(3), 726–737, March 2008.
25. A. Papachristodoulou, A. Jadbabaie, and U. Münz, Effect of delay in multi-agent consensus and oscillator synchronization, *IEEE Transactions on Automatic Control*, 55(6), 1471–1477, June 2010.
26. J.-C. Delvenne and R. Carli, Optimal strategies in the average consensus problem, *Proceedings of the 46th IEEE Conference on Decision and Control*, December 12–14, 2007, New Orleans, LA, pp. 2498–2503, 2007.
27. M. Rabbat and R. Nowak, Distributed optimization in sensor networks, *Proceedings of the Third International Symposium on Information Processing in Sensor Networks (IPSN 2004)*, April 26–27, 2004, pp. 20–27, Berkeley, CA, 2004.
28. S. Barbarossa and G. Scutari, Decentralized maximum likelihood estimation for sensor networks composed of nonlinearly coupled dynamical systems, *IEEE Transactions on Signal Processing*, 55(7), 3456–3470, July 2007.
29. R. Olfati-Saber and R.M. Murray, Consensus problem in networks of agents with switching topology and time-delays, *IEEE Transactions on Automatic Control*, 49(9), 1520–1533, September 2004.
30. D. Lucarelli and I.-J. Wang, Decentralized synchronization protocols with nearest neighbor communication, *SenSys'04: Proceedings of the Second International Conference on Embedded Networked Sensor Systems*, Baltimore, MD, pp. 62–68, ACM, New York, 2004.
31. IEEE Standard Power System test system configuration. Power systems test case archive. Available online at http://www.ee.washington.edu/research/pstca.

Section IV

Healthcare

10 Sensor Networks in Healthcare

Arny Ambrose and Mihaela Cardei

CONTENTS

ABSTRACT

Wireless sensor networks provide rapid, untethered access to information and computing, eliminating the barriers of distance, time, and location for many applications in national security, surveillance, healthcare, coverage, and many more. One of the most promising applications of wireless sensor networks technology is in healthcare. As the healthcare costs are rising, life expectancy is increasing, and the world population is aging, there has been a need to monitor patients and residents out of hospitals, in their own environment, and during emergency situations. Strategically placing a number of wireless sensors on the human body creates a wireless body area network that can monitor various vital signs that can provide real-time feedback to the user and medical personnel. In this chapter, we discuss the use of wireless sensor networks technology in emergency response applications, smart homecare applications, and mechanisms for continuous patient monitoring.

10.1 INTRODUCTION

A wireless sensor network is composed of sensors that are used to monitor a particular environment. The capabilities of such a network have advanced such that sensor networks can be used in healthcare systems. There has been much progress toward the integration of specialized medical technology with pervasive, wireless networks [1]. The use of sensors can enhance the care provided by a healthcare individual to a patient.

Continuous healthcare monitoring can become extremely expensive for a patient with chronic illness, and the use of sensors would alleviate the physical and financial burden of having a permanent caregiver. This would also give the patient a sense of independence, especially with elderly patients, while still having family and caregivers to monitor them and be alerted to any urgent situations. Wireless networks are the optimal method of providing continuous pervasive monitoring of not only the patient but also of the environment. This offers a variety of applications for the use of these networks in healthcare.

Medical systems can benefit from wireless sensor networks in a variety of applications. They can be used in disaster-response scenarios to allow for efficient tracking of patients and emergency response personnel [2–4], for continuous monitoring of patients in assisted living facilities [5,6], and for other mechanisms that require wearable sensors and provide efficient delivery of patient data.

10.2 EMERGENCY RESPONSE APPLICATIONS

During emergency response situations, it is important that the responders have a timely and accurate assessment of the health of the patients. Wireless sensor networks can be used to coordinate different teams of rescue personnel and also multiple organizations to create a cohesive and an efficient response effort. These networks usually comprise wearable sensors that can be placed on patients for continuous monitoring [4], a way of keeping track of response personnel and the patients, and a means of data collection and storage.

The emergency response software infrastructure CodeBlue [2] integrates wireless devices with a wide range of capabilities into a network that can be used for emergency care or disaster response. The infrastructure consists of wireless vital sign sensors, location beacons, and all the protocols and services required to make the information gathered by them useful to emergency medical technicians, police and fire rescue, and ambulance systems.

It is important that in coordinating the collection and transmission of data, there is proper discovery of the wireless devices, so that communication pathways are developed. In this architecture, data collection will be done using some kind of mobile device, for example, personal digital assistants (PDAs). It is important that the device name be specific to the application and not to use a low-level network address. This makes it easier for the collection device to request information from only sensors of a specific type or in a certain range. Also the discovery process is decentralized such that there is no concern of one failure making the entire network unusable.

When all sensors have been detected, it is important that communication between the devices is reliable since they may be communicating with other devices that are outside their communication range. The CodeBlue infrastructure (Figure 10.1) also contains ad hoc routing techniques that extend the effective communication range of the devices. These devices typically use a multicast communication method to allow one sensor to report its data to multiple receiving nodes.

During a disaster-response situation, it is also important that the locations of the rescuers and victims be monitored. CodeBlue uses a radio frequency (RF)–based location tracking system called MoteTrack [7] that operates using the low-power

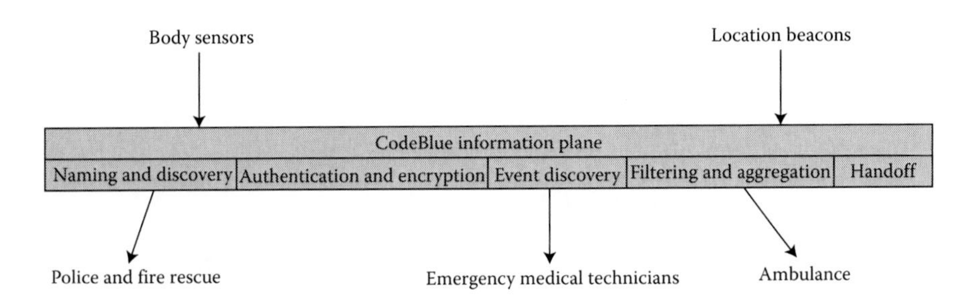

FIGURE 10.1 SATIRE architecture framework.

radio transceivers of the sensor nodes. MoteTrack uses beacon nodes that broadcast periodic messages that would contain the node's ID and the transmission power that is used to broadcast the message. Each beacon acquires a signature that will be sent to fixed beacons with known locations, serving as reference nodes in determining the location of the mobile nodes based on the power used to transmit the message.

Waterman et al. [4] developed the scalable medical alert response technology (SMART) that is designed to provide healthcare and patient monitoring. This system was developed to provide continuous monitoring of patients in an overcrowded hospital emergency room but can also be scaled up for use in disaster scenarios. Sensors are used to observe the patient's electrocardiogram (ECG) signals and oxygen saturation, and this information is transmitted to a PDA so that a quickly deteriorating patient can be immediately identified. The location system used is based on MIT Crickets [8], which merges RF and ultrasound to determine indoor locations where the global positioning system (GPS) would be ineffective.

Simulations [6] showed that the SMART system was able to manage 150 individuals. The system was operational after approximately 5 min and each patient was equipped with a monitoring equipment as soon as they arrived. It was found that the use of SMART enabled caregivers to work on other patients without having to continually monitor all patients.

Ko et al. [3] propose the medical emergency detection in sensor networks (MEDiSN), which is also designed for monitoring patients in hospitals and disasters. MEDiSN consists of patient monitors that are custom built, wearable motes that will collect and secure the data, relay points that will create a multihop wireless backbone for transmission of the data and a gateway, as shown in Figure 10.2. This mechanism differs from the others described as it uses a wireless mesh infrastructure of relay points that transmit the data from the patient monitors. This increases the scalability of the mechanism so that it can be used in situations with a large number of injured people.

Security and data aggregation are also discussed. MEDiSN secures patient data by performing end-to-end encryption and authentication of the data from the patient monitor. Since it is expected that there will be a large volume of data collected, the authors used a delta compression algorithm to condense the data as much as possible. This algorithm was found to have a very high compression ratio, while providing relatively low implementation complexity. The secured and compressed data are then delivered to the gateway via the collection tree protocol (CTP) [9]. This

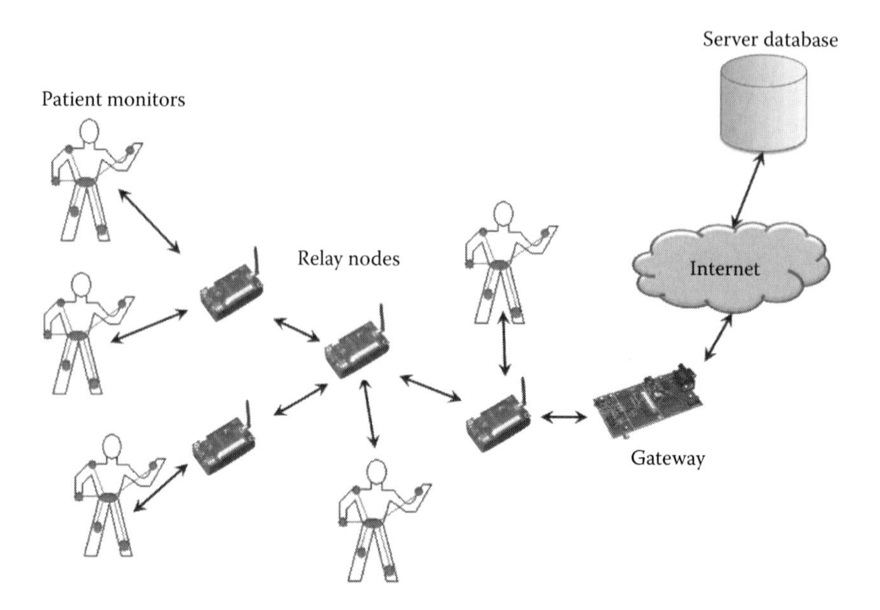

FIGURE 10.2 Data flow between patient and medical services.

protocol builds and maintains minimum cost trees to nodes that are designated as tree roots. Data are sent to the base station that requires the least amount of energy with no knowledge of the address of that node. The topology information is kept current using data path validations where data packets are used to query and validate the topology. For MEDiSN, this protocol is enhanced with a mechanism to deliver commands from the gateways to the patient monitors. Data will be retransmitted if there is an error between hops.

Simulations showed that the use of data aggregation combined with CTP cause a 20% increase in the number of messages that reach the gateway. MEDiSN was also compared with CodeBlue in an indoor environment with varying numbers of patient monitors. It was found that there was a significant improvement in the network utilization. This was due to the use of hop-by-hop transmission and retransmission of data. This can be used to mitigate the effects of interference and collisions better than the single-hop transmission mechanism used by CodeBlue.

10.3 SMART HOMECARE APPLICATIONS

The wireless network architecture for smart homecare requires that the technology be easily integrated with existing medical practices and technology, should provide real-time, long-term, and remote monitoring, the sensor should be small and unobtrusive and should provide assistance to elderly and chronic patients. Using a wireless sensor network for homecare provides doctors with a continuous stream of data that can provide a record of symptoms. The patients can also be monitored in cases where chronic life-threatening conditions may exist so that emergency personnel can be alerted if they require assistance.

Becker et al. [10] propose the SmartDrawer that would observe the medication of chronically ill patients. This drawer contains radio frequency identification (RFID) tags that enable the care provider to monitor the patient to ensure that the medication is being taken as prescribed. The authors expect that the system will have three users: the patient, the caregiver, and the maintainer. The SmartDrawer functionality for the patient is to give an alert when they deviate from the directions. The drawer will contain a record of the patient's medical data and an inventory of the different medication and the specifications on how to take them.

The caregiver will be able to retrieve historical data about the drugs that have been taken and will also be able to add new prescriptions or modify existing ones. The maintainer is described as a system administrator who would be able to add and remove different functionalities and access the data for manipulation so that they may be used in different applications.

The drawer operates only when it is in use. When the drawer is opened it begins to scan the medication that is being removed from it and to record the dosage that the patient is taking. When the drawer is closed, it updates the inventory and then stops scanning until it is in use again. This drawer would be especially useful to elderly patients who may have a tendency to forget to take their medication or forget the specific instructions on how to take them.

Wood et al. [5] extend the idea of a single drawer to consist of a wireless sensor network for assisted living and residential monitoring (ALARM-NET). ALARM-NET consists of a body network, which is made up of sensors that would be worn by the resident. These sensors would provide physiological sensing that would be customized to the medical needs of the resident. Environmental sensors would be deployed into the living space. These sensors would measure environmental conditions such as air quality, light, temperature, and motion. Motion detection is especially important as it can be used to track the resident's movements. These static sensors will also form a multihop network between the mobile body network and the AlarmGate. The AlarmGate manages all the system operations. These nodes enable interaction with the system. It serves as a gateway between the wireless sensors and the rest of the network that includes the back-end database that provides an area of long-term storage for data. This database also contains a circadian activity rhythm analysis program that learns the behavior of the resident and is able to adjust the configuration of the system, so that its operation is optimal based on that specific person. Finally, the architecture (shown in Figure 10.3) contains a user interface that could be a PDA or a computer that will allow authorized users to view data from the sensors or from a database.

ALARM-NET was designed not only to collect information but also to alert the healthcare provider if there are any emergency situations based on the patient's medical ailment. A resident, for example, has a condition where he should not remain sedentary for extended periods of time. Accelerometers can be placed in clothes and can alert the caregiver station if there is a prolonged period of inactivity.

Virone et al. [11] designed a network with an architecture that is similar to ALARM-NET. The architecture is multitiered, with heterogeneous devices such as lightweight body sensors and other mobile and stationary components. This architecture divides the devices into many layers:

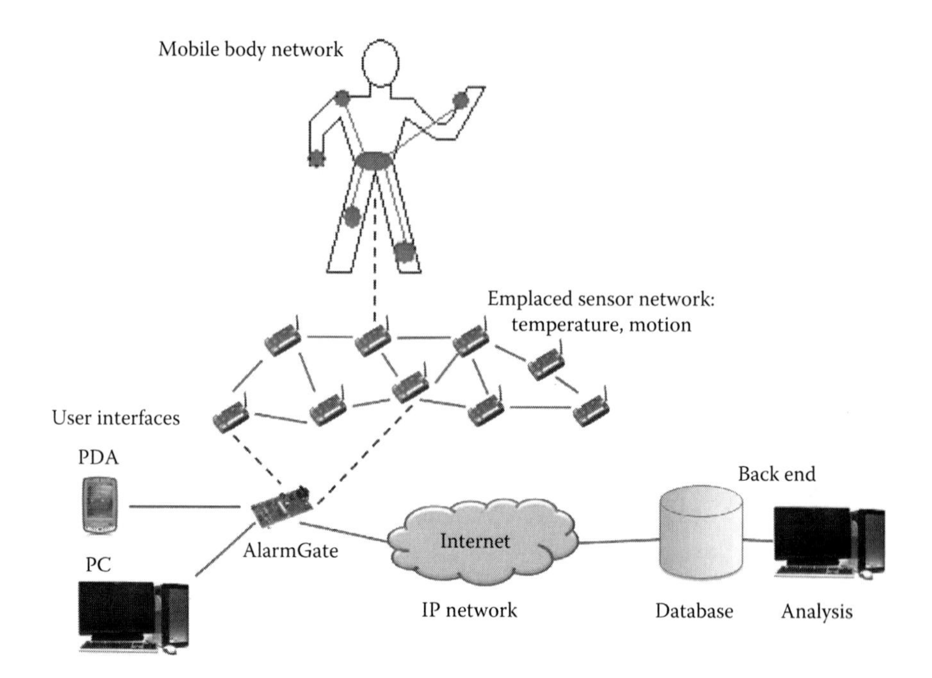

FIGURE 10.3 Telemedical system architecture.

Body Network and Subsystems: This network would include small devices that would contain different sensors such as accelerometers, heart rate, and temperature. These sensors will be necessary to monitor, identify, and locate the patient, as well as to perform a variety of other tasks based on the needs of the care provider. The authors determined that the sensors for this network should use kinetic energy to recharge their batteries. These sensors would also be able to alert the patient during abnormal circumstances, for example, an alert to remind a patient to move around if there are prolonged periods of inactivity.

Emplaced Sensor Network: This network would be made of sensors that have been installed into walls, furniture, etc. so that they can monitor the environment. These sensors are connected wirelessly to a backbone and do not carry out the data calculations or storage. They also serve as relay nodes for the sensors in the body network and the backbone.

Backbone: The backbone connects the stationary sensor network to other devices such as personal computers (PC), PDAs, and databases. It will also connect isolated sensors so that they can participate in more efficient routing.

Back-End Databases: This will be used for long-term storage of patient data. This may include archiving and data mining.

Human Interfaces: These include PDAs, PCs, and laptops that will be used by patients and caregivers to access the system. Access to the systems functions will be provided based on the role of the user, which will include data management, queries, and configuration.

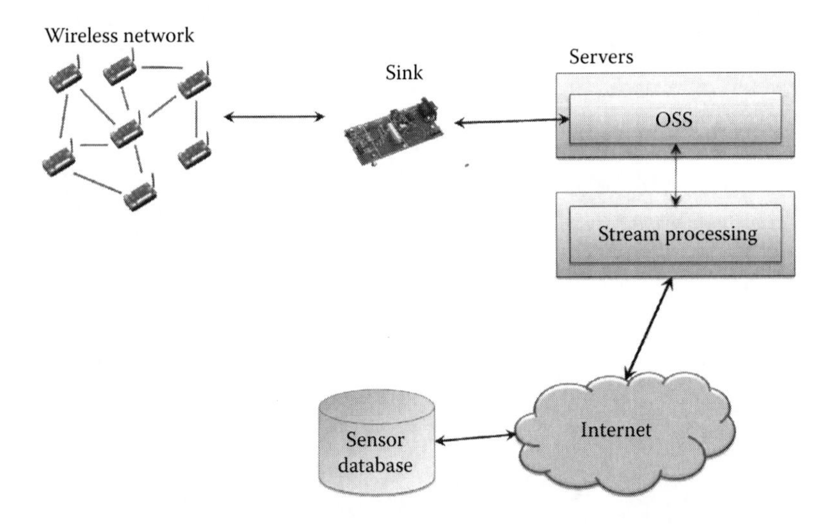

FIGURE 10.4 Smart Condo software architecture.

This system was designed to be a single-hop structure as the transmission range of the sensors is large enough to cover the entire facility being monitored. The authors recognize that for a multifloor facility, or in an effort to save energy by reducing the transmission range, a multihop protocol would be required.

Stroulia et al. [12] also devise a method of embedding sensors on the patient and in the environment in order to provide a continuous medical monitoring. The Smart Condo Project monitors the patient, similar to the other mechanisms, but attempts to use a more updated architecture than provided in ALARM-NET.

The architecture, shown in Figure 10.4, begins with a wireless network containing the relevant sensors for the patient. The sensors transmit data to the sink; data include a time stamp, network ID, node ID, sensor ID, sensor type, and event type. The sink transmits the packets it receives to a PC that contains the operation support system (OSS); this helps process the data and determines information about the resident based on the sensor data. The heart of the system is the server that runs the application-specific services; this controls all the functions of the network such as configuring the wireless sensors and interpreting the data received. The sensor database stores raw sensor data and data that have been processed for use by an application.

The authors have not had an opportunity to fully implement The Smart Condo project.

10.4 OTHER CONTINUOUS MONITORING MECHANISMS

Otto et al. [13] consider a body area sensor network that is used for health monitoring. This body network was designed such that it can be integrated with a larger telemedicine network for the continuous monitoring of patients. This system consists of individual monitoring networks for users that connect to the Internet where information can be transmitted to healthcare professionals for viewing, processing, and storage.

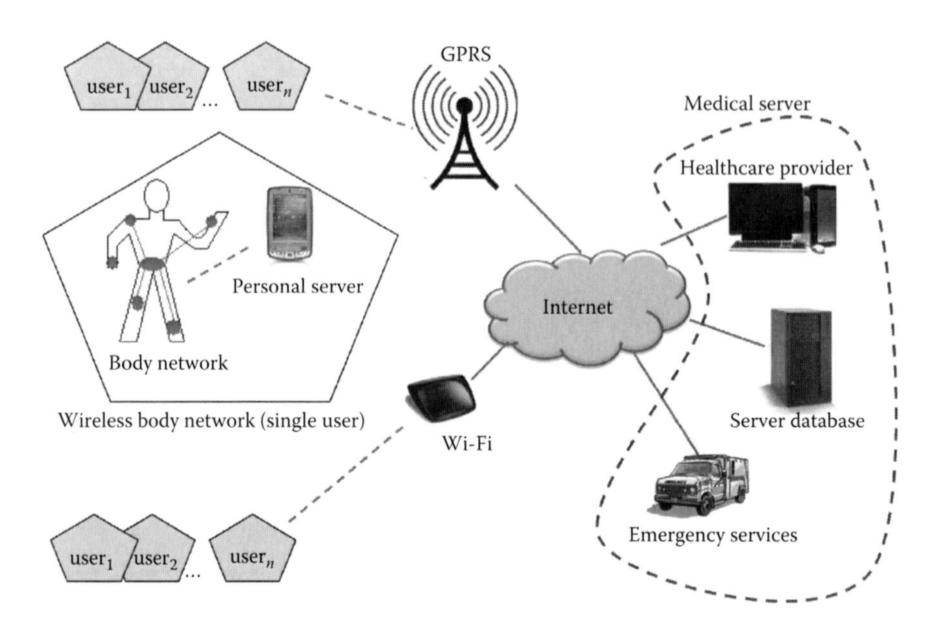

FIGURE 10.5 ALARM-NET architecture.

The architecture of such a system is shown in Figure 10.5 and comprises three layers. The top layer is the medical server, which is a network of medical personnel, emergency services, and healthcare providers. All persons in this layer are interconnected to enable the medical staff to provide services to thousands of individual users. Each user is equipped with a body area sensor network that will, based on the needs of that user, sample the vital signs and transfer the information to a personal server via a wireless personal network using either Bluetooth or ZigBee. The personal server can be any Internet-enabled device including a PDA, cell phone, laptop, or PC, which will manage the body network, provide an interface for the user and transmitted the information gathered from the sensors using an Internet connection or a cellular network such as general packet radio service (GPRS). The main functions of the medical server are to authenticate users when they connect to the system, download data that is transmitted from the users' personal network of sensors, parse the incoming data and store it in the matching medical records, analyze the data, identify serious irregularities in patient data and alert emergency medical technicians, and forward new care instructions to the user from the healthcare providers.

The personal server includes the authentication information and is configured to connect to the medical server automatically when a communication channel is available. In such cases, the information is uploaded so that it can be stored. If the channel is unavailable, then the personal server will store the data and make continual attempts to connect to the medical server. This ensures that the information being transmitted is as close to real time as practically possible.

The most critical element of the entire telemedical system is the wireless body area sensor network. This network is made up of sensor nodes that will be able to sense, process, and communicate the physiological signals of the user. These sensors

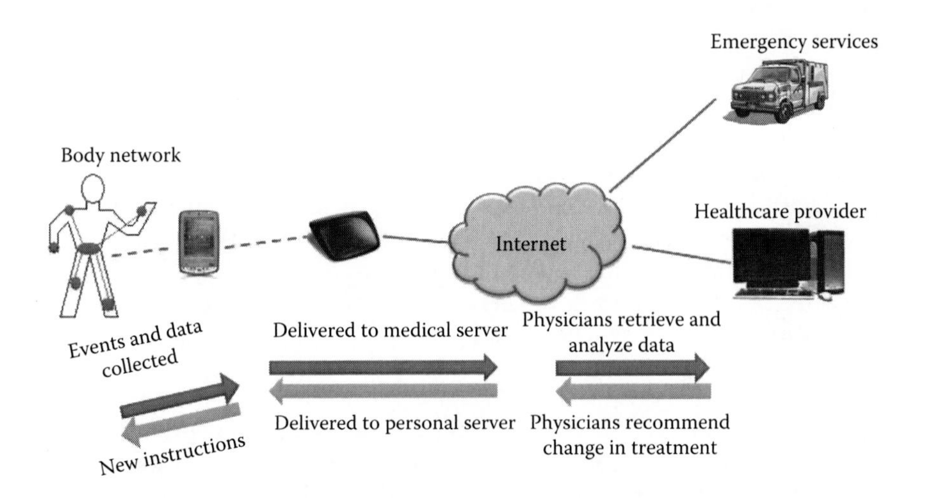

FIGURE 10.6 CodeBlue infrastructure.

include ECG sensor for monitoring heart activity, an electromyography (EMG) sensor for muscle activity, an electroencephalography (EEG) sensor for brain electrical activity, blood pressure sensors, motion sensors, and breathing sensors that can be used to measure the patient's activity level. Figure 10.6 shows the flow of data between patients and the healthcare providers and emergency services.

Jiang et al. [14] developed CareNet, which is an integrated wireless sensor environment for remote healthcare. CareNet is a two-tier wireless network [15] that is built on a heterogeneous networking infrastructure. The lower tier consists of a body network that employs wearable sensors that will obtain readings from the patient and transmit them via wireless communication to base-station sensors. The upper tier of the network consists of a backbone that would be used for transmitting these data to the healthcare provider. The backbone network offers a prompt and reliable delivery of the medical data. To support this claim, the authors designed specific features into the network:

Integrated Admission Control and Routing: This ensures that adding a new patient to the network would not compromise with the service of the existing patients. This is done using an admission control algorithm, which would approximate the capacity of the backbone network using its best case routing scenario.

Application-Level Routing: The routing protocol used by the backbone network routers has been implemented at the application level. This was done to make it reusable, that is, it can be used with many different operating systems without any modification to the protocol. The specific method of packet routing is determined based on the patients' activity patterns and so the routing table would be manually configured.

Multihop Transmission Control Protocol (TCP) Tunneling: Packet forwarding is also implemented at the application level. This is done by creating a TCP connection between backbone routers in each hop. This is done so that multiple threads can be used if the packet needs to be sent to more than one router at a time.

Mobile Sensor Hand-Off. Since the sensors are mobile, they would not be able to continuously communicate with the same backbone router at all times. Thus, sensors broadcast their message to all backbone routers within the transmission range. Duplicate packets are then removed based on a time stamp. The first packet received by the router is accepted and all subsequent duplicate packets are discarded.

The network was simulated using 90 patients where the traffic demand and patient location were randomly generated. The results showed that the network is unable to successfully support an increase in the number of patients when all patients are active at the same time. It was found, however, that if the patients maintain a stable mobility pattern, then the system would be able to support 10 more patients without a significant degradation of throughput.

Lupu et al. [16] propose the autonomic management of ubiquitous e-health systems (AMUSE), which uses multiple self-managed cells (SMCs) to implement a monitoring network. Each SMC manages a set of components such as the network of body and environmental sensors. The purpose of an SMC is to make that entire cell independent in that it is able to configure, optimize, and repair itself. The patient SMC should be able to support the addition and removal of sensors and then perform self-reconfigurations. SMCs are then expected to interact with each other and do so via composition and peer-to-peer interaction.

Composition interactions take place when there is a device inside the SMC that can manage the operation of that entire cell. In such a case, the device itself can be considered its own cell as it can function independently of the cell containing it. In such a case that cell will not advertise itself as an independent resource and will allow the cell containing it to govern its interactions with other devices. Peer-to-peer interactions occur between SMCs, for example, a nurse visiting a patient at home would need to interface with the patient SMC in order to interact with it. When the nurse SMC is detected, new policies such as alert behavior can be downloaded to the patient SMC and this will cause the patient cell to recalibrate itself without requiring the entire calibration to be stored in the nurse cell. Presently, the entire AMUSE network has not been simulated.

Zhu et al. [17] offer Vesta, which is a security improvement to the AMUSE system. Vesta integrates a variety of security protocols in order to secure the SMC. The AMUSE project was extended and secured in three major areas: secure sensor discovery, authentication module, and access control. In secure sensor discovery, it is ensured that only legitimate sensors can enter the network and that this new sensor will be paired with the intended patient. All wireless communication in the network has been encrypted using a key that is created when the new sensor gains access to the system. When users are authenticated, all communication between them is confidential.

The authentication protocol developed is based on a public key infrastructure where every user would be assigned a public key that identifies them. The nurse controller will broadcast a HELLO message to the patient. When the patient receives this message, the shared authentication process commences; in this process, each user is identified and authenticated before any information is shared between them. To guarantee that the process does not fail due to lost or corrupt messages during the exchange, a timer is initiated at the beginning of the authentication process and when the timer goes off, the process stops and is reset so that a new process can begin.

Access control is supported using the Ponder2 policy system [18] that manages authorization polices. Services and resources are characterized as objects. If an object is invoked, then access is permitted only if a policy exists that will allow access to this object. Otherwise, the request is denied. This ensures that only authorized users are allowed to view the patient information.

Ng et al. [19] present UbiMon, which is a five-component monitoring structure: the body sensor network (BSN) node, the local processing unit (LPU), the central server (CS), the patient database (PD), and the workstation (WS):

BSN Node: This wireless node was proposed to be a wireless module that either contains a very small battery or none at all. This node would be combined with a physiological sensor such as temperature or ECG. The context-aware sensors can then be used with the BSN to observe the patient activity.

LPU: The data transmitted by the body network are collected in the LPU. This can be a PDA or a cell phone, which will communicate with the network via Bluetooth, for short-range transmissions, and GPRS for long range. The LPU will collect data and recognize irregularities in patient data and provide warnings to the patient.

CS: From the data received from the LPU, the CS performs trend analysis and derives a pattern of the patient's condition. Based on this analysis, predictions can be made so that life-threatening anomalies can be detected.

PD: This is an area of long-term storage for the patient data, which is obtained from the CS. All data queries for patient data from WSs are satisfied by the PD.

WS: This may be a computer, laptop, or other handheld device that will be used by healthcare providers to access and analyze the data collected from the BSN. Historical data may also be accessed.

This system also is in its conception phase.

Chakravorty [20] introduces MobiCare that uses an architecture similar to others discussed. On the client side of this network, it consists of a BSN, which is made up of wearable sensors and actuators that communicate wirelessly and a BSN Manager, which will interact with the sensors and the healthcare provider servers. The BSN manager/MobiCare client monitors the sensors in the body network and aggregates the data collected by the sensors. This information is then uploaded to the servers using a secure wireless communication channel.

The authors suggest various devices that would meet the hardware requirements of a BSN manager, which should be user friendly, energy efficient, and contain wide area wireless connectivity, but it was decided that a cell phone would be the most suitable. Using a cell phone actually provides easy mobility for the MobiCare architecture. The MobiCare client will act as the server to connect to a gateway using the cellular 3G (generation) or 4G networks. This form of communication will allow the patient data to be transmitted in real time to the healthcare providers. This becomes especially important during emergency situations.

The healthcare provider side contains the services for the evaluation of the data. Different services will be provided to allow the patient to be continuously monitored and have their data reviewed by specialists. The data can also be stored for long-term health assessments.

Ganti et al. [21] introduced the idea of smart attire. Smart attire consists of sensors that have been embedded into the clothing that will be worn by the patient. These sensors will record the person's activities and location. The information gathered from the sensors can be stored in a database for subsequent analysis. The architecture to be used with this attire is called SATIRE.

SATIRE is based on a two-component structure, which is a PC or laptop and the mote. A data item is sensed at the sensor layer of the mote and passes to the application layer. The data then moves to the filter layer where data processing would occur. The results of the calculations move to the next layer, which contains the operating system and other important system functions. This layer provides an interface for communication with higher level applications. The final layer contains communication devices and drivers, which will send the data to the PC. On the PC side, the data are uploaded to the PC and then is passed to the next layer, which parses the raw data and stores them in a standard format, which may be based on a time stamp. The next layer is used for data processing. It deciphers the data using data mining and signal-processing techniques. This prepares the data for use in the application layer, which enables the user to perform queries on the information stored. This framework is illustrated in Figure 10.7.

SATIRE works with wearable sensors that are embedded into the users' clothing. When the patients' wearable sensor network is in the range of the rest of the network, the information can be transmitted in real time to the PC for processing. If the patient leaves this location and is out of range, then the information is stored within the sensors and can be uploaded when the body network is within the communication range of the rest of the network. Because the user is allowed to leave the range of the network and still be able to be monitored, albeit not in real time, he/she can be tracked over a variety of locations and so a GPS tracker is added to the sensors so that not only the location of the patient can be tracked but also the speed and duration of the activity.

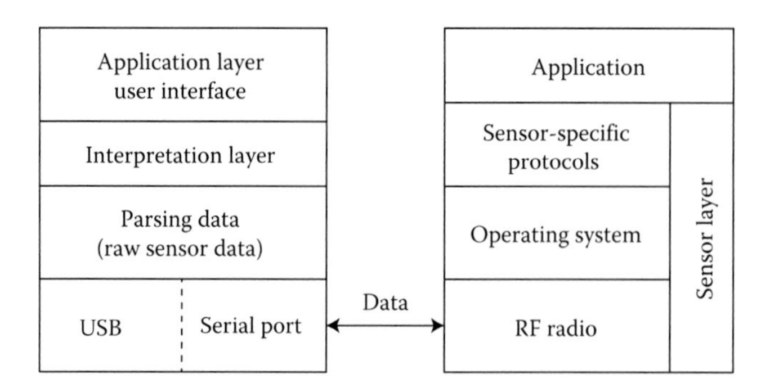

FIGURE 10.7 MEDiSN architecture.

10.5 ALGORITHMS COMPARISON

All the mechanisms surveyed in this paper have the common objective of improving the healthcare provided to patients by continuously monitoring them so that anomalies can be detected sooner. This premise was used in each article and adapted to different situations such as emergency response and smart homecare.

The algorithms discussed in Section 10.2 are compared in Table 10.1. The main objective of these schemes is to facilitate efficient communication among responding emergency services. This enables cooperation and a well-organized rescue effort. In Table 10.2, mechanisms from Section 10.3 are evaluated. These are methods that enable patients to live in their homes without assistance while retaining the ability to be monitored. Table 10.3 contains the mechanisms from Section 10.4, which are other schemes used to monitor patients in miscellaneous circumstances. These mechanisms are compared based on the following criteria:

Scalability: The ability of the system or network to accommodate increasing numbers of users with minimal reduction in the quality of service.

Self-Configuration: The network should be able to initialize itself and also adapt to nodes entering and leaving the network without having a loss of service.

TABLE 10.1
Emergency Response

Scheme	Scalability	Self-Configuration	Data Collection	Security	Location Tracking
CodeBlue	Large	Yes	PDA	No	RF tracking
SMART	Medium	No	PDA	No	MIT crickets
MEDiSN	Small	No	Multihop network	Yes	None

TABLE 10.2
Smart Homecare

Scheme	Body Sensor Network	Environmental Sensors	Context-Aware Sensors	Location Tracking	Data Collection
Alarm-Net	Yes	Yes	Yes	Motion detection	PDA and multihop network
Health Monitoring	Yes	Yes	Yes	Context aware	PDA and multihop network
Smart Condo	Yes	Yes	Yes	Motion detection	PDA/laptop
SmartDrawer	No	No	Yes	None	RF receiver

TABLE 10.3
Continuous Monitoring

Scheme	Body Sensor Network	Environmental Sensors	Architecture	Data Collection	Security
Telemedical system	Yes	No	Three-tier wireless network	PDA, cell phone, laptop, or PC	Yes
Carenet	Yes	No	Two-tier wireless network	Multihop network	Yes
AMUSE	Yes	Yes	Self-managed cells	PDA/laptop	Yes (enhanced by Vesta)
Vesta	Yes	No	Enhancement for AMUSE	NA	Yes
UbiMon	Yes	Yes	Five-component structure	PDA/cell phone	No
MobiCare	Yes	No	Three-component structure	cell phone	Yes
SATIRE	Yes	No	Two-component structure	PC/laptop	Yes

Data Collection: This refers to the method used to relay the data to the server for processing. This may be done by a PDA, which communicates directly with the server or using a multihop network.

Security: The scheme may or may not contain a mechanism that will be used to secure patients' sensor data as it is transmitted through the network.

Location Tracking: Some schemes have the ability to keep track of the patient's location through localization protocols.

BSN: The patient may be monitored using a BSN that will keep track of their vital information.

Environmental Sensors: Smart homecare schemes in particular make use of sensors that keep track of environmental conditions to keep the patient safe.

Context-Aware Sensors: These sensors keep track of the patient's body position to determine when a fall has occurred.

Architecture: This is the structure of the network's physical components and their operating organization and configuration.

10.6 CONCLUSION

Pervasive healthcare monitoring using wireless sensor networks is an important technology for the increasing demography of people with chronic illnesses. This type of monitoring is also extremely important for disaster response to take care and monitor the mass casualties when there are too few medical professionals for each patient to get continuous attention. Sensors can be placed on the body or in the homes of these patients and this can provide vital information that can alert the patient or medical professional when an emergency occurs.

It was shown that these networks can be used in a variety of applications. These include, but are not limited to, the disaster response, assisted living facilities, and smart homecare. These sensors can be used to make the lives of older patients more independent and to provide patients with chronic illnesses the benefit of having their condition monitored without enduring the expense of having a person to monitor them. In smart home care, many of the mechanisms employ environmental and context-aware sensors to keep track of the residents' location and to determine whether they are safe in their environment.

The architectures of these networks are designed to provide low congestion messages passing between sensors and the data collection unit. Some authors design a network that is only a single hop between the sensors and a cell phone that will use the cellular network to transmit information. Some schemes create multihop networks that use a gateway to connect to the Internet so that the information can be sent to a remote server for storage or viewing by the healthcare providers. Others describe an entirely local network where the sensors are able to communicate with the server directly or with the use of relay nodes if the sensors are out of the communication range.

Security is also a major concern when transmitting and storing medical information. Many of the mechanisms discussed, mention some form of securing data including Vesta [17], which is a mechanism that was proposed as a security enhancement for the AMUSE [16] mechanism. There are some mechanisms that do not refer to any mechanism of securing data but this enhancement would be necessary before successful implementation can occur.

Many of the schemes surveyed share a common need for some type of location tracking for the individual being monitored. This is especially important for those such as CodeBlue and MEDiSN that can be used for mass casualties where it would be impossible to locate a patient in need of assistance without some sort of protocol that could be used as a locator. It is advantageous that the mechanisms do not rely on a GPS as the means of tracking as this is not only more costly and significantly increases energy consumption but it has been found that GPS is not very accurate indoors. Most employ RF signals and motion detection to determine the patient position.

As the research continues into providing efficient pervasive healthcare monitoring, it is important that the architecture is low in congestion with a reliable connection to the Internet for remote access. The information should be completely secured and be processed so that it is available in a readable format for both patients and healthcare professionals to access.

REFERENCES

1. Stankovic, J., Cao, Q., Doan, T., Fang, L., He, Z., Lin, S., Son, S., Stoleru, R., and Wood, A. Wireless sensor networks for in-home healthcare: Potential and challenges, in *Proceedings of HCMDSS Workshop*, Philadelphia, PA, 2005.
2. Lorincz, K., Malan, D., Fulford-Jones, T., Nawoj, A., Clavel, A., Shnayder, V., Mainland, G., Welsh, M., and Moulton, S. Sensor networks for emergency response: Challenges and opportunities, *IEEE Pervasive Comput.* 2004; 2(3): 16–23.

3. Ko, J., Musaloiu, E. R., Lim, J. H., Chen, Y., Terzis, A., Gao, T., Destler, W., and Selavo, L. Demo Abstract: MEDISN: Medical emergency detection in sensor networks, in *ACM Conference on Embedded Networked Sensor Systems*, Raleigh, NC, 2008.
4. Waterman, J., Curtis, D., Goraczko, M., Shih, E., Sarin, P., Pino, E., Ohno-Machado, L., Greenes, R., Guttag, J., and Stair, T. Demonstration of SMART (scalable medical alert response technology), in *AMIA 2005 Annual Symposium*, Washington, DC, 2005.
5. Wood, A., Virone, G., Doan, T., Cao, Q., Selavo, L., Wu, Y., Fang, L., He, Z., Lin, S., and Stankovic, J. ALARM-NET: Wireless sensor networks for assisted-living and residential monitoring, Technical Report, Computer Science Department, University of Virginia, 2006.
6. Curtis, D., Pino, E., Bailey, J., Shih, E., Waterman, J., Vinterbo, S., Stair, T., Guttag, J., Greenes, R., and Ohno-Machado, L. SMART—An integrated wireless system for monitoring unattended patients, *J. Am. Med. Inform. Assoc.* 2008; 15(2): 44–53.
7. Lorincz, K. and Welsh, M. Motetrack: A robust, decentralised approach to RF-based location tracking, *J. Pers. Ubiquitous Comput.* 2007; 11(6): 489–503.
8. Wang, Y., Goddard, S., and Perez, L. C. A study on the cricket location-support system, in *IEEE International Conference Electronic/Information Technology*, Chicago, IL, 2007: 257–262.
9. Gnawali, O., Fonseca, R., Jamieson, K., Moss, D., and Levis, P. Collection tree protocol, in *Proceedings of the Seventh ACM Conference on Embedded Networked Sensor Systems (SenSys)*, Berkeley, CA, 2009.
10. Becker, E., Metsis, V., Arora, R., Vinjumur, J., Xu, Y., and Makedon, F. SmartDrawer: RFID-based smart medicine drawer for assistive environments, in *Proceedings of the Second International Conference on Pervasive Technologies Related to Assistive Environments*, Corfu, Greece, June 2009.
11. Virone, G., Wood, A., Selavo, L., Cao, Q., Fang, L., Doan, T., He, Z., Stoleru, R., Lin, S., and Stankovic, J. An advanced wireless sensor network for health monitoring, in *Transdisciplinary Conference on Distributed Diagnosis and Home Heathcare*, Arlington, VA, 2006.
12. Stroulia, E., Chodos, D., Boers, N., Huang, J., Gburzynski, P., and Nikolaidis, I. Software engineering for health education and care delivery systems: The smart condo project, in *SEHC '09, ICSE Workshop on Software Engineering in Health Care*, Vancouver, BC, 2009: 20–28.
13. Otto, C., Milenkovic, A., Sanders, C., and Jovanov, E. System architecture of a wireless body area sensor network for ubiquitous health monitoring, *J. Mobile Multimedia*, 2006; 1(4): 307–326.
14. Jiang, S., Cao, Y., Iyengar, S., Kuryloski, P., Jafari, R., Xue, Y., Bajcsy, R., and Wicker, S. CareNet: An integrated wireless sensor networking environment for remote healthcare, in *Third International Conference on Body Area Networks*, Tempe, AZ, 2008.
15. Jiang, S., Xue, Y., Giani, A., and Bajcsy, R., Robust medical data delivery for wireless pervasive healthcare, in *IEEE International Conference on Dependable, Autonomic and Secure Computing*, Chengdu, China, 2009.
16. Lupu, E., Dulay, N., Sloman, M., Sventek, J., Heeps, S., Strowes, S., Keoh, S. L., Schaeffer-Filho, A., and Twidle, K. AMUSE: Autonomic management of ubiquitous e-health systems, *J. Concurrency Comput. Pract. Exp.* 2008; 20(3): 277–295.
17. Zhu, Y., Sloman, M., Lupu, E., and Keoh, S. Vesta: A secure and autonomic system for pervasive healthcare, in *Third International Conference on Pervasive Computing Technologies for Healthcare (Pervasive Health 09)*, London, U.K., 2009.
18. Twidle, K., Dulay, N., Lupu, E., and Sloman, M. Ponder2: A policy system for autonomous pervasive environments, in *International Conference on Autonomic and Autonomous Systems*, Valencia, Spain, 2009.

19. Ng, J., Lo, B., Wells, O., Sloman, M., Toumazou, C., Peters, N., Darzi, A., and Yang, G., *Ubiquitous Monitoring Environment for Wearable and Implantable Sensors (UbiMon)*, UbiComp, Nottingham, U.K., 2004.
20. Chakravorty, R. MobiCare: A programmable service architecture for mobile medical care, in *Pervasive Computing and Communications Workshops*, Pisa, Italy, 2006.
21. Ganti, R., Abdelzaher, T., Jayachandran, P., and Stankovic, J. SATIRE: A software architecture for smart AtTIRE, in *Fourth ACM Conference on Mobile Systems, Applications, and Services (Mobisys 2006)*, Uppsala, Sweden, 2006.

11 Use of Body Sensor Networks in Clinical Settings and Medical Research

Carlo Alberto Boano, Felix Jonathan Oppermann, and Kay Römer

CONTENTS

ABSTRACT

Recent advances in wireless and embedded computing technologies have led to a radical transformation of healthcare. Wireless networks of low-power body sensors have been used to monitor the vital signs of postoperative patients outside of hospital facilities, for the daily care of elderly and chronically ill people, as well as to assist rehabilitation or to motivate people doing sports, allowing a significant reduction of healthcare expenses. Compared to these application areas, the number of applications of body sensor networks (BSNs) inside hospital structures and for scientific experimentation on patients has been, to date, rather limited, despite the potential improvements that they would offer in the management of hospitalized patients and in carrying out unobtrusive data collections for medical research. This chapter focuses specifically on the latter two application domains and carefully describes the technical and experimental challenges leading to such limited adoption. We give an overview of the most relevant applications that have been developed in the area and describe their noteworthy distinctive characteristics. We believe that this survey will serve as a reference to guide researchers and system designers working in these emerging areas.

11.1 INTRODUCTION

Fast progress in wireless technologies in conjunction with remarkable advances in sensor design and embedded computing technologies triggered a golden era for wireless sensor network (WSN) research and development. The resulting proliferation of miniature-sized low-power wireless devices revolutionized sensing in a wide range of civil, scientific, military, agricultural, and industrial applications [113].

The use of tiny, low-cost networked wireless sensors has also radically transformed healthcare. So-called body sensor networks (BSNs), that is, networks of noninvasive or invasive wireless biosensors worn or even inside the human body, enable a seamless collection of people's physical, physiological, and psychological information, which offers significant potential benefits at an economical, logistic, clinical, and scientific level. Wireless pervasive monitoring of patients offers indeed high degrees of flexibility and enhances patient mobility while addressing the typical weaknesses of traditional data collection, such as infrequent assessment and subjective observation [55].

With a *silver tsunami*, that is, a large number of retiring elders [147] that will soon exhaust the capacity of current hospitals, body sensor networks (BSNs) play a crucial role in shifting the monitoring of noncritical patients *outside* of hospital facilities, with a consequent reduction of healthcare expenses. As they offer the possibility to continuously monitor the vital signs of patients at their homes, enhancing prevention and allowing an early detection of diseases [157], BSNs have been widely used to build residential monitoring and assisted-living solutions for elderly care. BSNs have also been used to monitor patients with chronic conditions (i.e., patients with long-lasting or recurrent diseases) remotely, which helps in reducing the large expenditure of healthcare resources dedicated to chronic care [58].

Wireless sensor networking can also be used *inside* clinical and hospital structures to improve the management of inpatients, that is, patients that remain in the hospital structure overnight or for an indeterminate time. BSN can indeed simplify the transportation of patients across different hospital units [114] and avoid disconnections and failures occurring when handing off wired patients [75]. They can also enable a more frequent measurement of vital signs outside intensive care units: in step-down hospital units, indeed, where patients are not permanently wired and monitored, the vital signs of patients are measured sporadically (on average only 10 times during the first 24 h following an operation [163]), and a quicker detection of clinical deterioration is needed [30].

Closely related to the use of BSN inside hospital settings is their employment in specialized clinics or research laboratories for scientific experimentation on patients. Patients being visited and monitored in clinical environments do not behave as in their daily life, due to the stress and anxiety associated with the clinical visit and due to the exposure to a different environment, biasing the data collections [21]. An unobtrusive pervasive monitoring using miniaturized body sensors in more supportive environments would allow a more realistic quantitative assessment of the vital signs of patients, improving the quality and accuracy of data available to medical researchers.

Despite their potential, the number of applications of BSN inside hospital structures and for scientific experimentation on patients has been, to date, rather limited. On the one hand, wireless sensing systems used in such settings are safety critical, and hence require high reliability and accuracy. On the other hand, clinical trials broaden the requirements and challenges even further, as experimentation on patients requires the system to comply with several regulations and a significant overhead in obtaining all necessary permissions. It is therefore important to get a deeper understanding of the large amount of technical and nontechnical challenges introduced by experimentation on patients, as they largely reduce the adoption of BSN in clinical settings and for medical research.

Differently from existing surveys about BSNs for healthcare [2,33,35,56, 55,75,108,117,141], this chapter focuses specifically on BSN applications aimed for use inside clinical and hospital structures or for unobtrusive data collections to improve medical research. In the next section, we describe the vast scope of BSNs, showing that, in comparison to other application areas, their use in clinical settings and medical research has been rather limited. We carefully analyze and describe the technical and experimental challenges leading to such limited adoption in Section 11.3. We then review the most relevant BSN applications on hospitalized patients and for unobtrusive data collection to improve medical research in Sections 11.4 and 11.5, respectively. For both categories, we describe in detail exemplary applications, and we briefly highlight noteworthy distinctive characteristics. We conclude this chapter in Section 11.6 with an outlook on future BSN developments.

11.2 SCOPE OF BODY SENSOR NETWORK APPLICATIONS

Being healthcare their largest application domain, one would typically associate BSNs to the monitoring of vital signs of patients inside hospital facilities or to the monitoring of elderly people and postoperative patients in ambulatory environments. In reality, their scope is much broader and spans from human–computer interaction, leisure and gaming, over the management of soldiers in extreme environments, wellness and fitness applications, to the coordination of emergency squads during mass casualty disasters.

In this section, we provide a brief overview of the broad scope of BSN applications and point out that the use of BSN for building healthcare solutions on hospitalized patients and for unobtrusive data collections to improve medical research has been indeed rather limited, especially when compared to the number of applications in leisure and fitness or for ambulatory monitoring.

11.2.1 Leisure and Gaming

A rich body of research and commercial work has employed networks of body sensors to improve human–computer interaction and gaming experience. Hand-worn accelerometer and gyroscope sensors have been used to recognize gestures [15,121], with the Nintendo Wii console being probably the most famous example of a device allowing 3D motion sensing using three-axis accelerometers [127]. Other research prototypes have been proposed in the literature, including a heart-rate sensor to control an interactive game [107] or gloves integrating galvanic skin response (GSR) sensors to measure calm [144]. Another active area of research has been the use of BSNs to carry out interactive dance performance, in which the motion data of a dancer are used to control the generation of musical or graphical effects in real time [8,95,119].

11.2.2 Athletic Performance

The availability of low-cost wireless body sensors [90] increased the demand for systems assessing physical activity and performance, as several amateur and recreational athletes aim to maximize the efficiency of their individual training sessions [101]. The use of wearable motion sensors has been proposed for several sport activities, ranging from winter sports [24,63,103], running [145,146], and climbing [116] to water sports such as rowing [73] and swimming [11]. Other than purely measuring athletic performance, BSNs have also been used for real-time recognition tasks [59] such as counting of points in martial arts competitions [29].

11.2.3 Wellness and Fitness

The increasing availability of low-cost wireless body sensors, together with the advent of smartphones, has also revolutionized healthcare prevention. Applications that aim to motivate people doing sports [106,109] or that carefully check the caloric expenditures during the course of the day or during workouts [40,83,110,138,164] are now available at everyone's fingertips.

Similar to the use of BSNs to measure the athletic performance and to build interactive games, the new emerging mass market of wellness and fitness wireless solutions is also driven by the rather relaxed requirements from an engineering perspective, since these devices typically integrate only a few sensors, do not have strict requirements on accuracy or communication reliability, and do not require any certification as medical device [23].

11.2.4 Elderly Care and Ambulatory Monitoring

Triggered by the need of reducing the large expenditure of today's healthcare [58], the use of BSNs for elderly and chronic care and for postoperative monitoring in ambulatory settings has attracted, by far, the largest body of work. Ambulatory monitoring and assisted living of elderly people based on BSNs is nowadays a widely accepted solution [67,157]. From the earliest telecare projects [54,79,99,130,132], several research works have investigated fall detection systems to monitor elderly people using wearable sensors [3,84,85] or smart cane systems [158], as well as home nursing systems to monitor heart-rate variability [70] and to increase autonomy and independence of patients [14]. System designers have also gone beyond wearable sensors, by developing complete smart home solutions that include presence sensors, door contacts, microphones, and environmental sensors for classification of the patients' daily activities [16,46]. Similar solutions target chronically ill patients being remotely monitored at home [28]: wireless networks of body sensors have been proposed to monitor patients with cardiovascular problems and alert the caregivers as soon as the risk for heart attack increases [44,53] or to monitor diabetic patients and automatically inject insulin through a pump as soon as the insulin level in the body declines [52]. Furthermore, several solutions target patients that do not need a formal hospital admission or overnight stay after surgery, but can be monitored remotely: in particular, several research works have addressed postoperative recovery after abdominal surgery [9,89,154].

11.2.5 Rehabilitation and Therapy

Similar to solutions for postoperative monitoring, BSNs have been used also to follow rehabilitation after surgery and to improve the efficiency of therapies. Interactive games in combination with wireless wearable sensors have been employed to improve the efficiency of therapeutic and rehabilitation exercises by personalizing the exercises according to the patient's specific needs [125]. Solutions to assist rehabilitation after a cardiac disease [118] or to observe the recovery of motor abilities following stroke [61] or knee-replacement surgery [7] have shown the possibility to capture the progressively more regular gait patterns of subjects over time and to identify patients who fail to improve as expected, facilitating early clinical review and intervention. In [134], for example, wireless inertial sensors record and analyze the movements of patients and provide real-time feedback to the medical personnel.

11.2.6 Monitoring Vital Signs in Extreme Environments and Mass Casualty Disasters

BSNs have been largely employed to monitor the vital signs and stress level on soldiers and fire fighters working in extreme environments. Using wearable or textile sensors embedded into the inner garment of clothes to measure heart and respiration rates and additional environmental characteristics (e.g., temperature and the presence of toxic gas), several works aim to improve the safety of rescue workers and fire fighters [60] or to increase the safety in bomb disposal missions [71]. Similarly, systems recognizing activities of soldiers [104] and monitoring the health of astronauts have also employed wearable physiological wireless sensors. Another important application scenario for BSNs is their use in emergency scenarios and mass casualty disasters, with a quick deployment that would support coordination of the emergency recovery. Using wearable sensors to sense and record vital signs into an electronic patient record database, several researchers have shown how to enhance the capability to provide emergency care through automatic electronic triage [48,49,72,94,125].

11.2.7 Vital Sign Monitoring of Hospitalized Patients and for Medical Research

Vital sign monitoring in hospitals has great potential to improve decision making [23], simplify the transportation of patients [114], and enable a more frequent measurement on patients outside intensive care units [30]. Compared to the BSN systems described in the rest of this section, however, monitoring of vital signs on hospitalized patients has attracted only a few complete works [30,34,74,91] and has a much larger set of requirements. On the one hand, wireless sensor networking systems used in clinical settings are safety critical, and hence require high reliability and accuracy. They also have strong requirements with respect to the security and privacy of the collected measurements and to fault tolerance: a late detection of faults may cause harm to patients and threaten their lives. On the other hand, clinical trials broaden the requirements and challenges even further, as experimentation on patients requires the system to comply with the severe regulations of hospital environments and a significant overhead in preparing ethical applications and selecting the eligible patients. Similarly, the body of work targeting an unobtrusive data collection for medical research (see Section 11.5 for a complete survey) shares the same set of requirements, especially for what concerns experimentation on patients. In the next section, we will provide a deep analysis of these challenges.

11.3 CHALLENGES

Despite their potential benefits, to date, BSNs have not been widely deployed inside clinical and hospital structures. Compared to other application areas, BSNs deployed in these settings have a significantly larger set of challenging requirements. For example, ambulatory monitoring systems for the elderly and postoperative care must have a long operational lifetime, minimize false alarms, and maximize context awareness to better judge the relevance and significance of the recorded physiological data, but have no hard real-time constraints on the communication reliability and delay, which makes the design of communication protocols simpler. Similarly, systems used in emergency scenarios have real-time requirements and necessitate a highly reliable communication among the sensor nodes, and they further need to cope with the high degree of mobility of users, but do not have strong requirements on the operational lifetime of the system and on the privacy of the collected data [75]. When designing a new body sensor system to be used in clinical settings, these and several other requirements need to be addressed at the same time. In this section, we provide an analysis of these challenges, distinguishing between the technical challenges that need to be addressed when designing a new body sensor system to be used in clinical settings, and the intrinsic experimentation challenges that typically arise during the deployment in hospitals or during experimentation on patients.

11.3.1 Design Challenges

In the following paragraphs, we illustrate the technical challenges affecting the design of a BSN for monitoring hospitalized patients or for experimentation on patients for research purposes.

11.3.1.1 Operational Lifetime

The battery volume, and hence the energy budget available on body sensor nodes, is typically very limited, in order to decrease the size and weight of the devices and not to create discomfort to the final user. As wireless communication requires significant energy and as several sensing activities (e.g., electrocardiograms [ECGs] and accelerometer samplings) are carried out at a very high frequency, batteries are quickly depleted and need to be replaced. Most proposed systems have an operational lifetime that spans from a few hours [34] to a few days [30], hence requiring daily battery replacement: a nonnegligible burden on the already overworked hospital staff and a limitation for continuous long-term medical studies. Solutions reducing the battery consumption by means of aggressive duty cycling impact the prompt delivery of the measurements, whereas online data processing and sophisticated algorithms to reduce the volume of the data transfer require computational resources that may not be available in most body sensor nodes.

11.3.1.2 Resource Scarcity

In addition to scarce energy resources, body sensor nodes are indeed also constrained in their computational resources. In order to keep the size small and to keep the price per unit reasonably low, sensor nodes usually employ resource-constrained microcontrollers, with memories ranging from a few hundred bytes [21,22] to a few kilobytes. This makes it challenging to execute sophisticated algorithms, encryption mechanisms, or advanced data processing techniques on those resource-constrained devices.

11.3.1.3 Security and Privacy

BSNs are often used to measure the vital signs of hospitalized patients and may reveal information about their clinical deterioration. When used to build applications for scientific data collection on freely moving patients, BSNs may reveal the lifestyle of subjects as well as their daily activity. The potential abuse of this information by external entities such as insurance or advertisement companies is a major concern for the privacy of the individuals, and sensor data should be protected from external access. As a consequence, any medical system used in such settings needs to ensure the privacy of the recorded data and the patient's medical records, also to comply with the strict policies and laws for the protection of patient's personal data established in most countries. This task is quite challenging in the presence of wireless data transmission: a common route is to employ end-to-end encryption, but on resource-constrained battery-powered devices, this would affect energy efficiency [142,150]. Furthermore, to protect against eavesdropping attacks, it is necessary to implement suitable authentication procedures to ensure that only authorized personnel has access to the private data. Recent solutions have proposed several frameworks for preserving privacy [31,78], some of them based on biometric approaches that use an intrinsic characteristic of the human body as means for authentication [128]. In general, however, the number of deployed solutions embedding strong privacy measures has been rather limited to date.

11.3.1.4 Communication Reliability and Robustness

BSNs used in clinical settings are safety-critical systems that require reliable communications. When monitoring life-threatening deteriorations in vital signs of hospitalized patients, alarm procedures must be triggered immediately, and wireless systems need to guarantee the same reliability of wired systems [165]. Achieving a high communication reliability is not an easy task when using BSNs. On the one hand, the generic principle of body sensors is to restrict the communication radius to the body's periphery and to limit the transmission power in order to prolong the system lifetime [55]. Limiting the transmission power is especially important in clinical environments, as high-power transmissions may interfere with existing medical equipment, such as implanted drug delivery systems and stimulators. On the other hand, wearable systems are intrinsically affected by the attenuation of the transmitted signal from body shadowing, which may disrupt connectivity. Furthermore, wireless devices communicating using the freely available industrial, scientific, and medical (ISM) bands need to cope with the interference generated by other devices: in clinical environments, there is a high concentration of electronic devices that, together with off-the-shelf technologies used by patients such as mobile phones, are potential threats to low-power wireless communications [10]. For this reason, the number of systems adopting pure Institute of Electrical and Electronics Engineers (IEEE) 802.15.4 systems is rather limited [23], and there is a preference for frequency-agile Bluetooth systems and technologies such as smart textiles in which wires are embedded into clothing.

11.3.1.5 Real-Time Feedback

Reliable communications are essential to achieve real-time feedback, so that all relevant information are promptly delivered. This does apply not only to abnormal variations in the vital signs of hospitalized patients but also to the detection of faults, such as loose sensors and battery outages [34]. Within medical studies, researchers need to be constantly informed about the collected data, as well as about the behavior of the participants to the study. Even though test subjects are usually paid, it is common that they do not comply with unpleasant instructions [32], such as sleeping only for 4 h/day [21]. A proper handling of real-time requirements, however, increases the strain on the communication infrastructure in terms of bandwidth utilization and latency requirements.

11.3.1.6 High Data Rates

Besides the load introduced by sustaining real-time communications, systems tailored to medical applications often need to cope with the high data rates generated by physiological sensors such as ECGs and accelerometers. As low-power sensor devices are highly constrained with severe bandwidth limitations, several researchers have proposed the use of activity filters and on-node compression [93] or the creation of a backbone of relay nodes using optimized collection protocols [74].

11.3.1.7 Sensor Reliability

BSNs often have little room for redundancy; hence, the individual components of the system need to be highly reliable. This is especially true in clinical settings: the health and life of patients strongly depends on the recorded data of the BSNs, and a failure of a sensor could potentially be fatal. Similarly, medical researchers rely on the collected data to derive their conclusions, and a failure may compromise a whole experiment and imply a significant waste of resources.

11.3.1.8 Sensor Accuracy

In clinical applications, important decisions rely on the correctness of the collected measurements and may depend on tiny fluctuations in the vital signs of a patient. Similarly, also in medical research, different conclusions can be drawn for minimal differences in the physiological parameters that are monitored. Therefore, BSNs designed for clinical use in hospitals or for experimentation on patients need to provide a high accuracy, which needs to persist also when using flexible sampling rates and is expected to be comparable to traditional wired systems. This is not necessarily a straightforward task, as studies targeting extremely precise physiological measurements may require precise calibration and additional hardware complexity [21].

11.3.1.9 Context Awareness and Multisensor Data Fusion

Reaching a high accuracy is often not sufficient to explain a certain evolution of a given body parameter or to detect specific health conditions. In order to achieve a full understanding, a snapshot of the state of the surrounding environment and user activity is often required. Context data from ambient sensors enable the researcher to judge the relevance and significance of the recorded physiological data and to correctly classify outliers and false readings [30]. Also, the fusion of data from different sensors with diverse properties may be indispensable to achieve a full understanding of the collected measurements. In studies in which freely moving patients are involved, for example, it may be indispensable to integrate the measurements of several ambient sensors to understand the reasons of a sudden rise or decrease in temperature: this may occur if the participants to the study remain at the sunshine or take a walk outdoors during the winter [21]. Also, in case a large set of similar sensors is available, it is hard to make a proper sensor selection, as it may change over time. Furthermore, depending on the focus of the experiment or on the vital signs of interest, some sensors may be more informative with regard to certain activities than others.

11.3.1.10 Mobility of Users

An intrinsic challenge in the design of BSNs is how to cope with the mobility of users. Mobility introduces motion artifacts that may invalidate the measurements, and the system should provide means for their detection and (when possible) their compensation. Furthermore, mobility may impair communications. For example, in the presence of static relay points, freely moving users can move outside their radio range, leading to data loss in case persistent storage capabilities are not available.

11.3.1.11 Packaging

The design of a proper packaging is very challenging, as it encompasses several properties, ranging from durability to aesthetics, from portability to the easiness of cleaning and disinfection.

Firstly, the casing needs to be ergonomic, that is, easy to use and comfortable to be worn: this often comes together with a requirement of being lightweight and small. Ensuring comfort is intrinsically difficult as each user has a different perception on how comfortable a sensor is; therefore, different people might find a sensor comfortable and others may not. In medical research, it is indeed not uncommon that several patients quit an experiment because the device was irritating or uncomfortable [23,34,91]. The discomfort of patients is also often caused by an anxiety of being wired and triggers deliberate disconnections of sensors compromising the reliability of the system [23,75]. Fewer channels and consequently fewer electrodes may lead to more comfortable solutions, but require an efficient sensor selection to keep up the accuracy of measurements [140].

Secondly, the packaging needs to be durable. For example, it needs to resist rain, hand washing, showering, sweating, mechanical shock, and environmental temperature variations [19]. One of the biggest challenges is also the intrinsic mobility of users that may lead to motion artifacts especially when sensors are worn on the extremities [153].

Thirdly, the packaging needs to be easy to affix. Some notable examples show that even the easiest fixation mechanisms such as Velcro straps [23] or professional plasters [20] may not be optimal in all situations.

Fourthly, the package needs to be sufficiently aesthetically appealing, or there will be a high risk that the user refuses to wear it, especially in long-running medical experiments. As for the design of a comfortable packaging, to conceive an aesthetically appealing device is not straightforward, since aesthetic appeal is highly subjective. Furthermore, in order to be usable in clinical and sterilized environments, the packaging needs to be disinfectable and easy to clean; hence, the material employed to build the packaging needs to be resistant to common cleaning aids and disinfectants.

11.3.1.12 Sensor Biocompatibility

Sensors—especially implanted ones—need to be biocompatible, as they are usually in direct contact with the body for extended periods of time [161]. To reach this goal, it is necessary to select materials that are minimally irritating that are not likely to cause allergic reactions [105]. The presence of an extraneous material in contact to (or implanted inside) the body may cause an inflammatory reaction, that is, the injury of a tissue that may lead to acute and chronic inflammation, or a foreign body reaction that may lead to significant health risks for patients [112]. Designers can take inspiration from the lessons learned from the use of implantable sensors that have been largely applied on patients in the past years, for example, cardiac pacemakers and implantable cardioverter defibrillators [161].

11.3.1.13 Usability of the System

Irrespective of its final user, a BSN needs to be easy to use and resistant against errors of the operators. Fast decisions are required in the daily routine, and the presentation of the data should hence support the decision makers as much as possible. Setup procedures should be simple and hassle-free and provide sufficient feedback to immediately verify that the system operates as intended. To gain acceptance by the medical staff, new systems need to be easy to use, and the additional workload needs to be clearly justified by the added benefit. BSNs should hence not be more difficult to use than the traditional wired systems they replace [13] and should consist of simple and familiar interfaces; otherwise their adoption is quite unlikely.

11.3.1.14 Standardization and Interoperability

BSNs used in clinical environments need to synchronize and interoperate with a large number of existing measurement and clinical and medical record-keeping systems [55,143]. Integrating multiple devices that operate at different frequencies with different radio protocols is complex: in current prototypes, indeed, there is often a lack of seamless integration with infrastructure network and electronic medical records [67].

11.3.2 Experimentation Challenges

In the next paragraphs, we provide a detailed analysis of the intrinsic experimentation challenges that typically arise during the deployment in hospitals or during experimentation on patients.

11.3.2.1 Ethical Reviews

Every clinical study is covered by regulations that protect the health, safety, and dignity of the participants. In order to carry out a clinical study, researchers need to submit an application to an ethics committee, specifying the research questions the work is trying to answer, the criteria for the selection of participants, as well as the detailed procedure of the trial. Risks and benefits of the study need to be explicitly highlighted, and trials that address well-known problems without tackling novel research questions are considered to be unethical. An important criterion for the acceptance of an ethical application is that the number of participants to a given study is selected so that statistically significant results are obtained: failure to miss this requirement may invalidate the study. In addition to the detailed description of the trial, an information sheet describing in clear and simple words the study to be carried out is needed, so that the participants can autonomously decide whether to take part to the trial or not. The preparation of these documents may require to team up with people having medical expertise and is quite time-consuming. Furthermore, once an application is submitted, one may need to wait several weeks before getting an answer about its acceptance or rejection, as ethical committees do not necessarily meet frequently. Even more complex (at least in Germany) is an ethical application for a trial in which one wants to use newly designed uncertified devices, as the decision is not taken by a local committee within the university, but rather on a national level, with longer delays and even higher requirements.

11.3.2.2 Selection of (Eligible) Participants

The selection of the number of participants to a study can often be a catch-22 dilemma. On the one hand, studies that involve too few participants to achieve statistical significance are considered unethical and may not be approved by ethical committees (see earlier). On the other hand, one cannot simply involve an arbitrary amount of participants, as this would introduce remarkable costs. Indeed, only a few medical studies are carried out with volunteers, and it is more common that participants are paid to participate to a scientific trial and comply with a certain set of instructions. Furthermore, also the selection of which subjects are allowed to take part to a clinical trial is often challenging, as it needs to adhere to the strict inclusion and exclusion criteria of participants (for a detailed example, see [148]).

11.3.2.3 Acceptance by Medical Staff

Medical staff typically faces a high workload, and any additional task needs to be well justified. The operational systems should not create unnecessary work or hinder established procedures more than absolutely necessary. A common problem in hospital trials is the so-called alarm fatigue, that is, the situation in which the ward staff becomes desensitized to alarms produced by electronic devices such as infusion pumps, inflatable mattresses, and vital sign monitoring systems [23]. This requires the BSN to minimize the amount of false alarms as well as the amount of unimportant alarms triggered. Furthermore, it also requires proper training of the staff in order to avoid that relevant alarms are neglected, for example, battery status warnings.

11.3.2.4 Acceptance by Patients

Likewise, it is important to reach a high degree of acceptance among the patients (or participants to a given study). This is especially difficult as patients tend to have very diverse opinions about ergonomics, aesthetics, and painfulness: there is usually no one-size-fits-all solution and the system needs to be adaptable to different needs. It is not uncommon that patients complain that a device is uncomfortable [91] or even quit an experiment because the device is irritating [23,34]. For example, if a patient feels not comfortable while wearing a number of sensors, he or she may deliberately disconnect a portion of them or even suffer from anxiety attacks [148].

11.3.2.5 Uncooperative Patients

A critical issue during clinical experiments is the compliance of the study participants to a given protocol or prescribed regimen of treatments. Even though test subjects are typically paid, it is not uncommon that they do not comply with a number of unpleasant instructions [21], and this has been frequently observed in the literature [23,30,75]. Experimental designs should consequently include measures to record patient behavior in order to ensure reliability of the collected data.

11.3.2.6 Untrained Hospital Staff

Hospital staff usually has little to no technical knowledge about radio devices. If the system does not behave as it was described or is malfunctioning, the chances that they will step forward and fix the problem are close to zero. Instead, it is more likely that the hospital staff generates errors: existing deployments have shown that the corruption of measured data was caused by the medical staff misplacing the sensors [148]. This connects to the aforementioned *alarm fatigue* problem and may significantly affect the quality of measurements and data collections.

11.3.2.7 Suboptimal Sensor Placement

Depending on the application, the sensor can return significantly different results depending on the placement [67]. Skin temperature readings, for example, may significantly vary depending on where the sensor was located. The system should ideally provide an autonomous detection of suboptimal sensor placement, but this is rather complex to obtain, and one often needs to rely on the cooperation of the patient and on the ability of the medical staff.

11.3.2.8 Absence of Ground Truths

Any clinical study or investigation involving new systems or prototypes should, in principle, be carried out together with previous state-of-the-art devices in order to have a ground truth to compare and validate the results. In several cases, this step is not possible because BSNs enable measurements and data collections in settings that would not allow a direct comparison with wired technologies, for example, in studies on exercising athletes or long-term data collections in outdoor settings [20].

11.3.2.9 Data Selection and Analysis

The introduction of unobtrusive wireless BSNs produces a huge amount of data from several different physiological and ambient sensors. On the one hand, this requires tight synchronization among the different sensors composing a network so to better judge the relevance and significance of the recorded data. On the other hand, this requires efficient statistical methods in order not to overwhelm the mental capacity of medical doctors [51].

11.4 CLINICAL APPLICATIONS

As discussed in earlier sections, only a few systems based on BSNs targeting hospitalized patients have been actually evaluated in clinical settings. The body of work in this area can be categorized in two main classes: the design and evaluation of novel system prototypes and the evaluation of existing commercial wireless systems for patient monitoring. We consider both classes in turn.

11.4.1 System Prototypes

Most prototypes of BSNs that were evaluated in clinical settings have a short operational lifetime (from a few hours to a few days) and involve only a limited number of patients. All deployed systems focus on continuous vital sign monitoring and aim at replacing and augmenting existing wired systems and manual procedures, with the common goal of providing a more detailed undisrupted monitoring of the patients' vital signs. In the following, we distinguish three areas of application: monitoring of patients during transport, monitoring of unattended patients, and long-term monitoring of hospitalized patients.

11.4.1.1 Monitoring Patients during Transport

Hospitalized patients often need to be moved during their stay at the hospital. Today's wired monitoring systems are often a disturbance for the involved staff and are at risk of being intentionally or accidentally disconnected during the transport, leading to temporal disruptions of the monitoring. Wireless systems could reduce these problems and ensure a hassle-free continuous monitoring of the patients.

An early example of such a system is the wireless monitoring system for patient transports designed by Lin et al. [87]. The proposed system consists of one or more personal digital assistants (PDAs) connected to a central management system. Each PDA is equipped with sensors for the heart rate, the blood oxygen level, and ECG that are attached via a custom sensor board. A user-writable flash memory allows to cache or store measurement results locally. The management server is responsible for the persistent storage and visualization of the recorded data. The monitoring systems and the management system communicate with TCP/IP over an IEEE 802.11b wireless local area network (LAN), employing off-the-shelf access points. The standard wireless local area network (WLAN) security mechanisms are supplemented by advanced encryption standard (AES)-based end-to-end encryption, to protect the patient's sensitive data from eavesdropping. The system was demonstrated at the emergency department of the National Taiwan University Hospital (Taipei, Taiwan) over the course of 1 month. The demonstration was supplemented by a survey among the involved medical personnel, which yields an overall positive feedback on the usability of the system. Ratings of the system considerably outperformed those of the traditional monitoring system.

11.4.1.2 Monitoring of Unattended Patients

Nowadays, patients waiting for admission at a hospital unit are typically not monitored extensively for a sudden deterioration in their health status, and BSNs have been proposed to achieve a cost-effective unobtrusive monitoring. The scalable medical alert response technology (SMART) project [34] proposes a novel system for monitoring vital signs of unattended patients in hospitals. The system consists of wireless modules to monitor vital signs and the position of patients, a central monitoring system, decision and logistic support modules, and portable display units. The patient monitoring modules are based on commercial sensors for ECG and blood oxygenation attached to a PDA. The patient monitoring module is supplemented by a commercial indoor localization system. Both systems send their data to a central monitoring system. The monitoring system provides permanent storage of the recorded data, but also allows real-time visualization of the patient's vital signs. In addition, the system may also automatically monitor the recorded data for specific trigger conditions that might indicate a deterioration of the patient's health status. These decisions may incorporate information from the data streams of different sensors. In addition to critical medical decision making, the system is also able to detect a number of technical faults, such as disconnected sensors or depleted batteries. Based on the location of the patient as recorded by the localization subsystem, critical conditions are automatically reported to the closest appropriate caregiver currently available. If a caregiver does not respond on time, the alarm is relayed to an alternate receiver. The dispatch of alarms can be adjusted to the usual workflow of the hospital. To enable reception of alarms and remote access to relevant vital signs of patients, each caregiver is equipped with a caregiver module. Like the patient modules, these modules are also PDA based. To ensure the privacy of the patient's medical records, the use of the caregiver module requires authentication with a PIN and no data are permanently stored on the device. Both the patient and the caregiver module are communicated with the central system via a standard Wi-Fi network. While the communication with the caregiver modules is encrypted, data protection is not seen as necessary for the raw sensor data transmitted by the patient module as it does not contain any information that would allow to identify a patient. The system was evaluated in a pilot study at the waiting area of the emergency department of the Brigham and Women's Hospital in Boston, MA, United States. The study included 145 patients that were monitored between 5 min and 3 h. In contrast to the original system design, the data were not directly used by medical personnel due to regulatory constraints. Instead, the system was monitored by a specifically trained paramedic who evaluated all alarms. Significant events were communicated to the caregivers. In three cases, the system was able to detect an abnormal condition in a patient's vital signs. In all three cases, the alarm was confirmed by a manual examination, and the patient was immediately admitted to the emergency department. According to the authors, the system was met with an unexpectedly positive response by the patients. Only three patients withdrew from the experiment because of discomfort, and in a survey among 94 of the involved patients, 93% indicated that they would wear the system again [34].

11.4.1.3 Monitoring of Hospitalized Patients

The prevalent use case for wireless monitoring systems is continuous monitoring of already hospitalized patients. These systems are supposed to provide a more fine-grained monitoring of patient's vital signs than with current procedures. At the same time, these systems should reduce the workload of the involved hospital staff, by requiring less manual attendance.

An early system of this class is the MEDiSN system developed by Ko et al. [74]. The authors envision the system not only to be used at hospitals but also to monitor injured persons at disaster sites. Nevertheless, the system was only evaluated in clinical settings. The MEDiSN system consists of a number of monitoring units that record and transmit the patient's physical data, a number of relay nodes, and one or more gateways. A back-end server persistently stores the recorded data and provides access to it via representational state transfer (REST)-based web services. The sensors used in the deployment allowed pulse oximetry and pulse rate measurements. In addition to the sensors, each monitoring unit is also equipped with a display and four user buttons that enable limited direct interaction. The relay units form a routing tree employing the collection tree protocol [50] to forward the recorded data to one of the gateways. Monitoring units do not participate in routing, but simply send the data to the relay unit with the best link quality. The system is also capable of supporting downstream traffic and mobile nodes that roam between different relay nodes. Especially notable is the extensive use of encryption in order to ensure privacy of the sensitive medical data of the patients. All wireless communications are encrypted according to the advanced encryption standard. Two WSN pilot deployments in a clinical setting were carried out by Ko et al. [74] to evaluate MEDiSN. In the first deployment at the operating rooms and the postanesthesia care unit of the University of Maryland Shock Trauma Center, an area of approximately 1000 m^2 was covered by eight relay nodes. Up to eight patients at a time were equipped with measurement units that recorded blood oxygen levels and pulse rate. The patients wore the nodes between 90 min and several hours. These patients were successfully monitored over the extended period of time and despite their movements through the unit with an average reception rate of 98.25%. The second deployment at the emergency room of Johns Hopkins Hospital in Baltimore, MD, United States, covered an area of about 600 m^2 with eight relay nodes. This deployment was active for only 10 days during which 46 patients were equipped with monitoring units while waiting for their treatment. Like in the University of Maryland deployment, blood oxygen levels and pulse rate were sampled. For 14 of the patients, these measurements were also taken with a conventional instrument. The comparison indicates a high correlation to the records taken by the measurement nodes. Network reliability was comparable to the first deployment with an average reception rate of 95.43%.

Chipara et al. [30,96] developed a similar patient monitoring system for a step-down hospital unit. Such step-down units are usually responsible for patients that do not require intensive care, but are still at a higher risk than suitable for a regular hospital unit. Like MEDiSN, the system is optimized for low-data-rate sensors, like pulse sensors. The authors claim that low-data-rate sensors are sufficient for many medical monitoring tasks. This specialization on low data rates allows to use more power-efficient wireless communication technologies, resulting in extended battery life of the sensor nodes. Each sensor node was equipped with sensors for pulse and blood oxygen saturation. Both sensors were sampled at either 30 or 60 s intervals. The data were forwarded to a base station through a number of static relay nodes carefully placed in the step-down hospital unit. To improve the reliability of the system, sensor nodes did not participate in routing and only used a custom protocol to select the best relay node. The grid-powered relay nodes used a multihop routing scheme based on the collection tree protocol [50] to form a backbone network. By placing a sufficient number of relay nodes, the system could also support patient mobility. Patients could hence be monitored even when they left the unit for diagnostic testing. The architecture shares some striking similarities with the one employed in MEDiSN (see earlier). Both systems employ a relay network that is exclusively responsible for message routing. In both cases, message routing builds on top of the TinyOS collection tree protocol. The clinical monitoring system was deployed in a step-down unit of the Barnes-Jewish Hospital in Saint Louis, MO, United States, in the year 2009. Over a time period of 8 months, the system was evaluated on 46 patients. A distinctive feature of this study was the thorough analysis of the system's reliability. On the one hand, the network performed pretty well and delivered more than 99% of the data to the base station. Network disconnections only rarely lasted more than 2.5 min. On the other hand, the quality of the data provided by the sensor was affected by several factors, such as the mobility of the patient, the disconnection of the sensors, and the nonoptimal placement of the pulse oximeters. Consequently, the overall reliability of the system was lower than indicated by the high network reliability. Only for one-half of the patients, more than 80% of the scheduled sensor readings were successfully delivered to the base station. Sensor disconnections typically caused long-term failures and could hardly be noticed by the patients themselves. Complete sensor outages longer than 30 min were observed for more than 40% of the patients and lasted up to

14 h. To limit the effect of sensor disconnections, the authors propose an approach for quick and automatic detection of sensor disconnection. If a sensor disconnection is detected, an alarm is triggered to alert the hospital staff to check the sensor. In addition to long-term sensor outages, patient movements such as tapping or fidgeting also lead to short-term sensor failures resulting in incorrect readings. To mitigate the effect of short-term failures, the authors discuss the use of oversampling to increase sensing reliability. By taking samples at a higher frequency than required, the likelihood to record a valid sample in the original interval increases, but oversampling also significantly decreases the battery lifetime of the node. The deployment demonstrates that the reliability of medical BSNs does not only depend on the reliability of the network but is often predominantly affected by the reliability of the employed sensors. Nevertheless, the authors could reach a reliability that is comparable to commercial wired pulse oximeter systems. This demonstrates the feasibility of monitoring patients with a BSN in a clinical setting.

With LOBIN, López et al. [91] developed a generic monitoring system for patients in hospital environments that is especially noteworthy for the use of smart textiles and the integration of a patient tracking system. The LOBIN system allows to monitor a number of different vital signs, like heart rate and ECG. The use of smart textiles makes the creation of less-invasive and more comfortable sensors possible. The sensors do not need to be separately attached to the body, but are already included in the clothes worn by the patient. Each patient unit of the prototype system consists of a smart shirt with sensors that allow to measure the bioelectrical potential of the patient. The shirt is connected to a wearable data-acquisition device that transfers the measurement data wirelessly over an IEEE 802.15.4 network [64]. A three-axis accelerometer and a thermometer are integrated with the wearable data-acquisition device. The data are relayed by a number of distribution nodes employing dynamic source routing [66] and finally stored at a central management subsystem. The individual patient units do not participate in routing, but instead just associate with a single reachable distribution node. In addition to persistent storage, the central management subsystem also provides a graphical user interface to query and configure the system. To support localization and tracking, each patient wears a separate unit that communicates with beacon units distributed over the hospital in order to determine the current position of the patient. The individual patient units could reach a battery lifetime of up to 9 h during the evaluation. The complete system was deployed and evaluated at the Cardiology Unit at La Paz Hospital, Madrid, Spain. The evaluation involved five patients wearing the system for a 24 h period. The author considered the experiment successful. Packet losses stayed below 2%, and the quality of the visualized data was approved by the hospital doctors. Acquiring the clinically required signal quality with smart textiles turned out to be a challenge especially if patients were moving and required additional measures, such as the application of conductive gel.

11.4.2 Commercial Off-the-Shelf Systems

A number of wireless monitoring systems for hospital use are commercially available, and a few independent scientific studies have analyzed their reliability and accuracy during clinical trials. These commercial systems are typically simpler than the prototypes described in the previous section and typically consist of individual sensors with the capability of Wi-Fi or Bluetooth communication.

11.4.2.1 Evaluation of Arterial Blood Pressure Sensors

Øyri et al. [114] compared the performance of a wireless connected arterial blood pressure sensor with a conventional wired sensor under realistic conditions in an operation room. The examined WisMos wireless sensor employs Bluetooth to transmit the recorded data to a single laptop computer located in the operation room. The system was employed on four patients undergoing laparoscopic abdominal surgery. A conventional wired sensor was employed in parallel according to the medical protocol established at the hospital. The simultaneous use of both systems allowed to compare the measurements of both systems. An analysis of the data did not surface a significant difference between the two systems, and the authors conclude that the WisMos prototype could be a suitable replacement for traditional wired sensors in regard to data quality. Privacy is also identified as an important aspect to be considered when employing wireless sensors. Nevertheless, the built-in encryption functionality of the Bluetooth standard is deemed sufficient. In addition, the data recorded during the experimental study cannot be directly attributed to a specific patient, so that the effect of a possible privacy violation is limited.

11.4.2.2 Evaluation of Several ECG and Photoplethysmogram Systems

Bonnici et al. [23] evaluated four off-the-shelf wireless monitoring systems recording ECG and photoplethysmograms (PPGs). While several similar devices targeting the fitness market are readily available, only few devices meeting clinical standards are offered. In their experiment, four devices were tested under realistic conditions in a hospital unit. One device was a stand-alone ECG sensor, one was a stand-alone pulse oximeter, and two were combi devices that provide both functions. The two stand-alone devices were evaluated together as a single monitoring system comparable to the combi devices. The study does not reveal the exact types and manufacturer of the involved devices as it is only intended as a generic feasibility study. The recorded data were relayed to a central PC for later processing via Bluetooth and a Bluetooth-enabled smartphone. The smartphones were connected to the hospital-wide Wi-Fi network. For the remaining device, Bluetooth was activated during the experiments, but its software did not support direct streaming of the data. In addition to life streaming, all devices also recorded all data on local memory. The evaluation involved 31 patients. Each device type was worn by eight to twelve different patients for a period of 24 h. Only in three out of the 31 test runs that full ECG and PPG data could be collected for the full duration of 24 h. Five of the test runs were struck by severe data transmission losses. In six studies, the devices were preemptively removed by the patient due to discomfort. The remaining failures were due to low battery lifetime. The actual lifetime turned out to be significantly lower than proposed by the manufacturers. In addition, it turned out that the actual battery status was not indicated prominently enough and employing confusing units that are not clearly understandable for an average user. Based on the poor performance in realistic environments, the authors of the study conclude that commercial wireless monitoring devices are not ready for clinical applications. To be a viable choice for a broad set of applications beyond niche markets, like sleep laboratories, wireless monitoring devices need to be more reliable and easier to use. The additional benefit needs to be clear in order to justify the additional complexity.

11.5 APPLICATIONS IN MEDICAL RESEARCH

Medical researchers often struggle with the lack of realism in their routine data collections, often carried out in hospital environments or specialized research laboratories. Patients or volunteers being monitored in such settings suffer the so-called white coat syndrome [21] and do not behave as in their daily life both because of the anxiety and stress associated to the clinical visit and because of the different environments they are exposed to. Controlled lab settings for long-term monitoring of body parameters are indeed ultimately artificial environments often kept at constant temperature and isolated from external noise, in which subjects are requested to follow a prescribed list of activities, while having minimal social interaction and freedom of movement. Although this allows a direct supervision and control on the subjects and avoids external biases on the data, it involves high overhead and costs that significantly restrict the amount of tests and limit their duration, and it does not allow the same realistic profiling of body parameters that would be obtained when the subjects are living their daily life.

Within this context, the ongoing proliferation of wireless technologies that can be seamlessly worn or integrated into clothes plays an important role in improving the realism and quality of data collections. The development of tiny wireless physiological sensors measuring body temperature, as well as cardiovascular, respiratory, and muscle activity in freely moving people [5,6,26,43,65,82,111,120], enables accurate profiling of body parameters over several days in realistic environments and provides insights that cannot be replicated in controlled clinical and laboratory settings [51,75].

BSNs can hence find potential application in most branches of medical research: behavioral researchers have the possibility to study patients with medical or psychological disorders outside the clinic [155], as well as capture how stress influences smoking and alcohol consumption [42]. Chronobiologists can derive from long-term temperature measurements an accurate profile of the circadian-system activity on freely moving patients [21]. In sports medicine, the introduction of wireless systems allows to continuously collect information from training athletes for a better understanding of thermoregulation during exercise [20].

In this section, we report several applications of networked wireless body sensors that have revolutionized routine data collections of medical researchers.

11.5.1 Studies on Human's Thermoregulation

A large number of medical researchers are working towards a better understanding of human's thermoregulation, that is, of the mechanisms and control systems used by the human's body to keep its temperature certain boundaries, balancing thermal inputs and thermal losses. Thermoregulation affects human's behavior and varies depending on factors such as gender [69], age [162], and obesity [47], and it is highly complex to model accurately.

11.5.1.1 Thermoregulation during Exercise

In sports medicine, a precise knowledge of core body temperature (i.e., temperature in inner body and organs) during exercise can be used to optimize the athletic performance and to prevent injuries such as hypothermia, hyperthermia, and heat stroke. Although it is well known that core body temperature increases during exercise, factors such as climate, dehydration, (inappropriate) dressing, and high metabolic rate may significantly vary the body's responses, as well as increase or decrease the risk of circulatory collapse [41]. Hence, a thorough study on thermoregulation during prolonged exercise and extreme conditions, such as in marathons and iron-man races, needs to be carried out in realistic racing settings.

In this context, research has evolved significantly in the last century. Since the beginning of the twentieth century, researchers have shown that core body temperature can increase above 40° in marathon athletes immediately after the finish line [98,100,129], but no information was available about the internal body temperature during the race. Researchers have undertaken studies in which marathon runners were followed by moving vehicles matching their speed in order to connect the rectal probes of the runners to measuring devices every 9 min [98]. Runners were actively involved in the measurement process by manually checking the correct position of the probe and by placing thermistors in predefined body areas (thigh, finger, forehead, and upper arm) during the race to measure their skin temperature.

In later years, the introduction of ingestible pill telemetry systems [17,27,80] allowed the continuous monitoring of gastrointestinal temperature in several runners at the same time, without the need of wires or probes. Although the measured temperature is strongly affected by the ingestion of cold liquids or food [156,160], time of ingestion [27], limited lifetime (about 36 h between ingestion and expulsion), and continuous motion of the sensor that makes readings sensible to temperature gradients along the gastrointestinal tract [27], the ingestible wireless pill has become extremely popular among medical researchers. Communication-wise, ingestible pill telemetry systems are very limited in their connectivity range [27], that is, less than a meter in commercial platforms such as VitalSense (Mini Mitter Co., Inc.) and CorTemp (HQ Inc.). Studies have also revealed nonnegligible data loss during sleeping due to receiver falling out of range [102,160], as well as due to the presence of interference [27].

More recent studies have attempted to measure body temperature in an even less-invasive fashion [20,135]. Boano et al. have developed a telemetric system that continuously measures the core body temperature of marathon runners inside the ear by means of an infrared thermopile sensor pointing at the tympanic membrane [20]. Key feature of this telemetric system is the possibility to monitor the body parameters of the runners remotely during the race, thanks to the integration of the wireless wearable sensors with mobile phones that transmit the measurements to a remote database. Together with core body temperature measured in the ear, the system (that is similar to a common headphone and runs the Contiki operating system [39]) measures skin and environmental temperature using high-precision negative temperature coefficient (NTC) thermistors [20]. The integration of ambient and context sensors in a single system allows medical researchers to understand and correlate the impact of the environment on body temperature. For example, the system described in [20] can be used to understand the role of cold weather and winds on skin and tympanic temperature, aspects that in the past were investigated by artificially applying cold stimulus on the face [133] or by keeping subjects into cold or hot thermal chambers [38].

11.5.1.2 Monitoring of Circadian Rhythms of Body Temperature

In chronobiology, long-term measurements of body temperature are used to derive an accurate profile of the circadian-system activity, that is, the 24 h cycles in the physiological processes of humans [21]. Several researchers have analyzed the body temperature on subjects in specialized laboratories and showed that it follows strong circadian rhythms over the 24 h day, with a nocturnal decrease and a diurnal increase that reaches its maximum in the early evening [77]. The task of chronobiologists is indeed to acquire a deeper understanding of the phase, amplitude, and stability of circadian rhythms, as well as to study its correlation with factors such as age [97], gender, food intake, exercise, and sleep deprivation [21].

The commercialization of the first wearable data loggers has revolutionized experimentation in this field. The Maxim iButton, despite being a mere data logger with a limited lifetime that cannot provide immediate or remote feedback, has been used in several clinical studies to monitor the circadian rhythms of skin temperature for several days [57,76,86,136]. For example, iButtons were used to record the wrist temperature of the nondominant hand of 99 volunteers to derive their circadian rhythms [136]. Temperature was measured every 10 min, and none of the participants of the study reported discomfort in maintaining their usual lifestyle throughout the experiment.

Other commercial devices used to record body temperature in clinical settings are the Innovatec TherCom and the SenseWear BodyMedia. TherCom is a portable device that allows to measure temperature using several channels with different temperature ranges and motion using a three-axis accelerometer [139]. It has been used to monitor central and peripheral temperature on 30 patients, successfully measuring the circadian rhythms of temperature measured at the external auditory canal temperature and at the axillary line [152]. BodyMedia is a wearable device that measures the GSR, skin temperature, and caloric expenditure, and it has been used to assess the circadian rhythms of skin temperature on patients with acute coronary syndrome [1] and during fasting [12].

The high demand and vast use of these devices has triggered several research projects prototyping BSNs that could promptly provide either immediate feedback to the clinicians or remote monitoring capability during long-term data collections [37]. Rodrigues et al. [131] have developed an intravaginal temperature sensor based on the Shimmer platform [26] and show its usefulness in monitoring the ovulation period and detection of pregnancy contractions. The sensor logs all the temperature measurements in the local SD card and, whenever a Bluetooth connection to a base station is available, transfers the collected temperature measurements to the computer for real-time monitoring and analysis. Similarly, Boano et al. [21] have developed an IEEE 802.15.4 BSN to monitor the impact of sleep deprivation on the circadian temperature rhythms. Their system consists of low-power tiny sensor nodes measuring skin temperature using calibrated negative temperature coefficient (NTC) thermistors and transmitting the measurements wirelessly to a more powerful sink node that can be worn in a pocket. Whenever the sink node enters the range of a base station, it starts forwarding the collected samples to the laptop for a remote feedback to the medical team. Dementyev et al. [36] eliminate the need for a dedicated base station by designing a skin temperature monitoring system that provides immediate remote feedback through miniature wearable sensor nodes connected via Bluetooth to a cellular phone with Internet connectivity. This device has a lifetime of approximately 135 h when sampling at a rate of 0.2 Hz, which is comparable with [21] and longer than the average lifetime of an iButton (48 h) or other commercial devices [152]. This hints that the design of devices that seamlessly collect data from subjects in realistic settings privileges miniaturization and comfort at the price of a smaller battery size and system lifetime.

A precise knowledge of skin temperature rhythms can also give important information to chronotherapists in order to personalize and optimize the timing of drug delivery in cancer patients. Using VitalSense wireless skin temperature patches by Philips Respironics, Scully et al. have shown that the efficacy of anticancer medications improves significantly when personalizing the timing of chemotherapy delivery based on circadian timing [137].

11.5.2 Behavioral and Psychological Studies

Psychophysiologists and behavioral scientists investigate the relationships and links between psychological events and physiological and behavioral reactions of individuals. By carefully analyzing the behavioral, cognitive, affective, and physiological responses of subjects, they can study, among others, depression and anxiety disorders (the most prevalent class of psychiatric disorders nowadays [155]), as well as the influence of stress on smoking and alcohol consumption.

In this context, experiments in ambulatory settings offer several advantages over the laboratory. On the one hand, experiments in laboratory settings have a limited duration, making it difficult to detect specific events (e.g., spontaneous panic attacks). On the other hand, in psychological studies, the exposure to a novel environment induces anxiety that makes it difficult to generalize the obtained results [155]. Therefore, an increasing number of behavioral and psychological studies are carried out in supportive ambulatory environments and involve more and more the use of networks of portable wireless sensors designed to autonomously determine people's behavior, emotional state, adherence to medical recommendations, or interaction with objects in the environment [25,45,151].

11.5.2.1 Monitoring Depression and Anxiety or Stress Disorders

Several systems have been proposed to measure respiratory changes or peripheral responses (such as variations of skin temperature and resistance) as indicators of brain activity, state of mind, or psychological state, on subjects suffering depression, anxiety, or mental illness [18,45,68,149].

The most advanced systems are able to measure the three most significant emotional responses at the same time, namely, physiology, language, and motion. The LifeShirt [155], for example, is an example of textile sensor network worn on the body that can record respiratory, cardiovascular, and electrodermal activity, as well as motion and speech. Made of an array of sensors embedded in a sleeveless undergarment, the shirt has been proposed to monitor respiratory changes in patients with psychological disorders in ambulatory environments (e.g., breathing irregularities during panic attacks) [155]. Sung et al. [148,149] have designed the LiveNet system and shown its applicability in diagnosing and monitoring depression over long periods of time. In particular, they have shown that it is possible to classify depression states of subjects by processing data sampled by accelerometer sensors measuring body movements and that it is possible to accurately predict a state of depression using a combination of voice, physiological, and behavioral features.

Other researchers have attempted to monitor the activity of subjects and collect their behavioral patterns by enriching their home with several wireless sensors [16,46]. For example, Barger et al. [16] have created a smart home using simple and inexpensive motion sensors to identify the regular patterns of most activities of daily living, such as sleeping, cooking, or using the toilet, and capture any deviation in a subject's normal activity pattern.

Other studies have tried to measure and quantify stress on a large number of subjects by measuring their GSR [126] or to deduct its causes and consequences [42]. For example, Ertin et al. have developed AutoSense [42], a wireless suite that includes, among others, ECG, GSR and temperature sensors, and a respiratory inductive plethysmograph to measure relative lung volume and breathing rate. Their system has been used to collect hundreds of hours of data from subjects living their daily life and to derive models that capture the physiological and psychological effects of stress. Such models allow to continuously predict stress from physiological measurements and have shown to be fairly accurate, achieving up to 90% accuracy. AutoSense has also been used to relate how stress influences smoking and alcohol consumption using sensors that can collect the timing of smoking events, puff characteristics, as well as nicotine intake [42].

11.5.2.2 Adherence and Compliance Monitoring

A critical issue in noncontrolled settings is often to monitor the behavior of the participants to a given study and in particular to verify adherence and compliance to a given protocol or prescribed regimen of treatments. Even though test subjects are typically paid, it is frequent that they do not comply with a number of unpleasant instructions (e.g., sleeping only 4 h/day [21]). Furthermore, the effectiveness of a treatment is significantly reduced if prescribed regimens are not thoroughly followed, and studies have shown that average adherence rates are as low as 43% in patients receiving treatment for chronic conditions [32]. As noncompliance to a given protocol from the participants in a study leads to biased or incorrect data collections, it is extremely important to detect the occurrence of these events.

Pervasive wireless sensors used in long-term data collection in home environments and in freely moving people should indeed provide means to quantify the suboptimal adherence and compliance to specifications, which would otherwise be very difficult to model or derive [4]. In this regard, several solutions based on sensor networks and radio frequency identification (RFID) technologies have been proposed in the past years [2]. Ho et al. [62] have proposed to achieve remote medication intake monitoring by means of RFID tags identifying the medicine bottle and of a weight scale connected to a mote to verify the amount of medicine that was removed from the bottle. Similarly, Lopez et al. have also made use of smart RFID packaging to identify when a pill was removed, by means of a rupture of the electric flow into the RFID's integrated circuit [92]. Pang et al. [115] proposed instead the use of an intelligent medicine package sealed by controlled delamination material to control the dosage for a given medication (the medicine package can only be opened by the controlling device by applying a certain voltage). To monitor the exact nutritional intake in obesity and nutrition-related studies, Päßler and Fischer [122] have developed a method to monitor in a noninvasive fashion the food intake behavior by integrating in-ear microphone and a reference microphone in a hearing aid case and evaluating the chewing and swallowing sounds. Similarly, Liu et al. [88] have developed a headset embedding microphone and camera to autonomously provide detailed information about a subject's dietary habits.

11.5.3 Studies on Motor Diseases and Rehabilitation

Studies of motor diseases and rehabilitation progress after surgery are also often carried out in specialized laboratories. As their use is impractical, expensive, and labor intensive, a few researchers have proposed the use of BSNs to monitor postoperative recovery after surgery or to assess the severity of symptoms and motor complications in patients with Parkinson's disease.

11.5.3.1 Studies on Postoperative Rehabilitation

Atallah et al. [7] have followed the improvement in gait following knee arthroplasty surgery on eight patients. Using e-AR [89], a wearable miniaturized ear-worn sensor embedding a three-axis accelerometer, the authors observed activities such as sitting, walking, and standing 1, 3, 6, 12, and 24 weeks after the operation, respectively. They have collected measurements and shown that it is possible to capture the progressively more regular gait patterns of subjects over time, and that it would be possible to identify patients who fail to improve as expected, facilitating early clinical review and intervention. To analyze the walking pattern, the authors used discrete wavelet transform analysis for its compact representation of the signal in time and frequency and its computational efficiency [7].

Wu et al. [159] have designed MEDIC, a wearable sensor system that provides medical researchers with real-time streaming of high-bandwidth data from remote patients with motor disabilities. Wearable sensors are equipped with Bluetooth radios and include, among others, triaxial accelerometers and gyroscopes for gait analysis. A standard cellular phone or PDA equipped with Bluetooth forwards the sensor data directly to the servers at hospital or clinics for real-time monitoring of the patient. In their experiments, the authors have used the Nokia N770 PDA cellular phone because of its open software interface and have shown that they can reliably diagnose whether the gait is abnormal and if the patient is limping. In particular, through local signal processing, MEDIC can diagnose the conditions of the subject while selecting the minimum and optimal set of sensors and specify on which leg the patient is limping and which is the severity level of the limp.

Cellular phones are also exploited in [81] to develop a system that provides vibrotactile cues of body tilt. Using the embedded triaxial accelerometer of an iPhone and connecting a tactor bud (consisting of a controller, battery, and two tactors) through the headphone jack, Lee et al. have designed a balance trainer system producing real-time feedback to reduce body sway. Although the system is designed to provide real-time feedback, it can also store motion data for a subsequent analysis from the medical personnel. Preliminary tests on nine subjects have shown that participants had indeed a more stable body when the vibrotactile feedback was available, even in the most challenging postures, such as tandem Romberg.

Although the original goal of the three systems described earlier is to shift the monitoring of subjects outside hospital facilities and specialized laboratories for movement analysis, they can play an important role for medical research as well as they provide a simple and noninvasive way to compare, for example, the recovery of patients using different surgery techniques.

11.5.3.2 Studies on Motor Fluctuations

The study of Parkinson's disease requires the recording of high-bandwidth data, such as from accelerometers and electromyographs, for extended time periods, so BSNs may not represent the best choice since such high data rates exceed the capabilities of most low-power radio devices. To cope with this problem, Lorincz et al. [93] developed the Mercury system, which employs local feature extraction to reduce the amount of data that needs to be transmitted. Per default, the Mercury nodes continuously sample and store sensor data in a flash card. Using a reliable transfer protocol based on acknowledgment messages, the end user can extract selected raw data traces from each node and download them to a server for later analysis. The transfer is triggered remotely by the end user who needs to specify which specific set of data should be collected. In addition to the raw data, the user can also request filtered datasets. To support this mode of operation, Mercury supports local extraction of features from the collected data, which allows to save considerable bandwidth, storage, and energy. Mercury provides a suite of custom feature-extraction algorithms such as maximum peak-to-peak amplitude, mean, and root mean square of the time series that are computed on the fly as sensor data are being acquired [93]. Using Mercury, Patel et al. [123,124] have extracted six types of features from different body segments to verify if accelerometer data can be used to estimate the severity of symptoms and motor complications. The considered features include intensity, modulation, rate, periodicity, and coordination of movement and were shown to be useful to analyze bradykinesia (slowness of movement) and dyskinesia (presence of involuntary movements) using a support vector machine (SVM) classifier.

11.6 CONCLUSIONS

The early years of the twenty-first century have seen an authentic revolution in the provision of healthcare, with a steep rise in the number of systems that monitor the vital signs of patients in their homes or their progress in rehabilitation after surgery to improve the efficiency of therapies. The increasing popularity of WSN technology has indeed triggered the creation of a large number of BSN systems used in a wide range of settings, from human–computer interaction, leisure and gaming, over the management of soldiers in extreme environments, wellness and fitness applications, to the coordination of emergency squads during mass casualty disasters.

This survey has focused on the use of BSNs to monitor hospitalized patients and to carry out unobtrusive data collections for medical research. This field has received relatively little attention from the research community compared to other application areas, due to the high number of technical challenges that systems need to address at the same time and due to the high overhead caused by clinical trials and experimentation on patients. The huge body of work on communication protocols and optimized operations developed by the WSN research community in the last decade was sufficient to address some of these challenges, for example, how to deal with the limited computational resources of sensor nodes and how to achieve reliable communications. Most systems are also able to reliably support mobility, for example, among rooms in hospitals [23,30,34,74,87,91], or during long-term experimental studies in outdoor settings [20]. The pervasiveness of mobile phones and ubiquitous computing technologies has also simplified real-time feedback [20,34,87], and hence the prompt delivery of relevant information to the caregivers or medical researchers.

Several challenges, however, still remain open. Most systems deployed in hospitals and clinical settings have an extremely low operational lifetime, ranging from a few hours to a few days. Energy efficiency has always been a catch-22 dilemma for low-power embedded devices, but in the presence of reliable and encrypted communication, high data rates, and ergonomic lightweight body sensors (and hence the use of batteries with limited capacity), it is very difficult to achieve long-lasting operations. Similarly, security and privacy have been addressed only by some works [34,74], and further techniques are needed to ensure the privacy of the individuals and to protect sensor data from external access or malicious jamming. Systems used on hospitalized patients and to carry out unobtrusive data collections for medical research need still to improve the reliability of sensing. For example, reliably identifying the disconnection of sensors and motion artifacts is still an open problem [30,91]. Finally, one needs to address standardization and interoperability, especially for BSN systems used in hospitals: in current prototypes, indeed, there is often a lack of seamless integration with the network infrastructure and electronic medical records.

ACKNOWLEDGMENT

The research leading to these results has received funding from the European Union Seventh Framework Program (FP7/2007–2013) under grant agreement no. 317826 (RELYonIT: Research by Experimentation for Dependability on the Internet of Things) and 258351 (makeSense: Easy Programming of Integrated Wireless Sensor Networks).

REFERENCES

1. Al-Otair, H., Al-Shamiri, M., Bahobail, M., Sharif, M.M., and BaHammam, A.S.: Assessment of sleep patterns, energy expenditure and circadian rhythms of skin temperature in patients with acute coronary syndrome. *Medical Science Monitor* 17(7), 397–403 (2011).
2. Alemdar, H. and Ersoy, C.: Wireless sensor networks for healthcare: A survey. *Computer Networks* 54(1), 2688–2710 (2010).
3. Özgür Alemdar, H., Yavuz, G.R., Özen, M.O., Kara, Y.E., Incel, Ö.D., Akarun, L., and Ersoy, C.: Multi-modal fall detection within the WeCare framework. In: *Proceedings of the 9th International Conference on Information Processing in Sensor Networks (IPSN)*, Demo Session, ACM, New York, pp. 436–437 (2010).
4. Aloia, M.S., Goodwin, M.S., Velicer, W.F., Arnedt, J.T., Zimmerman, M., Skrekas, J., Harris, S., and Millman, R.P.: Time series analysis of treatment adherence patterns in individuals with obstructive sleep apnea. *Annals of Behavioral Medicine* 36(1), 44–53 (2008).
5. Anliker, U., Ward, J.A., Lukowicz, P., Tröster, G., Dolveck, F., Baer, M., Keita, F. et al.: AMON: A wearable multiparameter medical monitoring and alert system. *IEEE Transactions on Information Technology in Biomedicine* 8(4), 415–427 (2004).
6. Asada, H.H., Shaltis, P., Reisner, A., Rhee, S., and Hutchinson, R.C.: Mobile monitoring with wearable photoplethysmographic biosensors. *IEEE Engineering in Medicine and Biology Magazine* 22, 28–40 (2003).
7. Atallah, L., Jones, G., Ali, R., Leong, J., Lo, B., and Yang, G.Z.: Observing recovery from knee-replacement surgery by using wearable sensors. In: *Proceedings of the 8th International Conference on Body Sensor Networks (BSN)*, IEEE Computer Society, pp. 29–34 (2011).
8. Aylward, R. and Paradiso, J.A.: Sensemble: A wireless, compact, multi-user sensor system for interactive dance. In: *Proceedings of the International Conference on New Interfaces for Musical Expression (NIME)*, IRCAM—Centre Pompidou in collaboration with Sorbonne University, Paris, France, pp. 134–139 (2006).
9. Aziz, O., Atallah, L., Lo, B., Gray, E., Athanasiou, T., Darzi, A., and Yang, G.Z.: Ear-worn body sensor network device: An objective tool for functional postoperative home recovery monitoring. *Journal of the American Medical Informatics Association* 18(2), 156–159 (2011).
10. Baccour, N., Koubâa, A., Mottola, L., Youssef, H., Zúniga, M.A., Boano, C.A., and Alves, M.: Radio link quality estimation in wireless sensor networks: A survey. *ACM Transactions on Sensor Networks (TOSN)* 8(4), 1–33 (2012).
11. Bächlin, M., Förster, K., and Tröster, G.: SwimMaster: A wearable assistant for swimmer. In: *Proceedings of the 11th International Conference on Ubiquitous Computing (Ubicomp)*, ACM, New York, pp. 215–224 (2009).
12. BaHammam, A., Alrajeh, M., Albabtain, M., Bahammam, S., and Sharif, M.: Circadian pattern of sleep, energy expenditure, and body temperature of young healthy men during the intermittent fasting of Ramadan. *Appetite* 54(2), 426–429 (2010).
13. Baldus, H., Klabunde, K., and Müsch, G.: Reliable set-up of medical body-sensor networks. In: *Proceedings of the 1st European Workshop on Wireless Sensor Networks (EWSN)*, Springer-Verlag, Berlin, Germany, pp. 353–363 (2004).
14. Bamis, A., Lymberopoulos, D., Teixeira, T., and Savvides, A.: The BehaviorScope framework for enabling ambient assisted living. *Personal Ubiquitous Computing* 14(6), 473–487 (2010).
15. Bannach, D., Amft, O., Kunze, K.S., Heinz, E.A., Tröster, G., and Lukowicz, P.: Waving real hand gestures recorded by wearable motion sensors to a virtual car and driver in a mixed-reality parking game. In: *Proceedings of the 3rd IEEE Symposium on Computational Intelligence and Games (CIG)*, IEEE, pp. 32–39 (2007).

16. Barger, T.S., Brown, D.E., and Alwan, M.: Health-status monitoring through analysis of behavioral patterns. *IEEE Transactions on Systems, Man, and Cybernetics. Part A: Systems and Humans* 35(1), 22–27 (2005).
17. van Bladel, C.: Faster marathon times by measuring human performance. In: *Proceedings of the 7th Conference on Methods and Techniques in Behavioral Research*, ACM, New York, pp. 21–23 (2010).
18. Blum, J. and Magill, E.: M-psychiatry: Sensor networks for psychiatric health monitoring. In: *Proceedings of the 21st IEEE Conference in Biomedical Engineering and Medical Physics (PGBioMed)*, Liverpool John Moores University, School of Computing & Mathematical Sciences, Liverpool, United Kingdom, pp. 21–25 (2009).
19. Boano, C.A., Brown, J., Tsiftes, N., Roedig, U., and Voigt, T.: The impact of temperature on outdoor industrial sensornet applications. *IEEE Transactions on Industrial Informatics* 6(3), 451–459 (2010).
20. Boano, C.A., Lasagni, M., and Römer, K.: Non-invasive measurement of core body temperature in marathon runners. In: *Proceedings of the 10th IEEE International Conference on Wearable and Implantable Body Sensor Networks (BSN)*, IEEE Computer Society, pp. 274–279 (2013).
21. Boano, C.A., Lasagni, M., Römer, K., and Lange, T.: Accurate temperature measurements for medical research using body sensor networks. In: *Proceedings of the 2nd International Workshop on Self-Organizing Real-Time Systems (SORT) in Conjunction with the 14th IEEE International Symposium on Object, Component, Service-Oriented Real-Time Distributed Computing (ISORC)*, IEEE Computer Society, pp. 189–198 (2011).
22. Boano, C.A., Wennerström, H., Zúñiga, M.A., Brown, J., Keppitiyagama, C., Oppermann, F.J., Roedig, U., Nordén, L.Å., Voigt, T., and Römer, K.: Hot Packets: A systematic evaluation of the effect of temperature on low power wireless transceivers. In: *Proceedings of the 5th Extreme Conference on Communication (ExtremeCom)*, ACM, New York, pp. 7–12 (2013).
23. Bonnici, T., Orphanidou, C., Vallance, D., Darrell, A., and Tarassenko, L.: Testing of wearable monitors in a real-world hospital environment: What lessons can be learnt? In: *Proceedings of the 9th International Conference on Wearable and Implantable Body Sensor Networks (BSN)*, IEEE Computer Society, pp. 79–84 (2012).
24. Brodie, M., Walmsley, A., and Page, W.: Fusion motion capture: A prototype system using inertial measurement units and GPS for the biomechanical analysis of ski racing. *Sports Technology* 1, 17–28 (2008).
25. Bromundt, V., Wirz-Justice, A., Kyburz, S., Opwis, K., Dammann, G., and Cajochen, C.: Circadian sleep–wake cycles, well-being, and light therapy in borderline personality disorder. *Journal of Personality Disorders* 27(5), 680–696 (2013).
26. Burns, A., Greene, B., McGrath, M., O'Shea, T., Kuris, B., Ayer, S., Stroiescu, F., and Cionca, V.: SHIMMER: A wireless sensor platform for noninvasive biomedical research. *IEEE Sensors Journal* 10, 1527–1534 (2012).
27. Byrne, C. and Lim, C.L.: The ingestible telemetric body core temperature sensor: A review of validity and exercise applications. *British Journal of Sports Medicine* 41(3), 126–133 (2007).
28. Chakravorty, R.: MobiCare: A programmable service architecture for mobile medical care. In: *Proceedings of the 4th IEEE International Conference on Pervasive Computing and Communications Workshops (PERCOMW)*, IEEE Computer Society, pp. 532–536 (2006).
29. Chi, E.H.: Introducing wearable force sensors in martial arts. *IEEE Pervasive Computing* 4(3), 47–53 (2005).
30. Chipara, O., Lu, C., Bailey, T.C., and Roman, G.C.: Reliable clinical monitoring using wireless sensor networks: Experiences in a step-down hospital unit. In: *Proceedings of the 8th ACM International Conference on Embedded Networked Sensor Systems (SenSys)*, ACM, New York, pp. 155–168 (2010).

31. Choi, H., Chakraborty, S., Charbiwala, Z.M., and Srivastava, M.B.: SensorSafe: A framework for privacy-preserving management of personal sensory information. In: *Proceedings of the 8th Workshop on Secure Data Management (SDM)*, Springer-Verlag, Berlin, Germany, pp. 85–100 (2011).
32. Claxton, A.J., Cramer, J., and Pierce, C.: A systematic review of the associations between dose regimens and medication compliance. *Clinical Therapeutics* 23(8), 1296–1310 (2001).
33. Clifford, G.D. and Clifton, D.: Wireless technology in disease management and medicine. *Annual Review of Medicine* 63(1), 479–492 (2012).
34. Curtis, D.W., Pino, E.J., Bailey, J.M., Shih, E.I., Waterman, J., Vinterbo, S.A., Stair, T.O., Guttag, J.V., Greenes, R.A., and Ohno-Machado, L.: SMART—An integrated wireless system for monitoring unattended patients. *Journal of the American Medical Informatics Association (JAMIA)* 15(1), 44–53 (2008).
35. Darwish, A. and Hassanien, A.E.: Wearable and implantable wireless sensor network solutions for healthcare monitoring. *Sensors* 11(1), 5561–5595 (2011).
36. Dementyev, A., Behnaz, A., and Gorbach, A.M.: 135-hour-battery-life skin temperature monitoring system using a Bluetooth cellular phone. In: *Proceedings of the 7th IEEE International Meeting on Biomedical Wireless Technologies, Networks and Sensing Systems (BioWireless) in Conjunction with the IEEE Radio Wireless Week (RWW)*, IEEE Communications Society, Austin, TX (2013).
37. Dolgov, A.B. and Zane, R.: Low-power wireless medical sensor platform. In: *Proceedings of the 28th IEEE International Conference of the Engineering in Medicine and Biology Society (EMBS)*, IEEE, pp. 2067–2070 (2006).
38. Doyle, F., Zender, W.J., and Terndrup, T.E.: The effect of ambient temperature extremes on tympanic and oral temperatures. *American Journal of Emergency Medicine* 10(4), 285–289 (1992).
39. Dunkels, A., Grönvall, B., and Voigt, T.: Contiki—A lightweight and flexible operating system for tiny networked sensors. In: *Proceedings of the 1st Workshop on Embedded Networked Sensors (EmNetS)*, IEEE Computer Society, pp. 455–462 (2004).
40. Eisenman, S.B., Miluzzo, E., Lane, N.D., Peterson, R.A., Ahn, G.S., and Campbell, A.T.: The BikeNet mobile sensing system for cyclist experience mapping. In: *Proceedings of the 5th ACM International Conference on Embedded Networked Sensor Systems (SenSys)*, ACM, New York, pp. 87–101 (2007).
41. Emmett, J.: The physiology of marathon running: Just what does running a marathon do to your body? *Marathon and Beyond* 11(1), 20–36 (2007).
42. Ertin, E., Stohs, N., Kumar, S., Raij, A., al'Absi, M., and Shah, S.: Autosense: Unobtrusively wearable sensor suite for inferring the onset, causality, and consequences of stress in the field. In: *Proceedings of the 9th ACM Conference on Embedded Networked Sensor Systems (SenSys)*, ACM, New York, pp. 274–287 (2011).
43. Espina, J., Falck, T., Muehlsteff, J., Jin, Y., Adán, M.A., and Aubert, X.: Wearable body sensor network towards continuous cuff-less blood pressure monitoring. In: *Proceedings of the 5th IEEE International Symposium on Medical Devices and Biosensors (MDBS)*, IEEE, pp. 28–32 (2008).
44. Fensli, R., Gunnarson, E., and Gundersen, T.: A wearable ECG-recording system for continuous arrhythmia monitoring in a wireless tele-home-care situation. In: *Proceedings of the 18th IEEE International Symposium on Computer-Based Medical Systems (CBMS)*, IEEE, pp. 407–412 (2005).
45. Fletcher, R.R., Dobson, K., Goodwin, M.S., Eydgahi, H., Wilder-Smith, O., Fernholz, D., Kuboyama, Y., Hedman, E.B., Poh, M.Z., and Picard, R.W.: iCalm: Wearable sensor and network architecture for wirelessly communicating and logging autonomic activity. *IEEE Transactions on Information Technology in Biomedicine* 14, 215–223 (2010).

46. Fleury, A., Vacher, M., and Noury, N.: SVM-based multimodal classification of activities of daily living in health smart homes: Sensors, algorithms, and first experimental results. *IEEE Transactions on Information Technology in Biomedicine* 14, 274–283 (2010).
47. Froy, O.: Metabolism and circadian rhythms: Implications for obesity. *Endocrine Reviews* 31(1), 1–24 (2010).
48. Gao, T., Greenspan, D., Welsh, M., Juang, R.R., and Alm, A.: Vital signs monitoring and patient tracking over a wireless network. In: *Proceedings of the 27th IEEE International Conference of the Engineering in Medicine and Biology Society (EMBS)*, IEEE, pp. 102–105 (2005).
49. Gao, T., Massey, T., Selavo, L., Crawford, D., Chen, B., Lorincz, K., Shnayder, V., and Welsh, M.: The advanced health and disaster aid network: A light-weight wireless medical system for triage. *IEEE Transactions on Biomedical Circuits and Systems* 1, 203–216 (2007).
50. Gnawali, O., Fonseca, R., Jamieson, K., Moss, D., and Levis, P.: Collection Tree Protocol. In: *Proceedings of the 7th ACM Conference on Embedded Networked Sensor Systems (SenSys)*, ACM, New York, pp. 1–14 (2009).
51. Goodwin, M.S. and Velicer, W.F.: Telemetric monitoring in the behavior sciences. *Behavior Research Methods* 40, 328–341 (2008).
52. Greenemeier, L.: This really won't hurt a bit: Wireless sensor promises diabetics noninvasive blood sugar readings. http://www.scientificamerican.com/article.cfm?id=wireless-blood-glucose-diabetes, accessed on February 4, 2014 (2010).
53. Guo, D., Tay, F., Yu, L., Xu, L., Nyan, M., Chong, F., Yap, K., and Xu, B.: A wearable BSN-based ECG-recording system using micromachined electrode for continuous arrhythmia monitoring. In: *Proceedings of the 5th IEEE International Symposium on Medical Devices and Biosensors (MDBS)*, IEEE, pp. 41–44 (2008).
54. Halteren, A.V., Bults, R., Wac, K., Konstantas, D., Widya, I., Dokovsky, N., Koprinkov, G., Jones, V., and Herzog, R.: Mobile patient monitoring: The mobihealth system. *Journal on Information Technology in Healthcare* 2(5), 365–373 (2004).
55. Hanson, M.A., Powell Jr., H.C., Barth, A.T., Ringgenberg, K., Calhoun, B.H., Aylor, J.H., and Lach, J.: Body area sensor networks: Challenges and opportunities. *IEEE Computer* 42, 58–65 (2009).
56. Hao, Y. and Foster, R.: Wireless body sensor networks for health-monitoring applications. *Physiological Measurement* 29, 27–56 (2008).
57. Harper-Smith, A., Crabtree, D., Bilzon, J., and Walsh, N.: The validity of wireless iButtons and thermistors for human skin temperature measurement. *Physiological Measurement* 31, 95–114 (2009).
58. Hattangady, S.: Wireless medical devices: Endless possibilities! http://emblazeworld.com/Attachments-Articles/2009-June%20Emblaze%20Consulting%20-%20Wireless%20Medical%20Device%20Revolution.pdf, accessed on February 4, 2014, Emblaze Consulting LLC, McKinney, TX (2009).
59. Heinz, E.A., Kunze, K., Gruber, M., Bannach, D., and Lukowicz, P.: Using wearable sensors for real-time recognition tasks in games of martial arts an initial experiment. In: *Proceedings of the 2nd IEEE International Symposium on Computational Intelligence and Games (CIG)*, IEEE, pp. 98–102 (2006).
60. Hertleer, C., Langenhove, L.V., Rogier, H., and Vallozzi, L.: A textile antenna for off-body communication integrated into protective clothing for firefighters. *IEEE Transactions on Antennas and Propagation* 57, 919–925 (2009).
61. Hester, T., Hughes, R., Sherrill, D.M., Knorr, B., Akay, M., Stein, J., and Bonato, P.: Using wearable sensors to measure motor abilities following stroke. In: *Proceedings of the 3rd International Workshop on Wearable and Implantable Body Sensor Networks (BSN)*, IEEE Computer Society, pp. 5–8 (2006).

62. Ho, L., Moh, M., Walker, Z., Hamada, T., and Su, C.F.: A prototype on RFID and sensor networks for elder healthcare: Progress report. In: *Proceedings of the International Workshop on Experimental Approaches to Wireless Network Design and Analysis (E-WIND)*, ACM, New York, pp. 70–75 (2005).
63. Holleczek, T., Schoch, J., Arnrich, B., and Tröster, G.: Recognizing turns and other snowboarding activities with a gyroscope. In: *Proceedings of the 14th IEEE International Symposium on Wearable Computers (ISWC)*, IEEE, pp. 75–82 (2010).
64. IEEE 802.15.4 Working Group: IEEE standard for local and metropolitan area networks—Part 15.4: Low-rate wireless personal area networks (LR-WPANs), IEEE std 802.15.4-2011 edn (2011).
65. Jin, Z., Oresko, J., Huang, S., and Cheng, A.: HeartToGo: A personalized medicine technology for cardiovascular disease prevention and detection. In: *Proceedings of the 4th IEEE-NIH Life Science Systems and Application Workshop (LiSSA)*, IEEE, pp. 80–83 (2009).
66. Johnson, D.B., Maltz, D.A., and Broch, J.: DSR: The dynamic source routing protocol for multihop wireless ad hoc networks. In: *Ad Hoc Networking*, Addison-Wesley Longman Publishing Co., Inc., Boston, MA, pp. 139–172 (2001).
67. Jovanov, E., Poon, C.C.Y., Yang, G.Z., and Zhang, Y.T.: Body sensor networks: From theory to emerging applications. *IEEE Transactions on Information Technology in Biomedicine* 13, 859–863 (2009).
68. Jovanov, E., Raskovic, D., and Hormigo, R.: Thermistor-based breathing sensor for circadian rhythm evaluation. In: *Proceedings of the 38th Rocky Mountain Bioengineering Symposium (RMBS)*, Copper Mountain, CO (2001).
69. Kaciuba-Uscilko, H. and Grucza, R.: Gender differences in thermoregulation. *Current Opinion in Clinical Nutrition and Metabolic Care* 4(6), 533–536 (2001).
70. Karlsson, M., Forsgren, R., Eriksson, E., Edström, U., Bäcklund, T., Karlsson, J., and Wiklund, U.: Wireless system for real-time recording of heart rate variability for home nursing. In: *Proceedings of the 27th IEEE International Conference of the Engineering in Medicine and Biology Society (EMBS)*, IEEE, pp. 3717–3719 (2005).
71. Kemp, J., Gaura, E.I., Brusey, J., and Thake, C.D.: Using body sensor networks for increased safety in bomb disposal missions. In: *Proceedings of the IEEE International Conference on Sensor Networks, Ubiquitous, and Trustworthy Computing (SUTC)*, IEEE Computer Society, pp. 81–89 (2008).
72. Killeen, J.P., Chan, T.C., Buono, C., Griswold, W.G., and Lenert, L.A.: A wireless first responder handheld device for rapid triage, patient assessment and documentation during mass casualty incidents. In: *Proceedings of the AMIA Symposium*, American Medical Informatics Association, pp. 429–433 (2006).
73. King, R.C., McIlwraith, D.C., Lo, B., Pansiot, J., McGregor, A.H., and Yang, G.Z.: Body sensor networks for monitoring rowing technique. In: *Proceedings of the 6th International Workshop on Wearable and Implantable Body Sensor Networks (BSN)*, IEEE Computer Society, pp. 251–255 (2009).
74. Ko, J., Lim, J.H., Chen, Y., Musvaloiu-E.R., Terzis, A., Masson, G.M., Gao, T., Destler, W., Selavo, L., and Dutton, R.P.: MEDiSN: Medical emergency detection in sensor networks. *ACM Transactions on Embedded Computing Systems (TECS)* 10(1), 1–29 (2010).
75. Ko, J., Lu, C., Srivastava, M., Stankovic, J., Terzis, A., and Welsh, M.: Wireless sensor networks for healthcare. *Proceedings of the IEEE* 98(11), 1947–1960 (2010).
76. Kolodyazhniy, V., Späti, J., Frey, S., Götz, T., Wirz-Justice, A., Kräuchi, K., Cajochen, C., and Wilhelm, F.H.: An improved method for estimating human circadian phase derived from multichannel ambulatory monitoring and artificial neural networks. *Chronobiology International* 29(8), 1078–1097 (2012).
77. Kräuchi, K.: How is the circadian rhythm of core body temperature regulated? *Clinical Autonomic Research* 12(3), 147–149 (2002).

78. Kumar, P. and Lee, H.J.: Security issues in healthcare applications using wireless medical sensor networks: A survey. *Sensors* 12(1), 55–91 (2011).
79. Laerhoven, K.V., Lo, B., Ng, J., Thiemjarus, S., King, R., Kwan, S., Gellersen, H.W. et al.: Medical healthcare monitoring with wearable and implantable sensors. In: *Proceedings of the 3rd International Workshop on Ubiquitous Computing for Pervasive Healthcare Applications (UbiHealth)*, Tokyo, Japan (2004).
80. Laursen, P., Suriano, R., Quod, M., Lee, H., Abbiss, C., Nosaka, K., Martin, D., and Bishop, D.: Core temperature and hydration status during an ironman triathlon. *British Journal of Sports Medicine* 40(4), 320–325 (2006).
81. Lee, B.C., Kim, J., Chen, S., and Sienko, K.H.: Cell phone based balance trainer. *Journal of NeuroEngineering and Rehabilitation* 9, 10 (2012).
82. Leijdekkers, P., Gay, V., and Barin, E.: Trial results of a novel cardiac rhythm management system using smart phones and wireless ECG sensors. In: *Proceedings of the 7th International Conference on Smart Homes and Health Telematics (ICOST)*, Springer-Verlag, Berlin, Germany, pp. 32–39 (2009).
83. Lester, J., Hartung, C., Pina, L., Libby, R., Borriello, G., and Duncan, G.: Validated caloric expenditure estimation using a single body-worn sensor. In: *Proceedings of the 11th International Conference on Ubiquitous Computing (Ubicomp)*, ACM, New York, pp. 225–234 (2009).
84. Li, Q. and Stankovic, J.: Grammar-based, posture- and context-cognitive detection for falls with different activity levels. In: *Proceedings of the 2nd Conference on Wireless Health (WH)*, San Diego, CA. ACM, New York, NY (2011).
85. Li, Q., Stankovic, J., Hanson, M., Barth, A., Lach, J., and Zhou, G.: Accurate fast fall detection using gyroscopes and accelerometer-derived posture information. In: *Proceedings of the 6th International Workshop on Wearable and Implantable Body Sensor Networks (BSN)*, IEEE Computer Society, pp. 138–143 (2009).
86. Lichtenbelt, W.V.M., Daanen, H., Wouters, L., Fronczek, R., Raymann, R., Severens, N., and Someren, E.V.: Evaluation of wireless determination of skin temperature using ibuttons. *Physiology and Behavior* 88(4–5), 489–497 (2006).
87. Lin, Y.H., Jan, I.C., Ko, P.C.I., Chen, Y.Y., Wong, J.M., and Jan, G.J.: A wireless pda-based physiological monitoring system for patient transport. *IEEE Transactions on Information Technology in Biomedicine* 8(4), 439–447 (2004).
88. Liu, J., Johns, E., Atallah, L., Pettitt, C., Lo, B., Frost, G., and Yang, G.Z.: The development of an in-vivo active pressure monitoring system. In: *Proceedings of the 9th International Conference on Wearable and Implantable Body Sensor Networks (BSN)*, IEEE Computer Society, pp. 154–160 (2012).
89. Lo, B., Atallah, L., Aziz, O., Elhew, M.E., Darzi, A., and Yang, G.Z.: Real-time pervasive monitoring for postoperative care. In: *Proceedings of the 4th International Workshop on Wearable and Implantable Body Sensor Networks (BSN)*, IEEE Computer Society, pp. 122–127 (2007).
90. Lo, B., Thiemjarus, S., King, R., and Yang, G.Z.: Body sensor network—A wireless sensor platform for pervasive healthcare monitoring. In: *Adjunct Proceedings of the 3rd International Conference on Pervasive Computing (PERVASIVE)*, demo session. Munich, Germany (2005).
91. López, G., Custodio, V., and Moreno, J.I.: Lobin: E-textile and wireless-sensor-network-based platform for healthcare monitoring in future hospital environments. *IEEE Transactions on Information Technology in Biomedicine* 14(6), 1446–1458 (2010).
92. Lopez-Nores, M., Pazos-Arias, J., Garcia-Duque, J., and Blanco-Fernandez, Y.: Monitoring medicine intake in the networked home: The iCabiNET solution. In: *Proceedings of the 2nd International Conference on Pervasive Computing Technologies for Healthcare (PervasiveHealth)*, Tampere, Finland. IEEE, pp. 116–117 (2008).

93. Lorincz, K., Chen, Br., Challen, G.W., Chowdhury, A.R., Patel, S., Bonato, P., and Welsh, M.: Mercury: A wearable sensor network platform for high-fidelity motion analysis. In: *Proceedings of the 7th ACM International Conference on Embedded Networked Sensor Systems (SenSys)*, ACM, New York, pp. 183–196 (2009).
94. Lorincz, K., Malan, D.J., Fulford-Jones, T.R., Nawoj, A., Clavel, A., Shnayder, V., Mainland, G., and Welsh, M.: Sensor networks for emergency response: Challenges and opportunities. *IEEE Pervasive Computing* 3(4), 16–23 (2004).
95. Lynch, A., Majeed, B., O'Flynn, B., Barton, J., Murphy, F., Delaney, K., and O'Mathuna, S.C.: A wireless inertial measurement system (WIMS) for an interactive dance environment. *Journal of Physics: Conference Series* 15(1), 95–100 (2005).
96. Mao, Y., Chen, W., Chen, Y., Lu, C., Kollef, M., and Bailey, T.: An integrated data mining approach to real-time clinical monitoring and deterioration warning. In: *Proceedings of the 18th ACM International Conference on Knowledge Discovery and Data Mining (SIGKDD)*, ACM, New York, pp. 1140–1148 (2012).
97. Marin, R., Campos, M., Gomariz, A., Lopez, A., Rol, M., and Madrid, J.: Complexity changes in human wrist temperature circadian rhythms through ageing. In: *Proceedings of the 4th International Conference on Interplay Between Natural and Artificial Computation (IWINAC)*, Springer-Verlag, Berlin, Germany, pp. 401–410 (2011).
98. Maron, M.B., Wagner, J.A., and Horvath, S.: Thermoregulatory responses during competitive marathon running. *Journal of Applied Physiology* 42(6), 909–914 (1977).
99. Mathie, M.J., Coster, A.C., Lovell, N.H., Celler, B.G., Lord, S.R., and Tiedemann, A.: A pilot study of long-term monitoring of human movements in the home using accelerometry. *Journal of Telemedicine and Telecare* 10(3), 144–151 (2004).
100. Maughan, R., Leiper, J., and Thompson, J.: Rectal temperature after marathon running. *British Journal of Sports Medicine* 19(4), 192–196 (1985).
101. McIlwraith, D. and Yang, G.Z.: Body sensor networks for sport, wellbeing and health. In: G. Ferrari (ed.), *Sensor Networks: Where Theory Meets Practice, Signals and Communication Technology*. Springer-Verlag, Berlin, Germany, pp. 349–381 (2009).
102. McKenzie, J. and Osgood, D.: Validation of a new telemetric core temperature monitor. *Journal of Thermal Biology* 29(7–8), 605–611 (2004).
103. Michahelles, F. and Schiele, B.: Sensing and monitoring professional skiers. *IEEE Pervasive Computing* 4(3), 40–45 (2005).
104. Minnen, D., Westeyn, T.L., Ashbrook, D., Presti, P., and Starner, T.: Recognizing soldier activities in the field. In: *Proceedings of the 4th International Workshop on Wearable and Implantable Body Sensor Networks (BSN)*, Springer-Verlag, Berlin, Germany, pp. 236–241 (2007).
105. Morais, J.M., Papadimitrakopoulos, F., and Burgess, D.J.: Biomaterials/tissue interactions: Possible solutions to overcome foreign body response. *Journal of the American Association of Pharmaceutical Scientists* 12(2), 188–196 (2010).
106. Mueller, F., O'Brien, S., and Thorogood, A.: Jogging over a distance supporting a "jogging together" experience although being apart. In: *Proceedings of the ACM International Conference on Human Factors in Computing Systems (CHI)*, ACM, New York, pp. 2579–2584 (2007).
107. Nenonen, V., Lindblad, A., Häkkinen, V., Laitinen, T., Jouhtio, M., and Hämäläinen, P.: Using heart rate to control an interactive game. In: *Proceedings of the ACM International Conference on Human Factors in Computing Systems (CHI)*, ACM, New York, pp. 853–856 (2007).
108. Neves, P., Stachyra, M., and Rodrigues, J.: Application of wireless sensor networks to healthcare promotion. *Journal of Communications Software and Systems (JCOMSS)* 4(3), 181–190 (2008).

109. Nirjon, S., Dickerson, R.F., Li, Q., Asare, P., Stankovic, J.A., Hong, D., Zhang, B., Jiang, X., Shen, G., and Zhao, F.: Musicalheart: A hearty way of listening to music. In: *Proceedings of the 10th ACM International Conference on Embedded Networked Sensor Systems (SenSys)*, ACM, New York, pp. 43–56 (2012).
110. Oliver, N. and Flores-Mangas, F.: MPTrain: A mobile, music and physiology-based personal trainer. In: *Proceedings of the 8th International Conference on Human-Computer Interaction with Mobile Devices and Services (MobileHCI)*, ACM, New York, pp. 21–28 (2006).
111. Oliver, N. and Flores-Mangas, F.: Healthgear: Automatic sleep apnea detection and monitoring with a mobile phone. *Journal of Communications* 2(2), 1–9 (2007).
112. Onuki, Y., Bhardwaj, U., Papadimitrakopoulos, F., and Burgess, D.J.: A review of the biocompatibility of implantable devices: Current challenges to overcome foreign body response. *Journal of Diabetes Science and Technology* 2(6), 1003–1015 (2008).
113. Oppermann, F.J., Boano, C.A., and Römer, K.: A decade of wireless sensing applications: Survey and taxonomy. In: H.M. Ammari (ed.), *The Art of Wireless Sensor Networks, Signals and Communication Technology*, vol. 1. Springer-Verlag, Berlin, Germany (2014).
114. Øyri, K., Balasingham, I., Samset, E., Høgetveit, J.O., and Fosse, E.: Wireless continuous arterial blood pressure monitoring during surgery: A pilot study. *Anesthesia and Analgesia* 102(2), 478–483 (2006).
115. Pang, Z., Chen, Q., and Zheng, L.: A pervasive and preventive healthcare solution for medication noncompliance and daily monitoring. In: *Proceedings of the 2nd IEEE International Symposium on Applied Sciences in Biomedical and Communication Technologies (ISABEL)*, IEEE, pp. 1–6 (2009).
116. Pansiot, J., King, R., McIlwraith, D., Lo, B., and Yang, G.Z.: ClimBSN: Climber performance monitoring with BSN. In: *Proceedings of the 5th International Workshop on Wearable and Implantable Body Sensor Networks (BSN)*, IEEE, pp. 33–36 (2008).
117. Pantelopoulos, A. and Bourbakis, N.: A survey on wearable sensor-based systems for health monitoring and prognosis. *IEEE Transactions on Systems, Man, and Cybernetics, Part C: Applications and Reviews* 40(1), 1–12 (2010).
118. Paradiso, R., Loriga, G., and Taccini, N.: A wearable health care system based on knitted integrated sensors. *IEEE Transactions on Information Technology in Biomedicine* 9(3), 337–344 (2005).
119. Park, C., Chou, P.H., and Sun, Y.: A wearable wireless sensor platform for interactive dance performances. In: *Proceedings of the 4th IEEE Conference on Pervasive Computing and Communications (PerCom)*, IEEE Computer Society, pp. 52–59 (2006).
120. Park, C., Chou, P.H., Ying, B., Matthews, R., and Hibbs, A.: An ultra-wearable, wireless, low-power ECG monitoring system. In: *Proceedings of the IEEE Conference on Biomedical Circuits and Systems (BioCAS)*, IEEE, pp. 241–244 (2006).
121. Park, T., Lee, J., Hwang, I., Yoo, C., Nachman, L., and Song, J.: E-Gesture: A collaborative architecture for energy-efficient gesture recognition with hand-worn sensor and mobile devices. In: *Proceedings of the 9th ACM International Conference on Embedded Networked Sensor Systems (SenSys)*, ACM, New York, pp. 260–273 (2011).
122. Päßler, S. and Fischer, W.J.: Acoustical method for objective food intake monitoring using a wearable sensor system. In: *Proceedings of the 5th International Conference on Pervasive Computing Technologies for Healthcare (PervasiveHealth)*, IEEE, pp. 266–269 (2011).
123. Patel, S., Lorincz, K., Hughes, R., Huggins, N., Growden, J., Standaert, D., Akay, M., Dy, J., Welsh, M., and Bonato, P.: Monitoring motor fluctuations in patients with Parkinson's disease using wearable sensors. *IEEE Transactions on Information Technology in Biomedicine* 13(6), 864–873 (2009).

124. Patel, S., Lorincz, K., Hughes, R., Huggins, N., Growdon, J.H., Welsh, M., and Bonato, P.: Analysis of feature space for monitoring persons with Parkinson's disease with application to a wireless wearable sensor system. In: *Proceedings of the 29th IEEE International Engineering in Medicine and Biology Society Conference (EMBC)*, IEEE, pp. 6290–6293 (2007).
125. Patel, S., Park, H., Bonato, P., Chan, L., and Rodgers, M.: A review of wearable sensors and systems with application in rehabilitation. *Journal of NeuroEngineering and Rehabilitation* 9, 21 (2012).
126. Perala, C.H. and Sterling, B.S.: Galvanic skin response as a measure of soldier stress. Technical Report ARL-TR-4114, U.S. Army Research Laboratory, Aberdeen Proving Ground (APG), Aberdeen, MD (2007).
127. Pereira, O.R.E., Rodrigues, J.J.P.C., and Gomes, A.J.P.: Contributions of sensor networks to improve gaming experience. In: *Proceedings of the 24th IEEE International Conference on Advanced Information Networking and Applications Workshops (WAINA)*, Perth, Australia, pp. 1047–1052 (2010).
128. Poon, C.C.Y., Zhang, Y.T., and Bao, S.D.: A novel biometrics method to secure wireless body area sensor networks for telemedicine and m-health. *IEEE Communications Magazine* 44(4), 73–81 (2006).
129. Pugh, L., Corbett, J., and Johnson, R.: Rectal temperatures, weight losses, and sweat rates in marathon running. *Journal of Applied Physiology* 23(3), 347–352 (1967).
130. Reeves, A., Ng, J., Brown, S., and Barnes, N.: Remotely supporting care provision for older adults. In: *Proceedings of the 3rd International Workshop on Wearable and Implantable Body Sensor Networks (BSN)*, IEEE Computer Society, pp. 117–122 (2006).
131. Rodrigues, J.J., Caldeira, J., and Vaidya, B.: A novel intra-body sensor for vaginal temperature monitoring. *Sensors* 9(4), 2797–2808 (2009).
132. Rubel, P., Fayn, J., Nollo, G., Assanelli, D., Li, B., Restier, L., Adami, S. et al.: Toward personal eHealth in cardiology. Results from the EPI-MEDICS telemedicine project. *Journal of Electrocardiology* 38(4), 100–106 (2005).
133. Rustemeyer, J., Radtke, J., and Bremerich, A.: Thermography and thermoregulation of the face. *Head and Face Medicine* 3, 17 (2007).
134. Saini, P., Willmann, R., Huurneman, R., Lanfermann, G., te Vrugt, J., Winter, S., and Buurke, J.: Philips stroke rehabilitation exerciser: A usability test. In: *Proceedings of the International Conference on Telehealth and Assistive Technologies (Telehealth/AT)*, ACTA Press, Anaheim, CA, pp. 116–122 (2008).
135. Sanches, J.M., Pereira, B., and Paiva, T.: Headset Bluetooth and cellphone based continuous central body temperature measurement system. In: *Proceedings of the 32nd IEEE International Conference of the Engineering in Medicine and Biology Society (EMBS)*, IEEE, pp. 2975–2978 (2010).
136. Sarabia, J., Rol, M., Mendiola, P., and Madrid, J.: Circadian rhythm of wrist temperature in normal-living subjects: A candidate of new index of the circadian system. *Physiology and Behavior* 95(4), 570–580 (2008).
137. Scully, C., Karaboué, A., Liu, W.M., Meyer, J., Innominato, P., Chon, K., Gorbach, A., and Lévi, F.: Skin surface temperature rhythms as potential circadian biomarkers for personalized chronotherapeutics in cancer patients. *Interface Focus* 1(1), 48–60 (2011).
138. Seeger, C., Buchmann, A., and Laerhoven, K.V.: myHealthAssistant: A phone-based body sensor network that captures the wearer's exercises throughout the day. In: *Proceedings of the 6th International ICST Conference on Body Area Networks (BodyNets)*, Beijing, China (2011).
139. Innovatec Sensing and Communications. TherCom User's Manual: Ambulatory Body Temperature Monitor, Version 2012-10-26. http://innovatecsc.com/wp-content/uploads/2013/12/TherComUserManual.pdf, accessed March 1, 2014 (2012).

140. Shih, E.I., Shoeb, A.H., and Guttag, J.V.: Sensor selection for energy-efficient ambulatory medical monitoring. In: *Proceedings of the 7th International Conference on Mobile Systems, Applications, and Services (MobiSys)*, ACM, New York, pp. 347–358 (2009).
141. Shnayder, V., Chen, B.R., Lorincz, K., Fulford-Jones, T.R., and Welsh, M.: Sensor networks for medical care. Technical Report TR-08-05, Harvard University, Cambridge, MA (2005).
142. Srinivasan, V., Stankovic, J., and Whitehouse, K.: Protecting your daily in-home activity information from a wireless snooping attack. In: *Proceedings of the 10th International Conference on Ubiquitous Computing (UbiComp)*, ACM, New York, pp. 202–211 (2008).
143. Stankovic, J., Cao, Q., Doan, T., Fang, L., He, Z., Kiran, R., Lin, S., Son, S., Stoleru, R., and Wood, A.: Wireless sensor networks for in-home healthcare: Potential and challenges. In: *Proceedings of the High Confidence Medical Device Software and Systems Workshop (HCMDSS)*, Philadelphia, PA (2005).
144. Strauss, M., Reynolds, C., Hughes, S., Park, K., McDarby, G., and Picard, R.W.: The handwave Bluetooth skin conductance sensor. In: *Proceedings of the 1st International Conference on Affective Computing and Intelligent Interaction (ACII)*, Springer-Verlag, Berlin, Germany, pp. 699–706 (2005).
145. Strohrmann, C., Harms, H., Tröster, G., Hensler, S., and Müller, R.: Out of the lab and into the woods: Kinematic analysis in running using wearable sensors. In: *Proceedings of the 13th International Conference on Ubiquitous Computing (UbiComp)*, ACM, New York, pp. 119–122 (2011).
146. Strohrmann, C., Rossi, M., Arnrich, B., and Tröster, G.: A data driven approach to kinematic analysis in running using wearable technology. In: *Proceedings of the 9th International Conference on Wearable and Implantable Body Sensor Networks (BSN)*, IEEE Computer Society, pp. 118–123 (2012).
147. Summer, A.M.: The silver tsunami: One educational strategy for preparing to meet America's next great wave of underserved. *Journal of Health Care for the Poor and Underserved* 18(3), 503–509 (2007).
148. Sung, M., Marci, C., and Pentland, A.: Objective physiological and behavioral measures for identifying and tracking depression state in clinically depressed patients. Technical Report TR 595, Massachusetts Institute of Technology (MIT), Cambridge, MA (2005).
149. Sung, M., Marci, C., and Pentland, A.: Wearable feedback systems for rehabilitation. *Journal of NeuroEngineering and Rehabilitation* 2, 17 (2005).
150. Tan, C.C., Wang, H., Zhong, S., and Li, Q.: IBE-Lite: A lightweight identity-based cryptography for body sensor networks. *IEEE Transactions on Information Technology in Biomedicine* 13(6), 926–932 (2009).
151. Tapia, E.M., Marmasse, N., Intille, S.S., and Larson, K.: MITes: Wireless portable sensors for studying behavior. In: *Adjunct Proceedings of the 6th International Conference on Ubiquitous Computing (UbiComp)*, Nottingham, United Kingdom (2004).
152. Varela, M., Cuesta, D., Madrid, J.A., Churruca, J., Miro, P., Ruiz, R., and Martinez, C.: Holter monitoring of central and peripheral temperature: Possible uses and feasibility study in outpatient settings. *Journal of Clinical Monitoring and Computing* 23(4), 209–216 (2009).
153. Vogel, S., Hülsbusch, M., Hennig, T., Blazek, V., and Leonhardt, S.: In-ear vital signs monitoring using a novel microoptic reflective sensor. *IEEE Transactions on Information Technology in Biomedicine* 13, 882–889 (2009).
154. Wang, Z., Jiang, M., Zhao, H., Li, H., and Wang, Y.: A pilot study on evaluating recovery of the post-operative based on acceleration and sEMG. In: *Proceedings of the 7th International Conference on Body Sensor Networks (BSN)*, IEEE Computer Society, pp. 3–8 (2010).

155. Wilhelm, F.H., Roth, W.T., and Sackner, M.A.: The LifeShirt: An advanced system for ambulatory measurement of respiratory and cardiac function. *Behavior Modification (BMO)* 27(5), 671–691 (2003).
156. Wilkinson, D.M., Carter, J.M., Richmond, V.L., Blacker, S.D., and Rayson, M.P.: The effect of cool water ingestion on gastrointestinal pill temperature. *Medicine and Science in Sports and Exercise* 40(3), 523–528 (2008).
157. Wood, A., Stankovic, J., Virone, G., Selavo, L., He, Z., Cao, Q., Doan, T., Wu, Y., Fang, L., and Stoleru, R.: Context-aware wireless sensor networks for assisted-living and residential monitoring. *IEEE Network* 22(4), 26–33 (2008).
158. Wu, W., Au, L., Jordan, B., Stathopoulos, T., Batalin, M., Kaiser, W., Vahdatpour, A., Sarrafzadeh, M., Fang, M., and Chodosh, J.: The smartcane system: An assistive device for geriatrics. In: *Proceedings of the 3rd International Conference on Body Area Networks (BodyNets),* Tempe, AZ (2008).
159. Wu, W.H., Bui, A.A., Batalin, M.A., Au, L.K., Binney, J.D., and Kaiser, W.J.: MEDIC: Medical embedded device for individualized care. *Artificial Intelligence in Medicine* 42(2), 137–152 (2008).
160. Yamasue, K., Hagiwara, H., Tochikubo, O., Sugimoto, C., and Kohno, R.: Measurement of core body temperature by an ingestible capsule sensor and evaluation of its wireless communication performance. *Advanced Biomedical Engineering* 1(1), 9–15 (2012).
161. Yang, G.Z. (ed.): *Body Sensor Networks*, 1st edn. Springer-Verlag, London, U.K. (2006).
162. Yoon, I.Y., Kripke, D.F., Elliott, J.A., Youngstedt, S.D., Rex, K.M., and Hauger, R.L.: Age-related changes of circadian rhythms and sleep–wake cycles. *Journal of the American Geriatrics Society* 51(8), 1085–1091 (2003).
163. Zeitz, K. and McCutcheon, H.: Policies that drive the nursing practice of postoperative observations. *International Journal of Nursing Studies* 39(8), 831–839 (2002).
164. Zhan, A., Chang, M., Chen, Y., and Terzis, A.: Accurate caloric expenditure of bicyclists using cellphones. In: *Proceedings of the 10th ACM International Conference on Embedded Networked Sensor Systems (SenSys)*, ACM, New York, pp. 71–84 (2012).
165. Zhou, G., Li, Q., Li, J., Wu, Y., Lin, S., Lu, J., Wan, C.Y., Yarvis, M.D., and Stankovic, J.A.: Adaptive and radio-agnostic QoS for body sensor networks. *ACM Transactions on Embedded Computing Systems* 10(4), 1–34 (2011).

12 Prototype on RFID and Sensor Networks for Elder Healthcare

Melody Moh, Loc Ho, Zachary Walker, and Teng-Sheng Moh

CONTENTS

ABSTRACT

Radio-frequency identification (RFID) technology combined with sensor networks presents potential for a variety of applications including applications in healthcare. An RFID and sensor system consists of a tag and a reader. The tag is usually attached to a tracking device and the reader is used for tracking (sensing) and can collect necessary information. The healthcare applications for such a system are numerous, including patient monitoring and tracking and assisting individuals with physical challenges. This chapter describes a prototype project that integrates both sensor networks and RFID technologies. The project addresses in-home medicine monitoring for elderly patients.

12.1 INTRODUCTION

Radio-frequency identification (RFID) technology has recently become a viable replacement for the Universal Product Code (UPC) technology in many industries. Its fast growth and huge potential benefits have motivated a major move independently taken by Wal-Mart, the world's largest retailer, and the US Department of Defense (DoD), which required their suppliers to install RFID tags by 2005 [1]. In response, several major computer companies, including Intel, HP, IBM, and Sun, have announced their efforts and future plans to support RFID. RFID technology, however, has attracted relatively little attention in the network research community.

Meanwhile, sensors and sensor networks have in recent years been adopted as a major research focus by federal funding agencies. This has resulted in vast amount of research proposals, academic projects, and publications.

In an effort to bridge the gap between industry and academia focuses, we have worked on a prototype that utilizes both technologies and investigated the feasibility, technical challenges, and resulting capabilities of their integration.

An RFID system consists of two primary components—a tag and a reader. An RFID tag, like an UPC, is usually attached to a tracking object; a reader is then used to track tagged objects. While sensor network is used to sense and monitor physical, chemical, and biological environments through sensing of sound, temperature, light, etc., RFID tags allow any objects to be trackable or *sensible* as long as an RFID tag can be attached. Even though RFID technology has limitations, such as low tolerance to fluid or metal environments, tags can extend a sensor network by providing sensing/sensible property to otherwise unsensible objects, thus providing the last-hop connection of a sensor network.

RFID has been used in a number of biomedical and healthcare applications, such as artificial interocular pressure measurement [2], dental implants and molds [3], and hospital workflow including intrahospital patient and equipment tracking [4].

From a recent study, the population of age 65 and older in the United States will grow from 10.6 million in 1975 to 18.2 million in 2025, an increase of 72%, while the overall population increase is about 60%. The trend is global; the worldwide population over age 65 will be more than double from 357 million in 1990 to 761 million in 2025 [5]. Longevity has caused expensive age-related disabilities, diseases, and therefore healthcare. To help in addressing this aging population medication needs, we target our prototype on an in-home elder healthcare system. This is a continuation of our work on applying wireless technologies for biomedical applications [6–8].

The rest of this section presents major features of the RFID technology. Related studies on integration of RFID and sensor networks are described in Section 12.2. This is followed by a presentation of the two phases of the prototype system, in Sections 12.3 and 12.4, respectively. Section 12.5 discusses technical challenges and future improvements. Finally, Section 12.6 concludes this chapter.

12.1.1 RFID

Since sensor network has been a familiar topic in academia, we skip its introduction and focus only on RFID. An RFID system, more specifically, includes three components: (1) a tag or transponder located on the object to be identified, (2) an interrogator (reader) that may be a read or write/read device, and (3) an antenna that emits radio signals to activate the tag and read/write data to it.

At its simplest form, a tag is a beacon announcing its presence to a reader. These types of tags are often seen in retail stores used to prevent theft by announcing their presence when taken past a reader. RFID tag capabilities, however, extend well beyond a simple beacon. Tag can hold a unique identity (UID) of 8 bytes in length and can be used for inventory management at global scale, such as a UPC. More than just a UID, a tag can carry rewriteable persistent storage and accessible via a reader.

RFID tags are classified by its energy source as passive, semiactive (or semipassive), and active. A passive tag has no battery of its own and makes use of the incoming radio waves broadcast by a reader to power its response. An active tag uses its own battery power to perform all operations. A semiactive tag uses its own battery power for some functions but, like the passive tag, uses the radio waves of the reader as an energy source for its own transmission.

RFID readers employ tag-reading algorithms that are capable of identifying hundreds of tags per second. Once identified, a reader may read data from or write to tag memory, depending on the permissions granted by the tag. RFID readers generally fall into two categories—high frequency (HF) and ultrahigh frequency (UHF). Currently, HF RFID systems adhere to the International Organization of Standardization (ISO) standard while UHF RFID systems have yet to become standardized globally. Table 12.1 shows a comparison between HF and UHF RFID technology.

12.2 RELATED STUDIES

When an RFID tag is given sensing capabilities, the line between RFID and sensor network becomes blurred. Many active and semiactive tags have incorporated sensors into their design, allowing them to take sensor readings and transmit them to a reader at a later time. They are not quite sensor network nodes because they lack the capacity to communicate with one another through a cooperatively formed ad hoc network, but they are beyond simple RFID storage tags. In this way, RFID is converging with sensor networking technology. From the other direction, some

TABLE 12.1
Comparison of HF and UHF RFID Technology

	HF RFID	UHF RFID
Frequency	13.56 MHz	902–928 MHz North America 860–868 MHz Europe 950–956 MHz Japan
Read range	10–20 cm	3–6 m
Read rate	50 tags/s	400 tags/s
Memory size	64–256 bits read/write	64–2048 bits read/write
Power source	Inductive/magnetic field	Capacitive/electric field
Advantage	Low cost Standard frequency	High speed Longer read range

Source: SkyeTek Inc., SkyeRead Mini, http://skyetek.com/readers_Mini.html, downloaded April 16, 2005.

sensor nodes are now using RFID readers as part of their sensing capabilities. The SkyeRead M1-mini made by SkyeTek is an example of an RFID reader designed to mate directly with the Crossbow Mica2Dot sensor motes [9].

In the following, we describe several projects and prototypes taking place in industrial and federal research laboratories, as well as some products adopted by companies:

NASA: *Sensor Webs*—The project has the objective of using readily available technologies to create a wireless network with embedded intelligence [10]. In this way, instead of reporting to an external control system, sensed data can be shared throughout the network and be used by the embedded intelligence to act directly on any detected changes. RFID tagged objects, such as firefighters or astronauts, may be sensed and be guided by the intelligent sensor web; or product components and production flow may be sensed and be guided to slow down or to speed up.

Intel Labs: *Proactive Healthcare*—In addition to leading a major force on sensor network research, Intel Research Labs also initiated an effort to explore technology that can help in caring for the growing elderly population [11]. A joint project called *Caregiver's Assistant* and *CareNet Display*, developed by Intel Research Seattle and University of Washington, aims to provide elder care by monitoring elders' activities [12]. RFID tags are stuck on household objects. Combined with a sensor network, the system would collect information on which objects are touched and when. These data are used by an artificial-intelligent program, *Caregiver's Assistant*, to fill out a standard activities of daily living (ADL) form.

HP Labs: *Smart Rack and SmartLOCUS*—HP opened its US RFID Demo Center at HP Labs on October 2004. Two major research prototypes are

Smart Rack and SmartLOCUS [13]. Both prototypes gear toward integrating RFID and other types of sensors, such as video cameras or thermal sensors, into *multimodal sensor networks* that use more than one type, or mode, of sensors. Smart Rack uses thermal sensors and HF RFID readers to identify and monitor the temperature of servers sitting in large metal server cabinets. These sensors and readers are networked and the collected data are used to show, in real time, an inventory of the cabinets and temperature profile of each cabinet. It may become a commercial product and offered within HP's OpenView network management system.

Others: *DOD and BP Oil*—A few other DOD and private sectors are also using RFID with integrated sensor networks. The US Navy, working with Georgia Tech, has developed an RFID sensor network that monitors the temperature, humidity, and air pressure in containers where aircraft parts are stored [14]. The US military's Combat Feeding Program pilot uses active RFID tag-based sensor networks to provide real-time visibility of rations as they move from the manufacturer to units in the field [15]. The BP oil company uses RFID and sensor network to monitor assets and react quickly to changes in environmental conditions [16].

12.3 LEARNING PHASE: INTEGRATING OFF-THE-SHELF SENSOR NETWORK WITH SIMULATED RFID READER

The first phase of the project is to investigate the capability of sensors and RFID and how they may be integrated. In this section, we first illustrate an overview of the phase. Three major components are described in the three subsections. Finally, the last subsection presents performance results.

There are many choices of commercial products, and each costs from hundreds (sensors and HF RFID readers) to thousands of dollars (UHF RFID readers). Before making actual purchase decisions, in the learning phase, we develop a prototype consisting of some hardware and some software simulators.

There are a number of embedded platforms available. Adapting the sensor network platform is the most logical choice. The initial commercially available sensor network platform is the Berkeley mote. The Berkeley mote has been replaced by Mica, Mica2, Mica2Dot, and MCS Cricket manufactured by Crossbow Technology [17]. The Mica2 mote is selected for this phase to determine its capability, feasibility, and integration effect with RFID. Because of expensive hardware cost, in this phase, an RFID simulator reader is developed and used. On the basis of the experience learned, in the next phase, the developing phase, actual RFID readers are used.

In this prototype, there are four system components—two Mica2 motes, one simulated RFID reader, and a base station PC, as illustrated in Figure 12.1. The two Mica2 motes—named RFID reader mote and base station mote—are used for RF communication. The RFID reader simulator is used to simulate an actual RFID reader, communicating via a serial port. The base station PC is also used to perform statistic gathering as well as other required processing. It is connected to the base station mote via a serial port. The message flow of the entire system is also illustrated in Figure 12.1. Each component is described in the following subsections.

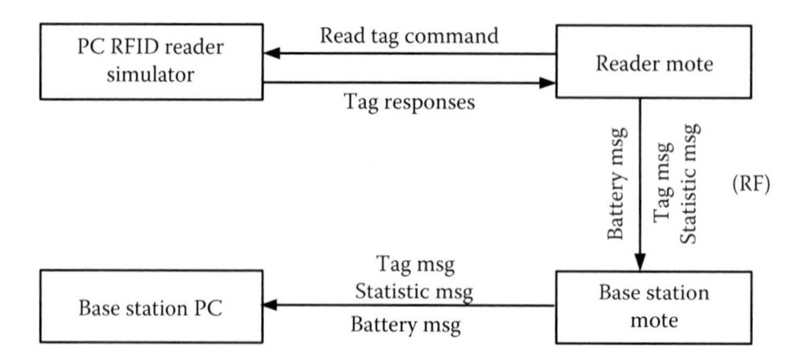

FIGURE 12.1 System component overview.

12.3.1 Software for RFID Reader Mote and Base Station Mote

The software developed for RFID reader mote is first described in detail, followed by a brief description of the software for base station mote.

The RFID reader mote software is developed using TinyOS and the nesC language [18]. This software interacts with the simulated RFID reader via the mote's serial port. The software module consists of RFID mote (control), RFID reader, battery, and communication modules, as illustrated in Figure 12.2. The control module provides control to all submodules and handles intermodule interactions. The RFID reader module handles interactions with the RFID reader. All RFID-specific details are hidden in the module. The battery module handles battery voltage measurement.

The communication module is divided into three submodules—packet management, serial communication, and RF communication—with an interface module. With limited memory resource, the packet management submodule manages a fixed memory size associated with each communication packet. The serial communication

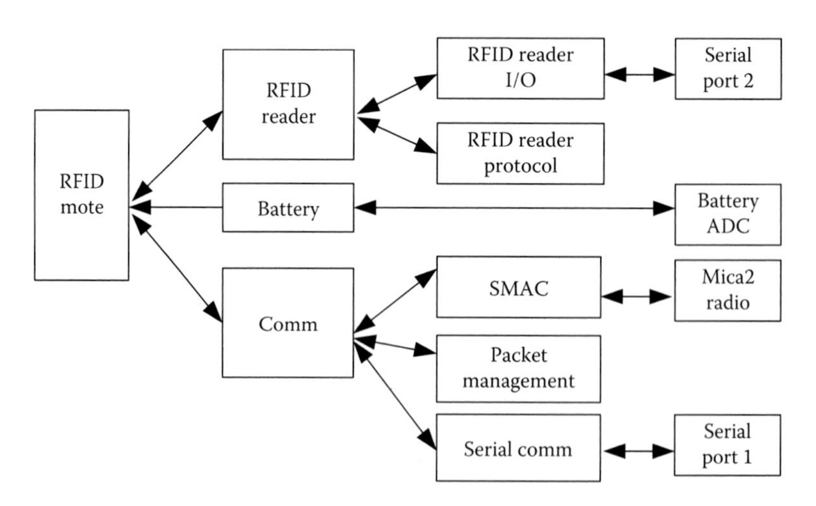

FIGURE 12.2 RFID reader mote software components (with SMAC RF communication submodule).

submodule is the TinyOS communication module over serial port. There are a few RF communication modules developed by various members of the open-source community. TinyOS has an RF communication module [18]. Crossbow Technology has a mesh RF communication module [17]. An SMAC module is also available for this platform [19]. The RF communication submodule is designed to allow for easy replacement. All three RF communication submodules are wrapped around a common interface and are selected at compile time.

The RFID reader mote software queries the RFID reader simulator every second for tag messages. In response, the RFID reader simulator sends a set of tag messages to the reader mote. Tag messages are queued by the RFID mote software and transmitted over RF to the base station mote. To allow efficient transmission, up to 12 tag messages are encoded into a single RF message. If there are fewer than 12 tag messages, the RFID mote waits for 300 ms before starting transmission. With a baud rate of 115,200 bps, 300 ms delay is sufficient. (To see this, each tag consists of 19 bytes, but each byte of 8 bits needs 1 stop bit; thus, transmitting would need to transmit 171 bits. A message of 12 tags needs transmission of 2052 bits. Given the aforementioned baud rate, it needs theoretically 17.8 ms.)

The base station mote software is similar to the RFID mote with run-time behavior changes based on the mote identity (ID). The base station mote software gathers the received packets and forwards them via the mote serial port to the base station PC.

12.3.2 RFID Reader Simulator

A simulator is developed to simulate an HF RFID reader. The HF RFID reader simulator emulates Texas Instrument HF Tag-it protocol [20]. It is written in Java using part of the existing TinyOS serial communication module. When the simulator receives a *Read Transponder Details Command*, it sends a series of simulated tag messages to the RFID reader mote via serial port. The number of tags is specified via its command line. The simulated tag IDs are fixed.

12.3.3 Software for Base Station PC

The base station PC is programmed to process data received from the base station mote. It is written in Java and makes use of existing modules from TinyOS. The architecture of this module is designed with component reuse for the next phase of the project. The base station PC software consists of seven modules—RFID station, RFID database, RFID station packet, RFID station statistic, RFID station graphical user interface (GUI), TinyOS Comm, and MySQL server—as shown in Figure 12.3.

The RFID station module is the main module and handles all the interactions among various submodules. The RFID database module handles all database-related interactions. Its main task is to store received tag messages to a persistent storage. The persistent storage is accomplished using MySQL server and interface via Open Database Connectivity (ODBC). The RFID station packet module handles message decoding. The RFID station statistic module gathers statistic information based on messages received or statistic message from each mote. With the help of the RF message header, it can determine RF message receive rate as well as lost packet.

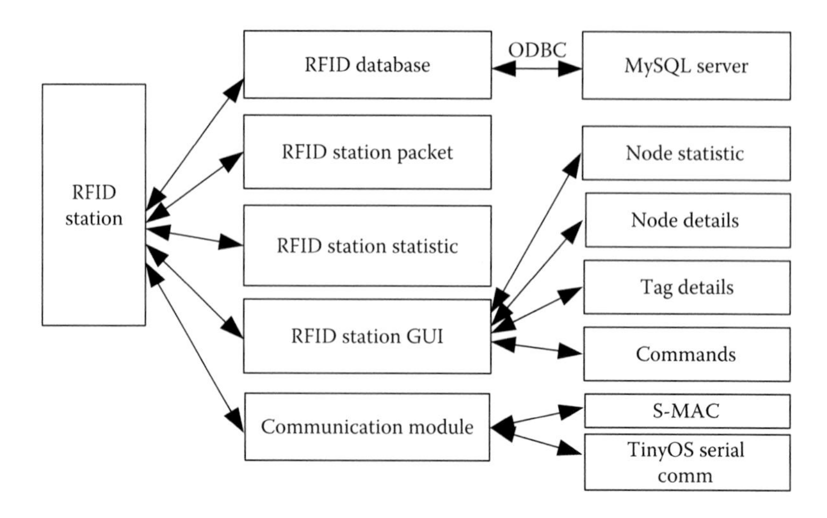

FIGURE 12.3 Base station PC module design overview.

The RFID station GUI module (obviously) handles GUI; the GUI screen includes node statistics, list of node details, list of tag details, and command input. Finally, the TinyOS Comm module is the TinyOS reliable communication module. This module handles all serial communications as well as network communication over a serial port.

12.3.4 Performance Results

The TinyOS RF communication module has a packet size limited to 29 bytes. This allows only one tag message with overhead to be transmitted in a single RF packet. The Crossbow RF mesh communication module is too slow with too much overhead. It is rated at 1 packet per second. The SMAC RF communication module has a packet size limit of 256 bytes; it is used for bandwidth test described later.

The test setup consists of two motes: an RFID reader simulator and a base station PC. The reader simulator and base station software both run on the same PC. The two motes are about 2 ft apart, both with battery power. The RFID reader mote (base station mote) communicates with the simulator reader (base station PC) by a USB serial port; the two motes communicate with each other via wireless communication.

The initial test achieved about 10 tag messages of 19 bytes in length. To achieve a better transfer rate, as mentioned earlier, the communication module is modified to queue up to 12 tag messages. This achieves about 25 tag messages or 500 application bytes per second with 100% reliable RF communication. This rate is sufficient in most embedded applications. Most commercial RFID readers can handle between 50 and 100 tags per second. On a pure ID-based application, this usually requires 12 bytes for a tag ID. Thus, this system can handle about 41 (i.e., 500/12) tags per second. If only 8 bytes are required for a tag ID, the support rate goes up to 62 (i.e., 500/8) tags per second.

12.4 DEVELOPING PHASE: SENSOR NETWORK WITH HF/UHF RFID FOR ELDER HEALTHCARE

In this section, we shall describe the development phase, including the details of the application prototype system, results, and future enhancements. As mentioned earlier, we developed a medicine monitoring and notification system for elder patients without personal care assistants. The high-level functionality may be described as follows:

> The system monitors, notifies, and assists an elder patient in taking the accurate amount of his/her medicines at the appropriate time. It notifies the elder patient when it is time to take his/her medicines. A buzzer (patient monitor system) mounted on the door of the patient's room beeps when there are medicines to be taken. When the patient walks to the medicine cabinet (medicine monitor system), the system guides the patient in taking the proper type and accurate amount of medicines using a GUI.

The rest of this section is organized as follows. Section 12.4.1 describes the prototype system, including its three subsystems and their interaction. Section 12.4.2 details three application mote software corresponding to these three subsystems. Section 12.4.3 presents the base station software including the GUI module.

12.4.1 Prototype System

The system utilizes the strengths of both HF RFID (lower cost) and UHF RFID (long distance) with three sensor network motes. The system is an extension to a prototype system researched by Intel Labs [21] but is completely built from bottom up with newer scale model, third-generation sensor network mote, and an additional UHF reader. There is no hardware or software reuse as they are not available to the public.

The system consists of three subsystems: the medicine monitoring subsystem, the patient monitoring subsystem, and the base station subsystem. Together, they use seven components—three Mica2 motes, an HF RFID reader, a UHF RFID reader, a weight scale, and a base station PC. In the following, we first describe each of the subsystems, followed by an illustration of their interactions in the system component configuration.

12.4.1.1 Medicine Monitoring Subsystem

In this subsystem, as shown in Figure 12.4, HF RFID tags are placed on each medicine bottle; each HF tag identifies a medicine bottle. The HF RFID reader (SkyeRead M1-mini [22]) is used in conjunction to track all medicine bottles within range of the reader. The medicine mote communicates with the HF RFID reader and with the scale to monitor HF tags and the total medicine weight. By performing reads of all tags at a regular interval, the system is able to determine *when* and *which* bottle is removed or replaced by the patient. (The short range of the HF reader is actually desirable for this aspect of the application.) The weight scale monitors the amount of medicine on the scale. Combining changes in weight and HF tag event, *which* medicine bottle, and the *amount* of medicine taken can be determined when the patient takes their pills. The medicine mote software is described in detail in Section 12.4.2.2.

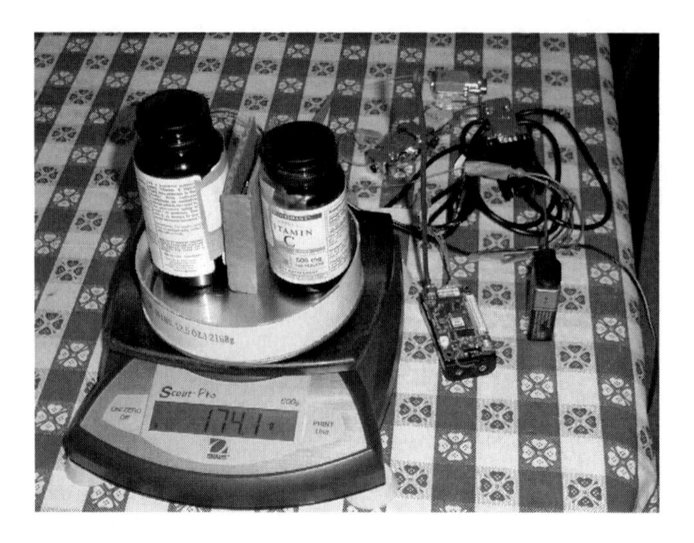

FIGURE 12.4 Medicine monitoring system.

12.4.1.2 Patient Monitoring Subsystem

This subsystem is shown in Figure 12.5. A UHF RFID system including a reader and one or more tags is used to track the elder patient who needs the medicines. This patient wears a UHF tag, which may be detected by the associated RFID reader within 3–6 m. The AWID UHF RFID reader [23] is chosen for this subsystem. The patient mote communicates with the UHF reader to monitor patient arrival at the door of a room or other areas where the system is installed. Thus, the system is able to determine that the patient is in the vicinity and alerts the

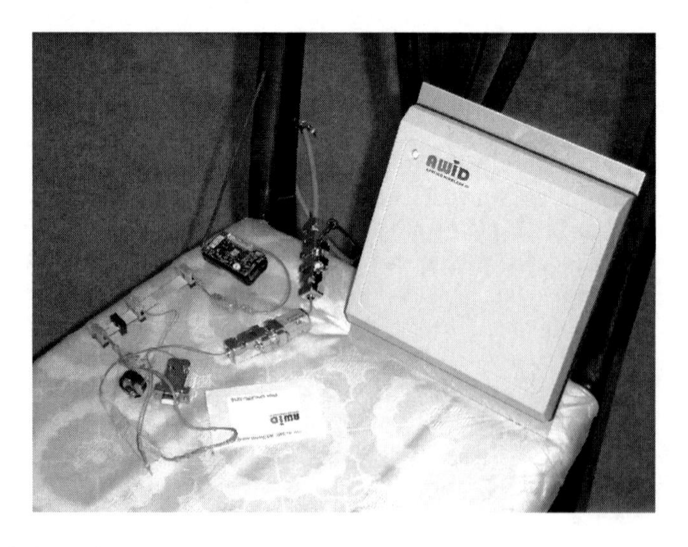

FIGURE 12.5 Patient monitoring subsystem.

FIGURE 12.6 Base station subsystem.

patient to take the required medicines via a buzzer. The patient mote software is described in Section 12.4.2.3.

12.4.1.3 Base Station Subsystem

This subsystem is shown in Figure 12.6. The base station mote provides message relay to the base station PC. The base station PC is a PC running a Linux operation system and it is redesigned from the first phase to accommodate our application. The base station software tasks include simulating a display and its GUI for the patient, determining when medicine is required, and maintaining various interactions between the medicine mote and patient mote. The base station mote software is described in Section 12.4.2.1.

12.4.1.4 System Component Configuration

Next, we describe major interactions among the seven system components, as shown in Figure 12.7. As explained before, motes are mainly used for communicating readers from RFID readers to the control system. The medicine mote communicates with the HF RFID reader and weight scale to monitor HF tags and medicine weight. (Recall that each HF RFID tag identifies a medicine bottle.) The patient mote communicates with the UHF RFID reader to monitor patient's arrival to the room or an area where the system is installed. The base station mote provides message relay to the base station PC (the control system).

12.4.2 Application Mote Software

The software for all three motes (medicine, patient, and base station motes) is identical with compile-time hardware assignment. Figure 12.8 shows the generic mote system software component, enhanced from that in the first phase (Figure 12.2). To support

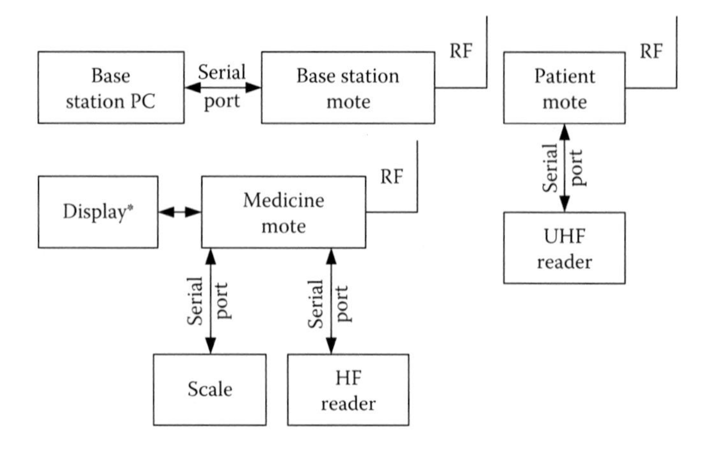

FIGURE 12.7 System component configuration. *The embedded display is simulated on the PC base station using appropriate messages.

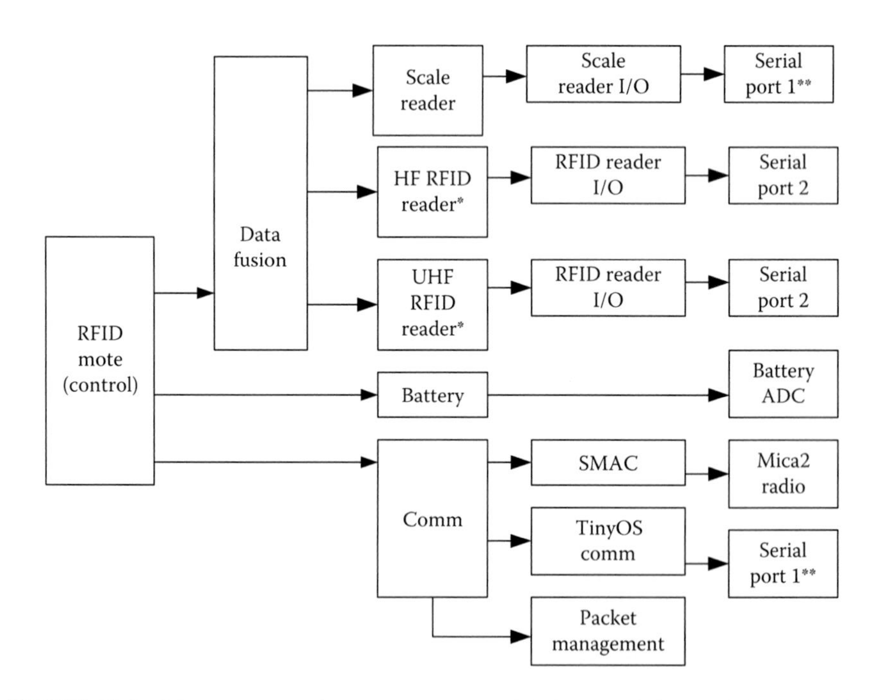

FIGURE 12.8 Mote system software components. *Only one of these modules is active; **only one of these modules is active.

the HF RFID reader, the internal of the RFID reader module developed in the first phase is replaced with the actual protocol interface. The weight scale module is added to handle the scale measurement. The RFID tag and weight data are fused into a single source of information for transmission. A set of data fusion messages is created to indicate the following: weight change, tag no longer detected, tag detected again, and patient detected (details in Section 12.4.3.1). This requires a number of error checkings and handlings to deal with unreliable serial port communications as well as scale weight instability. The UHF RFID reader module is added to communicate with the AWID UHF RFID reader. The scale reader module is added to communicate with the scale.

In correspondence to the three subsystems described in Section 12.4.1, there are also three pieces of mote software: the medicine mote software, the patient software, and the base station mote software. They are described in the following subsections.

12.4.2.1 Base Station Mote Software

The base station mote hardware is a Mica2 mote with breakout connectors and wire connectors to connect to the base station PC. Its software and functionalities are the simplest of the three motes. Its tasks are to receive wireless messages and relay them to the base station PC software application via serial port 1. The Comm module developed in the first phase is sufficient. The application message length and SMAC packet length changed from 250 to 50 bytes as our maximum message size is 50 bytes. This allows 30 messages to be queued for transmission instead of 12 messages from the first phase of the prototype system.

12.4.2.2 Medicine Mote Software

The medicine mote hardware is the same as the base station mote with an HF RFID reader and a scale. As illustrated in Figure 12.8, the scale reader and HF RFID reader modules have its only communication path. The design of the scale reader module encapsulates the scale communication protocol. When a scale reading is available, it notifies the data fusion module. Similarly, the HF RFID reader module encapsulates the HF reader communication protocol. All tags detected by the HF RFID reader module are passed to the data fusion module for processing.

The most complex module is the data fusion module, which performs aggregation on the HF RFID reader and scale data. One of its tasks is intelligent data aggregation, which aims to reduce the number of RF transmission. This process requires an additional message type with five states—medicine detected, medicine removed, medicine placed back, medicine taken, and medicine cleared. The medicine detected state message occurs when a new medicine (or tag) is received. The medicine removed state message occurs when a medicine is not detected after some elapsed interval. The medicine placed back state message occurs when a medicine is detected again after it has been removed. The medicine taken state message occurs when a medicine is placed back and the scale weight is stabilized. The medicine cleared state message is intended to handle the situation where weight changed because of the 100 mg resolution of the scale. In this case, the base station PC software will handle this situation by ignoring the changed weight.

The data fusion algorithm used in the medicine mote software is described as follows:

Data Fusion Algorithm (Medicine Mote)

1. On first receipt of each medicine message, perform the following operations:
   a. Record the medicine tag ID.
   b. Notify the base station PC by sending a medicine detected state message.
   c. Set detected counter to 1.
2. On subsequent receipt of each medicine message, perform the following operations:
   a. Increment its detected counter by 1.
   b. Reset the removed counter to 0.
   c. If the medicine is marked medicine removed, notify the base station PC by sending a medicine placed back state message and mark the medicine as medicine placed back.
3. On every half second timer, perform the following operations:
   a. Reset the detected counter to 0.
   b. If the detected counter is 0, increment the removed counter by 1.
   c. If the removed counter is greater than a threshold, notify the base station PC by sending a medicine removed state message and mark the medicine as medicine removed.
4. On each scale weight changed (and stabilized), perform the following operations:
   a. If a medicine is marked medicine removed, determine the scale difference and send a medicine taken state message to the base station PC.
   b. If no medicine is marked as medicine removed, send a medicine cleared state message to the base station PC.

To support reliable communication between the medicine mote and the base station PC application, each message is echoed back to its sender. Message is cleared if the sequence ID matches the expected sequence ID.

Finally, the scale reader module interfaces with the actual scale. At system start-up, the scale reader module will initialize the actual scale for proper operation. Every second, the scale reader module will send command to query the actual scale weight. This weight is passed to the data fusion module for processing. All communication errors with the scale will cause reinitialization.

12.4.2.3 Patient Mote Software

The patient mote hardware is the same as the base station mote with a UHF RFID reader. The UHF RFID reader is connected to the mote via a breakout board with various connectors and wire connectors. A passive UHF tag is worn on the patient's wrist to provide proper detection by the UHF RFID reader. (Because of limited access to UHF tags and interference of water concentration in human body, the UHF tag is required to be held on the finger for proper operation.) The patient mote software uses the same software modules as the medicine mote. The UHF RFID reader module is enabled while the scale reader and HF RFID reader modules are disabled via compiler switches. The UHF RFID reader module interfaces with the UHF RFID reader. All protocol-specific details are hidden in this module. (For more information on the AWID UHF RFID reader protocol, refer to LR-911 reader manual.) When a UHF tag is detected, it is forwarded to the data fusion module for processing. The design of the data fusion module is extended to support the patient mote functionalities.

The data fusion algorithm for the patient mote is described as follows:

Data Fusion Algorithm (Patient Mote)

1. On reception of a UHF tag ID message from the base station PC, enable the UHF RFID reader system. (This indicates medication needs to be taken.)
2. On reception of a cleared tag ID message, disable the UHF RFID reader system. (This indicates no medication required.)
3. On reception of a UHF tag ID from the UHF RFID reader, increment its detected counter if ID matches.
4. On every half second timer, perform the following operations:
   a. If counter is greater than 6, set counter to 6 (3 s).
   b. If counter is not zero, enable buzzer and decrement counter by 1.
   c. If counter is zero, disable buzzer.

12.4.3 Base Station Software

The base station PC software developed in the previous phase (Section 12.3.3 and Figure 12.3) is significantly expanded for the application. A data fusion module is added to process data aggregations. All data fusion messages are recorded in the persistent database. In supporting the application, a number of new database tables and schemes are created for recording additional information, patient, medicines, and log of events (from the patient mote and medicine mote).

The base station PC software component is shown in Figure 12.9. The RFID data fusion module tasks are to provide a list of medicines that need to be taken based on the patient data stored in the persistent database and handle response messages from the medicine mote and patient mote. The list of medicines is queried by the RFID patient GUI module for display. In the following section, we first discuss details of the data fusion module, which is the most complex among all modules; this is followed by the GUI module.

12.4.3.1 RFID Data Fusion Module

The module performs a number of crucial operations, which include interfacing with the database for patient data, computing the required medicines to be taken, handling messages from the medicine mote and patient mote, providing a list of medicines to be taken to the RFID Patient GUI module, and providing coordination logics to support the various application functionalities.

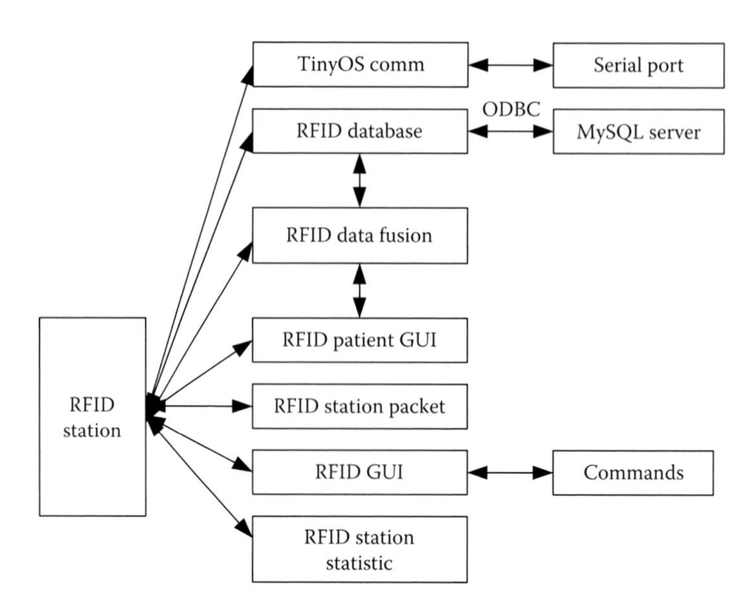

FIGURE 12.9 Application base station software component.

The RFID data fusion module algorithm is described as follows:

Data Fusion Algorithm (RFID)

1. On system start-up, perform these operations:
   a. Query the persistent database for the patient information such as user name and the UHF tag ID.
   b. Query the persistent database for a list of medicines to take.
   c. Query the persistent database for a list of medicines taken today.
   d. Compute the medicine list for the patient based on when a medicine is required to be taken and the current system clock.
   e. If there is medicine to be taken, send a UHF tag ID message to the patient mote if present.
   f. If there is no medicine to be taken, send a cleared message to the patient mote if present.
   g. If there is medicine to be taken, notify the RFID patient GUI module.
   h. If the RFID patient GUI module receives a notification, it retrieves the list of medicines required to be taken and displays them.
2. Perform the following operations once every minute:
   a. Compute the medicine list for the patient based on when a medicine is required to be taken and the current system clock.
   b. If there is medicine to be taken, send a UHF tag ID message to the patient mote if present.
   c. If there is no medicine to be taken, send a cleared message to the patient mote if present.
   d. If the RFID patient GUI module receives a notification, it retrieves the list of medicine required to be taken and displays them.

3. Upon reception of any messages from the patient mote or medicine mote, mark them as present.
4. Upon reception of a patient detected state message, log the event.
5. Upon reception of the medicine detected state message, perform these operations:
   a. If the tag ID matches the tag ID in the database, log the event.
   b. If the tag ID does not match the tag ID in the database, notify the RFID patient GUI to display invalid medicine.
6. Upon reception of the medicine taken state message, perform these operations:
   a. Log the event.
   b. Clear the medicine.
   c. Notify the RFID patient GUI to update its display.
7. Upon reception of the medicine cleared state message, perform these operations:
   a. Log the event.
   b. Add or subtract the additional weight to the last medicine taken record.
   c. Notify the RFID patient GUI to update its display.
8. Upon reception of any other messages, write to the persistent storage table accordingly.

12.4.3.2 Graphic User Interface Module

To provide user interactions with the system, a GUI is required with a display. An embedded display may be used for this purpose. With limited resource, an emulated display within the base station PC software is developed as a replacement. The display emulates and provides a GUI to assist the patient, as shown in Figure 12.10. Note the use of large font size for various medication/vitamins and of different

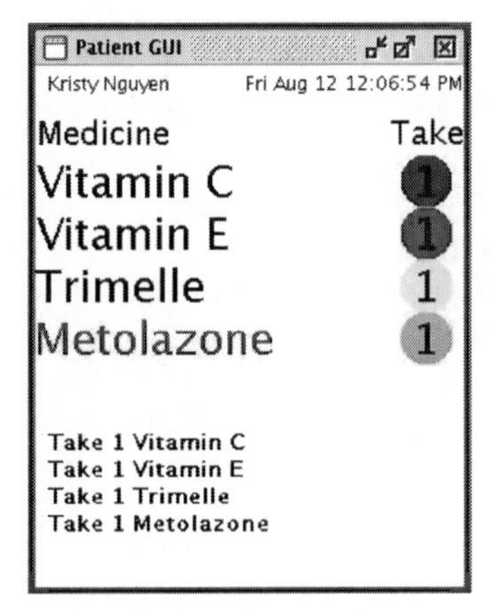

Before metolazone taken

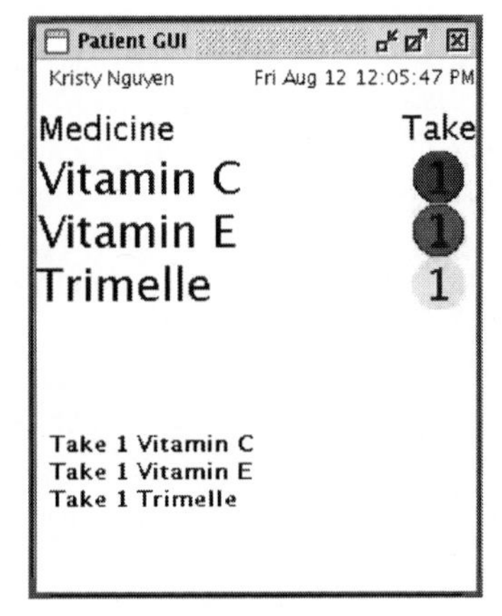

After metolazone taken

FIGURE 12.10 Patient GUI.

colors for pill quantity; this is to make it easier for old patients. Alternatively, pictures of various medicine brands/bottles may be used to replace medicine names.

12.5 TECHNICAL LIMITATIONS, CHALLENGES, AND FUTURE IMPROVEMENTS

In developing the prototype system, a number of technical challenges and limitations have been encountered. We feel that it is important to discuss them as well as suggest future improvements.

12.5.1 Technical Limitations and Challenges

The limitations and challenges included struggling with limited availability of hardware, lack of expertise and knowledge of RFID technology, finding and acquiring the proper hardware, immature RFID technology, and the nature of development on an embedded system. Unlike the traditional research of a pure software application or algorithm that requires only a PC or equivalent, our application required seven hardware components, software development, and interaction with each of them. Acquiring the HF and UHF RFID readers was quite difficult. As the UHF RFID reader is a relatively new technology with limited availability, a few UHF RFID reader manufacture companies would not sell us their hardware unless we attend their seminars, which could cost thousands of dollars. Although there are only a few UHF RFID readers, we still need to determine which reader would be appropriate for our application. Since we could not purchase all models, we chose the AWID LR-911 model as recommended by UCLA WINMEC [24] as the most reliable model available. For the HF and scale readers, we relied on Intel Labs' prototype as our guidelines. We chose newer models of the same HF RFID and scale readers used by Intel Labs, as newer models are better, smaller, and more accurate.

Because our application involved embedded platform, the development process was quite difficult. After our mote software was written, it was downloaded into the mote for testing. Without an in-circuit debugging capability, the only way to debug any part of the code was via the mote's three LEDs. Message printing capability was added after our mote software functioned properly with basic functionalities. Although the primary way to debug was still via toggling LEDs, we also carefully reviewed the mote source code when the system did not performed as expected.

As the UHF RFID technology is relatively new, we did not find an acceptable UHF tag. The preferred tag is a wrist UHF passive tag that can be worn on the patient's wrist without interference. With the current tag, the patient UHF passive tag is required to hold on the finger for proper detection.

On the HF reader end, the HF reader range was too short. Therefore, an external antenna is required. For this, a custom-designed external antenna may be more appropriate. At this stage, it is unlikely that we will acquire an external antenna and integrate in time. It is likely to be added in future phases of the prototype system.

12.5.2 Application Result and Future Improvements

The developed application can monitor and notify the patient to take their medication. The design of the system is fairly flexible. Additional patient motes can be added with minimal source changes on the base station PC software. As our first application prototype on RFID and sensors, the system is not a very commercially user friendly. Yet, it has proven the benefit of the integration of sensor network and RFID technologies, which is one major goal for this prototype.

There are a number of improvements and future extensions that can be added to the prototype system in order to be a feasible product. Improvements of the prototype system may include the following:

- Completely battery powered all system components with power management
- Higher precision scale as some medicines require 100 mg resolution
- Better UHF passive tag with less interference
- External HF antenna for more coverage area to detect medicine bottles
- Smaller HF tag in order to place on the bottom of medicine bottles
- An actual embedded display for the patient GUI
- A small form factor PC, such as mini-ITX or embedded platform, instead of a laptop
- Molding for all system components for better appearance
- Use or experiment with other MAC protocols

12.6 CONCLUSION

We have described a project that integrates both sensor network and RFID technologies. It includes an initial learning phase and a development phase. The learning phase investigates technology compatibility and capabilities through a sensor network interacting with a simulated RFID system. Simulating software modules are described as they provide excellent learning experiences and are needed before hardware purchases are possible. The development phase builds a system that consists of sensors and both HF and UHF RFID components. All the subsystems and the corresponding software modules are presented in detail. The project is targeted for in-home medication monitoring for elder patients. Future work may include, in addition of those improvements listed in Section 12.5.2, extending the prototype from medication monitoring to a broader elder home-care system, from one room to an entire house, featuring more sensors and RFID components distributed at various strategic places and on various household items. Another extension would be adding the capability of notifying family members via e-mail and networked with an external healthcare center monitor system for any assistance via the Internet.

ACKNOWLEDGMENTS

A preliminary version of the work appeared in the following paper: "A Prototype on RFID and Sensor Networks for Elder Healthcare: Progress Report," by L. Ho, Z. Walker, M. Moh, T. Hamada, and C.-F. Su, in *Proceedings of ACM Workshop on*

Experimental Approaches to Wireless Network Design and Analysis, in conjunction with *ACM SIGCOMM 2005*, Philadelphia, PA, August 2005. M. Moh is supported in part by Fujitsu Labs of America and by California State Univ. Program in Education and Research in Biotechnology (CSUPERB). T.-S. Moh is supported in part by CSU Research Grant.

REFERENCES

1. DOD, DOD RFID Website (Official Releases, Memos, etc.), http://www.acq.osd.mil/log/rfid/index.htm, http://www.dodrfid.org, downloaded April 16, 2005.
2. Finkenzeller, K., *RFID Handbook: Fundamentals and Applications in Contactless Smart Cards and Identification*, Wiley, England, 2003.
3. Ilic, C., Using tags to make teeth, *RFID Journal*, http://www.rfidjournal.com/article/articleview/1206/1/1/, downloaded January 22, 2005.
4. Exavera Technologies, eShepherd overview, http://www.exavera.com/healthcare/eshepherd.php, downloaded January 22, 2005.
5. Hooyman, N. and Kiyak, H., *Social Gerontology: A Multidisciplinary Perspective*, 6th edn., Allyn & Bacon, Boston, MA, 2002.
6. Culpepper, B., J.L. Dung, and M. Moh, Design and analysis of Hybrid Indirect Transmissions (HIT) for data gathering in wireless micro sensor networks, *ACM Mobile Computing and Communications Review (MC^2R)*, January/February 2004, 8(1), 61–83, A preliminary version, Hybrid indirect transmissions (HIT) for data gathering in wireless micro sensor networks for biomedical applications, appeared in *Proceedings of the IEEE Computer Communication Workshop*, Laguna Niguel, CA, October 2003, pp. 124–133.
7. Moh, M., B.J. Culpepper, L. Dung, T.-S. Moh, T. Hamada, and C.-F. Su, On data gathering protocols for in-body biomedical sensor networks, *Proceedings of the IEEE Globecom*, St. Louis, MO, November 2005.
8. Z. Walker, M. Moh, and T.-S. Moh. A development platform for wireless sensor networks with biomedical applications. *Proceedings of the IEEE Consumers Communication Networks (CCNC)*, Las Vegas, NV, January 2007.
9. Crossbow Inc., Mica2Dot Series, http://www.xbow.com/Products/productsdetails.aspx?sid=73, downloaded April 10, 2005.
10. Colllins, J., NASA creates thinking RF sensors, *RFID Journal*, http://www.rfidjournal.com/article/articleview/1146/1/47/, downloaded April 10, 2005.
11. Dishman, E., Inventing wellness systems for aging in place, *IEEE Computer Magazine*, May 2004.
12. Intel Research Seattle, Caregiver's assistant and carenet display: Making eldercare easier, http://seattleweb.Intelresearch.net/Projects/active/Dcdemo/, downloaded April 10, 2005.
13. O'Connor, M.C., HP kicks off US RFID demo center, *RFID Journal*, http://www.rfidjournal.com/article/articleview/1211/1/50, downloaded on April 10, 2005.
14. Roberti, M., Navy revs up RFID sensors, http://www.rfidjournal.com/article/articleview/990/1/, June 2004.
15. Roberti, M., Vendor to foxhole tracking, *RFID Journal*, http://www.rfidjournal.com/article/articleview/847, March 2004.
16. Roberti, M., BP leads the way on sensors, http://www.rfidjournal.com/article/articleview/1216/1/2/, November 1, 2004.
17. Crossbow Inc., Home page, http://www.xbow.com, downloaded April 10, 2005.
18. TinyOS, Tiny OS community forum, http://www.tinyos.net, downloaded December 4, 2004.

19. Ye, W., Download S-MAC source code for motes, http://www.isi.edu/ilense/software/smac/#Introduction, downloaded June 16, 2004.
20. Texas Instrument Inc., HF reader system series 6000, http://www.ti.com/rfid/docs/manuals/refmanuals/RI-STU-TRDCrefGuide.pdf, downloaded January 1, 2005.
21. Fishky, K. and Wang, M., A flexible, low-overhead ubiquitous system for medication monitoring, Intel Research Technical Report IRS-TR-03-011, October 2003.
22. SkyeTek Inc., SkyeRead Mini, http://skyetek.com/readers_Mini.html, downloaded April 16, 2005.
23. AWID Inc., Home page, http://www.awid.com, downloaded April 10, 2005.
24. WINMEC (Wireless Internet for Mobile Enterprise Consortium), Home page, http://www.winmec.ucla.edu/index.asp, downloaded February 20, 2005.

13 Exploiting Network Coding for Smart Healthcare

Scenarios and Research Challenges

Elli Kartsakli, Angelos Antonopoulos, and Christos Verikoukis

CONTENTS

ABSTRACT

The recent technological advances and the multiple challenges faced by the healthcare sector are strongly motivating the adoption and development of e-Health services. In this chapter, we provide a high-level framework for the design of end-to-end e-Health solutions and focus on the potential enhancements that can be achieved by the application of network coding (NC) techniques at the medium access control (MAC) layer. After a survey of existing solutions in the literature, we present a novel MAC scheme that employs NC techniques in a cooperative sensor network. The use of NC enables bidirectional communication in an efficient way, increasing the energy efficiency without compromising the quality of service (QoS).

13.1 INTRODUCTION

In the recent years, the healthcare sector is facing many challenges since it has to cope with budget restrictions and an increasing number of patients with long-term conditions. For example, chronic diseases such as asthma, heart and lung disease, and diabetes (expected to increase 252% by 2050) require frequent follow-ups and monitoring during extended periods of time, thus placing a significant burden on traditional healthcare provisioning [1]. This problem is further aggravated by the growing percentage of elderly population. Current estimations show that in Europe about 17% of the population is over the age of 65, and this percentage is expected to double by 2060 [2]. The rapid increase of the elderly population adds more challenges and demands on the public healthcare system and on medical services, in general.

The exploitation of information and communication technologies (ICT) to facilitate and improve healthcare and medical services, often referred to by the term e-Health, is bringing a shift in healthcare delivery. The e-Health paradigm involves the use of appropriate sensor devices on patients to enable the remote monitoring of vital signals, the early detection of critical conditions, and the remote control of certain medical treatments [3]. The medical sensors, placed in the vicinity of, or inside, the human body, are usually interconnected through a short-range wireless technology, thus forming a wireless body area network (WBAN). A gateway node collects all the sensory data from the WBAN and forwards them to a remote online server, where processing and integration with medical-related software applications take place. The connection of the gateway to the Internet is generally based on long-range communication access technologies for wireless local/metropolitan area networks (WLANs/WMANs).

The emerging application scenarios are numerous, including the active management of diseases such as diabetes (e.g., by measuring blood sugar levels and controlling the insulin dosage accordingly), the support for independent aging to the elderly (e.g., by tracking their medication intake and their activity level), and the monitoring of personal fitness activities to improve health and well-being (e.g., by logging health and fitness indicators during workouts) [3]. Overall, e-Health can offer significant benefits for both patients and healthcare providers, ensuring enhanced quality, efficiency, flexibility, and cost reductions in healthcare delivery.

From a wireless communications perspective, the fields of WBAN communications and wireless sensor networks (WSNs) for healthcare applications have been gaining a lot of attention by the research community. In particular, there is a fundamental need of designing novel efficient medium access control (MAC) protocols in order to tackle the specific challenges associated with the scenarios, the network topology, the constraints of the medical devices, and the particular requirements of e-Health applications. The MAC layer is responsible for the regulation of access among the system nodes to the shared wireless medium and the scheduling of transmissions, thus having a great impact on the network performance and the quality of service (QoS) provisioning.

Cooperation among nodes is a key MAC mechanism that can be employed to enhance diversity [4]. The broadcast nature of wireless communications enables the stations in the neighborhood of a transmitter to overhear its transmissions. As a result, these adjacent stations (*relays*, *helpers*, or *partners*) can actively contribute to the wireless communication, by forwarding the message from the source to other neighboring nodes or to the final destination, thus providing the latter with multiple copies of the same message, which can be locally combined using classical combining techniques to improve the reliability of the transmission.

The performance of conventional cooperative communication schemes has been further enhanced by the introduction of network coding (NC) [5]. In traditional cooperative networks, the intermediate relays simply forward the received packets to output links (store-and-forward method) in order to assist the communication between a source and a destination. However, NC has come into play to enable the intermediate nodes not only to forward but also to process the incoming information flows. In general, NC, along with error correction mechanisms such as automatic repeat request (ARQ), can significantly yield significant improvements in system throughput, robustness, complexity, security, and energy efficiency.

This chapter aims to present the potential of NC-based MAC layer protocols for WBANs and WSNs in the context of healthcare applications. The remainder of this chapter is organized in four sections. Section 13.2 presents two high-level scenarios that define the main architecture components for an end-to-end e-Health solution. Given this framework, Section 13.3 discusses existing works in the literature on end-to-end systems for e-Health provisioning, MAC protocol design for WBANs, and the application of NC in WBANs and WSNs as a means to achieve enhanced energy efficiency and reliability. Then, Section 13.4 focuses on a specific case study by presenting an NC-based MAC protocol for bidirectional communications in WSN networks. Finally, Section 13.5 closes this chapter with some concluding remarks.

13.2 e-HEALTH APPLICATION SCENARIOS

During the last few years, the research community has been motivated by the diversity of applications, the promising benefits and the potential market opportunities of e-Health solutions. Hence, the definition of a framework for the design of specific algorithms and protocols for pervasive healthcare applications is an important task. In this section, we present two application scenarios for single- and multi-patient monitoring, meant to illustrate the main architecture components of an end-to-end e-Health solution.

13.2.1 Single-Patient Scenario

The first scenario (depicted in Figure 13.1) focuses on the collection of multiple physiological data from a single patient and their delivery to a central unit. The network architecture can be divided into two parts: the WBAN and the distribution network.

The WBAN is formed among the multiple sensors deployed on the patient. The sensors can be divided into in-body devices (or implants) that generally have stricter requirements, especially in terms of energy consumption, and on-body detachable devices. In general, different sensors are employed in order to measure various metrics such as heart rate, respiratory rate, blood pressure, and oxygen saturation, thus generating heterogeneous traffic in the network. The WBAN includes a coordinator (also denoted as hub or cluster head) that is responsible for controlling the network and collecting all the sensory data. Then, a WBAN gateway is employed to relay the measured data to the distribution network. For the sake of simplicity and without loss of generality, it can be assumed that a single hardware device performs the roles of the coordinator and the gateway (e.g., a smartphone with multiple radio interfaces, interconnecting the WBAN with the distribution network).

The distribution network is responsible for the communication between the WBAN and the central control unit (e.g., the laptop of the treating physician). This communication can take place in both directions. On one hand, the patient data collected by the WBAN should be transferred in a prompt and reliable way to the central control unit in order to be revised by the responsible medical personnel. On the other hand, depending on the situation, the treating physician could adjust the

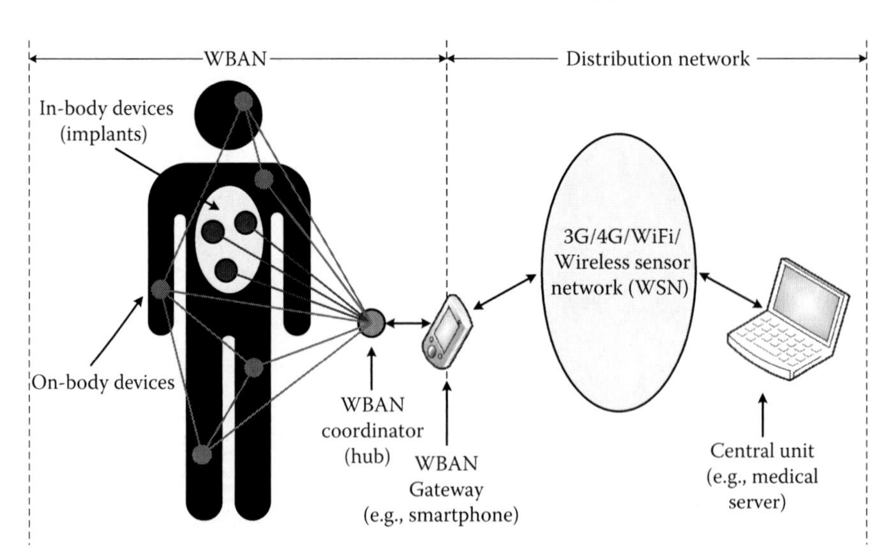

FIGURE 13.1 Single-patient scenario.

monitoring scheme (e.g., modify the sampling rate of some sensors) or update the treatment regime (e.g., optimize the dose of an insulin injection for a patient with diabetes). In this case, the distribution network should deliver the new treatment decisions to the WBAN coordinator, which, in turn, should forward the message to the corresponding sensors (the implanted insulin pump in the diabetes example). Hence, efficient bidirectional connectivity is often a desired feature in this type of scenarios.

Two different scenarios are identified for the communication in the distribution network: (1) a long-distance and (2) a local scenario. In the former case, the patient and the control unit are separated by a long distance (e.g., a home-bound patient monitored by a treating physician who is at the hospital). In this scenario, a long-range wireless technology is employed, such as 3G or 4G cellular networks or a broadband home connection to the Internet backhaul. In the second case, there is a relative proximity between the patient and the control unit (e.g., they are both located in different parts of the same hospital ward). In this scenario, a local network is deployed, such as an IEEE 802.11 WLAN or an ambient WSN, and the communication between the patient and the control unit can be established after one or more hops.

The main challenges associated with the single-patient scenario are summarized as follows:

- *Energy efficiency*: To provide comfortable and unobtrusive wearability, body sensor devices must have a very small form factor, which unavoidably has a direct impact on the size and the capacity of the available energy sources (i.e., the batteries). Furthermore, there is an imperative need for prolonged battery lifetime, sometimes in the order of years, since battery replacement in the context of healthcare applications is not a trivial issue, especially in the case of implantable devices.
- *QoS*: Low latency is the most stringent QoS requirement in WBANs, where real-time monitoring is the main application scenario. In addition, a critical issue in patient monitoring is the handling of alarm messages, issued by a sensor device when, for example, a sampled value exceeds some predefined limits. These unpredicted uplink data must be guaranteed timely delivery, since they can be closely associated to urgent, life-threatening conditions for the patient.
- *Reliability*: Closely connected to QoS guarantees, reliability is another important issue in WBANs. Reliability refers to the guaranteed delivery of the transmitted data that can be a challenge in the error-prone channels of WBANs, which must take into account the particular propagation characteristics of implanted and on-body wireless links and patient mobility.
- *Interference and coexistence*: WBANs operate on very low transmission power, given the short-range connectivity requirements and the strict regulations on acceptable specific absorption rate (SAR)* limits. Hence, on one

* SAR is a measure of the rate at which energy is absorbed by the body when exposed to a radio-frequency electromagnetic field and describes the potential for heating of the patient's tissue.

hand, interference caused by WBANs is relatively low. On the other hand, interference tolerance and coexistence with other wireless systems can be a challenging issue, especially for WBANs operating at the industrial, scientific, and medical (ISM) frequency bands and deployed in environments with many interfering sources.

- *Security*: The exchange of sensitive medical data in healthcare applications raises considerable security and privacy challenges. Hence, algorithms that guarantee security without compromising the network performance must be implemented.

13.2.2 Multipatient Scenario

The second scenario focuses on the collection of physiological data from multiple patients and their delivery to a central unit. The proposed scenario is shown in Figure 13.2.

In this topology, multiple patients are being monitored within a confined space whose size may vary from a single clinic room to a more extended hospital wing. Each patient is being monitored by a set of sensors, depending on the healthcare application. In the simplest case, a homogeneous scenario where all patients wear the same type of sensors can be assumed. A typical example would be the electrocardiogram (ECG) monitoring of multiple patients in a cardiology wing. A more complex scenario could include a variety of sensors measuring multiple physiological data,

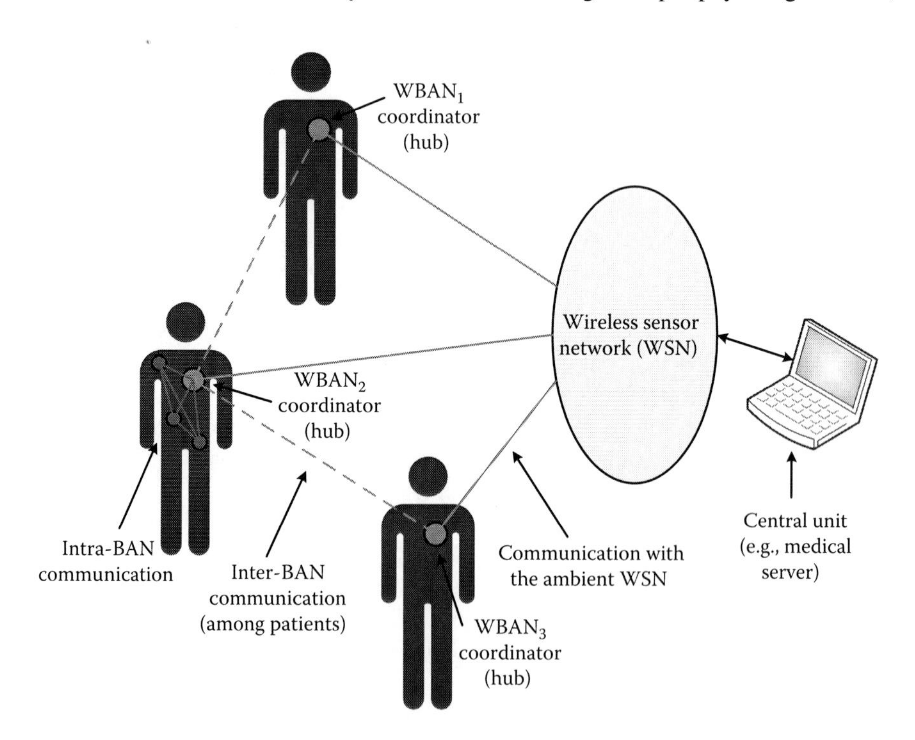

FIGURE 13.2 Multipatient scenario.

similar to the heterogeneous traffic case considered in single-patient scenario. In any case, the sensors deployed on each patient form a WBAN (or a cluster) with a single coordinating node per patient. The coordinator is responsible for collecting all the data and forwarding it to the central unit, through an ambient WSN.

This scenario can be considered as an extension of the single-patient case, introducing two additional challenges, apart from the energy efficiency, QoS, reliability, interference, and security issues that must still be tackled:

- *Cooperation*: Communication among coordinators is also possible, opening opportunities for cooperation between WBANs, in order to facilitate the delivery of data to the ambient WSN and eventually to the central care unit. Different frequency channels should be employed by each WBAN to avoid interference, whereas a common channel should be selected for inter-WBAN communication.
- *Mobility*: The impact of patient mobility should also be studied, given that patients may be allowed to move within the considered scenario (e.g., walk within the corridors of the hospital ward). As a result, the topology of the WBANs is not static (as opposed to the fixed ambient sensor network), and links among WBAN coordinators may be formed dynamically. Under scenarios with mobility, the design of localization and tracking mechanisms can be particularly important, not only to provide information on the patient's location (e.g., to track the patient in case of emergency) but also for diagnostic purposes (e.g., fitness monitoring or tracking of behavior patterns).

13.3 SOLUTIONS FOR E-HEALTH APPLICATIONS

13.3.1 End-to-End e-Health Solutions

In this section, we present some end-to-end solutions available in the literature for e-Health applications, closely related to the scenarios described in Section 13.2. End-to-end solutions for ubiquitous e-Health typically combine a short-range wireless technology, such as Bluetooth, ZigBee, IEEE 802.15.4 (WSNs), and IEEE 802.15.6 (WBANs), with a long-range wireless technology, such as Wi-Fi, WiMAX, or cellular 3G/4G.

In [6], a two-tier network architecture is considered. The lower tier consists of the WBAN, where multiple sensor devices worn by a single patient are connected to a coordinating node by employing the carrier sense multiple access with collision avoidance (CSMA/CA) access mode of IEEE 802.15.4 standard for WSNs. At the upper tier, multiple WBAN coordinators (corresponding to multiple patients) located within a specific area, for example, a hospital ward, communicate with an access point through Wi-Fi. End-to-end packet delay and access time have been modeled as a function of the number of coexisting WBANs. An extension of this work in [7] has introduced service differentiation to prioritize high-rate data streams (e.g., electroencephalogram [EEG] data) over low-rate streams (e.g., ECG data). The idea is to provide contention-free access to the high-priority data flows, while maintaining CSMA/CA access for the lower priority nodes.

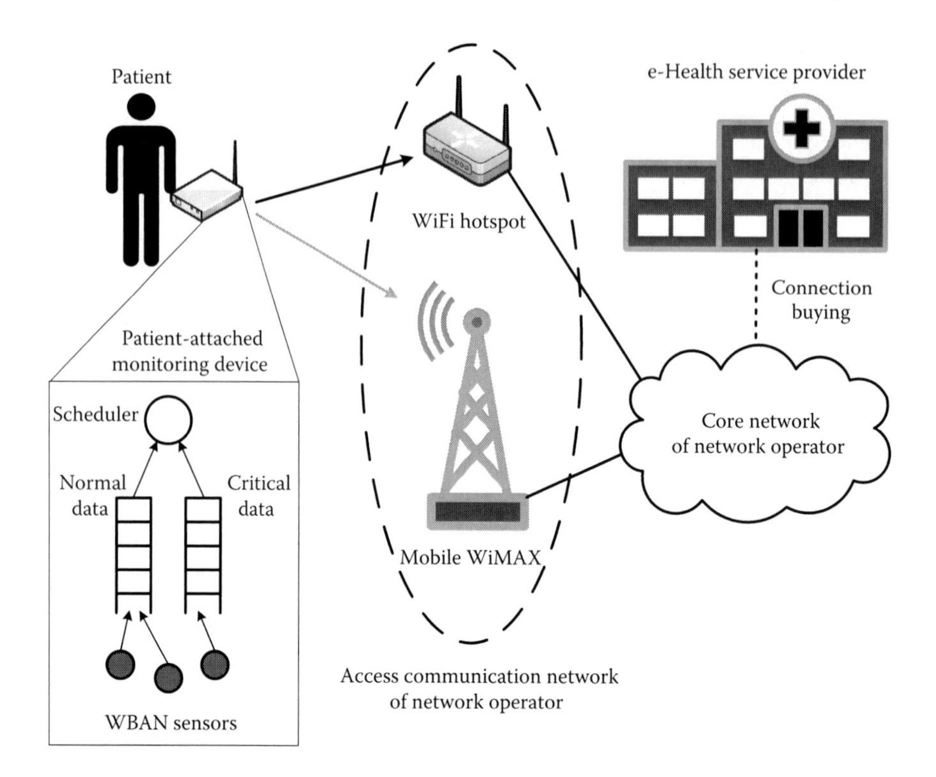

FIGURE 13.3 Architecture of remote patient monitoring system for Wi-Fi/WiMAX heterogeneous scenario. (Adapted from Niyato, D. et al., *IEEE J. Sel. Areas Commun.*, 27, 412, 2009.)

Another remote monitoring scheme that provides ubiquitous connectivity for mobile patients has been presented in [8]. In the proposed scheme, depicted in Figure 13.3, a patient-attached monitoring device collects the WBAN data, classifies them as high priority (e.g., critical data such as blood pressure, pulse rate, and heart rate) or normal priority (e.g., ECG signal), and forwards them to the e-Health provider through an heterogeneous Wi-Fi/WiMAX access communication network. The access technology is selected depending on the patient's location, considering that Wi-Fi hotspots cover only specific (mainly indoor) locations and WiMAX has a wider (outdoor) coverage. In addition, two types of connections are provided by the network operator: (1) low-cost reserved connections, allocated to patients for given amounts of time (e.g., weeks), and (2) high-cost on-demand connections, employed when the available bandwidth for reserved connections is not enough to cover the traffic load. The authors approach this e-Health scenario from the service provider's side, who has to buy in advance a certain number of reserved connections from the network operator to serve a given number of patients. Stochastic programming techniques are used to determine the optimal number of reserved connections for each wireless technology in order to minimize the provider's cost.

In some works, ambient sensor networks for environmental monitoring are employed in conjunction with WBANs, in order to provide additional information on

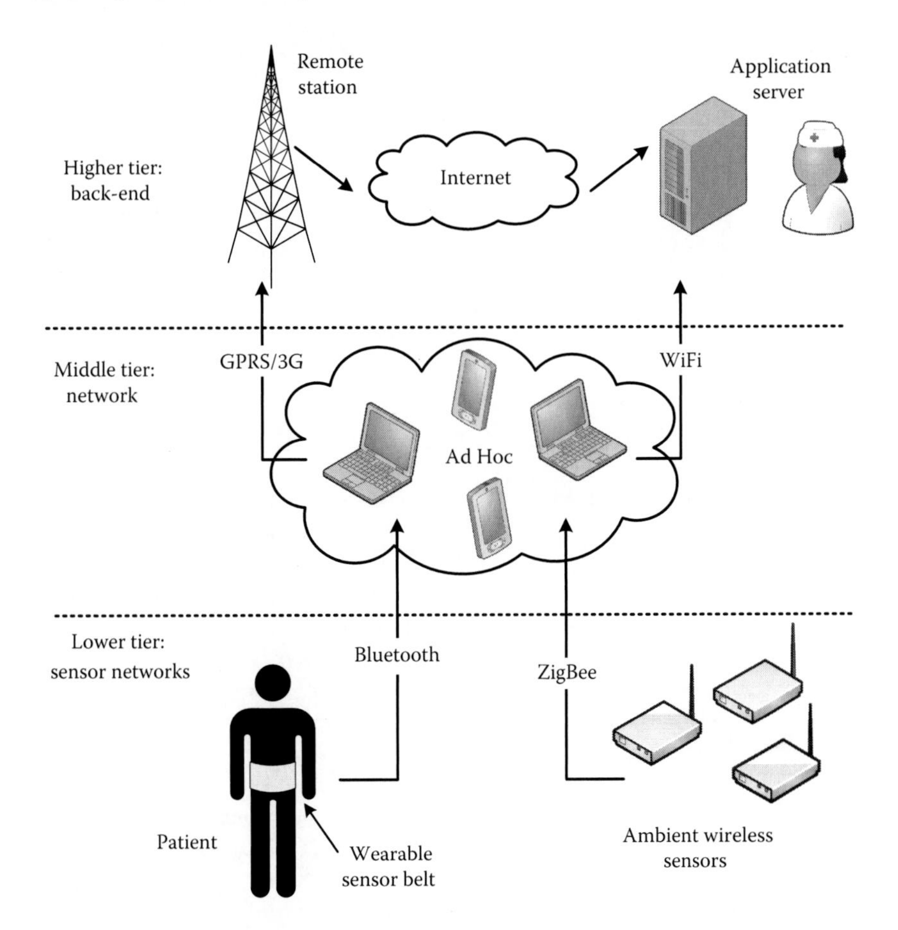

FIGURE 13.4 Example of three-tier network architecture. (Adapted from Huang, Y. et al., *IEEE J. Sel. Areas Commun.*, 27, 400, 2009.)

the patient's whereabouts, such as temperature, humidity, and light conditions. Along this line, a three-tier network architecture, depicted in Figure 13.4, is proposed in [9], for the remote monitoring of elderly or chronic patients in their residence. The lower tier consists of two systems: (1) a patient-worn fabric belt, which integrates the medical sensors and is equipped with a Bluetooth transceiver, and (2) the ambient wireless sensors that form a ZigBee network and are deployed in the patient's surroundings (e.g., in the patient's home or in a nursing house). In the middle tier, an ad hoc network of powerful mobile computing devices (e.g., laptops, PDAs) gathers the medical and ambient sensory data and forwards them to the higher tier. The middle-tier devices must have multiple network interfaces: Bluetooth and ZigBee to communicate with the lower tier and WLAN or cellular capabilities for connection with the higher layer. Finally, the higher tier is structured on the Internet and includes the application databases and servers that are accessed by the healthcare providers. The study involves a real implementation of the proposed architecture and tackles several security issues that arise along the three tiers.

13.3.2 MAC Protocols for WBANs

Focusing on short-range wireless technologies for healthcare applications, this section provides an overview of MAC layer protocols and mechanisms for WBANs/WSNs.

Many proposals guarantee collision-free data transmission by introducing low-cycle time division multiple access (TDMA) scheduling schemes where channel access is managed by a central coordinating node. An example of a TDMA-based solution for periodic traffic that takes into consideration packet retransmission due to channel errors has been proposed in [10]. In the proposed scheme, nodes are granted a fixed slot allocation in every superframe, long enough to transmit a data packet and receive an acknowledgment (*ACK*) by the coordinator. The nodes generally remain in a sleep mode and only wake up at predefined intervals to receive beacons and transmit data.

Another mechanism for collision-free data transmission has been proposed in cascading information retrieval by controlling access with distributed slot assignment (CICADA) [11]. In CICADA, the network is organized in a tree topology, with the coordinator placed at the top of the tree. The protocol operation defines two cycles for control and data transmission, further divided into time slots. During the control cycle, a slot assignment scheme is forwarded from parent to child node, starting from the coordinator. During the data cycle, the nodes wake up during their allocated slots to transmit data toward the coordinator, while data are flowing upward from the bottom toward the coordinator.

An example of a hybrid access mechanism has been implemented in BodyMAC [12]. The protocol introduces a superframe structure divided into three parts: a synchronization beacon, a downlink part reserved for the transmission of data from the coordinator, and an uplink part for data transmitted by the nodes to the coordinator. The uplink is split into a contention access part (CAP), during which the nodes can access the channel through CSMA/CA rules, and a contention-free part (CFP), during which time slots are allocated dynamically to the nodes by the coordinator. In order to gain slot allocation, the nodes must notify their bandwidth requirements to the coordinator during the CAP period.

A different implementation of the same principle is adopted in Distributed Queuing Body Area Network (DQBAN) [13]. The superframe structure in DQBAN is again divided into three parts: (1) the access part, further divided into a number of access minislots; (2) the scheduling part, further dividing into a number of scheduling minislots; and (3) the data part, in which collision-free data transmission takes place. Nodes compete for channel access by transmitting small control packets in randomly selected access minislots, and a collision resolution algorithm is employed to resolve any collisions among access requests. After a successful access request, nodes enter a transmission queue and are enabled to transmit their packet in the data part in a collision-free manner.

Another approach has been adopted in traffic-aware dynamic MAC (TAD-MAC) [14], in which the coordinator tries to learn the sleeping pattern of the nodes. TAD-MAC operation takes place in two phases. During the evolution phase, the coordinator polls the nodes at different times to see if they have packets to transmit, thus gathering statistics on their traffic load that are stored in a register bank. Different policies are employed for in-body (implantable) and on-body (wearable) nodes. In the first case, a star topology is considered and the register bank is maintained at the coordinator, whereas in the second case, a mesh topology is assumed, and nodes keep statistics for their neighbor's traffic. The second operation phase begins after several cycles, when the protocol converges to a steady state and the beacon interval is adapted to the wakeup times of each node.

13.3.3 NC for WBANs

Recently, the application of NC techniques on the MAC layer has been studied as a potential way to improve energy efficiency and enhance reliability in wireless networks. The impact of NC in *green* communications is currently an active research subject, especially in broadcast and multicast scenarios [15–18]. The recent research work that investigates the energy aspect of NC applications deals mostly with the network layer. Cui et al. [19] introduced CORP by using a suboptimal scheduling algorithm that exploits NC opportunities, thus achieving a significant power saving over pure routing. More recently, Miao et al. [20] proposed an energy-efficient broadcast algorithm using NC for gradient-based routing in wireless networks. Their algorithm aims to reduce the network traffic and, consequently, the energy consumption in order to prolong the network lifetime.

In the domain of WSNs, Munari et al. [21] have introduced NC-PAN in order to enhance the throughput gain in TDMA high data-rate scenarios. However, NC-PAN is not compatible with the non-beacon-enabled mode of IEEE 802.15.4 that adopts CSMA. In the context of WBANs, the work presented in [22] shows the potential throughput improvement under error-prone channels through the application of NC, as a function of the number of redundancy packets, the employed relays, and the number of sink nodes. More advanced schemes, such as cooperative diversity coding where both coded and uncoded data packets are transmitted, can yield ever better results [23]. A tree topology has been assumed in [24], where multiple nodes communicate with the coordinator through a small number of relays. The nodes send their uncoded data to the relays, which in turn forward a set of both coded and uncoded packets to the destination. The obtained results show that NC can offer higher reliability with respect to traditional redundancy schemes where multiple retransmissions of uncoded packets take place. Finally, similar conclusions are drawn in [25], where the coding decisions take into account the reliability requirements of different traffic flows.

13.4 CASE STUDY: NC-AIDED COOPERATIVE ARQ PROTOCOL FOR WSNs

In this section, we focus on a specific case study that illustrates the potential benefits of NC on sensor networks. In particular, we present an NC-aided cooperative ARQ MAC protocol for WSNs (NCCARQ-WSN) that coordinates the retransmissions among a set of relay nodes, which act as helpers in a bidirectional communication.

The proposed scheme could be applied to any scenario in which bidirectional communication is desired.

For instance, in the single-patient scenario, depicted in Figure 13.1, NCCARQ-WSN could be employed for bidirectional connection in the distribution WSN, connecting the WBAN hub with the central unit. In this case, the hub would transmit the medical data collected from all the sensors to the central unit, while the central unit would transmit downlink control information (e.g., based on previous measurements) that should be forwarded to the sensors.

The main features of NCCARQ-WSN are given as follows:

1. NC techniques are used in order to enhance the system performance.
2. Less control packets—and consequently less overhead—are inserted in the network.
3. Our protocol operates in CSMA scenarios, hence being compatible with the IEEE 802.15.4/IEEE 802.15.6 standards.
4. We present an analytical model for the energy consumption in the network from the MAC layer point of view.

13.4.1 System Model

Figure 13.5 depicts the considered network, consisting of two nodes (S and D) that have data packets to exchange in a bidirectional communication and a set of n nodes ($R_1, R_2, \ldots, R_n$) that act as relays in this particular network setup. We further assume that node D is located marginally in the transmission range of node S, and vice versa, resulting in a weak direct link with high packet error rate (*PER*). However, the erroneous direct transmissions are compensated by employing network cooperation after the transmission of special control packets (i.e., request for cooperation [*RFC*]) either by S or D.

In our system model, all nodes in the network operate in a promiscuous mode in order to be able to listen to every ongoing transmission and cooperate if requested, while the relays always store a copy of any captured data packet (regardless of its destination address) until it is acknowledged by the intended destination. In addition, all relays are equipped with NC capabilities, thus being able to apply XOR techniques to the packets before any further forwarding takes place. Apparently, the existence of many relay nodes in the cooperation phase causes conflicting situations that need to be handled by novel MAC protocols. To this end, NCCARQ MAC has been designed to coordinate the retransmissions among the relay set in the network.

13.4.2 Protocol Description

In NCCARQ-WSN, a cooperation phase is initiated once a data packet has not been received correctly by the destination.* Various error detection mechanisms such

* The terms *source* and *destination* are used with regard to the initial transmission. Throughout the communication, both nodes can act as source or destination of data flows.

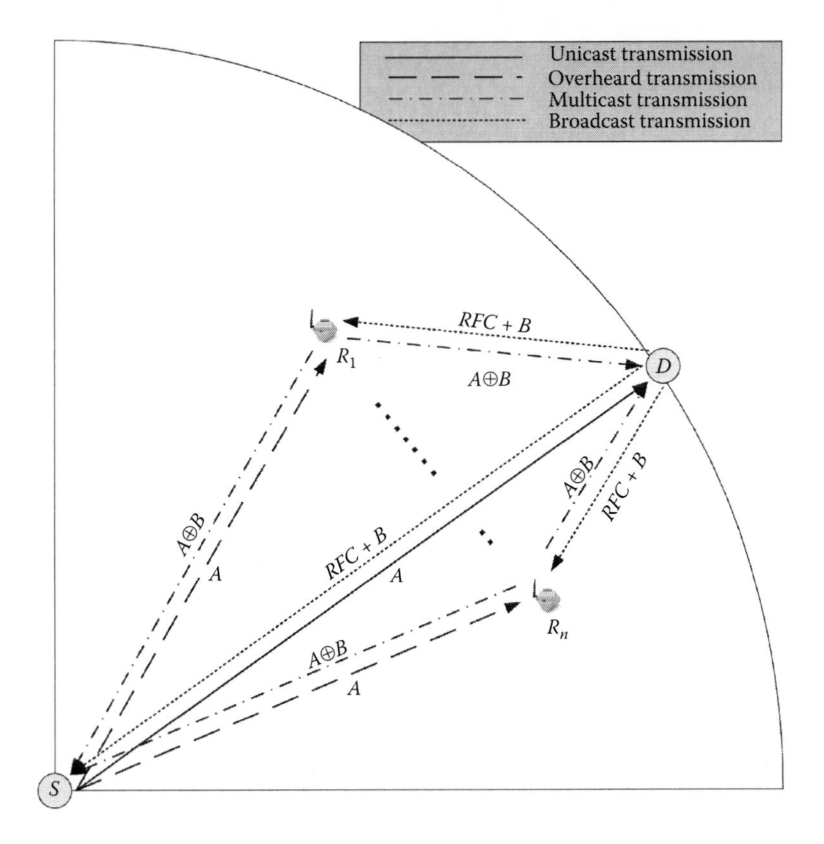

FIGURE 13.5 NCCARQ-WSN system model.

as cyclic redundancy check (CRC) [26] can be applied to perform error control to the received messages. Therefore, the destination station initiates the retransmission phase by broadcasting an *RFC* packet to the network, indicating the need for cooperation. Furthermore, in case of bidirectional traffic, that is, when the destination has data for the source, the data packet is broadcasted piggybacked on the *RFC* message.

The stations that receive the *RFC* packet are potential candidates to become active relays for the communication process. Therefore, the relay set is formed upon the reception of the *RFC*, and the participants (relays) get ready to forward their information. Since the relays have already stored the packets addressed both to the destination (so-called cooperative packet) and to the source (so-called piggy-backed packet), they create a new coded packet by combining the two native data packets, using the XOR method. Accordingly, the active relays will try to gain access to the channel in order to persistently transmit the network-coded packet. Once the source and the destination receive the network-coded packet from the relays, they are able to decode it (by applying again the XOR operation to their respective native packet and the received network-coded packet) and extract the original data packets. Subsequently, they confirm the received data packet by transmitting the respective

ACK packets, thus terminating the cooperation phase. Upon successful *ACK* reception, the relays are informed that the particular communication has been completed and, hence, may erase the respective packets from their buffers. Otherwise, in case that either one or both of the received coded packets cannot be decoded after a certain maximum cooperation timeout due to transmission errors, the relays are obliged to forward again the coded packet following the same rules.

An important feature of NCCARQ-WSN is backward compatibility with the non-beacon-enabled mode (CSMA) of IEEE 802.15.4. In this mode, the stations maintain backoff counters and sense the channel for a clear channel assessment (*CCA*) period before transmitting their data. In addition, the transmission of *ACK* packets commences a turnaround time (T_{ACK}), which enables the station to switch mode between reception and transmission. It is also worth mentioning that long frames are followed by a long interframe space (*LIFS*), while short frames are followed by a SIFS period of time. Moreover, there is a common transmission bit rate for both data and control packets. Nevertheless, despite the backward compatibility, NCCARQ-WSN has undergone some modifications in order to efficiently exploit the advantages of both cooperative and NC techniques:

1. A reliable multicast communication scheme is guaranteed by employing *ACK* packets for the reception of the network-coded data.
2. For bidirectional traffic, the data packets are transmitted along with the *RFC* packets, without taking part in the contention phase.
3. Since the subnetwork formed by the relay set operates in saturated conditions, as the relays store the packets of the overheard transmissions, it is necessary to execute a backoff mechanism at the beginning of the cooperation phase to minimize the probability of an initial collision.

An operation example of NCCARQ-WSN is given next, in order to clarify our proposed access protocol. A simple network topology with four stations is considered, all of them in the transmission range of each other. A source station (*S*) transmits a data packet (*A*) to a destination station (*D*) that also has a packet (*B*) destined to the source station. Furthermore, there are two relay nodes (R_1 and R_2) that support this particular bidirectional communication. The entire procedure is depicted in Figure 13.6 and explained as follows:

1. At instant t_1, node *S* transmits the data packet *A* to node *D*.
2. At instant t_2, node *D* fails to demodulate the received data packet. Hence, it waits for T_{ACK} to switch mode in order to broadcast an *RFC* message asking for cooperation by the neighboring sensor nodes (R_1 and R_2 in this example), along with the data packet *B*, destined to node *S*. It is worth mentioning that the transmission takes place after a *CCA* time, while the *LIFS* time is omitted to provide the cooperation phase with higher priority.
3. The reception of both *RFC* and *B* (t_3) triggers the stations R_1 and R_2 to become active relays and set up their backoff counters (CW_1 and CW_2, respectively) in order to participate in the contention phase for the transmission of the network-coded packet $A \oplus B$. Moreover, before starting the

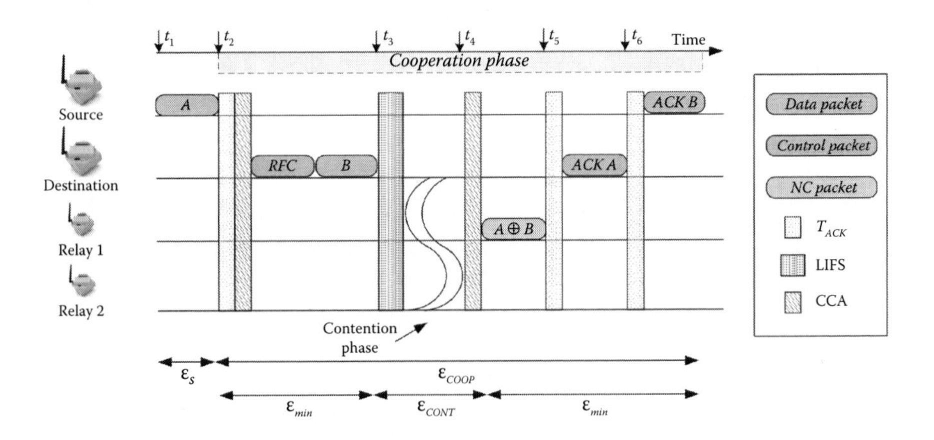

FIGURE 13.6 NCCARQ-WSN example of frame sequence.

transmission, they have to sense the channel idle for *LIFS* time, as the standard dictates.

4. At instant t_4, CW_1 expires and R_1 transmits the coded packet $A \oplus B$ to the nodes S and D simultaneously.
5. At instant $t_5(t_6)$, the station $D(S)$ decodes properly the XOR-ed packet and transmits an *ACK* packet to verify the correct reception of packet $A(B)$.

13.4.3 Protocol Analysis

In this section, taking into account the NCCARQ-WSN operation, we develop a theoretical probabilistic model to analytically evaluate the protocol performance in WSN scenarios. The design of these models leads to closed-form expressions for the packet delay, the system throughput and the energy consumption in the network.

13.4.3.1 Delay Analysis

The NC techniques applied in NCCARQ-WSN imply the simultaneous transmission of more than one packet in the network. Therefore, we analytically estimate the expected time needed for two packets to be exchanged in our protocol.

The total time elapsed from the initial transmission from the source until the correct reception of the coded packet at the destinations (S and D) can be defined as

$$\mathbf{E}[T_{total}] = \mathbf{E}[T_S] + \mathbf{E}[T_{COOP}] \tag{13.1}$$

where

$\mathbf{E}[T_S]$ represents the average time for the direct transmission of a single data packet from the source to the destination

$\mathbf{E}[T_{COOP}]$ corresponds to the average time required for a cooperative transmission via relays to be completed

Since the estimation of $\mathbf{E}[T_S]$ is straightforward depending on the network configuration (i.e., packet length and transmission data rate), our analysis is focused on the term $\mathbf{E}[T_{COOP}]$ in order to derive a closed-form expression for the packet delay. The average time spent during the cooperation phase can be defined as

$$\mathbf{E}[T_{COOP}] = \mathbf{E}[T_{min}] + \mathbf{E}[T_{CONT}] \tag{13.2}$$

where $\mathbf{E}[T_{min}]$ is the minimum average delay in case of perfect scheduling among the relays, that is, when no contention takes place, while the term $\mathbf{E}[T_{CONT}]$ is used to denote the additional delay caused due to the contention phase.

In order to obtain $\mathbf{E}[T_{min}]$ and $\mathbf{E}[T_{CONT}]$, we must first calculate the expected number of retransmissions ($\mathbf{E}[r]$) required to properly demodulate the coded packet at the destination nodes. In general, ($\mathbf{E}[r]$) is directly connected with the *PER* of the link between the relays and the destination ($PER_{R\to D}$). However, in our scheme, two packets are sent at the same time via different channels (relay to source and relay to destination), and as a result, the number of retransmissions can be expressed as

$$\mathbf{E}[r] = \frac{1 + \dfrac{(1 - PER_{R\to S}) \cdot PER_{R\to D}}{1 - PER_{R\to D}} + \dfrac{(1 - PER_{R\to D}) \cdot PER_{R\to S}}{1 - PER_{R\to S}}}{1 - PER_{R\to S} \cdot PER_{R\to D}} \tag{13.3}$$

Therefore, the term $\mathbf{E}[T_{min}]$ can be calculated as

$$\begin{aligned}\mathbf{E}[T_{min}] = {} & T_{T_{ACK}} + T_{CCA} + T_{RFC} + T_B + T_{ONC} + \mathbf{E}[r] \cdot (T_{LIFS} + T_{CCA} + T_{A\oplus B} + T_{T_{ACK}}) \\ & + T_{ACK} + T_{T_{ACK}} + T_{ACK}\end{aligned} \tag{13.4}$$

where

$\mathbf{E}[r]$ corresponds to the expected number of retransmissions until the correct reception of the network-coded packet by S and D

T_{RFC}, T_{ACK}, and T_B represent the transmission times for the *RFC*, *ACK*, and *B* packet, respectively

Furthermore, $T_{A\oplus B}$ is the time required to retransmit a coded packet, while T_{ONC} is the time that a relay needs for applying NC techniques in the two native packets. Finally, T_{LIFS}, $T_{T_{ACK}}$, and T_{CCA} are the duration of a *LIFS* silence period, a T_{ACK} period, and a *CCA* period in IEEE 802.15.4.

Moreover, the term $\mathbf{E}[T_{CONT}]$ can be defined as

$$\mathbf{E}[T_{CONT}] = \mathbf{E}[r] \cdot \mathbf{E}[T_c] \tag{13.5}$$

where $\mathbf{E}[T_c]$ represents the average time required to transmit a single packet during the contention phase among all the relays. In order to compute this value, we need to model the backoff counter of each of the relays with the Markov chain presented in [27] (Figure 13.7), since the formed subnetwork behaves as a saturated IEEE 802.15.4 non-beacon-enabled network despite the modifications in the access rules.

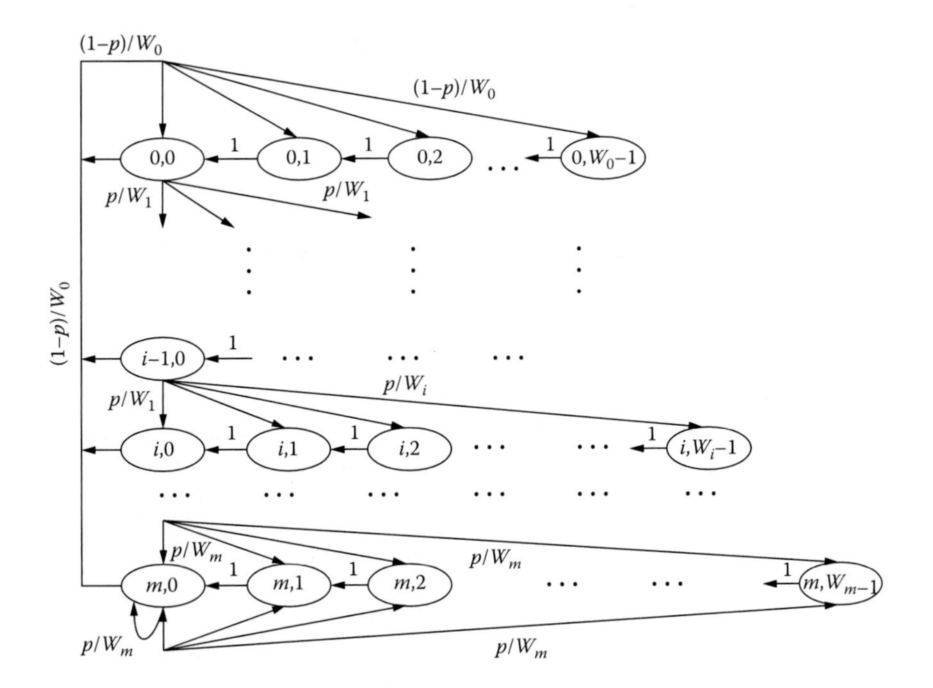

FIGURE 13.7 Modeling the backoff counter operation with a 2D Markov chain. (From Fujii, Y. et al., Saturation throughput analysis of unslotted CSMA-CA networks, *Proceedings of the 25th International Technical Conference on Circuits/Systems, Computers and Communications*, Pattaya, Thailand, pp. 688–691, July 2010.)

According to this model, the probability τ that a station transmits in a randomly chosen slot is given by

$$\tau = \sum_{i=1}^{m} b_{i,0} = \frac{b_{0,0}}{1-p} = \frac{2(1-2p)}{(1-2p)(W+1)+pW(1-(2p)^m)} \tag{13.6}$$

where

$$b_{0,0} = \frac{2(1-2p)(1-p)}{(1-2p)(W+1)+pW(1-(2p)^m)} \tag{13.7}$$

and the probability of a collision p as a function of τ is given by

$$p = 1-(1-\tau)^{n-1} \tag{13.8}$$

In formulas (13.6) and (13.7), $b_{i,k}$ represents the steady-state probability of the state $\{i, k\}$, W is the size of the contention window, m denotes the number of the backoff stages, and n corresponds to the number of the relays in the network.

Furthermore, the probability that at least one relay attempts to transmit can be expressed as

$$p_{tr} = 1 - (1 - \tau)^n \tag{13.9}$$

and the probability of a successful transmission, that is, one station transmits conditioned on the fact that at least one station transmits, is given by

$$p_{s|tr} = \frac{n\tau(1-\tau)^{n-1}}{1-(1-\tau)^n} \tag{13.10}$$

Moreover, the probabilities of having an idle (p_i), successful (p_s), or collided (p_c) slot can be written as

$$p_i = 1 - p_{tr} \tag{13.11}$$

$$p_s = p_{tr} \cdot p_{s|tr} \tag{13.12}$$

$$p_c = p_{tr}(1 - p_{s|tr}) \tag{13.13}$$

Considering the aforementioned probabilities and given that the average number of slots that we have to wait before having a successful transmission can be represented as

$$\mathbf{E}[N] = \sum_{k=0}^{\infty} k(1-p_s)^k p_s = \frac{1}{p_s} - 1 \tag{13.14}$$

the total contention time can be written as

$$\mathbf{E}[T_c] = \mathbf{E}[N] \cdot \mathbf{E}[T_{slot|non\_successful\_slot}] \tag{13.15}$$

Applying Bayes' theorem, we are able to estimate the average duration of a slot, given that the specific slot is either idle or collided:

$$\mathbf{E}[T_{slot|non\_successful\_slot}] = \left(\frac{p_i}{1-p_s}\right)\sigma + \left(\frac{p_c}{1-p_s}\right)T_{col} \tag{13.16}$$

with σ representing the duration of an idle slot, while T_{col} corresponds to the time of a collision and in our scheme is equal to

$$T_{col} = T_{LIFS} + T_{CCA} + T_{A \oplus B} + T_{TACK} \tag{13.17}$$

Therefore, using Equations 13.14 through 13.16, formula (13.5) can be rewritten as

$$\mathbf{E}[T_{CONT}] = \mathbf{E}[r] \cdot \left(\frac{1}{p_s} - 1 \right) \left[\left(\frac{p_i}{1 - p_s} \right) \sigma + \left(\frac{p_c}{1 - p_s} \right) T_{col} \right] \tag{13.18}$$

Finally, we are able to derive a closed-form formula and compute the total delay for two packets to be exchanged in the system by exploiting Equations 13.2, 13.4, and 13.18.

13.4.3.2 Throughput Analysis

The total throughput of the network can be defined as the sum of the throughput that is produced by the successful direct transmissions plus the throughput derived by the cooperation phase after erroneous packet receptions. This can be mathematically expressed as

$$\mathbf{E}[S_{total}] = \mathbf{E}[S_D] + \mathbf{E}[S_{COOP}] \tag{13.19}$$

where

$$\mathbf{E}[S_D] = (1 - PER_{S \to D}) \cdot \frac{\mathbf{E}[P]}{\mathbf{E}[T_D]} \tag{13.20}$$

and

$$\mathbf{E}[S_{COOP}] = 2 \cdot PER_{S \to D} \cdot \frac{\mathbf{E}[P]}{\mathbf{E}[T_{total}]} \tag{13.21}$$

In the preceding expressions, the parameters $\mathbf{E}[T_D]$ and $\mathbf{E}[T_{total}]$ have been already defined. Furthermore, the PER between the source and the destination is given by $PER_{S \to D}$, while $\mathbf{E}[P]$ denotes the average packet payload. In this point, it must be clarified that the coefficient 2 in formula (13.21) is mandatory, since NC enables the simultaneous transmission of two packets.

Therefore, having obtained a closed-form expression for $\mathbf{E}[T_{total}]$ and since $\mathbf{E}[P]$, $\mathbf{E}[T_D]$, and $PER_{S \to D}$ are known parameters, we are able to compute the theoretical system's throughput.

13.4.3.3 Energy Performance Analysis

Following the same line of thought as in the throughput calculation, the average total energy consumption in the network is described by the following expression:

$$\mathbf{E}[\mathcal{E}_{total}] = \mathbf{E}[\mathcal{E}_S] + \mathbf{E}[\mathcal{E}_{COOP}] \tag{13.22}$$

where $\mathbf{E}[\mathcal{E}_{COOP}]$ and $\mathbf{E}[\mathcal{E}_S]$ represent the energy consumption during the cooperative phase and the initial transmission from the source, respectively. These terms have also been depicted in Figure 13.6.

In our attempt to clarify the aforementioned equation, we consider three discrete power modes:

1. *Transmission mode*, when the node is transmitting data/control packets
2. *Reception mode*, when the node is receiving data/control packets
3. *Idle mode*, when the node is sensing the medium without performing any action

The power levels associated to each mode are P_T, P_R, and P_I, respectively. Furthermore, the relationship between energy and power is given by $\mathcal{E} = P \cdot t$, where the terms $\mathcal{E}$, P, and t represent the energy, the power, and the time, respectively. Let us recall that the network consists of a source, a destination, and a set of n relays. Therefore, taking into account the network topology, we have

$$\mathbf{E}[\mathcal{E}_S] = P_T \cdot T_A + (n+1) \cdot P_R \cdot T_A \tag{13.23}$$

where T_A corresponds to the transmission time for packet A. On the other hand, the term $\mathrm{E}[\mathcal{E}_{COOP}]$ could be further expressed as

$$\mathbf{E}[\mathcal{E}_{COOP}] = \mathbf{E}[\mathcal{E}_{min}] + \mathbf{E}[\mathcal{E}_{CONT}] \tag{13.24}$$

where

$\mathbf{E}[\mathcal{E}_{min}]$ denotes the energy consumption in a perfectly scheduled cooperative phase

$\mathbf{E}[\mathcal{E}_{CONT}]$ is the energy consumed during the contention phase (i.e., idle slot and collisions)

The minimum time is computed as follows:

$$\begin{aligned}\mathbf{E}[\mathcal{E}_{min}] = {} & (n+2) \cdot P_I \cdot (T_{T_{ACK}} + T_{CCA}) + P_T \cdot (T_{RFC} + T_B) + (n+1) \cdot P_R \cdot (T_{RFC} + T_B) \\ & + (n+2) \cdot P_I \cdot T_{ONC} + \mathbf{E}[r] \cdot ((n+2) \cdot P_I \cdot (T_{LIFS} + T_{CCA}) + P_T \cdot T_{A\oplus B} + 2 \cdot P_R \cdot T_{A\oplus B} \\ & + (n-1) \cdot P_I \cdot T_{A\oplus B} + (n+2) \cdot P_I \cdot T_{T_{ACK}}) + 2 \cdot P_T \cdot T_{ACK} + (n+2) \cdot P_I \cdot T_{T_{ACK}} \\ & + 2 \cdot (n+1) \cdot P_R \cdot T_{ACK}\end{aligned} \tag{13.25}$$

Equations 13.23 through 13.25 are based on the following principles:

- All stations remain idle during the *LIFS, CCA*, and T_{ACK} times.
- The relays that lose the contention phase turn in idle mode.
- When a station transmits a packet (control or data), the rest of the stations are in promiscuous mode, thus capturing the packets.

The total energy consumption during the contention phase is derived from the energy spent during both idle and collided slots. Hence, formulating an analytical model for the energy consumed during this phase constitutes a very challenging task, mainly due to the uncertainty of the average number of nodes involved in a collision. Let us start by defining that

$$\mathbf{E}[\mathcal{E}_{CONT}] = \mathbf{E}[r] \cdot \mathbf{E}[\mathcal{E}_C] \tag{13.26}$$

where $\mathbf{E}[\mathcal{E}_C]$ represents the average energy required to transmit an NC packet during the contention phase among all the relays. In order to calculate the energy consumption during the collisions, we have to estimate the average number of stations that transmit a packet simultaneously. The probability p_k that exactly k stations are involved in a collision is

$$p_k = \frac{\binom{n}{k} t^k (1-t)^{n-k}}{1-(1-t)^{n-(k-1)}} \tag{13.27}$$

Therefore, the expected number $\mathbf{E}[K]$ of stations that are involved in a collision is

$$\mathbf{E}[K] = \sum_{k=2}^{n} k \cdot p^k = \sum_{k=2}^{n} k \cdot \frac{\binom{n}{k} t^k (1-t)^{n-k}}{1-(1-t)^{n-(k-1)}} \tag{13.28}$$

During the idle slots, all the stations in the network remain idle. On the other hand, during the collisions, more than one relay is in transmission mode, two stations (the source and the destination) are in reception mode, while the rest of the relays are in idle mode. Considering the probabilities that we have derived regarding the contention phase (p_c, p_i), the preceding assumptions can be mathematically expressed as

$$\mathbf{E}[\mathcal{E}_C] = p_i \cdot ((n+2) \cdot P_I \cdot \sigma) + p_c \cdot (\mathbf{E}[K] \cdot P_T \cdot T_{col} + 2 \cdot P_R \cdot T_{col} + (n - \mathbf{E}[K]) \cdot P_I \cdot T_{col}) \tag{13.29}$$

where all the parameters have been already defined. Thus, combining Equations 13.22 through 13.26 and 13.29, we are able to estimate the total amount of the energy that is consumed in our protocol.

13.4.4 Performance Evaluation

In order to validate our analysis and further evaluate the performance of NCCARQ-WSN, we have developed a time-driven C++ simulator that executes the rules of the proposed protocols. In the following sections, we present the simulation setup along with the results of our experiments.

13.4.4.1 Simulation Scenario

The network under simulation consists of a pair of transmitter–receiver (both nodes transmit and receive data) and a set of relay nodes that facilitate the communication, all of them in the transmission range of each other. In our experiments, we consider saturated conditions, that is, the nodes have always packets to send in their buffers. Additionally, the relay nodes are capable of performing NC techniques to their buffered packets before relaying them. In order to focus on the impact of both NC and cooperative communication, we have made the following assumptions:

1. The traffic is bidirectional, that is, the destination node has always a packet to transmit back to the source node.
2. Original transmissions from source to destination are always received with errors, as we consider a highly noisy channel with $PER_{S \to D} = 1$.
3. The channel between the source and the destination is error symmetric, that is, $PER_{S \to D} = PER_{D \to S}$.
4. The channel between the source and the relays is error-free, that is, $PER_{S \to R} = 0$.

The configuration parameters of the network are summarized in Table 13.1 considering the IEEE 802.15.4 physical layer. The relay set consists of five nodes, each of them implementing a backoff counter starting with a contention window $CW_{min} = 8$. The time for applying NC (T_{ONC}) to the data packets is considered to be negligible, since the coding takes place between only two packets. Based on hardware specifications and since different power modes are allowed, we have chosen the following power levels for our scenarios: $P_T = 15$ mW,* $P_R = 35$ mW, and $P_I = 712$ μW [28]. A packet size of 100 bytes has been assumed, given that medical sensors usually employ small size data packets.

In order to evaluate our approach, we compare our scheme with a simple cooperative ARQ scheme (so-called CARQ), where the bidirectional communication takes place in two steps. In the first step, the source sends a packet to the destination and, upon the erroneous reception, the destination broadcasts the *RFC* packet, thus triggering the relays to retransmit the packet. In the second step, the destination transmits its own packet to the source, and the same procedure as in the first step is repeated, thus consuming valuable network resources. In both steps, the relays take part in the contention phase in order to access the medium and transmit their packets.

TABLE 13.1
System Parameters

Parameter	Value (Bytes)	Parameter	Value	Parameter	Value
Data packets	100	*Data rate*	256 kbps	*Backoff unit*	320 μs
MAC header	9	T_{ACK}	192 μs	P_T	15 mW
PHY header	6	*CCA*	128 μs	P_R	35 mW
ACK, RFC	11	*LIFS*	640 μs	P_I	712 μW

* In our example, we employ the maximum transmission power levels that are offered in IEEE 802.15.4.

The delay and the throughput, as defined in Sections 13.4.3.1 and 13.4.3.2, respectively, are the metrics that we use in order to evaluate the QoS performance of our protocol. Moreover, in order to evaluate the energy performance of our proposed protocol, we use the energy efficiency metric η [29], defined as

$$\eta = \frac{\text{Total amount of useful data delivered (bits)}}{\text{Total energy consumed (J)}} \tag{13.30}$$

Before proceeding to the simulation results, it is worth mentioning that the definition in Equation 13.30 inherently implies that NC benefits the energy efficiency of a protocol, as the number of the delivered bits increases by combining multiple data packets.

13.4.4.2 Performance Results

Figure 13.8 shows the throughput enhancement that NCCARQ-WSN achieves compared to simple cooperative protocols without NC capabilities. While, in the case of simple cooperative schemes, throughput values saturate around 66 kbps, and NCCARQ-WSN achieves 99 kbps, exhibiting a 50% throughput increase. Evidently, it can be observed that as the number of required retransmissions (x-axis) grows due to channel impairments, the relative throughput gain increases as well, reaching a difference of 80% in networks where five retransmissions are

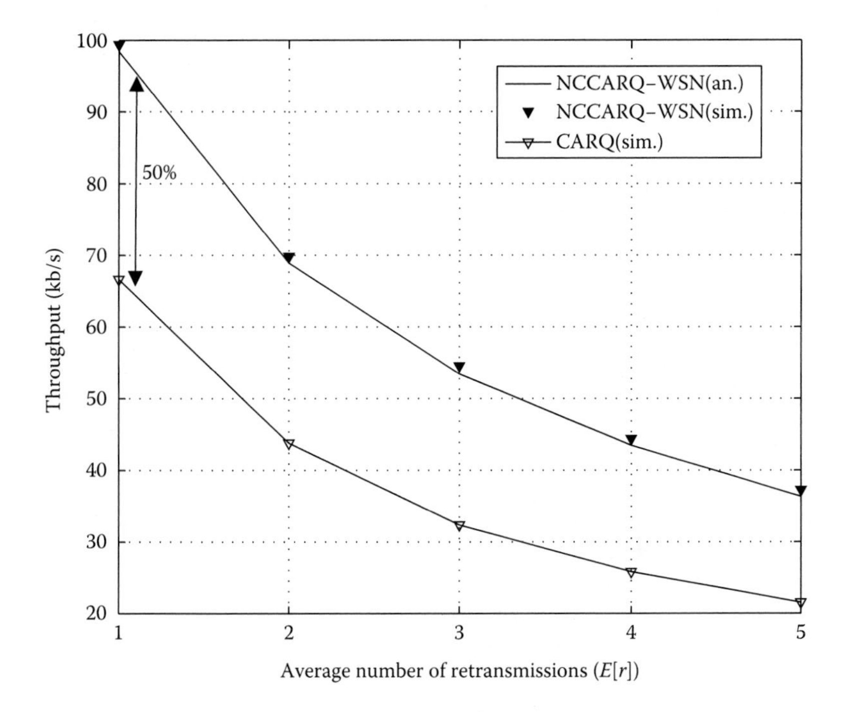

FIGURE 13.8 System throughput (NCCARQ-WSN vs. CARQ).

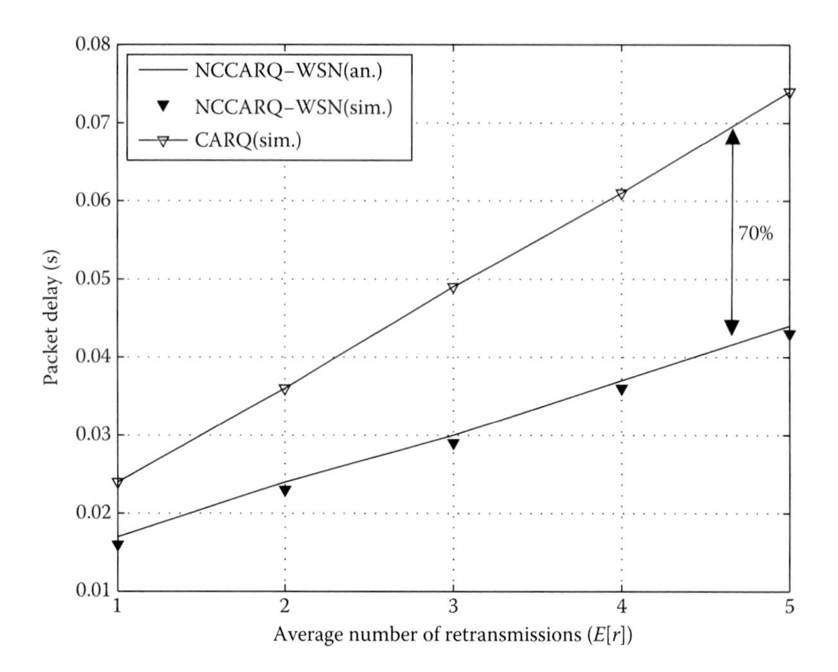

FIGURE 13.9 Packet delay (NCCARQ-WSN vs. CARQ).

expected in order for the packets to be properly delivered. This significant improvement makes sense for three main reasons: (1) the total number of transmissions in NCCARQ-WSN protocol is lower compared to CARQ, (2) the number of *RFC* packets is also decreased, and (3) for bidirectional traffic, in NCCARQ-WSN, the cooperation phase is initiated only once, thus saving valuable time compared to other cooperative schemes where the cooperation takes place upon every erroneous packet reception.

Figure 13.9 presents the packet delay in both NC-based and simple cooperative ARQ MAC protocols. In this point, we must recall that two packets are delivered to their respective destinations in each transmission cycle of NCCARQ-WSN. Hence, in order to be accurate, we compare the delay in NCCARQ-WSN with the time required for two packets to be exchanged in CARQ.

As it can be observed, we can achieve significantly lower packet delay by using NC techniques. Specifically, the average time that is required for two packets to be transmitted using CARQ is 0.024 s in channels where one retransmission is necessary, reaching up to 0.074 s when five retransmissions are required. On the other hand, the delay values in NCCARQ-WSN are 0.016 and 0.042 s for one and five retransmissions, respectively. This difference can be rationally explained considering the operation of our proposed NCCARQ-WSN scheme, where some data packets are sent to the relay (attached to the *RFC* message), thus avoiding the erroneous channel. Furthermore, in our proposed scheme, we manage to reduce the backoff phases by sending two packets simultaneously, while in simple cooperative protocols, the relays have to participate in the contention phase for each packet that has to

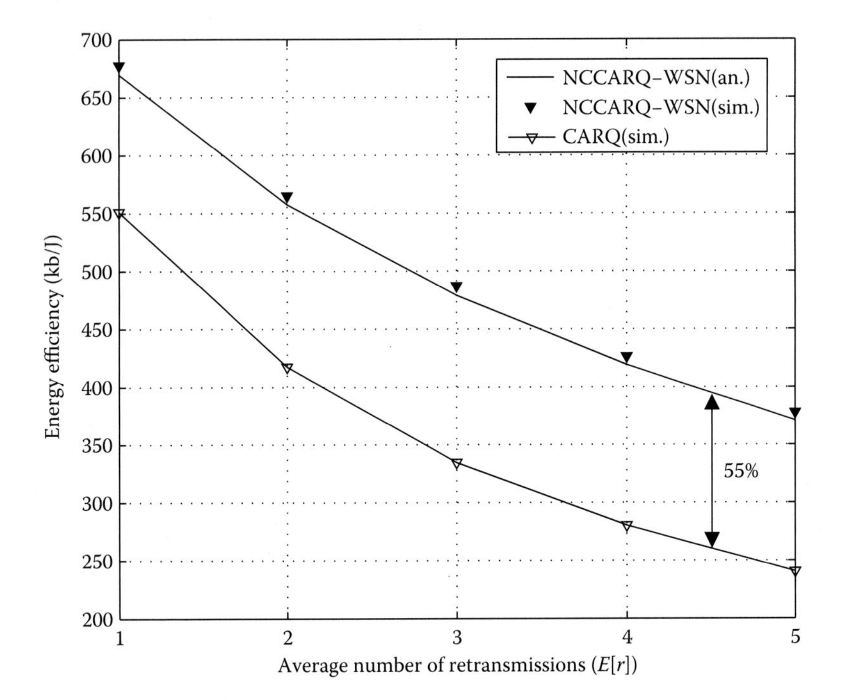

FIGURE 13.10 Energy efficiency (NCCARQ-WSN vs. CARQ).

be retransmitted. Therefore, we are able to enhance the packet delay, since the time that is spent in idle slots and collisions is significantly reduced, especially as the number of required retransmissions grows.

Figure 13.10 shows that our analysis verifies the simulation results with regard to the energy performance. Comparing our proposed NC-based scheme with simple cooperative protocols for different number of retransmissions (and consequently different *PER* between the relays and the destinations), we observe that our scheme is more energy efficient than non-network-coding-based schemes, since more bits are delivered over the same amount of energy consumed. Keeping constant the data packet length (100 bytes), the energy efficiency of NCCARQ-WSN decreases as the number of relay retransmissions grows. However, the difference with simple cooperative schemes remains steadily over 30%.

13.5 CONCLUSIONS

This chapter has focused on e-Health services from a wireless communications perspective and has studied the potential enhancements that can be achieved through NC-based MAC layer protocols for WBANs and WSNs in the context of healthcare applications. Two high-level healthcare scenarios for single-patient and multiple-patient monitoring have been presented, setting the framework for the design of end-to-end e-Health solutions. In continuation, an overview of the existing state-of-the-art proposals has been given, with respect to end-to-end wireless communication systems, MAC protocols for WBANs, and the application of NC techniques in WBANs and WSNs. In order to better illustrate the performance gains that can be achieved through NC techniques, a novel CSMA-based distributed MAC protocol for cooperative wireless networks, NCCARQ, has been presented. NCCARQ achieves significant diversity and performance gains by allowing users to cooperate among them and enabling the nodes to perform NC for bidirectional transmissions.

There are several open lines of investigation in the context of MAC layer protocols for e-Health applications. MAC layer optimization for energy harvesting is an issue that must be tackled in order to further increase battery lifetime. Efficient scheduling mechanisms to jointly achieve QoS provisioning and energy efficiency are a rewarding but very challenging task, since low-power modes of operation (e.g., low-duty-cycle MAC protocols) often have a negative impact on performance. Another promising direction for future work involves the design of MAC layer mechanisms to support and exploit cooperative NC techniques, since most relevant works in the field adopt simplified TDMA-based access methods. Finally, to optimize the energy management in the network, it is important to consider cross-layer approaches that combine MAC schemes with energy-aware routing solutions.

ACKNOWLEDGMENTS

This work has been funded by the Research Projects CO2GREEN (TEC2010-20823), WSN4QoL (PIAP-GA-2011-286047), KINOPTIM (PIAP-GA-2012-324491), ESEE (ENIAC/324284-2) and the ENIAC ARTEMOS Project by UE and Spanish Government (EUI2010-04252 and EUI2011-4349).

REFERENCES

1. NHS, Department of Health, QIPP Long Term Conditions, Supporting the local implementation of the Year of Care Funding Model for people with long-term conditions, 2011. Available online at: https://www.gov.uk/government/uploads/system/uploads/attachment_data/file/215060/dh_133652.pdf, accessed on February 10, 2014.
2. European Economy 2/2012, The 2012 Ageing Report: Economic and budgetary projections for the EU27 Member States (2010–2060), Available online at: http://ec.europa.eu/economy_finance/publications/european_economy/2012/pdf/ee-2012-2_en.pdf, accessed on February 10, 2014.
3. ETSI, Machine to Machine Communications (M2M): Use Cases of M2M Applications for eHealth, Draft TR 102732 v0.4.1, March 2011.
4. T. Cover and A. Gamal, Capacity theorems for the relay channel, *IEEE Transactions on Information Theory*, 25(5), 572–584, 1979.
5. R. Ahlswede, N. Cai, S. Li, and R. Yeung, Network information flow, *IEEE Transactions on Information Theory*, 46(4), 1204–1216, 2000.
6. J. Misic and V. Misic, Bridging between IEEE 802.15.4 and IEEE 802.11b networks for multiparameter healthcare sensing, *IEEE Journal on Selected Areas in Communications*, 27, 435–449, May 2009.
7. J. Misic and V. Misic, Bridge performance in a multitier wireless network for healthcare monitoring, *IEEE Wireless Communications*, 17, 90–95, February 2010.

8. D. Niyato, E. Hossain, and S. Camorlinga, Remote patient monitoring service using heterogeneous wireless access networks: Architecture and optimization, *IEEE Journal on Selected Areas in Communications*, 27, 412–423, May 2009.
9. Y. Huang, M. Hsieh, H. Chao, S. Hung, and J. Park, Pervasive, secure access to a hierarchical sensor-based healthcare monitoring architecture in wireless heterogeneous networks, *IEEE Journal on Selected Areas in Communications*, 27, 400–411, May 2009.
10. S. Marinkovic, E. Popovici, C. Spagnol, S. Faul, and W. Marnane, Energy-efficient low duty cycle MAC protocol for wireless body area networks, *IEEE Transactions on Information Technology in Biomedicine*, 13, 915–925, November 2009.
11. B. Latré, B. Braem, I. Moerman, C. Blondia, E. Reusens, W. Joseph, and P. Demeester, A low-delay protocol for multihop wireless body area networks, in *Proceedings of the Fourth Annual International Conference on Mobile and Ubiquitous Systems: Networking Services (MobiQuitous 2007)*, Philadelphia, PA, pp. 1–8, August 2007.
12. G. Fang and E. Dutkiewicz, BodyMAC: Energy efficient TDMA-based MAC protocol for Wireless Body Area Networks, in *Proceedings of the Ninth International Symposium on Communications and Information Technology (ISCIT 2009)*, Incheon, Korea, pp. 1455–1459, September 2009.
13. B. Otal, L. Alonso, and C. Verikoukis, Highly reliable energy-saving MAC for wireless body sensor networks in healthcare systems, *IEEE Journal on Selected Areas in Communications*, 27, 553–565, May 2009.
14. M. Alam, O. Berder, D. Menard, and O. Sentieys, TAD-MAC: Traffic-aware dynamic MAC protocol for wireless body area sensor networks, *IEEE Journal on Emerging and Selected Topics in Circuits and Systems*, 2, 109–119, March 2012.
15. C. Fragouli, J. Widmer, and J.-Y. Le Boudec, A network coding approach to energy efficient broadcasting: From theory to practice, in *Proceedings of the 25th IEEE International Conference on Computer Communications (INFOCOM 2006)*, Barcelona, Spain, pp. 1–11, April 2006.
16. Y. Wu, P. Chou, and S.-Y. Kung, Minimum-energy multicast in mobile ad hoc networks using network coding, *IEEE Transactions on Communications*, 53, 1906–1918, November 2005.
17. H. Hosseinmardi and F. Lahouti, Online multicasting using network coding in energy constrained wireless ad hoc networks, in *Proceedings of the Third International Symposium on Wireless Pervasive Computing (ISWPC 2008)*, Santorini, Greece, pp. 545–549, May 2008.
18. S. Kim, T. Ho, and M. Effros, Network coding with periodic recomputation for minimum energy multicasting in mobile ad-hoc networks, in *Proceedings of the 46th IEEE Annual Allerton Conference on Communication, Control, and Computing*, Allerton, IL, pp. 154–161, September 2009.
19. T. Cui, L. Chen, and T. Ho, Energy efficient opportunistic network coding for wireless networks, in *Proceedings of the 27th IEEE Conference on Computer Communications (INFOCOM 2008)*, Phoenix, AZ, pp. 361–365, April 2008.
20. L. Miao, K. Djouani, A. Kurien, and G. Noel, Network coding and competitive approach for gradient based routing in wireless sensor networks, *Ad Hoc Networks*, 10, 990–1008, August 2012.
21. A. Munari, F. Rossetto, and M. Zorzi, Hybrid cooperative-network coding medium access control for high-rate wireless personal area networks, in *Proceedings of the 2010 IEEE International Conference on Communications (ICC 2010)*, Cape Town, South Africa, pp. 1–6, May 2010.
22. G. Arrobo and R. Gitlin, Improving the reliability of wireless body area networks, in *Proceedings of the Annual International Conference of the IEEE Engineering in Medicine and Biology Society (EMBC 2011)*, Boston, MA, pp. 2192–2195, September 2011.

23. G. Arrobo and R. Gitlin, New approaches to reliable wireless body area networks, in *Proceedings of the IEEE International Conference on Microwaves, Communications, Antennas and Electronics Systems (COMCAS 2011)*, Tel Aviv, Israel, pp. 1–6, November 2011.
24. S. Marinkovic and E. Popovici, Network coding for efficient error recovery in wireless sensor networks for medical applications, in *Proceedings of the First International Conference on Emerging Network Intelligence (EMERGING 2009)*, Sliema, Malta, pp. 15–20, October 2009.
25. A. Taparugssanagorn, F. Ono, and R. Kohno, Network coding for non-invasive wireless body area networks, in *Proceedings of the 21st IEEE International Symposium on Personal, Indoor and Mobile Radio Communications Workshops (PIMRC Workshops 2010)*, Istanbul, Turkey, pp. 134–138, September 2010.
26. W. Peterson and D. Brown, Cyclic codes for error detection, *Proceedings of the IRE*, 49, 228–235, January 1961.
27. Y. Fujii, D. Umehara, S. Denno, M. Morikura, and T. Sugiyama, Saturation throughput analysis of unslotted CSMA-CA networks, in *Proceedings of the 25th International Technical Conference on Circuits/Systems, Computers and Communications*, Pattaya, Thailand, pp. 688–691, July 2010.
28. B. Otal, Optimization of wireless ambient and body sensor networks for medical applications, PhD dissertation, Universitat Politècnica de Catalunya (UPC), Barcelona, Spain, March 2010.
29. M. Zorzi and R. Rao, Energy constrained error control for wireless channels, in *Proceedings of the Global Telecommunications Conference (GLOBECOM'96)*, London, U.K., Vol. 2, pp. 1411–1416, November 1996.

14 Monitoring Health and Wellness Indicators for Aging in Place

Kevin Bing-Yung Wong, Tongda Zhang, and Hamid Aghajan

CONTENTS

ABSTRACT

The field of personal informatics has become progressively mainstream recently with the rise of groups such as Quantified Self and the introduction of commercially available user-friendly fitness trackers and visualization software. The users of personal informatics technologies monitor aspects of their daily lives and habits for the purpose of self-reflection. For example, by recording workout habits, sleep patterns, and medicine intake, a person can generate concrete health-related statistics that can drive personal change toward healthier habits. We propose a solution that seeks to offer the same type of self-reflective statistics but also supply more detailed wellness indicators for caregivers and medical professionals. Our target population is the rapidly aging segment of many developed countries, who have also been the impetus for much research in the area of assistive technologies for older adults. We intend to monitor users for extended durations in their homes in order to extract and track various wellness indicators and behavioral patterns. These indicators and patterns of behaviors can then be tracked over time to look for trends and sudden changes that can indicate incipient health problems. Our processing framework is based on motion sensor data as a privacy-preserving alternative to camera data while still extracting similar observations of relevant health-related indicators. We will derive from these motion sensor data parameters such as the user's position and trajectory, wellness indicators such as sedentary motion periods and sleep patterns, and social indicators such as guest visitations and out-of-home activities. Each of these indicators can eventually be further processed to look for anomalous behavior in the form of outliers, which can be used in conjunction with other indicators to look for potential correlations that may indicate causes of the anomalous behavior. We intend to expand our work to include data from more sensor types, including cameras and smartphones to evaluate the potential increased utility compared to a system that uses only motion sensors. In the following sections, we present our initial results in extracting indicators that we believe to be important in tracking a person's health and well-being as well as present the various mechanisms we use to process motion sensor data.

14.1 INTRODUCTION

14.1.1 Motivation

Population aging is one of the growing societal concerns for developed countries, due to both rising life expectancy and decreased birth rates. This shift in population demographics is troubling, since eldercare is traditionally very labor intensive and costly. Organizations such as the US Department of Health and the World Health Organization have predicted that the population of people older than 60 will triple between 2000 and 2050, from 600 million to 2 billion. In the United States alone, the number of citizens older than 65 is expected to grow from 40.3 to 72.1 million in 20 years [1,2]. As a result of this growing population demographic, many assistive technologies are being researched to reduce the cost for assistive living, to allow the aging population to retain a measure of independence by enabling them to continue to live in their own homes, and to allow family members and other caregivers to monitor the wellness and safety of older adults [3–7].

To aid in the care of the rapidly aging population, we are developing a prototype system to analyze the long-term behavior, interactions, and possible indicators of health of a user. We first intend our system to be able to extract meaningful indicators of well-being, such as sleep durations, eating habits, social activities, and movement statistics. We believe that by measuring and analyzing these well-being indicators over an extended period of observation, we can observe trends, such as a shift or gradual change. From these trends, we can conceivably infer the cause of the change, for example, a lower mobility over time, change in the duration of sleep, or change in TV watching patterns can all be possible indications of a change of a user's physical or mental condition or a sign of depression. Using such a long-term monitoring framework, we can also capture trends and representative behaviors over a long time span and present them to a medical professional for further analysis. Possible uses of such a system include monitoring of older adults who choose to live alone and monitoring of large assistive living facilities to more effectively utilize staff members. The proposed system would allow family and caregivers to see the patient's status and remotely monitor their well-being and also generate alerts in case of detected accidents or negative well-being trends.

There has been much work in fall detection using cameras and other sensing modalities; however, such technologies are only useful to respond after a patient has already fallen and possibly injured themselves. Although a quick response to a fall can bring significant peace of mind both to the patient and caregivers, it does not address the significant costs and health-related consequences of a fall on the patient. Authors of a survey of such fall-related costs [8] note that preventative measures should be explored. We believe that long-term monitoring is a possible solution to this problem, as we hope to extract behavioral and health-related indicators that can be used to infer mobility or movement disorders that could signal an increase in fall likelihood. The use of long-term health and well-being indicators is also beneficial for self-reflection as a motivational tool to change possibly unhealthy trends in lifestyle. Continuous monitoring can also be used by doctors, since in-person patient visits are only an infrequent and potential skewed view of a person's current health. The Quantified Self movement is a part of this recent trend for self-reflection and health monitoring; however, using environmental sensors, such as motion sensors and cameras, would allow a monitoring system to unobtrusively monitor a patient continuously while they are in the home with no active compliance on the patient's behalf [9]. Wearable monitors, such as the Fitbit and Nike+ FuelBand, would be able to gather more patient-specific information, but they require the user to remember to use the device [10,11].

14.1.2 Related Work

Among recent developments in assistive technologies for the elderly, there has been an increasing interest in developing systems that can autonomously sense and identify potential conditions and trends that could indicate changes in health or safety that would be of concern to caregivers. One such system developed by Shin et al. uses a network of five passive infrared (PIR) sensors to track the motion activity of nine elderly subjects in a government-sponsored housing [12]. The goal of their system was to look for abnormal patterns in the PIR data by using higher-level derived indicators, such as the percentage of time the motion sensors were triggered, how often the user would move between motion sensors, and the period of time that the user was not moving. The authors then used 24 different support vector data description (SVDD) classifiers, one for each hour of the day, to classify normal activities.

However, their system would often generated false abnormality warnings due to the irregular behavior patterns of the users, such as waking up late and performing cleaning activities during different times of day. Other work by O'Brien et al. also used PIR sensors in the home as a primary input, but their work was focused on visualization techniques to potentially identify movement disorders for the older adults [13].

Cook at the Washington State University (WSU) has been studying smart environments and their applications for a number of years, and her research group focuses on data mining and machine learning approach to recognize activities of daily life or ADL, in order to automatically identify important activities through finding the most common pattern in motion sensor data and treating those common occurrences as activities [14–16]. This work was motivated by both assisted living applications and health monitoring applications to track daily routines and look for deviations. Some of the recent work by her student Jakkula focuses on using one class of support vector machines (SVM) to classify anomalous behavior using an annotated dataset based on motion and door sensors in a home setting [17]. Anomaly detection by monitoring drifts and outliers of detected parameters were also studied by Jain et al.; however, that research focused less on ambient sensors and more on wearable health monitors [18]. More theoretical approaches to activity detection, such as work by Kalra et al., have focused on machine learning and statistical models for ADL detection [19].

There have been many other proposed approaches toward detecting, representing, and analyzing activity and behavioral patterns in a home setting. One such approach by Lymberopoulos et al. uses a home sensor network to track a user's motion and presence throughout the home [20]. Using region occupancy and the associated occupancy time, they create a set of symbols that encode the location, duration of the user's presence, and the time of day the user was present in a specific area in the home. From these symbols, they discover frequent sequences of symbols and their likelihoods to extract the user's activity patterns from their 30-day dataset.

More vision-centric approaches by Gómez-Conde et al. [21] focus on detecting deviations from a model of normal behavior using cameras and computer vision techniques such as motion detection and object segmentation to develop tele-assistance applications for the elderly. Their system mainly focuses on the sensing and classification of events, not recognizing longer-term patterns or behaviors. Other research projects such as [22,23] also focus more on shorter-term monitoring and focus on techniques for person tracking and fall detection. Due to the general lack of long-term monitoring data and privacy issues with data collection, there has not been a significant body of work dedicated to long-term, on the order of months or years, monitoring of behavior and health.

14.2 DATASET DESCRIPTION AND PROPOSED WELLNESS INDICATORS

Ideally, we would have liked to use data from an existing camera network deployment, since cameras are passive and do not require user participation or compliance, and with the proper computer vision algorithms, it is possible to extract very nuanced data regarding a person's behavior, activities, and movements around a monitored space. Unfortunately, long-term video data is difficult to obtain due to privacy and other considerations; however, dense motion sensor data is widely available since many sets were gathered in smart home testbeds. These PIR-based motion sensor deployments have the same fundamental problem as cameras: they depend on a fixed infrastructure and can only monitor a person inside of a fixed coverage area. Wearable sensors are another option for long-term monitoring, but they depend heavily on patient cooperation, as they have to be worn correctly and continuously in order to be effective. Additionally, accurately localizing wearable devices indoors can be complicated, since GPS often is unavailable and GPS chipsets can have a large power draw, which would reduce the device's maximum run time, or require larger batteries, which would make the device more cumbersome.

Due to the lack of long-term camera or tracking data of users in their homes, we developed algorithms to process long duration motion or PIR sensor data to extract meaningful indicators of well-being. We intend to develop a general framework to take long-term observations of people; extract indicators that would relate to mobility, social behavior, eating habits, and other activities; and then make inferences about the person's health by observing how these indicators fluctuate or change over time. These inferences can then be used as feedback for both the person being monitored and to caregivers.

14.2.1 Description of Dataset

The dataset that we used for processing was obtained from WSU's CASAS project. The original use of the CASAS dataset involved extracting common activities through data mining and machine learning techniques to automatically find ADL, defined as common activities that a person performs to care for themselves [24]. The CASAS project's ultimate goal was to eventually characterize a person's behavior and develop predictive and assistive technologies to aid the elderly and the cognitively impaired accomplish various ADL tasks [14]. The dataset features 220 days of PIR motion, door, and temperature sensor data stored by a central networked data logger. The dataset is partially user annotated with a fixed set of events. Door and motion sensors are event driven by nature, so the dataset only recorded when a door or motion sensor changed state. As a result, the data was nonuniformly sampled and did not lend itself to make efficient range queries that compared the states of different sensors in the same time. To make processing easier, we mapped the raw data by first generating a uniform 1 s time scale and mapping events to the closest time in our new uniform scale. In the following sections, we describe the ideal set of indicators that we would like to track and describe the current set of algorithms that we used to process our motion sensor data and some of our initial results showing the extracted parameters.

14.2.2 Wellness Indicators

The parameters in Table 14.1 represent an initial set of what we considered to be useful health and behavior indicators that we believed could be derived from the limited information provided to us by binary motion and door sensors. The work presented in this chapter represents our current progress in extracting these indicators.

TABLE 14.1
Table of Health and Behavioral Indicators

Parameter	Importance to Wellness	Proposed Algorithm
Mobility (walking pace/ area coverage)	Changes in mobility could indicate a lack of energy or physical ailments.	Track motion detection events over time and infer position.
Trajectory anomalies	Frequent changes in trajectories from place to place could indicate that the user frequently becomes disoriented.	Compare current trajectories to look for deviations from typical trajectories.
Sleep duration	Changes in sleep duration and interruptions could indicate conditions including depression, sleep apnea, and bladder disorders.	Track motion level in the bedroom at night; high levels in and around the bed would indicate disruptions.
Social activities (visitors and outdoor activities)	Decreasing amounts of social activities could have a negative impact on well-being and be an indicator of social isolation.	Use multiuser tracking to track visitors and door sensors to track time spent outdoors.
Eating times	The duration, frequency, and regularity of meals can be used to infer dietary intake.	Track time spent in the dining area.
Sedentary activities	Long periods of sedentary activity are generally regarded as unhealthy.	Track periods of stationery low motion activity.

14.3 PROCESSING METHODOLOGY

Our first processing steps in using the CASAS dataset involved extracting movement and behavioral-based indicators. We initially sought to obtain an understanding of how the user moves through the instrumented apartment by analyzing the spatial and temporal relationship between motion sensor activations. To accomplish this, we looked at both the temporal correlation between motion sensor activations and the concurrent sensor activations and explored graphs that repressed how the sensors were spatially and temporary linked. We then used other methods, such as position averaging and particle filtering, to obtain an estimate of the users position at every time point during the CASAS dataset. The following sections describe how we extracted mobility information from the CASAS dataset along with our results.

14.3.1 Temporal Link Transition Analysis

A temporal analysis of the motion sensor activation times can yield the user's transition time between sensors by obtaining the cross correlations between motion sensor activations between every sensor to every other sensor in the network. For this portion of the processing, we do not make any strict assumptions of the detection range of the motion sensors; other than that, they are not all overlapping, since in that case, the user would simply activate all sensors simultaneously. Previous work done in this area used similar binary proximity sensors to determine the topology and transition

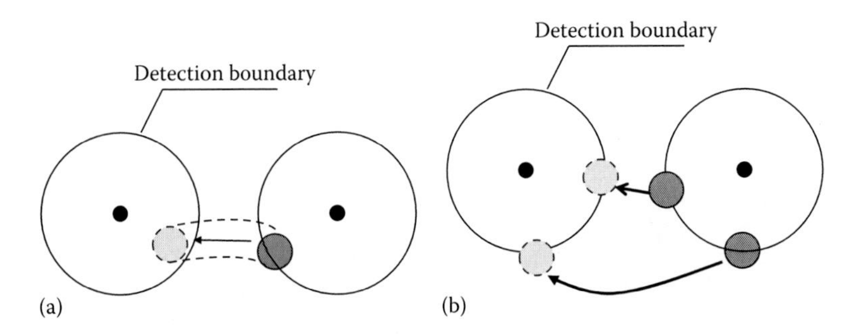

FIGURE 14.1 Illustrations of factors that could cause variations in the transition time histogram: (a) variable speed and (b) variable path.

times between sensors; however, they were frequently deployed in hallways and other areas where movement directions were restricted [25–27].

In the case of the CASAS dataset, as users walk through the apartment, they trigger motion sensors as they move into each sensor's detection range. The transition time between these trigger events is dependent on both the user's walking speed and the path they take between the sensor's respective detection ranges. This causes a variability in the transition time caused by both variations in the walking speed and path chosen by the user. These two cases are illustrated in Figure 14.1. Since we know the distance between sensors from the CASAS deployment maps, we can use the transition time to infer walking pace or even the presence of physical obstructions if the transition time is significantly higher than normal for sensors of a certain spacing.

More formally, we consider the trigger times of each sensor as a sequence of timestamps

$$S_i = \{t_1, t_2, \ldots, t_n\}$$

We then take each pair of sensors (S_i, S_j) and find the time difference, τ, between triggers of sensor S_i and sensor S_j. We then accumulate these time differences in a histogram with 0.25 s bins. In practice, we limit the search window for the triggers of S_j to 60 s to limit the size of the histograms generated and apply a moving average filter to smooth out the histogram for later model fitting. We also subdivide the 220-day experimental data into day-long segments to be processed individually, so the transition times of each day can be compared.

The bar plot in Figure 14.2a shows the time differences of a sensor to itself, which is useful in characterizing the detection delays of a sensor. Figure 14.2b shows the transition time histogram using 90 days of accumulated data; however, the distinctive peak representing a distinctive transition time can also be seen with only 2 days of accumulated data, as shown in Figure 14.2c. It is also interesting to note that the sensors are not able to cycle from on to off states quickly, most likely due to some internal averaging or threshold present in the PIR motion sensors. This delayed deactivation effect was observed during the PIR deployment in previous work by Cho et al., as the sensors would indicate the presence of motion for 1–2 s after all motion has ceased [25]. This effect was noted in the CASAS data as well, observing

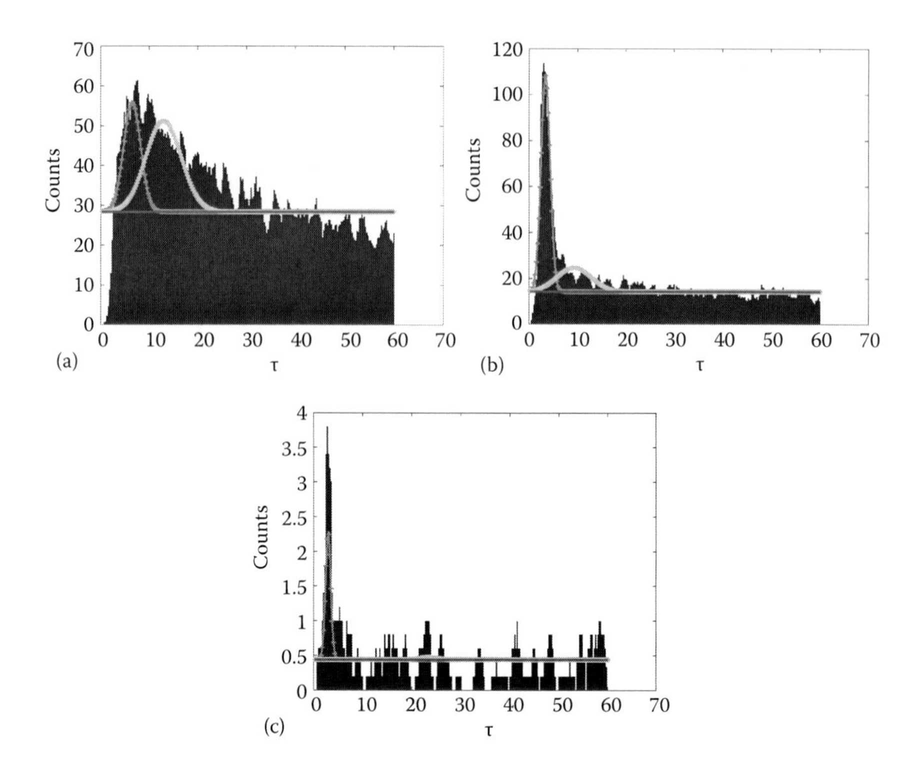

FIGURE 14.2 Time transition histograms with fitted Gaussian link estimates and uniform noise floor: (a) self-sensor transition times, 90-day accumulated data; (b) adjacent sensor transition times, 90-day accumulated data; and (c) adjacent sensor transition times, 2-day accumulated data.

the transition times of a sensor to itself (S_i, S_i), as shown in Figure 14.2a. While there are some trigger events that occur in the first (0–0.25) bin, there is a lack of activations spaced less than 2 s apart, which indicates that the CASAS sensors displayed a similar delayed deactivation.

We assumed that the transition times between sensors are approximately Gaussian, since the walking speed of individuals is approximately Gaussian, which has been experimentally demonstrated by Young [28]. We also assumed that the histograms of time differences between two sensors would be composed of shifted and scaled Gaussians, each representing a different walking speed or path. Variations in walking speed would be most evident in the spread of each Gaussian, while multiple peaks in the histograms would most likely be due to path diversity between sensors. Thus, to model the transition times, we used a model that was based on a mixture of Gaussian model that we fitted to the normalized transition time histograms, which is the same approach used by Cho et al. [25]:

$$P\left(\tau,\pi,\mu,\sigma^2\right)=\left(\sum_{k=1}^{N}\pi_k\mathcal{N}\left(\tau,\mu_k,\sigma_k^2\right)\right)+\pi_{N+1}\mathcal{U}(\tau),$$

where

$$\sum_{k=1}^{N+1} \pi_k = 1$$

Following a similar approach to [25], we used the iterative expectation maximization method to fit the earlier model to our data, using only two Gaussians to model up to two possible transition paths and one uniform component to model indirect paths and other sensor activations that are not a result of movement between two sensors. An example of an indirect path adding histogram noise is if the users move from $S_1 \rightarrow S_2 \rightarrow S_3 \rightarrow S_2$ within our 60 s search window. Then the time delay histogram for (S_1, S_2) would contain both the first $S_1 \rightarrow S_2$ transition and also include another longer link as the user returns to S_2 after first walking in range of S_3.

The plotted Gaussian distributions in Figure 14.2c were generated using only 2 days of accumulated data; although the peak's mean was captured, the estimate Gaussian distribution does not capture the histogram peak's height due to the limited data. When we use 90 days of data to perform the fitting, the primary transition time peak was better fitted, as shown in the Gaussian plots in Figure 14.2b. This shows some inherent trade-off between the quality of our transition model fitting and the amount of data that we must accumulate. We did noticed that some of the sensor pairs had a time difference distribution that was longer tailed; this can either result from an incorrect assumption about the Gaussian nature of the transition times or could be the result of significant path variability between two sensors.

After the models are fitted, we can use the most likely transition time for each pair of sensors, determined by looking at which of the fitted Gaussian distributions had a higher mix component π_i, to generate an adjacency matrix and display a graph of motion sensor connectivities. We can also capture the link strengths by integrating each link's fitted histogram and multiplying by both its mix coefficient *pi* and by the inverse factor used to normalize the transition time histogram prior to fitting. These link strengths can indicate how frequently a path between two sensors was used. Both the adjacency and link strengths vary throughout the experimental period; thus, we posit that they can be used to determine a pattern of behavior for the user by observing how they vary over time.

A graph representation of the extracted adjacency matrix is shown in Figure 14.3; edges or links are only shown for transition times less than 5 s and with a minimum of 100 transitions. Additionally, the darkness of the links indicates their strength, with stronger links rendered in a darker shade. Plots of the matrices representing the pairwise transition times and link strengths are shown in Figures 14.4a and 14.5a, each cell representing the strength between a pair of motion sensors. These two figures illustrate that the strong and short links are relatively sparse, which is due to the limited movement speed of the user and the environmental obstacles that prevent the user from walking through walls. Figures 14.4b and 14.5b show the differences in these transition time and link strength matrices from month to month, which indicate that there is a measurable difference in mobility

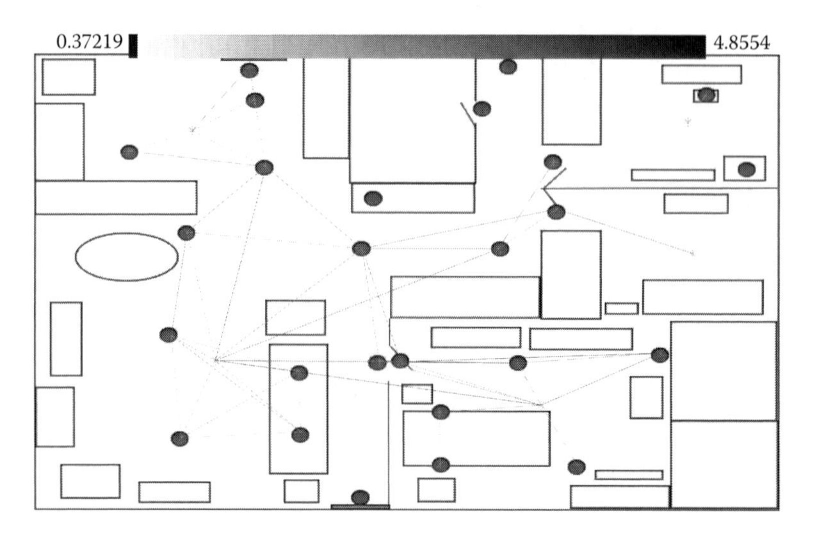

FIGURE 14.3 Thresholded transition connectivity graph for 3 consecutive months.

patterns between the first and second months, which can indicate changes in movement and behavior.

14.3.2 Concurrent Sensor Activation Analysis

Much research has been devoted to networks of simple binary motion sensing devices, but much of the work focuses on very specific tasks including detection, tracking, and localization [29–31]. As a result, the study of the user's activity location distribution and the spatial relationship between the sensor's detection regions has not been thoroughly studied.

In this section, we want to use concurrency events, defined as the periods of time when two or more sensors are activated simultaneously, to analyze the overlap of the CASAS sensors' coverage and user's activity location distribution. The basic idea is that if two sensors are activated at the same time, it most likely means that the user is moving in the overlapping area of the two sensors' detection regions. Therefore, the number of concurrency events among sensors reflects the user's activity frequency in the overlapping regions in the sensor network.

Given that the motion sensor network consists of N binary motion sensors $(s_1, s_2, \ldots, s_N)$ and that the coordinates of the sensors are $\{(x_1, y_1), (x_2, y_2), \ldots, (x_N, y_N)\}$, each sensor's state $f(s_i)$ $(i \in \{1, 2, \ldots, N\})$ has only two possible values $\{0, 1\}$, where 0 means the sensor is not activated and 1 indicates the sensor is triggered.

For sensor s_i $(i \in \{1, 2, \ldots, N\})$, assuming the detection region of every sensor is a circular area, the relation between the distance from user to sensor s_i and the probability of s_i being activated can be described by a parameter tetrad $(\delta_1^{(i)}, \delta_2^{(i)}, \beta_1^{(i)}, \beta_2^{(i)})$, where $\delta_1^{(i)} > \delta_2^{(i)}$ and $0 < \beta_2^{(i)} < \beta_1^{(i)} < 1$.

Figure 14.6a shows the meaning of the parameter tetrad, where $P(f(s) = 1)$ is the probability the sensor s being activated and $d(u, s)$ refers to the distance between the user and the sensor s. So for a given sensor s_i, we have

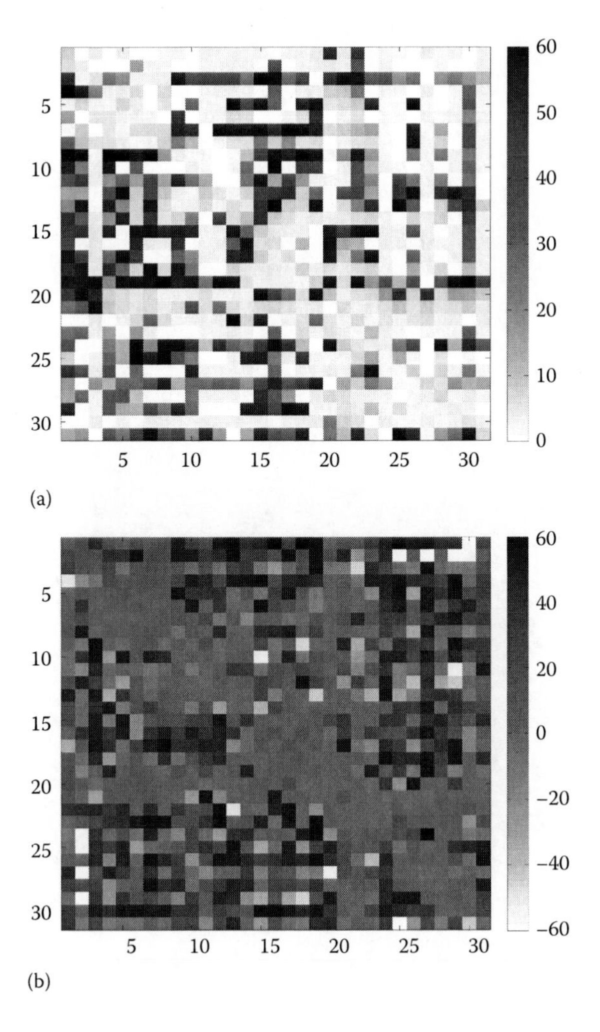

FIGURE 14.4 Transition time visualization of (a) the first month of data and (b) the changes from month 1 to month 2.

$$P(f(s_i)=1)\begin{cases}\leq \beta_1^{(i)} & d(u,s_i)<\delta_1^{(i)}\\ \geq \beta_2^{(1)} & d(u,s_i)<\delta_2^{(i)}\end{cases},$$

where $d(u,s_i)=\sqrt{(x_0-x_i)^2+(y_0-y_i)^2}$ is the distance between user and sensor s_i.

The intuition of this sensor activation model can be demonstrated through Figure 14.6b. Since the sensor is placed on the ceiling, the detection region is a cone in space, a circular area in 2D space. From the perspective of sensor (looking down from the ceiling), the user can be regarded as a circle. If the user is inside the detection boundary, with large probability, the sensor s_i would be activated; oppositely, if the user is outside the detection boundary, there will be a small chance that the

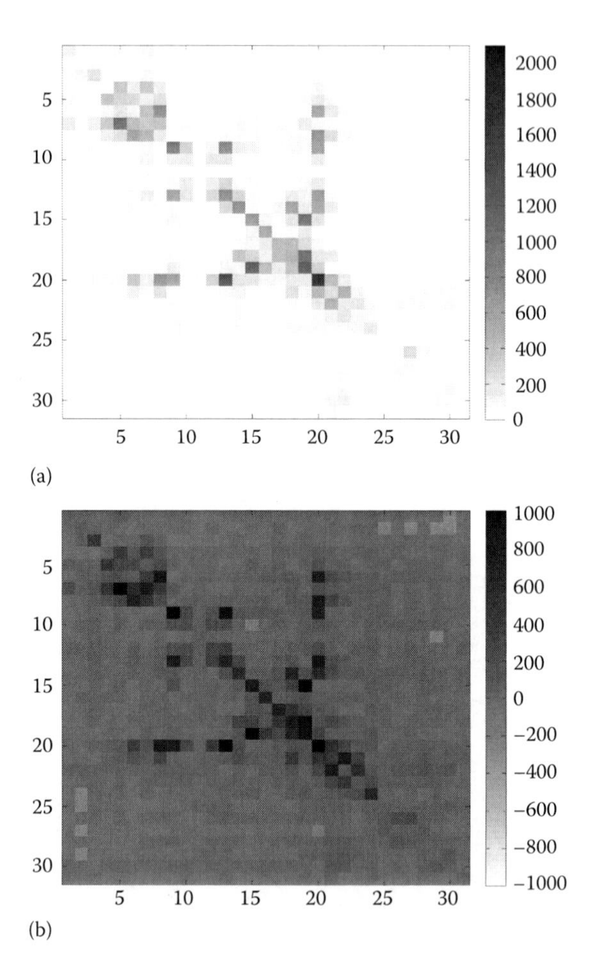

FIGURE 14.5 Link strength visualization of (a) the first month of data and (b) the changes from month 1 to month 2.

sensor is triggered. In other cases, when the user is on the boundary of detection, or the user is moving quickly through the detection boundary, sensor's activation possibility is in between the prior two cases.

To simplify our subsequent analysis, we assume that all the motion sensors have the same parameter tetrad (δ_1, δ_2, 1, β_2) (see Figure 14.7a). We then split events in the transition area (caused by fast user movement and activity near the detection boundary) into two pieces: one is regarded as activity inside the motion sensor detection boundary, and the other is regarded as outside of a particular sensor's detection boundary, reflected in the parameter tetrad, $|\delta_1 - \delta_2| \ll 1$ (see Figure 14.7b), based on which the concurrency probability distribution function can be deducted.

Let the whole area A be divided into $\{B_i\}_{i=1,\ldots,k}$ such that $A = B_1 \cup B_2 \cup \cdots \cup B_l$, where B_v is all the overlapping areas of v sensors' detection regions and l is the biggest number of sensors that can be activated together. Then, we can define $P(u \in B_i)$

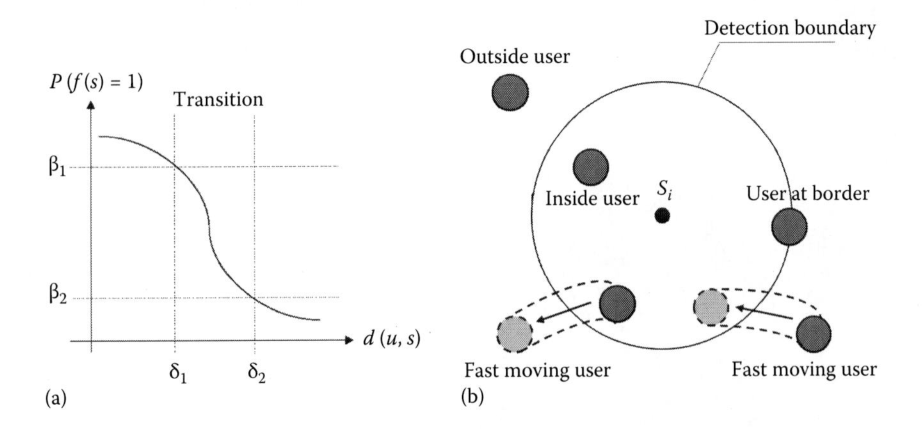

FIGURE 14.6 Sensor's activation: (a) sensor's activation probability and (b) intuition of sensor activation.

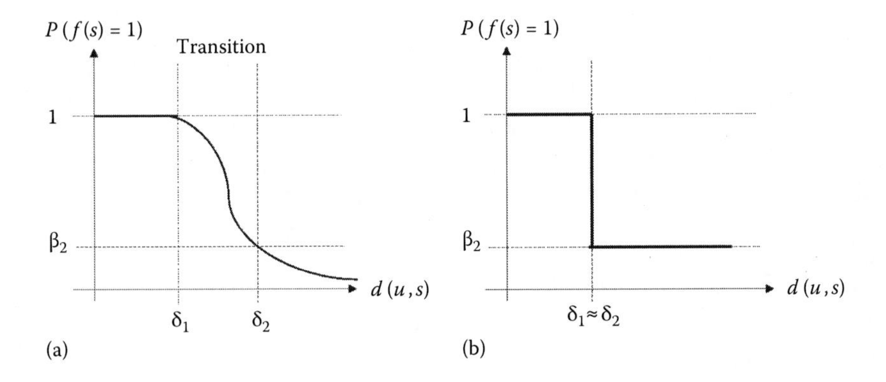

FIGURE 14.7 Simplified sensor's activation: (a) simplified activation probability and (b) further simplified activation probability.

as the probability user u has activities in region B_i. Thus, the probability the sensor network has exactly k sensors triggered at the same time can be expressed as follows:

$$P(k) = \sum_{i=1}^{\min\{l,k\}} P(u \in B_i) \cdot \binom{N-i}{k-i} \cdot \beta_2^{k-i} \cdot (1-\beta_2)^{N-k} \quad (14.1)$$

The previous equation is the concurrency probability distribution function based on our concurrency model, which is deducted from a simple sensor activation curve assumption. Thus, the concurrency distribution can be represented by a $1 \times N$ vector $P = [P(1), P(2), \ldots, P(N)]_{1 \times N}$.

To verify our concurrency model and estimate the parameters in the model, we take the CASAS dataset from January 1, 2011, to June 1, 2011. The data involves

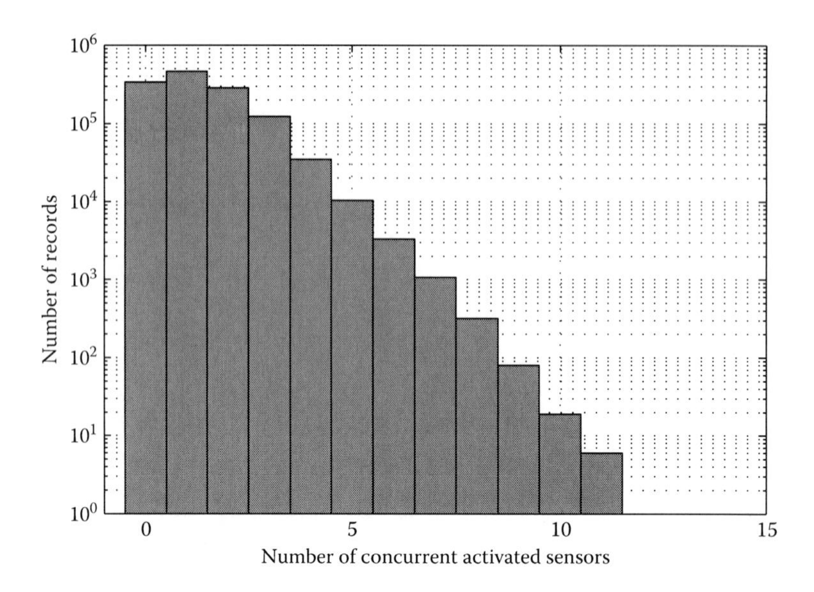

FIGURE 14.8 Concurrency statistic vector H, a histogram of concurrent sensor activations.

$N=31$ motion sensors' states. Let's use a $1\times N$ vector $H=[h_1, h_2, \ldots, h_N]_{1\times N}$ to represent the statistics of the concurrency in the dataset, where h_i ($i \in \{1, 2, \ldots, N\}$) is the number of times that i sensors are activated at a same time, as shown in Figure 14.8. Assuming the total number of events in the dataset is λ, we can estimate the statistics vector H, a histogram of the number of concurrent activations, using the probability distribution provided by the concurrency model in Equation 14.1 as follows:

$$H \approx P \cdot \lambda \tag{14.2}$$

The parameter estimation and model fitting can be interpreted as an optimization problem. In our model, the unknown parameters are $(l, b_1, b_2, \ldots, b_l, \beta_2)$, where l means at most l sensors have a intersect detection region, $b_i=P(u \in B_i)$ ($i \in \{1, 2, \ldots, l\}$) is the probability that the user is moving within i sensor's detection range, and β_2 is the sensor's false-positive activation rate. The object function or cost function can be defined as the difference between H, the normalized statistic vector of concurrency, and the normalized concurrency distribution vector P.

Figure 14.9a shows the optimized objective values for different value of l. Based on the observation of Figure 14.9a, after l taking the value 2, the objective value does not change too much. The intuition is that the probability that the user locates at the overlapping area of k sensors will decrease rapidly as k becomes bigger. So b_1, b_2, b_3 contribute most part of the objective function in the optimization problem; as l becomes bigger, the change of the entries $b_4, \ldots, b_l$ would have little influence on the objective value.

To continue the model fitting, we can set l to any value in 3, 4, …, 11. Here, we set $l=4$. The optimization problem gives the estimated parameters $\tilde{b}_1 = 0.9435, \tilde{b}_2 = 0.0415, \tilde{b}_3 = 0.0091, \tilde{b}_4 = 0.0053, \tilde{\beta}_2 = 0.0406$. Just as expected, b_i decrease rapidly with an increasing i. Equation 14.2 offers a way to evaluate the

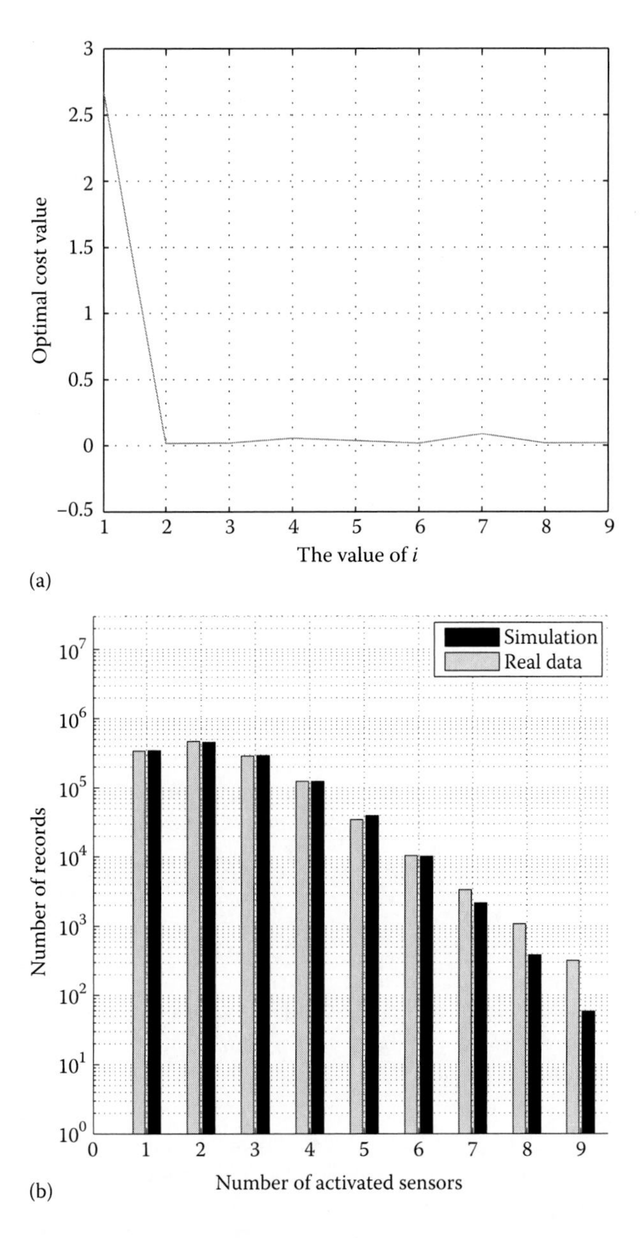

FIGURE 14.9 Concurrency model fitting: (a) optimized objective values under the change of l and (b) real data compared to concurrency model.

performance of the model. The comparison between the estimated statistic vector of concurrency $\tilde{H} \approx P \times \lambda$ and the statistical result from the real dataset is shown in Figure 14.9b. The estimation results based on the model are overlaid with the real data.

To obtain a better understanding of the information provided by sensors' concurrent activation events, we then attempt to use the concurrency graph to visualize

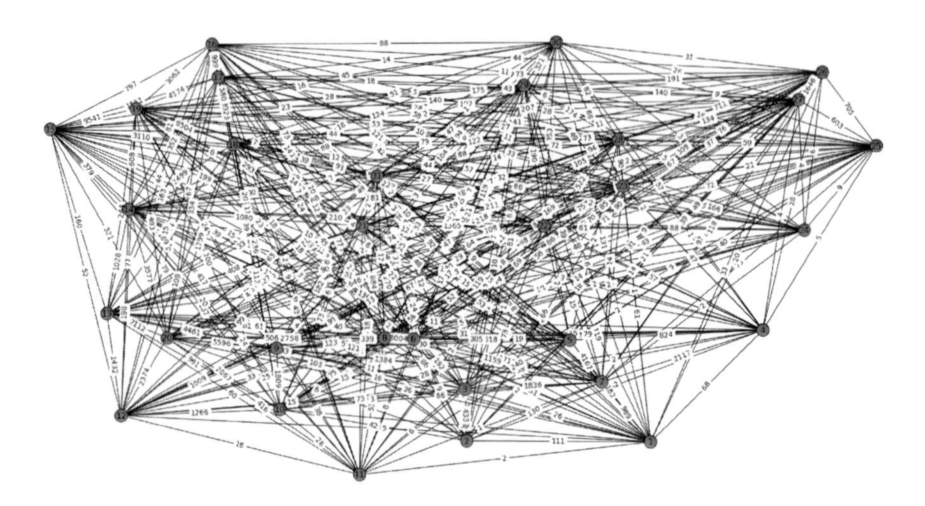

FIGURE 14.10 Concurrency graph for 2010–2011.

the concurrency statistic data. Basically, the concurrency graph is a weighted graph $G=(V, E)$, which consists of a set V of motion sensors as nodes and a set E of edges. Two nodes (sensors) have an edge between them only if they have been activated together at least once. The weight of an edge represents how many times those two sensors have been triggered at the same time. Therefore, to build the concurrency graph, we initialized the graph by adding all sensors as nodes. Then we go throughout the dataset, and for every concurrency event, we increase the weight for corresponding edges.

According to the model built in the previous part, the weight of an edge is proportional to the probability that the user generates a moving activity in the overlapping area of two sensor's detection region, or the user is moving fast through two sensors' detect regions. Figure 14.10 shows the concurrency graph built on CASAS dataset from November 1, 2010, to December 1, 2010. The circles in the figure are the sensor nodes; the lines are the edges between nodes; and numbers on the edges are the weights of corresponding edges.

As can be seen, there are too many spurious low-weight edges that mask the important highly weighted edges, since most of the edges have relative small weights. From the previous model fitting part, the false-positive rate β_2 has already been estimated $\tilde{\beta}_2 = 0.0406$. Knowing the total event number λ, the expectation of the weight that is caused by false-positive activation is $T = \lambda \cdot \tilde{\beta}^2$, because the increase of the weight of an edge only depends on the two connected nodes. For each concurrency event, two sensors can be falsely triggered together with probability $\tilde{\beta}^2$. To reduce the influence from the false trigger event, a threshold can be set to filter out the edges with small weight. Choosing threshold $= T = \lambda \cdot \tilde{\beta}^2$, a new concurrency graph (Figure 14.11a) can be established.

With the aid of thresholding, the concurrency graph becomes closer to matching the topology of the apartment. From the description of CASAS dataset, sensors {7, 19, 20, 24, 27} are configured as area sensors that have a large activation region

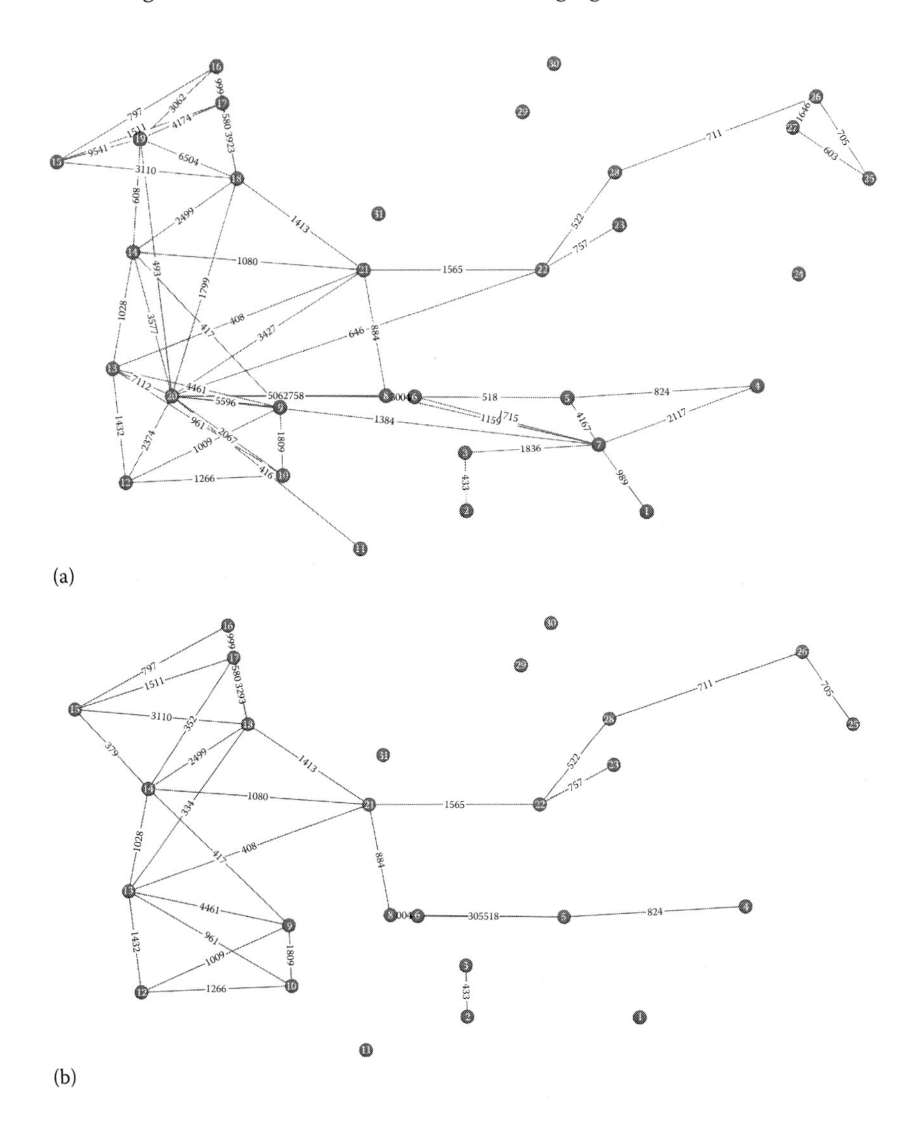

FIGURE 14.11 Improved concurrency graph: (a) concurrency graph for 2010–2011 with thresholding and (b) concurrency graph for 2010–2011 with thresholding and area sensor deleted.

compared to the ceiling-mounted sensors. To further refine the graph, all the edges connected to those area sensors can be deleted. Figure 14.11b shows the result after deleting nodes {7, 19, 20, 24, 27}. It becomes clear that the concurrency graph can be viewed as a skeleton of the apartment.

There are many possible ways to interpret the concurrency graph. One possible interpretation is that the user's moving activity frequency between two sensors is represented by the weight of the edge between those two sensors. So the bigger the weight is, the user may have activities between corresponding two node sensors with more probability. Another way is to interpret the concurrency graph as a whole, as it can be interpreted as a 2D pattern of a user's moving activities during a certain time period.

Therefore, using different interpretations, the concurrency graph can be used in different tasks. For example, in the task of estimating the user's location, the graph can act as a given prior location distribution map, which can help to improve the location estimation precision. Another example is the task to analyze the change of user's lifestyle over time; the concurrency graph for different time periods may give information about how the user's activity distribution changes over time. Similar interpretations can also be made using the graphs generated from temporal analysis of the sensor activations, as presented in Section 14.3.1.

14.3.3 User Position/Trajectory Estimation

14.3.3.1 Position Averaging

Our first attempt to estimate the user's position was to average the locations of the activated motion sensors by using the CASAS sensor deployment map to localize the individual motion sensors. Most of the PIR sensors used in the instrumented apartment were ceiling-mounted units with a relatively small detection radius of 8 ft. With these pieces of information, we are able to determine the position and time of all motion detection events. The layout of the apartment used to collect the dataset can be seen in Figure 14.12. The small circle represents the ceiling-mounted sensors, and the large diffuse circles represent the rooms covered by a wall-mounted sensor. Figure 14.13a and b shows a close-up of the kitchen and living room areas with the motion sensors highlighted.

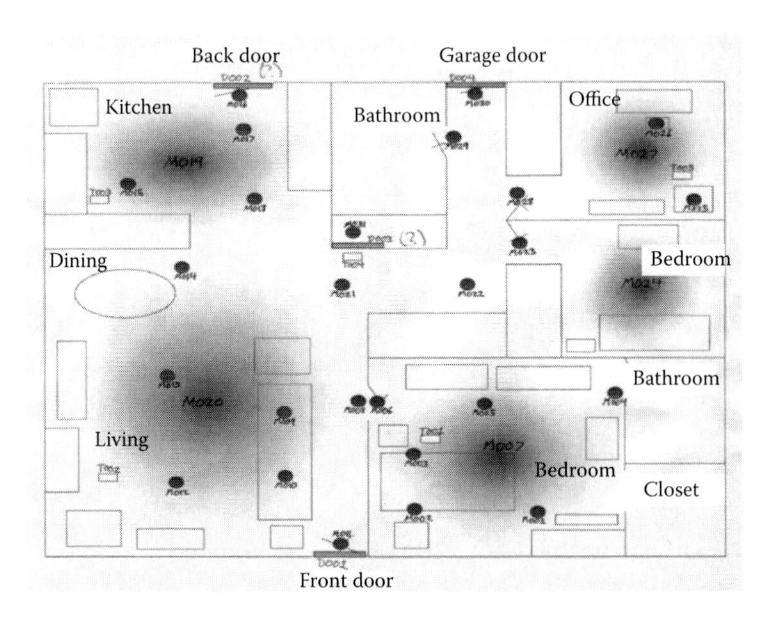

FIGURE 14.12 Apartment layout in WSU CASAS dataset.

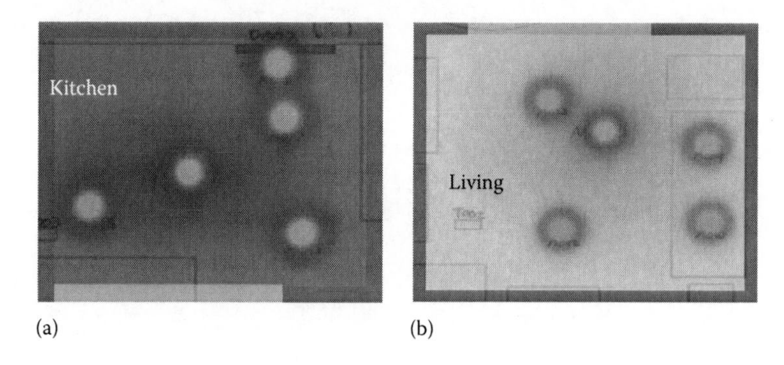

FIGURE 14.13 Sensor layouts for two rooms in the test apartment: (a) kitchen layout and (b) living room layout.

To estimate the user's position, we iterated through every motion sensor event and averaged the location of the motion event with the locations of all other motion events that occurred within a small, 2 s window around the current event. This has the effect of a moving average filter for the user's position based on the locations of motion sensor's reporting movement. The positions used in this averaging operation were estimated to be the center of each motion sensor, and the window size was chosen so that only adjacent sensor's activations would be averaged but large movements across rooms would not. The end result is a nonuniform sampling of a person's position for every motion detection event in the dataset. We then use a nearest neighbor interpolation to interpolate this position information to a 1 s granularity and then smooth it using a moving average filter.

From this position information, we can then visualize approximately how far the user moves in a given day and for a given week. Figure 14.14 plots the distance moved as a heat map. The top portion of the figure is a stacked plot of all 220 days of the experiment, each vertical bar representing a measure of movement for every hour in a day in pixels, as measured from the deployment map in Figure 14.12. The bottom figure plots the cumulative distanced moved every day for every week of the experiment, also in the arbitrary unit of pixels. The darker the pixel, the more movement was recorded from the user for a particular time instant. The most immediately useful portion of this figure is the low mobility periods during the late night and early morning periods, which corresponds to periods when the user is most likely sleeping and thus times that they are least likely to move around the apartment.

14.3.3.2 Particle Filtering

Position averaging was adequate for estimating user positions during our initial experiments when we simply wanted to know how often the user would move from one room to another. However, the averaging approach yields trajectories that are essentially piecewise linear and lack the smoothness necessary to extract useful mobility parameters such as the user's pace and distance traveled, since frequent sensor activation in an area with multiple overlapping sensors tended to make the average user position jump around the apartment.

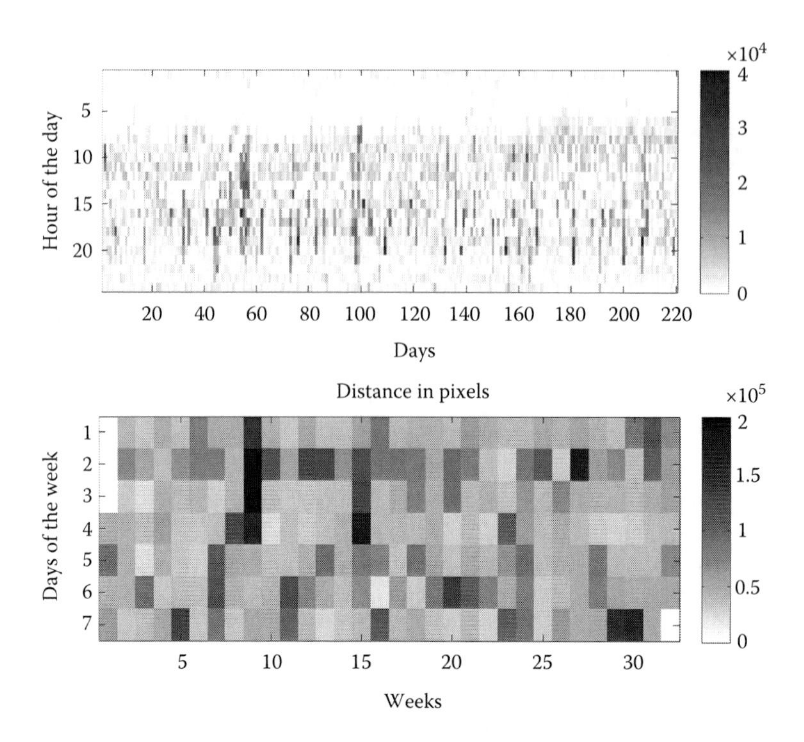

FIGURE 14.14 Estimated movement distances by day and by week.

As a result, we decided to explore particle filtering as another approach to discovering movement and behavior patterns using a user's trajectory. Our approach is based on earlier work in the robotics localization and control space with some modifications inspired by the particle swarm optimization [32–34].

The general approach to applying a particle filter on the CASAS dataset was based largely on the approach developed by Thrun et al. [35]. However, many assumptions and simplifications had to be made in order to define the probability distributions required to update the particle filter. A particle filter uses an iterative process to update a set of particles that represent samples of the estimated user position. These particles are represented by the set χ_t, which encodes the positions of the particles:

$$\mathcal{X}_t := x_t^{[1]}, x_t^{[2]}, \ldots, x_t^{[N]}$$

We used a relatively small number of particles, 100, to limit computation time. As an input to our particle filter, we used a resampled version of the motion sensor activations, since we wanted a uniform temporal spacing between the particle filter update steps.

Ordinarily, two distributions are needed to update the particles during each iteration. The first distribution $p(x_t|u_t, x_{t-1})$ represents the distribution of likely new positions based on the previous particle position and a control variable u_t. Since we do not have control over the user's position, we estimated the simplified $p(x_t|x_{t-1})$ as a Gaussian distribution with a small variance to let the particle's position drift slightly when there are no sensor activations. The intuition behind this is that if no sensors are activated, then there would be an increased uncertainty as to the location of the user, so the variance of the user's estimated position would increase.

The second distribution, $p(z_t|x_t)$, is used to determine the importance factor, which is the likelihood of a sensor measurement z_t, given a particle positioned at x_t. To model this distribution, we used two cases; if there were no sensor activations, we used a uniform distribution for all particles, giving them equal importance. However, if there were sensor activations, we calculated a sensor likelihood map by placing shifted copies of a 2D Gaussian with a variance corresponding to the detection radius of the motion sensors at the location of every sensor activated during a specific time. We then used this sensor likelihood map to estimate $p(z_t|x_t)$ for every particle. Lastly, we used a low-variance sampler to resample the particles based on their importance factors [35].

To account for the restrictive environment of the apartment, we modified the standard particle filter algorithm so that the particles could neither leave the apartment nor leave a room if there were no sensor activations. An example of the particle filter output when there are no sensor activations can be seen in Figure 14.15a; in this example, the particles are confined to the room, since that was the location of the last sensor activation. An example of the particle filter output with sensor activations can be seen in Figure 14.15b; in this case, the particles will be effectively attracted to the locations of the motion sensor activations.

14.4 INFERENCE OF INDICATORS

Beyond studying the user's trajectory and mobility characteristics through the CASAS data, we wanted to also extract relevant activities and behaviors that are derived from the user's position, along with the level of activity in the apartment. The previous section described our efforts to extract aggregate movement patterns and user position estimates. The work presented in this section is more closely tied to the well-being indicators we presented in Table 14.1. In the following sections, we discuss how we extracted indicators of the user's behavior, from how sedentary they were, how much they slept, how often they had visiting guests, and lastly how often they spent time outside of their apartment.

14.4.1 Sedentary Activity Classification

With our existing motion sensor dataset, we can only infer low motion activities, such as sedentary behavior and sleeping, by the absence of motion detection events. Thus, we make the assumption that if the user is stationary, and the level of motion detection events is low, then the user is most likely sitting. To make this determination, we calculate the active periods when the user is in the apartment and triggers more than two motion detectors in any given region. The motion detectors are largely nonoverlapping, so if only one motion sensor is triggered, it is likely that the user is stationary/sedentary. Periods when the user is not in the apartment and when visitors are present were masked out, since we currently cannot associate detection events with multiple users. This information is used to determine the amount of time in each day that the users were either in motion or still, as a fraction of time they were in

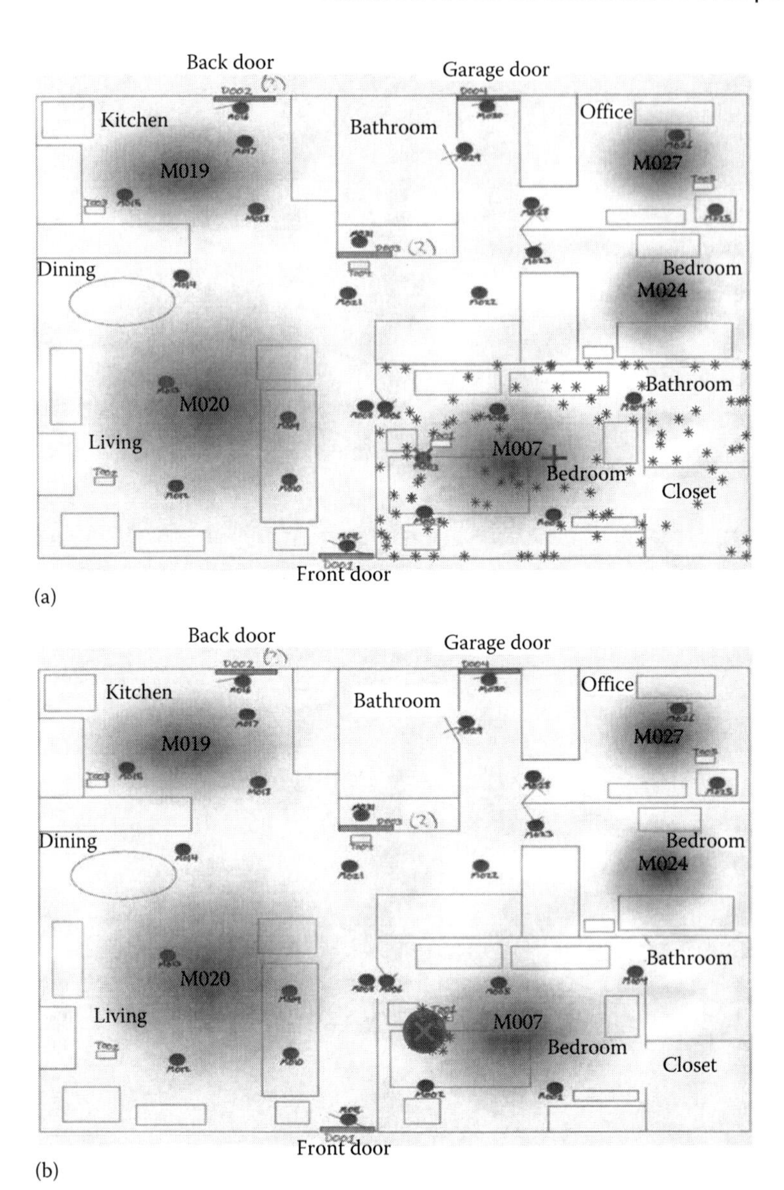

FIGURE 14.15 Particle distributions with and without detection events: (a) no motion events, particles will spread out, only bounded by the room's walls and (b) motion event, particles will be effectively attracted toward detection events.

the apartment. Figure 14.16 shows the ratio between the active motion times and the sedentary motion times, as well as potential outliers that are more than one standard deviation from the mean motion ratio. We can ultimately use the times associated with these outliers to correlate to other potential anomalies of other indicators to attempt to find a causal relationship between indicators.

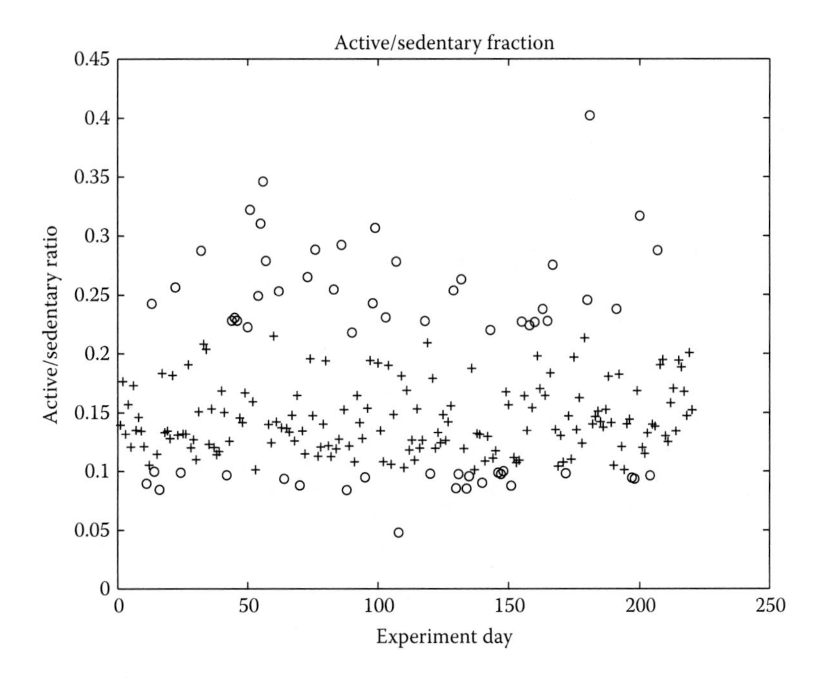

FIGURE 14.16 Active/sedentary ratio with outliers (circles).

14.4.2 Sleep Detection

To detect when the user is sleeping, we use a combination of previously derived user trajectory information and the motion detector traces for the area around the user's bed, as defined in the deployment map shown in Figure 14.12. We then calculate the distance of the user to the center of the bed as well as the motion activity levels of the sensors in the bedroom; if both are sufficiently low, we can assume that the user is both in the bed and relatively stationary or asleep. With this technique, we can determine when the user is likely to sleep at any point during any experiment day, as well as any interruptions, such as bathroom visits. We can also highlight abnormal sleep patterns, such as taking long naps during the day. Such a visualization is shown in Figure 14.17c, which shows typical nighttime sleeping periods in gray and nap periods that are separated by more than 1 h from other sleep periods in black. These long nap times are uncommon in the dataset that we studied and could indicate illness or tiredness from previous lack of sleep. We performed a manual verification for certain days of the experiment to verify if anomalous sleep times were genuine or a result of erroneous processing. This verification was done by viewing the raw motion detection traces and trajectories in sliding 15 min windows to confirm when the user went to bed and by inspecting the full motion detection trace to look for singles of global instrumentation errors. An example of the motion detection events and trajectories is shown in Figure 14.17a; the motion detections are shown as hot spots overlaid on a deployment map. An example of a global motion trace for the same experiment day is shown in Figure 14.17b; motion detection events are shown as black bar segments. We inspected this particular motion trace because our sleep detector indicated that the user had an abnormally low amount of sleep for this day. However, the detection of a low amount of sleep was due to the fact that all motion sensors in the apartment were triggered simultaneously, which can be seen by the thick vertical black bars

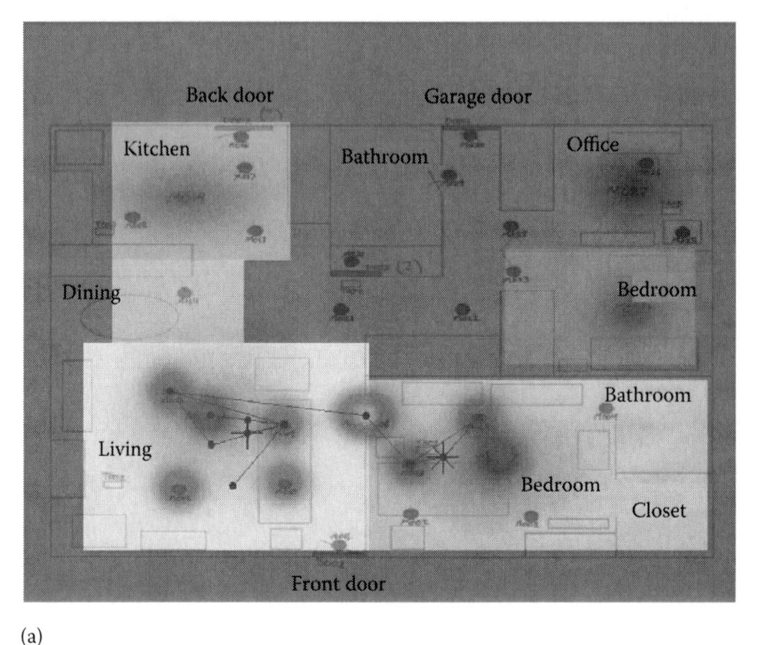

(a)

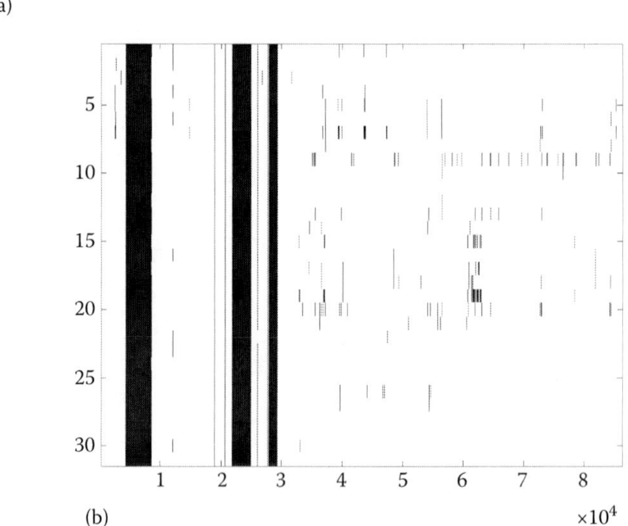

(b)

FIGURE 14.17 Motion detection visualizations for sleep detection and estimated sleep periods: (a) motion detection events in a 15 min window, (b) motion detection trace for a 1-day period.

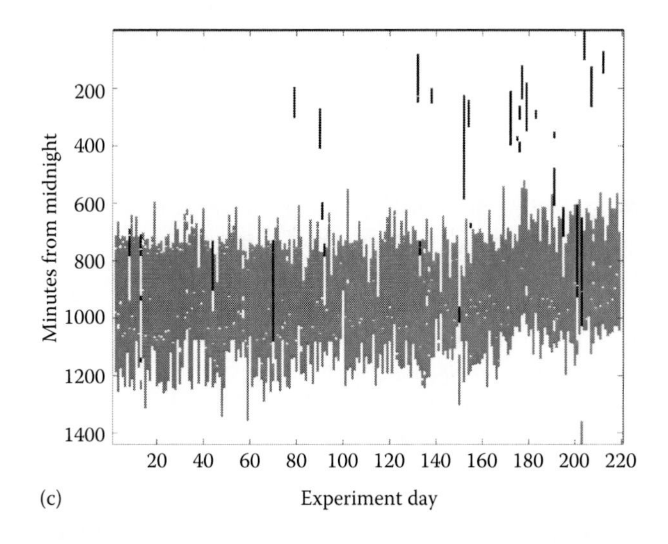

FIGURE 14.17 (continued) Motion detection visualizations for sleep detection and estimated sleep periods: (c) visualization of estimated sleep periods for the experiment.

indicating that all motion detectors were active for extended periods of time, which was most likely a failure in the logging hardware.

14.4.3 Visitor Detection

Since motion detectors are incapable of distinguishing between detection events caused by different people, the data must be postprocessed to see if the detection events in the apartment were caused by a single user or multiple users for later processing. As a first step, we mask out any periods where multiple people are detected and make the assumption that if a single user is detected, it is the primary occupant of the apartment. The method that we used to classify if multiple users are present is based on the idea that if the users are moving throughout the apartment and are sufficiently far apart, there should be two *clusters* of motion detection events. We search for two or more *clusters* of motion activity events that are too close together in time to be caused by a single person. This search is performed by looking at all motion events around small time 2 s window around every motion detection event and then clustering the locations of the motion detection events in that window. The value of 2 s was chosen through experimentation to be large enough to see simultaneous motion detections from multiple users, yet small enough, so that the motion detection events do not blur together and become indistinguishable.

If there are two or more distinct clusters, we can assume that there are more than two users in the space. This method will not work well if two occupants are moving together, as they would not generate two motion *clusters*. Additionally, due to the windowing approach, if one user stops moving, then only one of the users would generate a *cluster*. This limitation is a result of the type of motion sensor used, as PIR sensors can only detect motion, not presence, so it is difficult to track slow-moving or stationary users. As a result, our multiple-person detections can be quite noisy; however, since our initial use for this information is to mask out sections of data, it is adequate for our initial processing. We further refine this mask by using door detection events combined with global motion information to determine if a door opening event is the result of people entering or leaving the apartment. Both events are classified based on observing motion detection events during time windows immediately before and after and door event. If there is motion before but not after a door event, then it is assumed that the occupants have left the apartment. Likewise, if there is no motion before and only motion after a door event, we assume that someone entered the apartment. This method can also be used to determine door events in which the user opens and closes the door without leaving the apartment for a significant amount of time. In the case of guests, we take any multiple-person detection events and extend the time range to the door opening events that occur both immediately before and after the detection event. Since the door sensors are magnetic reed sensors, they have a very low false detection rate and thus can be a good indicator of entry and exit events and make good points to use to extend and fill in gaps of the multiuser detections.

Figure 14.18 shows instances where the clusters are sufficiently far apart so that we can be more certain that the motion detection events are caused by more than one person.

14.4.4 Out-of-Apartment Activities

The durations for the time spent outside and the time spent indoors with a guest or other visitors were derived from entering and leaving events and multiuser detection events, respectively. Figure 14.19 shows a time series plot of both outdoor and visitor activity durations as a function of experiment day, as well as a histogram of the respective activity durations. The longer visitor durations are mostly likely due to overnight visits by family members, which can be inferred from the notes provided with the dataset.

14.5 EVALUATION

We were only able to evaluate a subset of extracted well-being indicators, specifically sleep detection and bathroom interruptions detection, since they were the only indicators representing events that were annotated in the CASAS dataset. We evaluated our sleep detector by comparing sleep/wake intervals that were detected by our indicator extractor with the annotated *sleep* events and *bed-to-toilet* events that were contained in the dataset. It is important to note that we do not explicitly detect *bed-to-toilet* events; we simply detect sleep disruptions in our current version of our algorithm, which can be viewed as more general than *bed-to-toilet* events. We compared each detected *sleep* or *bed-to-toilet* event to the corresponding CASAS annotated events to see if there was an overlap. If an overlap existed between the detected events and the annotated events, and the overlap duration was at least 75% of the predicted event duration, we classify it as a correct classification. The confusion matrices representing the performance of our *sleep* and *bed-to-toilet* classifiers

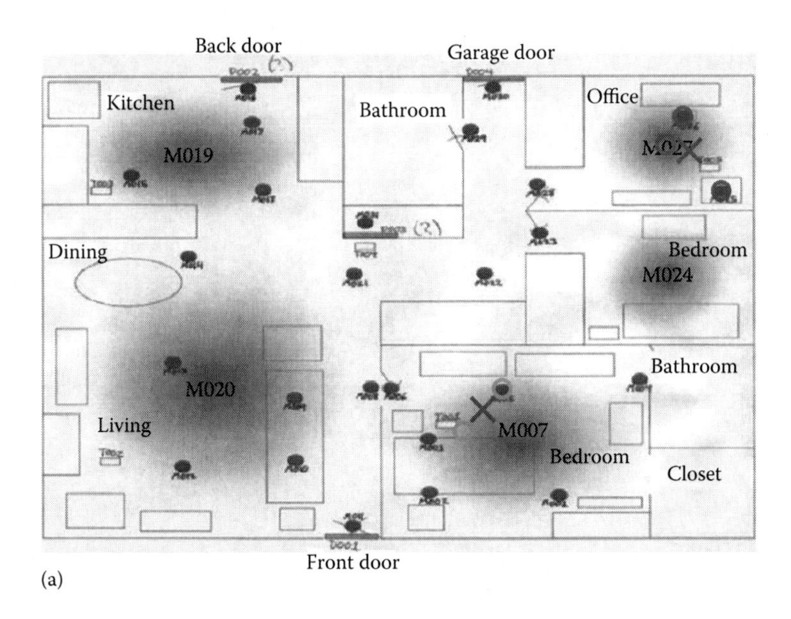

(a)

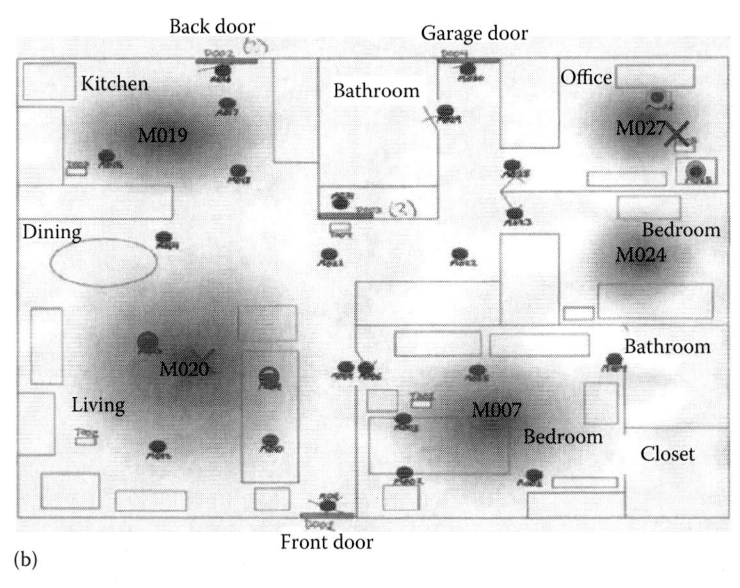

(b)

FIGURE 14.18 Positive multiuser detection events: (a) positive office and bedroom multiuser detection and (b) possible living room and bedroom clusters.

are presented in Table 14.2. Using the multiperson detection algorithm described in Section 14.4.3, we can evaluate our *sleep* and *bed-to-toilet* classifiers only during periods when a single user is in the apartment and when multiple people are in the apartment. The single-user performance is presented in Table 14.3, and the multiuser performance is presented in Table 14.4.

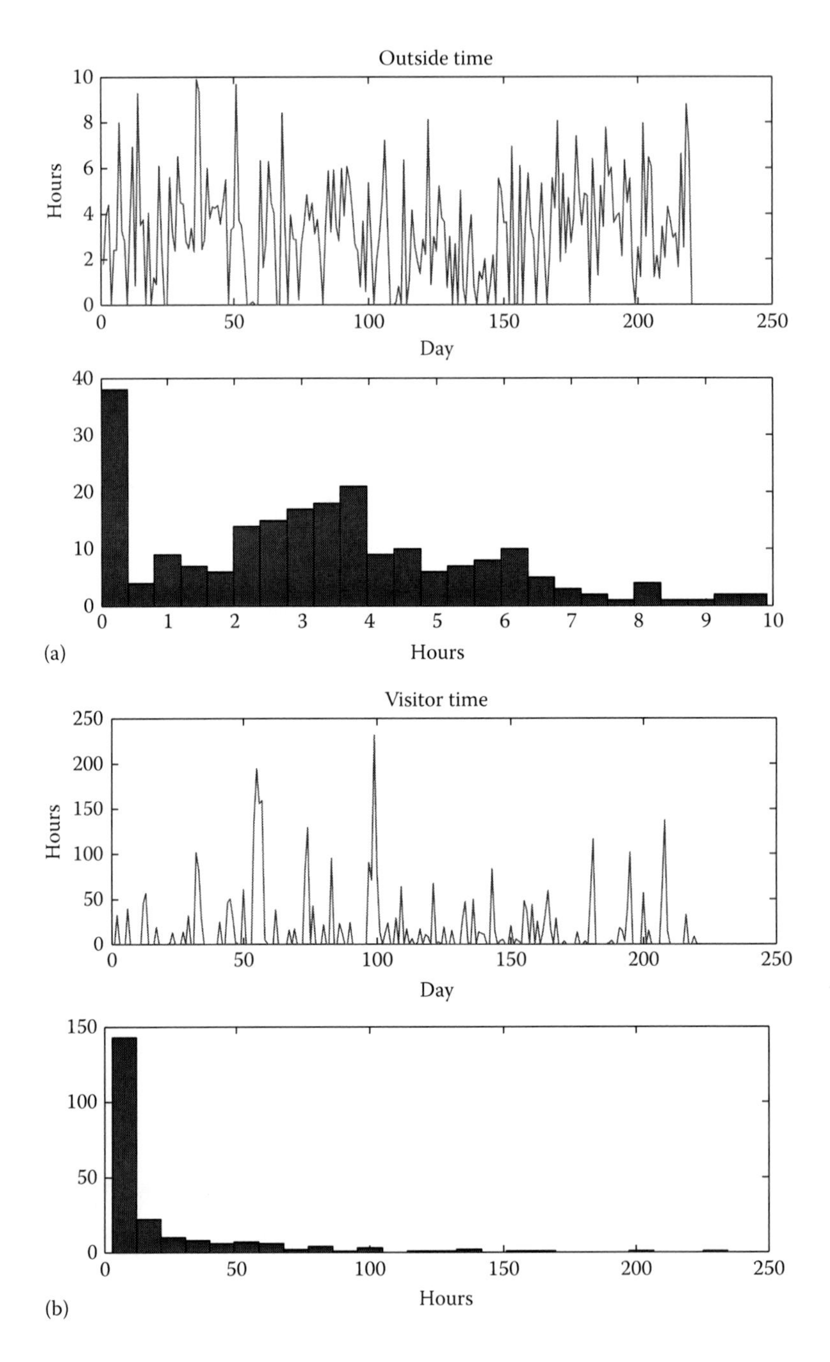

FIGURE 14.19 Extracted outside activity durations: (a) outside activity duration and (b) visitor activity duration.

TABLE 14.2
Confusion Matrix of Sleep/Wake and Bed-to-Toilet Detectors Using Interval Comparisons

(a) Sleep/wake classification performance

		Actual	
		Awake	**Sleep**
Predicted	**Awake**	336	159
	Sleep	25	469

(b) Bed-to-toilet performance

		Actual	
		Negative	**Positive**
Predicted	**Negative**	275	3
	Positive	258	19

TABLE 14.3
Confusion Matrix of Sleep/Wake and Bed-to-Toilet Detectors Using Single-User Data Only

(a) Sleep/wake classification performance

		Actual	
		Awake	**Sleep**
Predicted	**Awake**	332	144
	Sleep	24	439

(b) Bed-to-toilet performance

		Actual	
		Negative	**Positive**
Predicted	**Negative**	262	3
	Positive	240	18

When comparing event intervals, we find that our sleep detector frequently classifies periods that were annotated as *sleep* events as *wake* events. This is most likely due to the fact that our sleep detector also uses motion sensor activity levels in the bedroom in its classification. Periods of high motion were not regarded as sleep; rather, we treated them as possible sleep interruptions. Since each interval can be of an arbitrary length, the false wake events generated a relatively high number of intervals, which in turn may have exaggerated the false negatives in

TABLE 14.4
Confusion Matrix of Sleep/Wake and Bed-to-Toilet Detectors Using Multiple-User Data Only

(a) Sleep/wake classification performance

		Actual	
		Awake	Sleep
Predicted	Awake	4	15
	Sleep	1	30

(b) Bed-to-toilet performance

		Actual	
		Negative	Positive
Predicted	Negative	13	0
	Positive	18	1

TABLE 14.5
Confusion Matrix of Sleep/Wake Detector Using Uniform Partition Comparisons

		Actual	
		Awake	Sleep
Predicted	Awake	211,896	2,131
	Sleep	1,907	91,676

our sleep classifier. When we compared our sleep classifier to the annotated *sleep* events on a minute-by-minute basis, our performance is much better, as seen in Table 14.5, with a similar number of false negatives and positives, some of which are due to differences in determining the onset of sleep and the wake times, since the data was manually annotated by CASAS. The single-user performance was similar to the performance in the general case, since the user has very few overnight guests; however, the multiple-user performance was worse, since we considered the user to be awake if there was any activity outside the bedroom, including the guest bedroom. Previous work performed on the CASAS dataset was focused on the automated detection and extraction of common activities, some of which include sleeping and *bed-to-toilet* events; however, depending on the dataset used, they obtained a recognition rate for sleep of 99.3% for one dataset and 0% for another, since the *sleep* activity was not properly extracted for the second dataset [15]. In both cases, they were unable to properly extract *bed-to-toilet* events.

Our sleep classification results compare favorable to earlier CASAS when compared on a minute-by-minute basis, as we have a 98.7% success rate. Our results in classifying *bed-to-toilet* events were much worse, correctly identifying only 19 of the 157 annotated events. Since we were equating sleep disruptions with these *bed-to-toilet* events, our classifier was far to general and triggered on every sleep interruption.

We also evaluated the performance of our multiperson detection algorithm by visualizing a sequence of short 5 s time windows of motion sensor activations overlaid on the sensor deployment map to manually detect the presence of multiple clusters of motion sensor activations. If we could see two distinct clusters of motion sensor activations, we regarded this as a likely case of multiple people. We used our manual labeling to classify the intervals between external door openings, since these are the only times when guests can enter the apartment, and matches our algorithm's mechanism to propagate a multiple person detection to the nearest entering and exiting door events. For each interval, if we saw activity in two different rooms, but no activity between, we classified the apartment as having multiple people. It is possible for multiple people to move together throughout the apartment during a visit, but we are not able to detect this type of movement via our detector or via manual labeling. Since the complete dataset is quite long, we only manually evaluated approximately 10% of the data.

Figure 14.20 shows two examples from the motion sensor visualizations used to detect multiple users. The visualizations used the multiple-person detection results to indicate if there are possible visitors by showing the current apartment status using

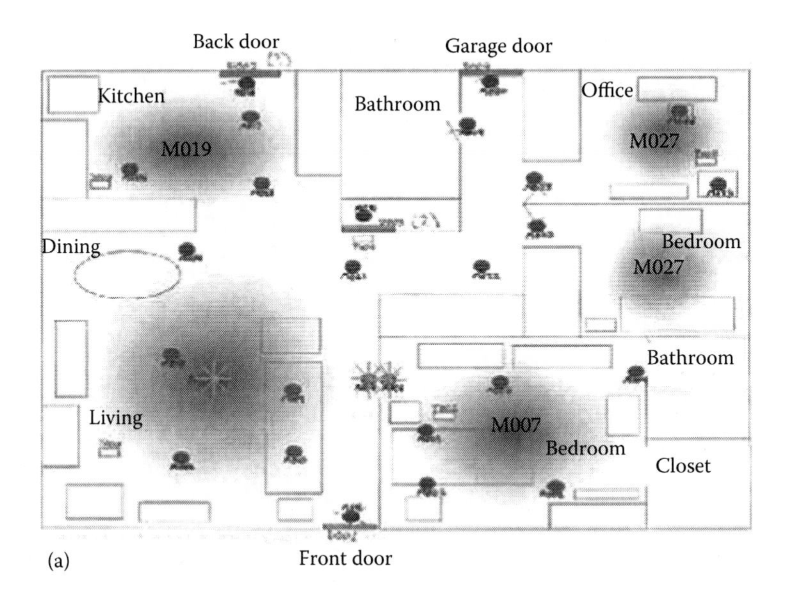

FIGURE 14.20 Example of difficulties of using a sliding window approach to detect multiple people: (a) only one of the two occupants is moving.

(continued)

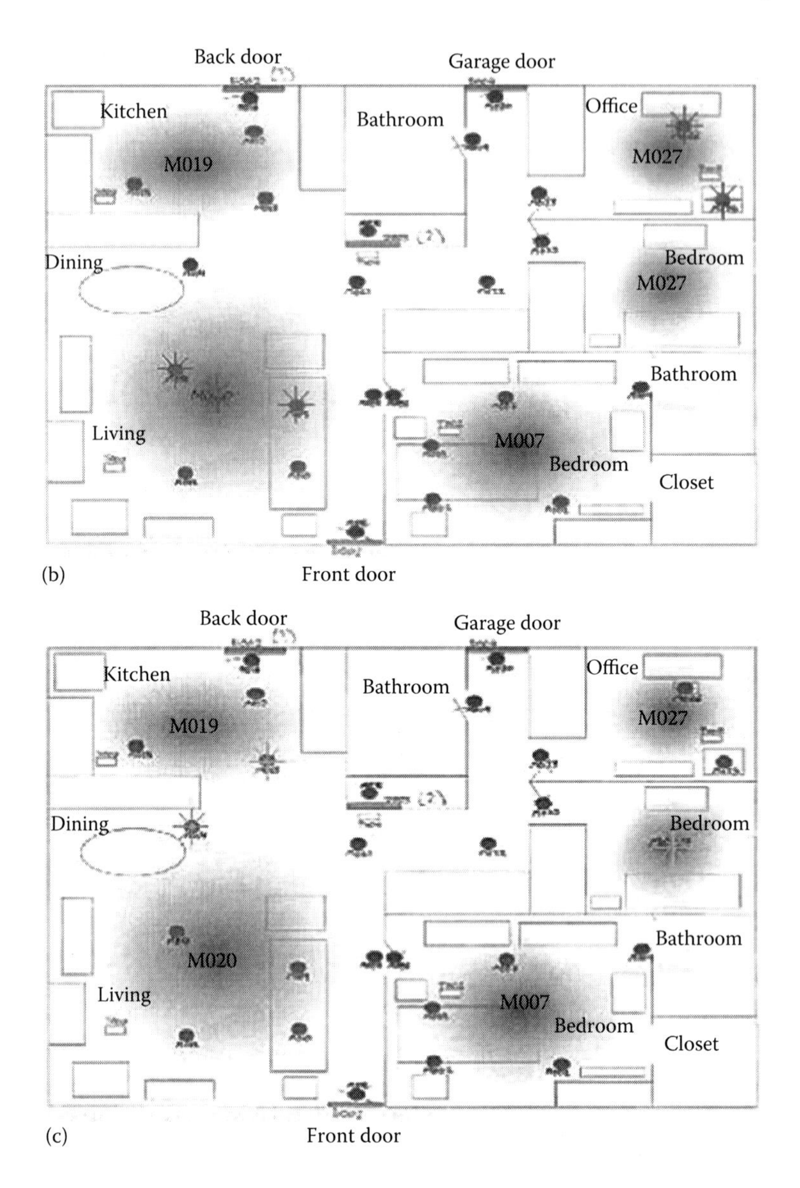

FIGURE 14.20 (continued) Example of difficulties of using a sliding window approach to detect multiple people: (b) both occupants are moving in distinctive locations, and (c) case of a multiple-detection failure.

status text and markers. The markers indicate a triggered motion detector during the 5 s window, and the text indicates if a visitor has been detected for the period between two door entry/exit events. Figure 14.20a shows the case when only one of the two users is moving, so during that time instant, our system cannot determine if a visitor is present; but using knowledge from adjacent time instances, as in Figure 14.20b, we propagate this information backward and forward in time to the previous

TABLE 14.6
Confusion Matrix of Multiple-Person Detector

		Actual	
		Single	**Multiple**
Predicted	Single	54	3
	Multiple	1	12

and future entry events, respectively. As a result, our system does not have to detect all cases where multiple people are present. There are several degenerate situations when our algorithm is unable to detect multiple people; one such example is presented in Figure 14.20c. Our algorithm does not work in this case since the far right cluster of activity only has a single event, so the clustering algorithm most likely failed to assign it to a unique cluster.

The confusion matrix representing the performance of our multiple-person detector is presented in Table 14.6. In general, our detector does a good job of classifying intervals with multiple persons present with an 94% success rate; however, there were some instances when multiple people were not properly detected. Upon reviewing these misclassified intervals, we found that most were due to the lack of instrumentation in the guest bedroom, which only had two sensors, one of which was directed at the door, so our clustering algorithm frequently failed to assign a cluster to the lone detection event in the guest room. We did have some false-positive detections, all of which were for very short intervals that involved very fast motion of the occupant. The fast motion generated a higher than normal number of motion sensors to trigger across a wide area compared to a normal movement. Our detector seems to work best when the areas where users are present are heavily instrumented. For the CASAS testbed, the multiple-person detector is frequently triggered when the primary occupant and guests were spread across the master bedroom, the office, and the kitchen and living room.

14.6 FUTURE WORK

The work on our processing framework to date has focused on extracting useful indicators of health, such as sleep times and active versus sedentary activities, as well as movement and trajectory parameters. We intend to use these indicators as input into our next stage to recognize useful patterns and make inferences using our extracted indicators and parameters as inputs. A block diagram of our current processing framework and possible future models for health inferences is shown in Figure 14.21.

One useful inference that we could make is to determine if long-term changes, or rate of change, of the indicators we extract indicate a change in health or wellness where the caregiver should be notified. However, short-term variability along with measurement noise can complicate identifying these long-term trends, especially

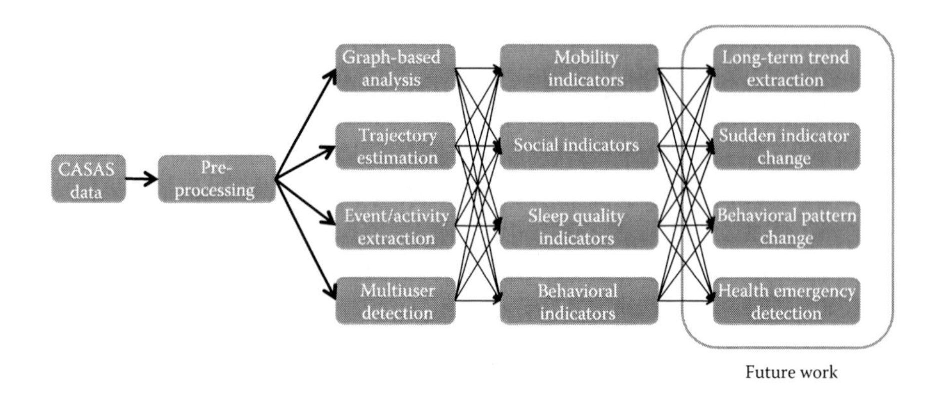

FIGURE 14.21 System block diagram including future health inference modules.

since natural aging can cause changes to the users' mobility that, while interesting, may not be diagnostically relevant to caregivers.

Sudden changes in wellness indicators are another area that we intend to explore in the future. Unfortunately, we have limited knowledge of the health status of the CASAS apartment occupant, so this stage of the research may depend on an alternate dataset or simulated motion sensor activations. Examples of sudden changes could be a marked deviation from previous movement patterns and trajectories, which could indicate a change in mental state, to abnormal sleeping habits and increased sedentary activity. The challenge in this area is to distinguish normal variability in a user's behavior from changes that could be the result of a deteriorating medical condition.

Ultimately, we would like to look for relationships between indicators, to look for causal relationships between them to see if changes in one indicator could predict future changes in other indicators. This would allow us to make inferences as to the cause of wellness indicator changes and possibly identify causes or events that precipitated them.

14.7 CONCLUSION

We presented our initial work in developing a system to extract movement and behavior indicators from long-term data from smart home sensors with the eventual goal of using these indicators to track the wellness of a home's occupants. We hope to use these extracted indicator time series to form user-specific behavioral models for use in detecting abnormal behavior to enable caregivers or relatives to be informed of any health or behavioral changes. We were able to extract sleep patterns that were verified by comparing with the CASAS projects annotations. We were also able to develop a multiple-person detection technique and verify its performance using manual labeling of the visualized sensor data. Since motion sensors have traditionally been accepted in residential environments for lighting controls and security applications, further research into the utility of such data for long-term monitoring seems relevant, as other sensing modalities, such as cameras, have considerable privacy implications.

REFERENCES

1. World Health Organization, What are the public health implications of global ageing? 2012 [Online]. Available: http://www.who.int/features/qa/42/en/index.html, accessed on March, 2012.
2. Administration for Community Living, A profile of older Americans, 2012 [Online]. Available: http://www.aoa.gov/aoaroot/aging_statistics/Profile/2011/4.aspx, accessed on March, 2012.
3. A. O'Brien and R. Mac Ruairi, Survey of assistive technology devices and applications for aging in place, in *Second International Conference on Advances in Human-Oriented and Personalized Mechanisms, Technologies, and Services* (*CENTRIC'09*), September 2009, Porto, Portugal, pp. 7–12.
4. B. Tran, Home care technologies for promoting successful aging in elderly populations, in *Proceedings of the Second Joint 24th Annual Conference and the Annual Fall Meeting of the Biomedical Engineering Society EMBS/BMES Conference on Engineering in Medicine and Biology*, October 2002, Houston, TX, Vol. 3, pp. 1898–1899.
5. M. E. Pollack, Intelligent technology for an aging population: The use of AI to assist elders with cognitive impairment, *AI Magazine*, 26(2), 9–24, 2005.
6. H. Hashimoto, T. Matsunaga, T. Tsuboi, Y. Ohyama, J. She, N. Amano, S. Yokota, and H. Kobayashi, Comfortable life space for elderly—Using supporting systems based on technology, in *SICE, 2007 Annual Conference*, September 2007, Kagawa University, Takamatsu, Japan, pp. 3037–3042.
7. L. Wang, Z. Wang, Z. He, and X. Gao, Research of physical condition monitoring system for the elderly based on ZigBee wireless network technology, in *2010 International Conference on E-Health Networking, Digital Ecosystems and Technologies* (*EDT*), April 2010, Shenzhen, China, Vol. 1, pp. 32–35.
8. S. Heinrich, K. Rapp, U. Rissmann, C. Becker, and H.-H. Knig, Cost of falls in old age: A systematic review, *Osteoporosis International*, 21, 891–902, 2010 [Online]. Available: http://dx.doi.org/10.1007/s00198-009-1100-1, accessed on March, 2012.
9. Quantified self: Self knowledge through numbers, Available: http://quantifiedself.com/, accessed on March, 2012.
10. Fitbit®, Force, Flex, One and Zip Wireless Activity and Sleep Trackers. Available: http://www.fitbit.com/, accessed on March, 2012.
11. Nike+ Fuelband, Tracks your all-day activity and helps you do more, March 2012 [Online]. Available: http://www.nike.com/us/en_us/c/nikeplus-fuelband, accessed on March, 2012.
12. J. H. Shin, B. Lee, and K. S. Park, Detection of abnormal living patterns for elderly living alone using support vector data description, *IEEE Transactions on Information Technology in Biomedicine*, 15(3), 438–448, May 2011.
13. A. O'Brien, K. McDaid, J. Loane, J. Doyle, and B. O'Mullane, Visualisation of movement of older adults within their homes based on PIR sensor data, in *2012 Sixth International Conference on Pervasive Computing Technologies for Healthcare* (*PervasiveHealth*), May 2012, San Diego, CA, pp. 252–259.
14. D. Cook, Learning setting-generalized activity models for smart spaces, *IEEE Intelligent Systems*, 27(1), 32–38, January–February 2012.
15. P. Rashidi and D. J. Cook, Mining and monitoring patterns of daily routines for assisted living in real world settings, in *Proceedings of the First ACM International Health Informatics Symposium* (*IHI'10*), 2010, Arlington, VA. New York: ACM, pp. 336–345.
16. P. Rashidi, D. J. Cook, L. B. Holder, and M. Schmitter-Edgecombe, Discovering activities to recognize and track in a smart environment, *IEEE Transactions on Knowledge and Data Engineering*, 23(4), 527–539, April 2011.

17. V. R. Jakkula and D. J. Cook, Detecting anomalous sensor events in smart home data for enhancing the living experience, in *AAAI Workshops on Artificial Intelligence and Smarter Living*, 2011, San Francisco, CA, pp. 33–37, Vol. WS-11-07.
18. G. Jain, D. Cook, and V. Jakkula, Monitoring health by detecting drifts and outliers for a smart environment inhabitant, in *Proceedings of the International Conference on Smart Homes and Health Telematics*, Belfast, Northern Ireland, UK, 2006, pp. 114–121.
19. L. Kalra, X. Zhao, A. J. Soto, and E. E. Milios, A two-stage corrective Markov model for activities of daily living detection, in *ISAmI'12*, 2012, Salamanca, Spain, pp. 171–179.
20. D. Lymberopoulos, A. Bamis, and A. Savvides, Extracting spatiotemporal human activity patterns in assisted living using a home sensor network, in *Proceedings of the First International Conference on PErvasive Technologies Related to Assistive Environments* (*PETRA'08*), 2008, Athens, Greece. New York: ACM, pp. 29:1–29:8.
21. I. Gómez-Conde, D. Olivieri, X. Vila, and L. Rodriguez-Liñares, Smart telecare video monitoring for anomalous event detection, in *2010 Fifth Iberian Conference on Information Systems and Technologies* (*CISTI*), Santiago de Compostela, Spain, June 2010, pp. 1–6.
22. C. Rougier, J. Meunier, A. St-Arnaud, and J. Rousseau, Robust video surveillance for fall detection based on human shape deformation, *IEEE Transactions on Circuits and Systems for Video Technology*, 21(5), 611–622, May 2011.
23. F. Cardile, G. Iannizzotto, and F. La Rosa, A vision-based system for elderly patients monitoring, in *2010 Third Conference on Human System Interactions* (*HSI*), May 2010, Rzeszow, Poland, pp. 195–202.
24. S. S. Roley, J. V. DeLany, C. J. Barrows, S. Brownrigg, D. Honaker, D. I. Sava, V. Talley et al., Occupational therapy practice framework: Domain & practice, 2nd edition, *American Journal of Occupational Therapy*, 62(6), 625–683, 2008.
25. E. Cho, K. Wong, O. Gnawali, M. Wicke, and L. Guibas, Inferring mobile trajectories using a network of binary proximity sensors, in *2011 Eighth Annual IEEE Communications Society Conference on Sensor, Mesh and Ad Hoc Communications and Networks* (*SECON*), June 2011, Salt Lake City, UT, pp. 188–196.
26. J. Singh, U. Madhow, R. Kumar, S. Suri, and R. Cagley, Tracking multiple targets using binary proximity sensors, in *2007 Sixth International Symposium on Information Processing in Sensor Networks* (*IPSN'07*), April 2007, Cambridge, MA, pp. 529–538.
27. Q. Le and L. Kaplan, Target tracking using proximity binary sensors, in *2011 IEEE Aerospace Conference*, Big Sky, MT, March 2011, pp. 1–10.
28. S. B. Young, Evaluation of pedestrian walking speeds in airport terminals, *Transportation Research Record: Journal of the Transportation Research Board*, 1674(1), 20–26, 1999.
29. P. Karras and N. Mamoulis, Detecting the direction of motion in a binary sensor network, in *2006 IEEE International Conference on Sensor Networks, Ubiquitous, and Trustworthy Computing*, June 2006, Taichung, Taiwan, Vol. 1, p. 8.
30. R. Brooks, P. Ramanathan, and A. Sayeed, Distributed target classification and tracking in sensor networks, *Proceedings of the IEEE*, 91(8), 1163–1171, August 2003.
31. D. De, W.-Z. Song, M. Xu, C.-L. Wang, D. Cook, and X. Huo, Finding HuMo: Real-time tracking of motion trajectories from anonymous binary sensing in smart environments, in *2012 IEEE 32nd International Conference on Distributed Computing Systems* (*ICDCS*), June 2012, Macau, China, pp. 163–172.
32. S. Thrun, D. Fox, W. Burgard, and F. Dellaert, Robust Monte Carlo localization for mobile robots, *Artificial Intelligence*, 128(1–2), 99–141, 2000.

33. D. Fox, S. Thrun, W. Burgard, and F. Dellaert, Particle filters for mobile robot localization, in *Sequential Monte Carlo Methods in Practice*, A. Doucet, N. de Freitas, and N. Gordon, eds. New York: Springer Verlag, 2001, pp. 499–516.
34. J. Kennedy and R. Eberhart, Particle swarm optimization, in *Proceedings of the IEEE International Conference on Neural Networks*, November/December 1995, Piscataway, NJ, Vol. 4, pp. 1942–1948.
35. S. Thrun, W. Burgard, and D. Fox, *Probabilistic Robotics* (*Intelligent Robotics and Autonomous Agents*). Cambridge, MA: The MIT Press, 2005.

15 Architecture of Sports Log Application Using Nine-Axis MEMS

Rana E. Ahmed, Imran A. Zualkernan, Nada Ibrahim, Rashid Al Hammadi, and Farhan Hurmoodi

CONTENTS

ABSTRACT

The adoption of MEMS-enabled devices in health industry is on the rise due to their incorporation of various functionalities and sensors on the devices. The devices are highly portable and dissipate less energy. This chapter presents the architecture, design, and test results of a sports log application using MEMs. The architecture is based on nine-axis MEMS system consisting of three-axis digital gyroscope, accelerometer, and compass. The system uses Bluetooth wireless communication to communicate information from the sensors to the user's smartphone. Using a simple and efficient algorithm, the system computes different types of information, such as distance covered and calories burnt by the person while walking or jogging. The information is displayed on the user's smartphone according to the profile selected by the user. Various tests are performed to check the validity and statistical accuracies of our results.

15.1 INTRODUCTION

The adoption and application of microelectromechanical systems (MEMS)–enabled devices in sports and health industry have increased tremendously thanks to improved functionality and portability of the devices along with their decreasing prices. The wearable MEMS devices that wirelessly link with the smartphones have modernized the classical pedometers. The popularity of app stores for software downloads for mobile phones has also contributed toward quick and cost-effective adoption of mobile applications and related hardware. Sports applications have witnessed an increase of 113% during the year 2010 [1], and this trend is expected to stay in the near future. One factor linked with the popularity of applications is a certain class of mobile platforms (e.g., Android and iOS [2,3]). There has been a strong need for a user-friendly sports log applications for athletes and weight loss patients to monitor their exercise progress under controlled conditions.

Functioning by sensing a body's motion, pedometers are used by regular walkers and sports enthusiasts. They are used while walking or running to count the number of steps taken as well as the distance traveled. Most pedometer applications rely on the information provided by an accelerometer or global positioning system (GPS) or a combination of both. One of the mobile applications currently dominating the market is from a famous shoe manufacturer, which depends on the use of the accelerometer to calculate calories consumed in addition to distance traveled and speed. Using GPS on its own has its disadvantages. It cannot accurately record position indoors, in areas with building obstructions or under heavy tree canopy [4]. In addition, atmospheric conditions can also affect the measurements made by GPS devices, and so affecting the accuracy of speed and position [4]. Furthermore, GPS accuracy fails on curved paths [5]. Accelerometers, on the other hand, may seem more accurate than GPS systems [5], but they too have their disadvantages, especially inaccuracies caused on uphill or downhill paths [5]. Moreover, efforts are underway to communicate information gathered by sports log application to cloud-based computers for analytics, and to social media sites to share information with friends and fellow athletes [6].

MEMS are tiny devices that have elements ranging in size from 1 to 100 μm, about the thickness of human hair. MEMS devices are generally divided into two categories: sensors and actuators. Sensor devices gather information from their surroundings, and actuators execute giving command. There has been a recent trend in the industry to integrate several sensors/actuators in MEMS [7]. The MEMS used in the present research work includes three major components: an accelerometer, a gyroscope, and a compass.

An accelerometer can be visualized as a mass on springs. As long as there is no rotation, accelerometers can measure linear acceleration in three axes and tilt. The velocity is obtained by a single integration; however, the accelerometers cannot distinguish between the acceleration due to linear movement and the acceleration due to gravity.

Gyroscope senses angular velocity. This sensor also has three axes (making total nine axes) and measures its own rotation using the well-known Coriolis effect.

Compasses in the MEMS are basically magnetic field sensors that pick up every possible surrounding magnetic field. Compasses have three axes as well.

The lack of distinction between the linear acceleration and the acceleration due to gravity in accelerometer can be solved when a gyroscope is added to it. The relative distance can be obtained by a double integration; nevertheless, the integration leads to errors in the long run. As a compass can pick up unrelated surrounding magnetic fields, a gyroscope sensor is needed to solve the field corruption problem [8–10].

The placement of the sensor on human body is quite critical as it can affect speed, distance, the number of steps, and calorie expenditure calculations. According to Liu and Won [11], after testing the placement of the gyroscope on the waist, knee, and foot, the placement on the knee gave the most accurate distance traveled. This is due to the fact that the placement on the knee takes advantage of the periodic angular displacements at each step taken by the person, and by using this information, the system will keep the angular displacement errors bounded.

This research work presents the design of a sports log application that uses the information produced by an external nine-axis motion sensor. The system uses a simple, yet efficient algorithm to compute desired results.

15.2 SYSTEM ARCHITECTURE

15.2.1 Brief Overview of System Architecture

The architecture of the system is made up of three levels: presentation layer, data layer, and data source layer. Figure 15.1 shows all the views containing the different data presented to the user, whereas Figure 15.2 shows how each layer interacts with the other and the user.

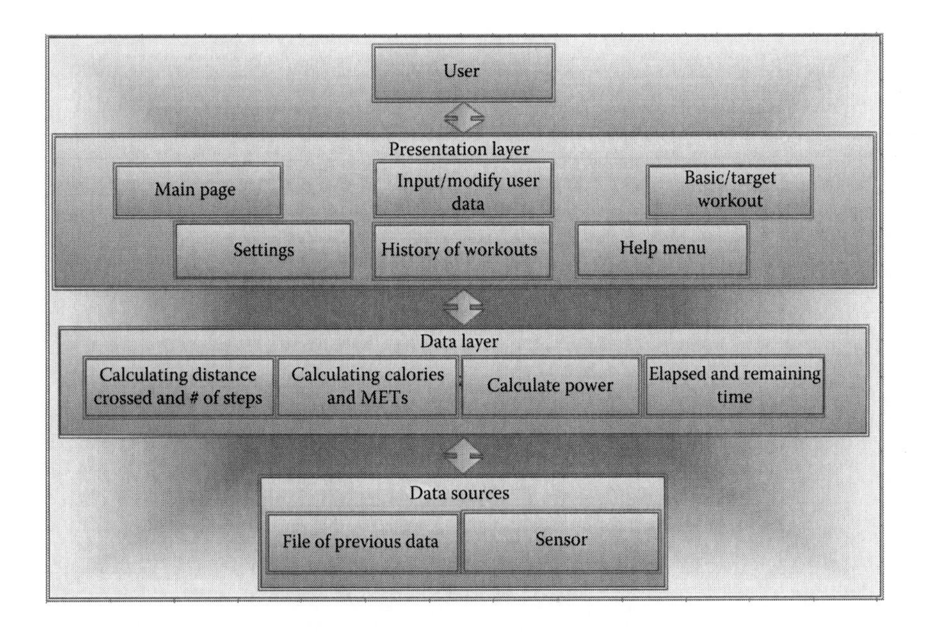

FIGURE 15.1 Overall system architecture.

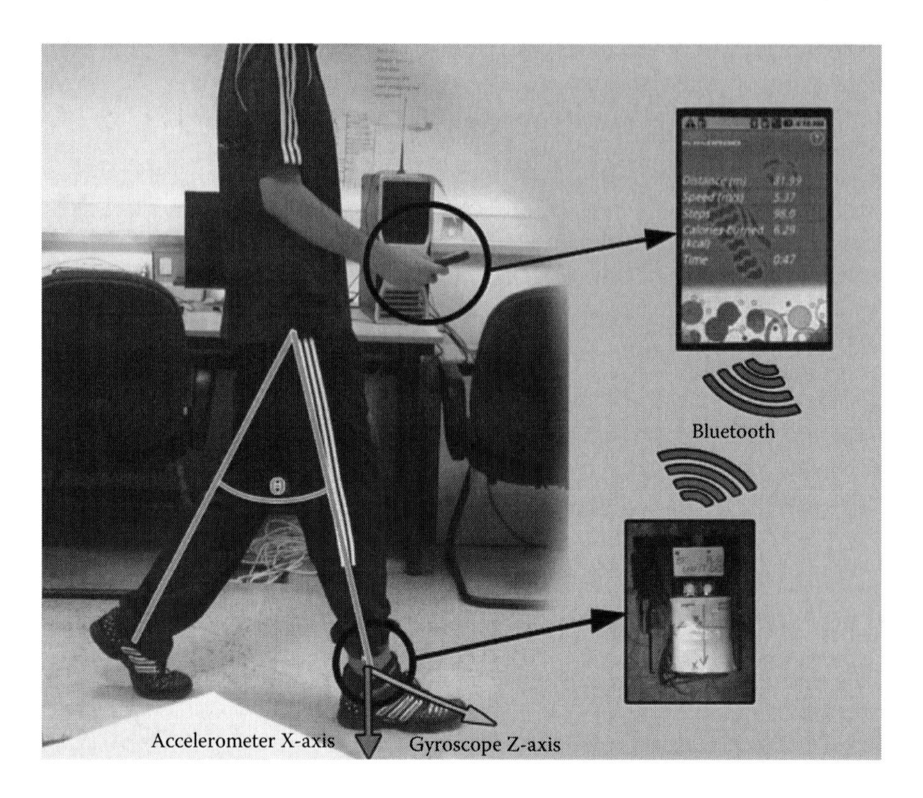

FIGURE 15.2 Interaction between the user and layers.

The system uses information produced by a sensor board containing nine-axis MEMS. The sensor board is attached to the user's right ankle. The information is sent wirelessly using Bluetooth radio to a smartphone. The sensor information is used in an algorithm running on smartphone that calculates distance, speed, the number of steps taken, and calories burned while walking. The user is expected to input the needed algorithm parameters (weight and leg length) to get the required information as accurately as possible.

The software application developed is also responsible for interacting with the user, maintaining a history of workout, user settings for different types of workouts (Basic, Target), and *power song* option.

15.2.2 Hardware Components and Graphical User Interface

The hardware of the motion sensor consists of a universal serial bus (USB)/battery power board, a sensor board, and a radio frequency (RF) board. The 1.1 in. × 0.9 in. (approximately) USB/power board basically provides the sensor board with the supporting functions, serving mainly to power the sensor board with a rechargeable battery. The USB power board also includes a USB connection for charging the lithium ion battery and allowing serial debug port connection to a PC and a battery charger.

The sensor board, which is approximately 1.3 by 0.9 in. in size, contains a three-axis digital gyroscope, accelerometer, and compass (creating a total of nine axes) [12].

Accelerometers measure linear acceleration in three axes and tilt. The velocity is obtained by single integration, whereas relative distance is obtained by double integration. Compasses pick up every possible surrounding magnetic field; therefore, a gyroscope sensor was added to solve the field corruption problem. Gyroscopes sense angular velocity by measuring its own rotation using the Coriolis effect.

The three components (USB/power board, sensor board, and RF board) are connected to an 8 bit microprocessor that performs sensor sampling and motion sensor calculations. The buck/boost efficient switching regulator allows wide battery inputs (approximately 1.8–3.6 V), allowing the sensor board to operate from any power source from 2.0 to 3.6 VDC (such as 2 AA cells). The RF board used operates in Bluetooth SPP mode and is approximately 1.35 by 0.9 in. in size and is intended for communicating data to the host console or PC [12].

The application graphical user interface (GUI) has different views: *Main* page, *Basic Workout* page, *Target Workout* page, *Settings* page, *History of Workouts* page, and the *Help Menu* page, as shown in Figures 15.3 through 15.6.

15.2.3 Data Layer

In this section, the computations done by the application algorithm for both the accelerometer and the gyroscope are discussed. The accelerometer x-axis value is

FIGURE 15.3 Main page window screen.

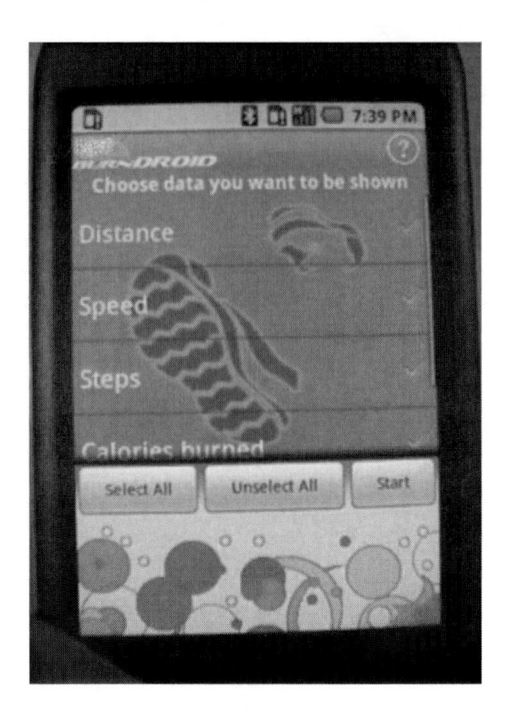

FIGURE 15.4 Basic workout screen.

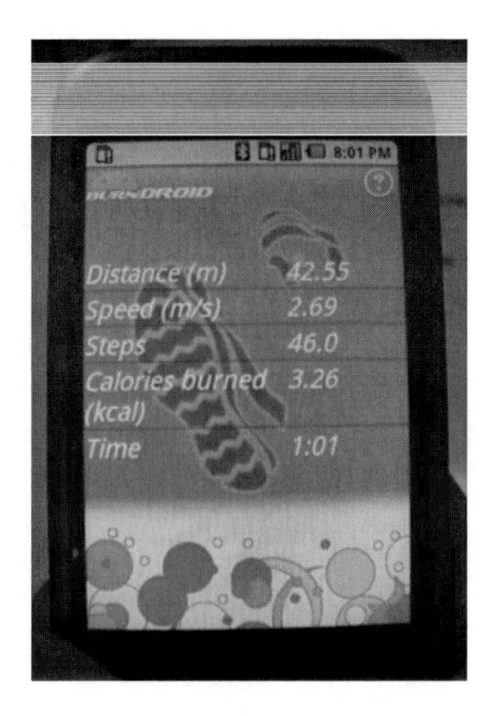

FIGURE 15.5 Basic workout exercise screen.

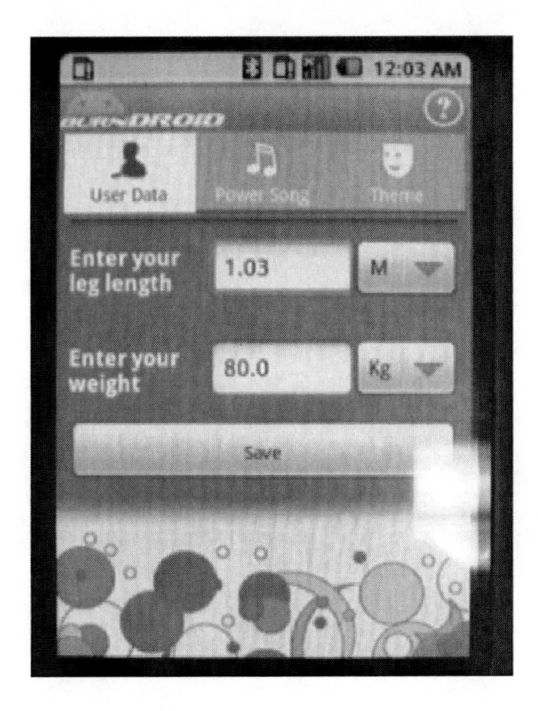

FIGURE 15.6 Setting screen.

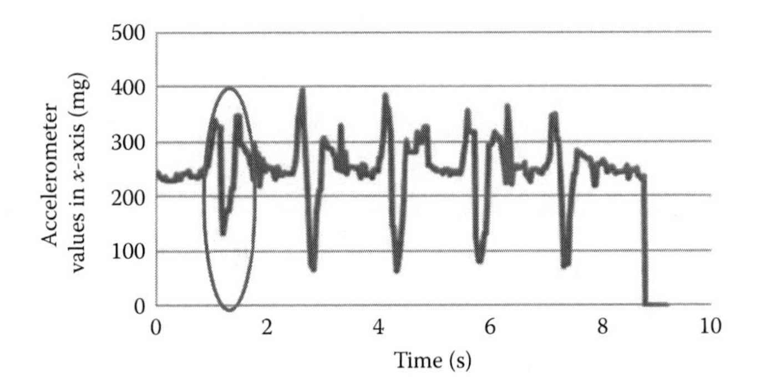

FIGURE 15.7 Accelerometer x-values vs. time (unstable).

used to count the number of strides by detecting the fluctuating areas of the acceleration values produced, as shown in Figure 15.7. The stable point of the accelerometer values is detected to determine when the user is not making a step, as shown in Figure 15.8.

The angular velocity (in degrees/s) produced by the gyroscope is used to calculate the distance traveled (in m), calories burned (in kcal), speed (in m/s), and the number of steps taken by the user. The gyroscope is used to calculate the number of strides taken using the values of the z-component, since the z-axis is the axis at which the foot rotates [13], as shown in Figure 15.10. Due to a clear pattern in the

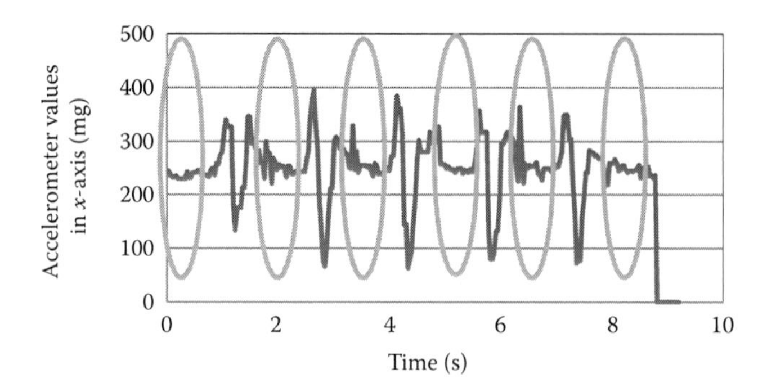

FIGURE 15.8 Accelerometer x-values vs. time (stable).

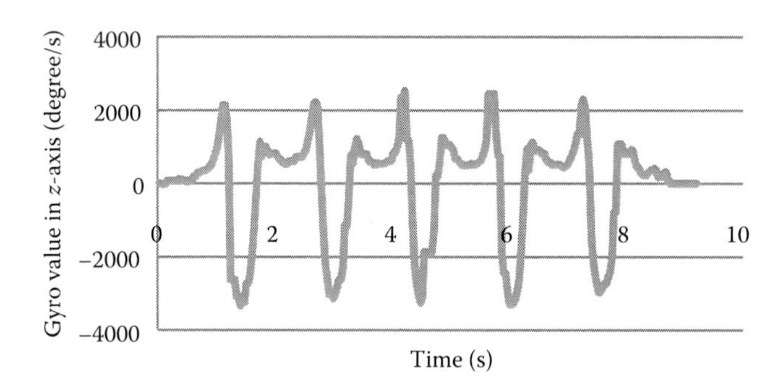

FIGURE 15.9 Gyroscope z-values vs. time.

values retrieved from the z-component, as indicated in Figure 15.9, each negative drop represents a stride made by the right leg. Nevertheless, these negative values are also obtained while a person is standing still. Therefore, the accelerometer values are used to complement the gyroscope values to differentiate between a stride and standing still, resulting in a more accurate value for the number of strides. Since each stride is equal to a step, the number of strides is multiplied by 2 to get the number of steps carried out by both feet.

In order to compute the distance traveled, the value of the angular velocity in each of the x-, y-, and z-directions is read from the gyroscope in the motion sensor strapped around the user's ankle, as shown in Figure 15.10.

The displacement angle (Θ) is calculated by integrating the angular velocity's z-component (in degrees/s) as shown in the following equation [11]:

$$\Theta = \int_{t1}^{t2} \omega(t)_z \, dt \tag{15.1}$$

where

$t1$ is the time (in seconds) at which the leg is having a 0° with respect to the z-axis
$t2$ is the step ending time

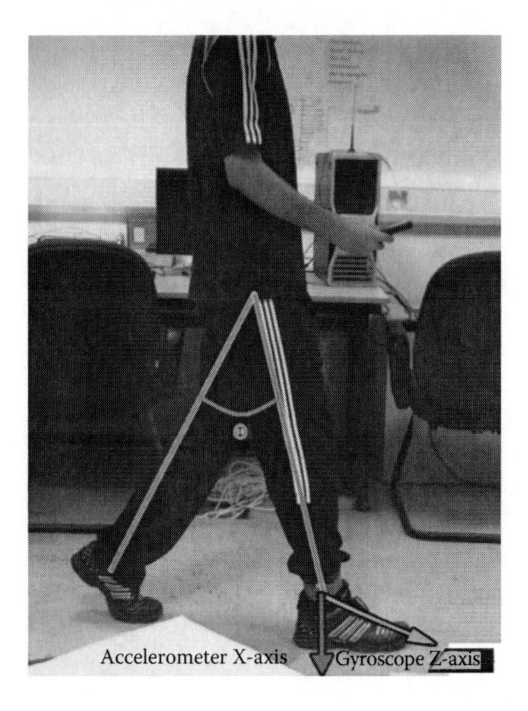

FIGURE 15.10 Human stride model.

In order to find the distance covered by one stride (in m), the following formula [11] is used:

$$\text{Stride length} = \text{leg length} \times \sqrt{2 \times [1 - \cos(\Theta)]} \tag{15.2}$$

The distance traveled is simply the distance covered by all the strides the user took. Therefore, the total distance is

$$\text{Distance traveled} = \sum \text{Lengths of strides} \tag{15.3}$$

The walking speed (in m/s) is calculated as follows:

$$\text{Walking speed} = \frac{\Delta\text{Distance traveled}}{\Delta\text{Time}} = \frac{\text{Length of current stride}}{\text{Current time}} \tag{15.4}$$

In order to calculate the calories burnt, the following formula [11] is used:

$$\text{Calories burnt} = 0.7 \times \text{weight (lbs)} \times \text{distance (m)} \tag{15.5}$$

The value of 0.7 is the coefficient for walking. Each activity has its own coefficient depending on how rigorous the activity is.

MEMS sensor communicates with the mobile phone via Bluetooth, sending one reading every 33 ms. This layer includes a text file, to be used later, that saves previous MEMS data in addition to any previous data entered by the user.

15.3 TESTING AND ANALYSIS OF RESULTS

In order to keep the study under controlled conditions, all physical walking and running experiments were carried out by the same person at a time so that the leg length and weight remained constant, with the sensor strapped to the ankle at the same position. With every batch of experiment, the tester was changed. The tests were carried out for walking in a straight line, running in a straight line, curves, and slopes (both uphill and downhill).

15.3.1 Experimental Design

Four people with different leg lengths (100, 102, 95, and 105 cm) and weights (65.6, 63, 65.6, and 72 kg) were asked to walk briskly for exactly 100 m in a straight line on a flat surface. Measurements about the number of steps taken, speed, distance traveled, and calories burned were recorded using four different devices:

- A leading manufacturer's application where a motion sensor is present in the shoes
- Application involving pedometer attached to the shoulder
- Application involving pedometer in pocket
- Our application

For the application, the combination of gyroscope and accelerometer (*gyro*) was tested, and the accelerometer (*acc*) was tested separately as well.

15.3.2 Results

Tables 15.1 through 15.5 show the results of running the *Anderson–Darling normality test* using the statistics package *Mini Tab*. This test indicates if the test results acquired from our experiments follow a normal distribution. To determine if the results are normal, we need to look at the *P*-value. If the *P*-value is below 0.05, then it is considered to be not normal, and we take the median value; otherwise, we take the mean value. The mean is the average value of all the experimental data where the data would be clustered around. The median on the other hand is simply the middle value, and if we have an even number of data, then it is the average of the two middle points.

As shown in Table 15.1, the only application with a *P*-value above 0.05 is the shoulder application, and so we looked at its mean instead of median. The results show that the median and 95% confidence interval for median are exactly on the mark of the 100 m. We have our gyroscope results as second best, in terms of both median and range of % confidence interval for median. Taking the leading manufacturer application's results into consideration, in Table 15.2, we have the Anderson–Darling

TABLE 15.1
Anderson–Darling Normality Test

	Gyro (m)	Acc (m)	Pocket App (m)	Shoulder App (m)	Leading App (m)
A-squared	0.78	1.11	4.15	0.35	5.14
P-value	0.038	0.006	0.005	0.457	0.005
Median distance	100.72	66.10	93.88	—	100
Mean distance	—	—	—	97.09	—
95% confidence interval for median	99.54–104.03	61.14–72.46	90.22–99.48	—	100.00–100.00
95% confidence interval for mean	—	—	—	92.97–100.12	—
95% confidence percentage difference	4.49%	11.32%	9.26%	7.15%	0%

TABLE 15.2
Leading Application vs. Other Applications

	Leading App—Gyro (m)	Leading App—Acc (m)	Leading App—Pocket (m)	Leading App—Shoulder (m)
A-squared	4.48	3.98	3.16	2.72
P-value	0.005	0.005	0.005	0.005
Median	−3.04	29.42	8.41	1.03
95% confidence interval for median	−5.72 to 1.02	27.57–34.25	4.63–14.37	−3.50 to 6.56
95% confidence percentage difference	6.74%	6.68%	9.74%	10.06%

TABLE 15.3
Sign Test for Median

Sign Test of Median = 0.1000 vs. not = 0.1000

	N	Below	Equal	Above	*P*	Median
Leading app distance (m)	34	8	18	8	1.000	100.0
Shoulder app distance (m)	34	23	11	11	0.0576	97.37
Pocket app distance (m)	34	23	0	11	0.0576	97.37
Acc distance (m)	34	34	0	0	0.000	97.37
Gyro distance (m)	34	14	0	20	0.3915	100.7

TABLE 15.4
Individual 95.0% Confidence Interval

```
                                  Individual 95.0% Confidence Interval
Leg length (cm)   Median (m)   ------+---------+---------+---------
 95                113.4                          (-------*--)
100                 95.1       (-----*--)
102                103.9                  (---*---------)
105                100.2        (---------*---)
                               ------+---------+---------+---------
                                   96.0      104.0     112.0
```

TABLE 15.5
Limitations

	Straight (Walk)	Straight (Run)	Round-About (Walk)	Upslope (Walk)	Downslope (Walk)
Number of tests	30	11	29	33	33
Distance (m)	100	100	66.6	15.7	15.7
A-squared	0.27	0.51	0.25	0.36	0.81
P-value	0.639	0.161	0.711	0.431	0.033
Mean	98.65	67.20	66.514	14.794	—
Median	—	—	—	—	16.45
95% confidence interval for mean	97.96–99.34	62.99–71.41	65.983–67.045	14.42–15.16	—
95% confidence interval for median	—	—	—	—	16.08–16.65
95% confidence percentage difference	1.38%	8.42%	1.062%	0.74%	0.57%

normality test for all applications. Using the leading application as a benchmark, we calculated the error compared with the leading application.

As shown in Table 15.1, we are looking at the median since all *P*-values are below 0.05. By focusing on the 95% confidence interval for median, we notice that while having the leading application as our benchmark, because of its value, range size, and range values, the gyroscope results are the best.

The *sign test* for median tests the null hypothesis with the median equals to a specific value, in this case being 0.1000.

The null hypothesis states that if the data are very improbable (referring to the *P*-value when it is equal to or less than 0.05), then the null hypothesis can be determined to be false.

Based on the Table 15.3, we notice that both gyroscope and the leading application are statistically the same, as they are both above the *P*-value of 0.05. The *P*-values for pocket and shoulder applications are considered to be equal to 0.05 and thus prove the null hypothesis to be wrong.

The *Mood's median test* simultaneously compares all the results of the gyroscope distance results with the leg length. As seen in Table 15.4, each person had a different median for distance covered. When two areas in the individual 95.0% confidence interval overlap, they are considered to be the same. For example, the 95 cm leg length is considered to be the same as the 102 cm leg length. Not all areas overlap; however, no area is separated/different from at least one other area. Therefore, we conclude that all areas/leg lengths are the same and yield the overall median of 100.7 m.

In order to determine how well our end algorithm works, we tested it by walking in a straight line for a 100 m, around a 66.6 m circumference round-about, 15.7 m upslope, and 15.7 m downslope and run in a 100 m long straight line. Based on the Anderson–Darling normality test results, our gyroscope application seems to function well for all earlier-mentioned scenarios/exercises except for the straight-line run. This is mainly due to limitation in our algorithm to compute distances.

15.4 SUMMARY AND CONCLUSIONS

This work presented a sports log application based on nine-axis MEMS consisting of three-axis digital gyroscope, accelerometer, and compass (creating a total of nine axes). The system uses Bluetooth wireless communication to communicate information from sensors to the user's smartphone. The main objective was to compute, as accurately as possible, different type of information, such as distance covered and calories burnt while walking or jogging. Using a simple but efficient algorithm, the distance traveled and the number of steps were found to be within a small percentage of error when compared with other similar state-of-the-art sports log applications. Various tests were carried out, and they proved that a gyroscope, rather than an accelerometer, is more accurate, especially in the case of using the MEMS sensor. Moreover, the analysis carried out shows that our application provides results that are statistically as accurate as offered by the leading manufacturer's applications.

ACKNOWLEDGMENT

The authors would like to thank Mr. Yasser Al-Jundi and Dr. Hamid Najafi (InvenSense Inc.) for helping them in conducting experiments and gathering results and providing them with the necessary equipment to develop this project.

REFERENCES

1. Retrieved from website: http://www.internetekg.com/?p=9, March 1, 2012.
2. Fusion on android devices: A revolution in motion processing, retrieved from website: http://invensense.com/mems/videolibrary.html, January 1, 2011.
3. Retrieved from website: http://mobithinking.com/stats-corner/global-mobile-statistics-2011-all-quality-mobile-marketing-research-mobile-web-stats-su, May 18, 2011.
4. Reed, R.G., Using fitness devices and global positioning system (GPS) technology in measuring energy expenditure and distance walked over flat and inclined surfaces, PhD thesis, University of Arizona, Tucson, AZ.

5. C. Mallula, Comparing Garmin Forerunner 405CX and Nike+iPod to accurately measure energy expenditure, distance and speed of over ground running, Master thesis, Cleveland State University, Cleveland, OH, 2010.
6. MEMS in the World of Sports, *EE Times Asia*, January 25, 2012. Available at http://www.eetasia.com/ART_8800660096_499495_NT_b79e23b6.HTM.
7. MEMS Industry Group, retrieved from website: http://invensense.com/mems/videolibrary.html, January 1, 2011.
8. K. Takata, M. Tanaka, J. Huang, A. Runhe, and N. Shiratori, A wearable system for outdoor running workout state recognition and course provision, *Proceedings ATC*, Hong Kong, China, 2007, pp. 385–394.
9. Y. Lim, I. Brown, and J. Khoo, An accurate and robust gyroscope-based pedometer, *30th Annual International IEEE EMBS Conference*, Vancouver, British Columbia, Canada, August 2008.
10. A critical review of the market status and industry challenges of producing consumer grade MEMS gyroscopes, retrieved from website: http://www.invensense.com/mems/whitepapers.html, January 1, 2011.
11. Z. Liu and C.-H. Won, Knee and waist attached gyroscopes for personal navigation: Comparison of knee, waist and foot attached inertial sensors, *2010 IEEE/ION Position Location and Navigation Symposium (PLANS)*, May 4–6, 2010, pp. 375–381.
12. InvenSense Inc., *9-Axis 8-Bit Motion Processing Reference Design:* Hardware Functional Description, retrieved from http://www.cdiweb.com/datasheets/invensense/Hardware%20Functional%20Description%20V22.pdf, accessed January 1, 2011.
13. Development of high-performance high-volume consumer MEMS gyroscopes, retrieved from website: http://www.invensense.com/mems/whitepapers.html, January 1, 2011.

Section V

Transportation

16 Social Sensor Networks for Transportation Management in Smart Cities

Francesco Chiti, Romano Fantacci, and Tommaso Pecorella

CONTENTS

ABSTRACT

Internet of Things represents a promising approach to improve the context awareness and smart interactions with surrounding environment. In this area, both wireless sensor networking and social networking paradigms are gaining momentum. These trends suggest a collaborative integration of sensor and social data, which result to be mutually compensatory in various data processing and analysis. This chapter, therefore, deals with mechanisms for an effective coordination to improve the quality of published information, with particular reference to the smart cities vision. In particular, the focus is on transportation management, where an integrated approach could be applicable to a large variety of urban scenarios, as it is flexible, reactive, and robust. In the following, a brief overview of the most suited enabling technologies is provided, together with addressing a practical case study to point out the advantages provided by the proposed approach.

16.1 INTRODUCTION

The Internet of Things (IoT) represents an approach (Palattella et al., 2012) recently envisaged to deliver constant information flow from the physical objects in our world and to make the sensory information about our environment available online. Simultaneously, mobile phones are becoming multisensor devices, accumulating large volumes of data related to our daily lives, while they also serve as a major channel for recording people activities at social-networking services in the Internet. These trends obviously raise the potential of collaboratively integrating sensor and social data: these two popular data types are, in fact, mutually compensatory in various data processing and analysis. Therefore, it is desirable to provide mechanisms for wireless sensor networks (WSNs) collaboration and coordination to improve the quality of published information. Participatory sensing, for instance, enables to collect people-sensed data via social-network services (e.g., Twitter) over the areas where physical sensors are unavailable. Simultaneously, sensor data are capable of offering precise context information, leading to effective analysis of social data (Yerva et al., 2012). In addition to this, it is interesting to provide a framework for employing social networks to facilitate the collaboration and coordination of distributed sensor networks. Seamlessly exchanging sensory measurements among geographically dispersed sensor networks is anticipated to enhance our capability to capture the status of the physical world and to realize a *smart environment*, once sensory information is delivered to various machineries in the world. Additionally, sensor networks may utilize social networks for distributing the sensing responsibilities among sensor networks and, as a result, prolonging the lifetime of the networks while maintaining constant monitoring and detection of the environment (Baqer, 2010).

One of the main areas for applying WSN monitoring paradigm is related to environmental control especially in urban regions, considering that about 80% of the population of developed countries currently live in towns, so that it represents one of the main causes affecting the quality of life. The collection of environmental data is a basic block for the climate changing tracking and, consequently, for environmental protection. *Climate* is usually defined as the *average* of the atmospheric conditions over both an extended period of time and a large region. Small-scale patterns of climate resulting from the combined influence of topography, urban buildings structure, watercourses, and vegetation are known as *microclimates*, which refer to a *specific* site or location. The microclimate scale may be at the level of a settlement (urban or rural), neighborhood, cluster, street, or buffer space in between buildings or within the building itself. Specifically, the dispersion and dilution of air pollutants emitted by vehicles is one of the most investigated topics within urban meteorology, for its fundamental impact on the environment affecting cities of all sizes. These issues, which arise in urban planning, concern the average and peak values of various air pollutants as well as their temporal trends and spatial variability. The accurate detection of these values might be advantageously exploited by public authorities to better plan the public and private transportation by evaluating the impact on people's health while controlling the greenhouse phenomenon.

As the unpredictable nature of a climate variation requires an incessant and ubiquitous sensing, WSNs represent a key technology for environmental monitoring, hazard detection, and, consequently, decision making (Martinez et al., 2004). A WSN is designed to be self-configuring and independent from any preexisting infrastructure, being composed of a large number of elementary sensor nodes (SNs) that can be large-scale deployed with small installation and maintenance costs.

VSN has been widely studied in the literature: some are related to routing, traffic management, and road safety (Gao et al., 2010; Chen et al., 2011). In addition to this, several examples of frameworks for evaluating the urban air quality with WSNs have been proposed, as it is reported in Santini et al. (2008). An infrastructure-less framework composed of a distributed warning system and a location-based back-off scheme is proposed in Chen et al. (2011) to avoid the read-end collision. A real-time environmental monitoring system is proposed in Resch et al. (2009), which aims to combine the live measurement data with the historic data for monitoring the temporal and spatial changes of environmental status. A vehicular WSN (VWSN) is introduced in Wong et al. (2009) to monitor the air quality and environment in urban areas. Tsugawa and Kato (2010) introduced the contribution of intelligent transportation system (ITS) on the energy saving and the prevention of global warming. An environmental air quality sensing system is proposed in Aoki et al. (2009) using the street sweeping vehicles. In addition, in Cordova-Lopez et al. (2007), the monitoring of exhaust and environmental pollution through the use of WSN and GIS technology is addressed. As microclimate monitoring usually requires deploying a large number of measurement tools, in Shu-Chiung et al. (2009), the VWSN approach to reduce system complexity while achieving fine-grained monitoring is adopted. However, many existing works on the application of VWSNs in the environmental protection focus on using the collected data to generate the environmental report. The increasing popularity of social networks motivates the interests in providing more innovative applications by integrating the VWSN with social networks, as it is discussed in Tse et al. (2011) together with addressing some challenging issues.

Another aspect strictly correlated with microclimate establishment is represented by the ecologic footprint of traffic congestion due to inefficient traffic management. As a consequence, an increasing number of cities are going to develop ITS as an approach to harmonize roads and vehicles in optimized and green paths. ITSs involve several technologies as advanced informatics, data communications and transmissions, electronics, and computer control with the aim of real-time traffic reporting and alerting. Such a framework allows remote operation management and self-configuration of traffic flows, as well as specific information delivering to vehicles concerning, for instance, traffic congestion or the presence of accidents (Pinart et al., 2009).* Thus, the research on data acquisition scheme has become a key point to enable effective ITSs.

* These goals could be summarized in two main fields as traffic flow *forecast* and traffic *congestion control*.

At present, the acquisition of real-time traffic data is by means of installation and use of *wired* monitoring equipment in most cities. However, several concerns are associated with this choice: firstly, with the continuous expansion of the city size and the increasing number of traffic roads, the more the number of wired monitoring equipment increases, the more the cost grows (*scalability*). Further, the installation of wired monitoring equipment does not have the flexibility, being difficult to (re) deploy. Finally, as urban traffic congestion has a certain degree of space–time randomness, then it is inappropriate to install monitoring equipment in *fixed* locations. On the other hand, large-scale universal installation will cause larger waste. To solve these problems, a promising approach is currently represented by WSNs applicable to all types of urban environment (Laisheng et al., 2009), as they have no space constraints, flexible distribution, mobile convenience, and quick reaction.

In the following, a brief overview of the most suited enabling technologies is provided, together with addressing a practical case study to point out the advantages provided by the proposed approach.

16.2 BACKGROUND

16.2.1 Social Sensors

Social-network services facilitate their users to share their ideas, opinions, pictures, videos, news, or actually any form of contents in the web. Such social data typically contain highly valuable information, aiding a wide range of applications, for example, allowing social scientists to understand human behaviors, companies to figure out their customers' preferences, and news agencies to identify significant news. Previously, it was difficult to obtain the rich set of social information or required large amounts of laborious human efforts like conducting surveys and interacting with the users.

A key issue concerns the integration of distributed sensing to a communications infrastructure able to provide pervasive capabilities. A promising approach is focused on Internet features, specifically on Web 2.0: the so-called *sensor-to-web* model, such as SensorWeb (Kansal et al., 2007), CenceMe (Miluzzo et al., 2008), and CitySense (Murty et al., 2008), is in essence designed to share sensory information and infrastructure. Whether they are sensing human or environmental parameters, they all provide means of delivering sensory information to end users. The privacy of data and thus the user privileges to access the sensory information depend generally on what is being sensed and their respective applications. Integrating mobile phones with sensors provides sensory measurements about the owner of the mobile phone. Social networks such as Facebook and Twitter are being employed to publish sensory information (Kansal et al., 2007; Baqer and Kamal, 2009). In essence, social networks enable the creation of communities based on their common interests (Java et al., 2007). Similarly, in sensor networks, a community may include SNs of the same sensory application or SNs deployed to monitor the same target. For instance, in CenceMe, project users with mobile phones seamlessly uploaded their status onto Facebook (Kansal et al., 2007). Specifically, Twitter—a microblogging web service that enables users to publish their current status by sending short messages of about 140 characters—suits the requirements of sensor networks (Java et al., 2007). Any SN or sensor network may be associated with a dedicated Twitter page where information about the monitored event is published. Sensory data are processed and filtered en route within the network to avoid overwhelming the Twitter page with sensory measurements. Thereby, current sensor-to-web models are limited to conveying sensory readings of particular sensors or sensor networks to their dedicated web portal. The author in S-Sensors (Baqer and Kamal, 2009) employed Twitter to publish temperature readings.

At the light of these considerations, a wide variety of research directions are open, which might be summarized as follows:

- *Mood analysis*: To this purpose, one popular research line aims at extracting and analyzing mood information from Twitter messages. In Marcus et al. (2011), microblogs are used for mood analysis, presenting a method for associating a mood to a certain event. Another study (Bollen et al., 2010) tries to predict the impact of public mood expressed in Twitter messages on the stock market; they make use of mood tracking tools, namely, OpinionFinder (which measures positive vs. negative mood) and Google-Profile of Mood States (GoPMS), which measures mood in terms of six dimensions. The authors of Thelwall et al. (2011) focus on Twitter messages in order to study why certain events resonate well with the population. In particular, the impact of an important event is measured by checking if the associated *sentiment* is higher than the Twitter average or by assessing whether this event is associated with increased sentiment strength. Finally in Pak and Paroubek (2010), considering its widespread diffusion, Twitter is used as the reference media for training the sentiment classifier, which classifies tweets as expressing positive, negative, or neutral sentiment.
- *Social sensing*: In the light of the increasing importance of sensor networks in our everyday activities, some studies (Nagarajan et al., 2011; Rosi et al., 2011) went ahead and consider the people participating in microblogs or social networks as *social sensors* providing the rich social context, which are hard to infer using physical sensors. For instance, the work in Sakaki et al. (2010) monitors the flows of Twitter messages for quickly detecting an earthquake that occurs in an area where seismic sensors are unavailable. Another study (Weng and Lee, 2011) mines the Twitter messages to identify relevant events to given monitoring conditions.
- *Social data fusion*: This field has been opened by the ubiquitousness of the mobiles and sensing devices. Specifically, in Nagarajan et al. (2011), the paradigm of CitizenSensing is introduced, enabled by Mobile sensing and Human Computing, according to which humans acting both as citizens on the ubiquitous web and as sensors and sharing their observations and view through Web 2.0. Likewise, SocialFusion (Beach et al., 2010) proposes the use of sensor networks to enable context-aware social applications, analyzing the data generated by the users of the applications. In addition, SocialSensors (Rosi et al., 2011) describes the need for fusing social data with pervasive sensors for providing improved

services. Finally, in Lovett et al. (2010), heuristic methods for data fusion that combine the user's personal calendar with his social network posts are presented, in order to produce a real-time multisensor interpretation of the real-world events while achieving significant improvement through data fusion.

16.2.2 Internet of Things

The *future* Internet, designed as an IoT, is foreseen to be a worldwide network of interconnected objects uniquely addressable, based on standard communication protocols (Internet of Things in 2020: Roadmap for the Future, 2008). Identified by a unique address, any object might be able to dynamically join the network and collaborate and cooperate efficiently to achieve different tasks. In fact, the everyday objects will become *proactive* actors of the Internet, generating and consuming information. The elements of the IoT comprise not only those devices that are already deeply rooted in the technological world (such as cars or fridges) but also objects foreign to this environment (garments or perishable food) or even living beings (plantations, woods, or livestock). By embedding computational capabilities in all kinds of objects and living beings, it is actually possible to provide a qualitative and quantitative leap in several sectors: healthcare, logistics, domotics, entertainment, and critical monitoring, to name a few. Comprehensively, the European Commission envisions IoT as the approach where things are endowed with identities and virtual personalities operating in smart spaces using intelligent interfaces to connect and communicate within social, environmental, and user contexts. In particular, the use of standard technologies in the World Wide Web to instrument the IoT is frequently referred to as the Web of Things (Guinard et al., 2010b).

In this disruptive scenario, WSNs are able to open new perspectives. Covering a wide application field, WSNs, indeed, can play an important role by collecting surrounding context and environment information. The benefits of connecting both WSN and other IoT elements go beyond remote access, as heterogeneous information systems can be able to collaborate and provide common services. This approach is effectively supported by several international companies' initiatives, such as: (i) IBM A *Smarter Planet* strategy, which considers sensors as fundamental pillars in intelligent water management systems and intelligent cities, and (ii) the HP *Labs CeNSE*, which focused on the deployment of a worldwide sensor network in order to create a central nervous system for the Earth.

However, deploying WSNs configured to access the Internet raises novel challenges, which need to be tackled before taking advantage of the many benefits of such integration. Essentially, the technologies that will enable the integration are being developed and tested. For example, the 6LowPAN standard, defined by IETF (Kushalnagar et al., 2007), allows the transmission of IPv6 packets through computationally restricted networks. Moreover, it is actually possible to link the data produced by the elements of a WSN (SNs) with web services based on SOAP and REST (Guinard et al., 2010a), messaging mechanisms (such as e-mails and SMS), or social networks (e.g., Twitter) and blogs (e.g., WordPress) (Libelium: Interfacing the Sensor Networks with the Web 2.0, 2010).

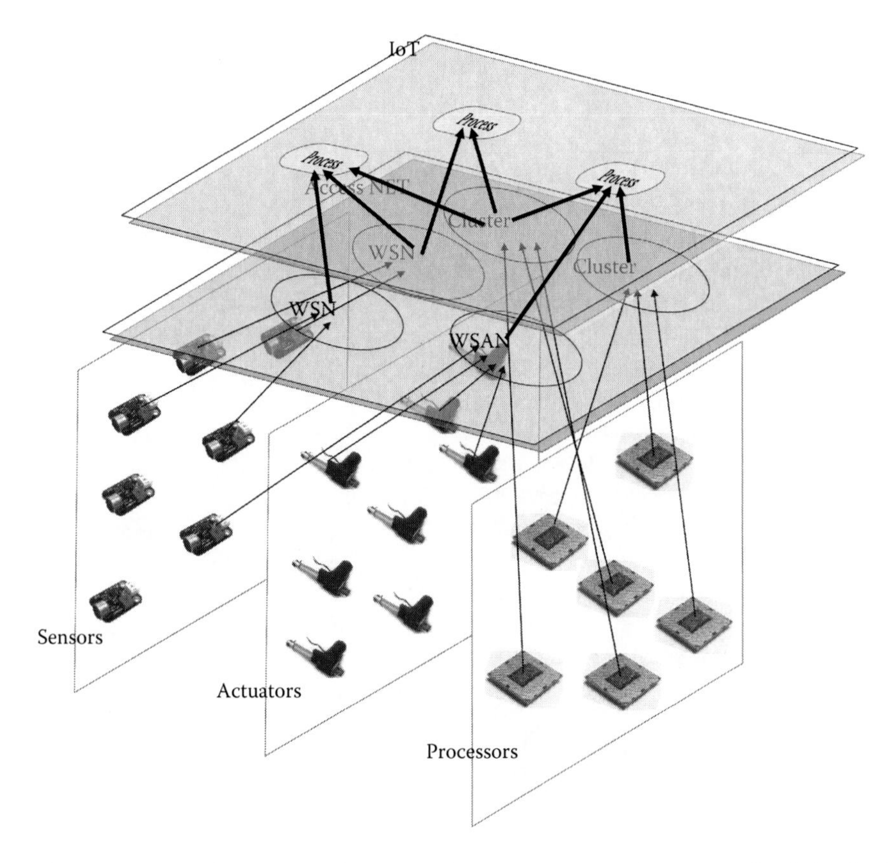

FIGURE 16.1 Reference functional architecture for dynamic setup of IoT-inspired applications.

Besides the technological challenges, IoT is understood to bring along multiple economical and societal challenges. The latter result from the high pervasiveness of the technological artifacts realizing IoT and the introduction of machine to machine (M2M) applications with no or minimal human involvement. Among the various technological developments supporting the IoT, wireless networking, embedded computing, and scalable computation through virtualization are considered the key driving forces for transforming the IoT vision into reality. Herein, a comprehensive functional architecture for IoT enabling collaboration among generic WSNs and wireless sensor and actuator networks (WSANs), together with cluster of devices dedicated to distributed data storing and processing, is presented in Figure 16.1, pointing out the role of access networks and IoT/M2M technologies.

16.2.3 ITS Communications Paradigms Overview

ITS services availability relies on the presence of an infrastructure usually comprising fixed devices interconnected by an underlying network, either wired or wireless.

Data exchange toward or among mobile terminals is inherently wireless, since information should directly reach the drivers through PDAs or onboard transceivers; in evidence, the IEEE 802 committee has activated 11p Task Group to define a Wi-Fi extension for wireless access in vehicular environments (WAVE) (Jiang and Delgrossi, 2008). Moreover, wireless connections are needed also for data gathering, according to the WSN paradigm, comprising a large number of devices in charge of sensing and relay informations to the core network (Tubaishat et al., 2009).

Within the previous scenario, several communications paradigms are possible (Yousefi et al., 2006). The case in which fixed access points (APs) allow mobile nodes to join the network is usually referred as *infrastructure-to-vehicle* (I2V) communications and can support advanced applications such as web surfing, multimedia streaming, remote vehicle diagnostics, and real-time navigation, to name a few; on the other side, *vehicle-to-vehicle* (V2V) communications represent the option in which mobile nodes can directly communicate to each other without any need of infrastructure. Although V2V and I2V communications are both prominent research fields, this chapter is mainly focused on the latter, as it aims to efficiently exchange short amounts of data, collected and aggregated by an in-field deployed WSN, to nomadic users, while keeping the complexity of onboard circuitry as low as possible. It is worth noticing that a reliable I2V scheme is extremely valuable even for V2V communications since, whenever a direct link among vehicles is not available, message exchange can leverage on the infrastructure instead of being successively relayed by few less-reliable mobile nodes, as addressed in Gerla et al. (2006).

The urban environment is usually composed of a large number of mobile terminals that are likely to quickly change their reference AP, therefore facing frequent disconnection and reconnection procedures, so that it may be not viable to deliver the total amount of required data within a single session. Moreover, the urban channel is affected by long- and short-term fading that introduces additional delays for data retransmissions (in the case of TCP traffic) or sensibly lowers the data reliability (in the case of UDP traffic); these issues are addressed in details in Bychkovsky et al. (2006) and Ott and Kutscher (2004), providing a practical case study involving IEEE 802.11b.

In general, content distribution through overlay networks is more efficient when compared to traditional solutions using multiple unicasts. In order to achieve higher throughput and failure resilience, parallel downloading from multiple overlay nodes represents a typical approach in most recent proposals (Wu and Li, 2007). However, the same content may be unnecessarily supplied by multiple nodes, raising the problem of the so-called content reconciliation, which usually is a time- and bandwidth-consuming operation (Byers et al., 2002a).

16.3 PROPOSED APPROACH

16.3.1 System Requirements and Architecture

The reference system model is derived from a real-world case study, inspired by the Tuscany Region project *Metropolitan Mobility Agency Supporting Tools* (SSAMM),

FIGURE 16.2 Reference network topology for general I2V applications.

devoted to enhance the quality of urban transportation system introducing innovative paradigms. The addressed urban communications scenario is modeled as a two-level network, as illustrated in Figure 16.2. In particular, the lower level is composed of a large number of SNs, positioned in such way that suitable and effective sampling of the road traffic is achieved within the area of interest (Tanner, 1957). Whenever possible, SNs are deployed in correspondence with road infrastructures such as posts, lamps, and traffic lights, typically arranged in a square grid fashion. Their purpose is to collect traffic flow information* and relay it to the higher layer consisting of interconnected network of APs. In addition to these fixed SNs, also mobile sensors are introduced; it could be the case of a public vehicle equipped with gas analyzers for the classical air pollutants NO, NO_2, O_3, and NO, in order to record air pollution and meteorological data within different urban zones. In the meanwhile, vehicles can deliver information regarding the interarrival time between adjacent APs, which is useful in estimating the congestion level. Finally, APs deliver gathered data toward a mobile collector (MC) usually referred to as *data mule*.

As the proposed application scenario is concerned with fast and efficient information retrieval, these drawbacks could be faced by introducing an appropriate *data dissemination* algorithm, enhancing the information *persistence* throughout the network without an excessive overload in terms of total packet transmissions.

* Different types of information, for example, average crossroad waiting time and presence of roadworks or accidents, could be of interest.

To face MC inherent mobility, a *distributed data gathering* protocol has been introduced (Stefanovic et al., 2011) to efficiently collect all the sensed data by visiting only an arbitrary subset of the SNs; this general requirement is extremely important in urban scenarios, since paths are usually space–time *constrained*. This has been achieved by resorting to a distributed implementation of rateless codes (Byers et al., 2002b), that are a particular class of erasure correction codes which rely on sparse binary coefficient data combining, being suitable for the I2 V data dissemination application, in which devices exhibit low computational capabilities. Moreover, it has been introduced as an adequate *data dissemination protocol*, which has been integrated with an MC *data gathering* scheme specifically designed for an urban-wide area monitoring WSN in order to allow reliable and accurate sensing collection.

16.3.2 Communications Scheme

Each homogeneous subset of SNs is connected in a star-wise or tree topology to an AP; APs encode and exchange packets received from SNs, then broadcast the information to MCs. MCs usually join the network without need of an association with a specific AP by adopting a *passive operation mode* and continuously collecting information regarding the surrounding environment broadcasted by APs.* Nevertheless, whenever MCs are involved in disseminating their own information, they explicitly associate with the best AP and operate both in *transmitting* and *receiving* modes. However, intervehicle communications are not hereafter considered. Finally, we assume that MCs have onboard capabilities to process the downloaded data according to suitable applications in order to interpret current traffic information. Specifically, the collected real-time data provide opportunity for onboard computer to perform optimal route calculation, delay estimates, and present driver with visual map representation of critical locations where accidents, high pollutant concentrations, or severe road congestions took place. Although the push mode is possible, data processing is usually implemented in an automatic (i.e., periodic) manner in order to guarantee as much as possible real-time monitoring of the traffic-load conditions. In particular, the latter mode of operation makes it possible for MCs to be informed about dangerous situations (e.g., accidents) in a short time span, hence allowing for increased safety of people and vehicles.

The communication between SNs (even mounted on board of a MC) and APs, and between MCs and APs, is assumed to be based on wireless technology. As recently shown, IEEE 802.11b/g standards demonstrated significant potential for vehicular applications (Bychkovsky et al., 2006). Another candidate could be IEEE 802.11p (still in the draft stage), whose one of the aims is to support efficient data exchange between roadside infrastructure and vehicles.

Regarding the communications between APs, it is accomplished by leveraging on a preexisting infrastructure deployed in an urban area, that is, connecting APs to wired metropolitan area network (MAN), such as the one adopted by Florence Municipality,

* This is adopted in order to lower the implementation complexity and cost of the mobile equipment and minimize the downloading time, avoiding access contentions and complex handover procedure.

called *FI-Net*, comprised of a double fiber-optics ring with a 2×2.5 Gbps full-duplex capacity. Full-mesh wireless interconnections are not considered as the adoption of an IEEE 802.11 unique radio interface could pose several limitations in terms of coverage or, equivalently, scalability.

However, the communication scenario described earlier fits in the infrastructure mode of the IEEE 802.11 standard in which several APs are interconnected using an external distributed system, forming an *extended service set* (ESS).

16.3.3 Rateless Codes

Rateless (or digital fountain) codes are a recently introduced class of forward error correction codes with a universally capacity-approaching behavior over erasure channels with arbitrary erasure statistics. The first practical rateless codes, called LT codes (Luby, 2002), are based on a simple encoding process where the source message of length k information symbols is encoded into a message containing a potentially infinite number of encoded symbols. Each encoded symbol is an independently and randomly created representation of the source message and, as soon as the receiver correctly receives any set of k' encoded symbols, where k' is only slightly larger than k, it is able to recover the source message.

LT encoding process is defined by the LT code degree distribution $\Omega(d)$, which is a probability mass function defined over the interval $[1, k]$. To create a new encoded symbol, the degree d is randomly sampled from the degree distribution $\Omega(d)$, d information symbols are uniformly and randomly selected from the information message, and the encoded symbol is obtained by XOR-ing d selected information symbols. Usually, information and encoded symbols are equal-length binary data packets and the XOR operation is the bitwise XOR. Encoded symbols are transmitted over an erasure channel and decoded at the receiver using the iterative belief-propagation (BP) algorithm. BP algorithm for erasure channel decoding iteratively recovers information symbols from the degree-one encoded packets and cancels out the recovered information symbols in all the remaining encoded packets (which may result in a new set of degree-one encoded packets). The iterations of this simple process can lead to the complete message recovery; otherwise, the receiver will have to wait for additional encoded packets in order to complete the decoding process. The key problem of LT code design is the design of the degree distribution $\Omega(d)$ that will enable source message recovery from any slightly more than k received encoded symbols using the iterative BP decoding algorithm. This problem is solved asymptotically in Luby (2002), where it is shown that using the so-called robust soliton degree distribution, it is possible to recover the source message from any k' encoded symbols, where $k' \to k$ asymptotically, with encoding/decoding complexity of the order $O(k \cdot \log k)$.

Rateless codes are usually applied in multicast scenarios, where the source message is entirely available to the source node. However, in many practical systems such as wireless ad hoc networks, WSN or p2p networks, the message of interest might be distributed over many or all network nodes. As shown recently—see Vukobratovic et al. (2010) and references therein—distributed rateless coding may be performed as efficiently as its centralized counterparts and may provide a number of benefits in distributed network scenarios for applications such as data gathering, data persistence, and distributed data storage.

16.3.4 Distributed Data Gathering, Encoding, and Dissemination

The system application, residing in APs, periodically performs the following three procedures, according to Stefanovic et al. (2011): (1) data gathering from SNs, (2) encoding, and (3) disseminating encoded data to MCs. We refer to these three stages as upload, encoding, and download phase, respectively, and the period encompassing all of them as *data refreshment period*. According to the IEEE 802.11 standard, the link time in every AP coverage zone is divided in superframes (IEEE, 2007), and the data refreshment period in each zone is aligned with superframe boundaries (see Figure 16.3).

During the upload phase, every AP polls all SNs in its domain* and collects the most recent measurements. As typically foreseen by most of the IEEE 802.11 standards (IEEE, 2007), superframes are divided into the contention-free period (CFP) and contention-based period (CBP), where the former is used to avoid MAC collisions and deliver prioritized information to MCs. The polling phase can be accomplished within the CFP part of a typical frame. CFP always starts after a beacon with a delivery traffic information map (DTIM) field sent by AP to STAs (which is SNs in our case). STAs associated with AP learn when the CFP should begin and automatically set their NAV to MaxCFPDuration when the CFP is expected to begin. Then AP individually polls each STA with a CF-poll message waiting for DATA and CF-ACK messages from it, where messages are separated by a short interframe space (SIFS) period. We assume that APs are globally synchronized,† so the actual upload takes place in the *first* superframe period following the start of the data refreshment period. Each SN uploads its

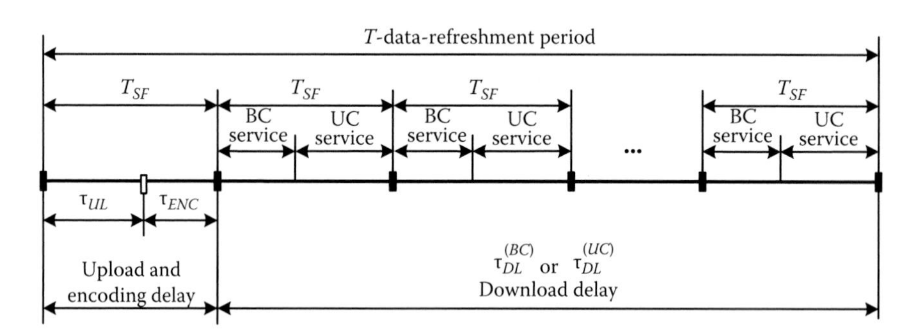

FIGURE 16.3 Data refreshment period of the proposed application.

* SNs can be either fixed, that is, infrastructured, or mobile, that is, on board of MCs. In the latter case, it is necessary that the sojourn time of MC is comparable with superframe. This condition is easily satisfied in the case of public transportation means close to a regular or temporary stop where an AP has been placed.

† It is easily provided by the preexisting MAN communications infrastructure which arrange a sort of *APs network*.

measurements within a single data packet of length L bits. Since SNs and APs form an *infrastructured* network, it has been supposed that nodes have been previously deployed in a line-of-sight (LoS) fashion in order to optimize link quality; however, possible packet losses are managed by means of Automatic Repeat reQuest (ARQ) scheme, so that from an application point of view data, delivering could be considered reliable. In particular, to match the constraint of polling completion within the first frame, a maximum ARQ retransmission persistency equal to NA attempts has been selected, in order to yield a negligible residual packet error probability.

On reception, AP stores and uniquely indexes each received data packet, where the indexing scheme is known to all APs. The total number of stored data packets in APs network per data refreshment period is k, which is equal to the total number of SNs. These k data packets represent a single data generation, upon which the rateless coding is performed. The differentiation among data generations can be achieved using appropriate field in packet header, allowing MCs to maintain global time references.

After the upload phase, the system application, distributed over all APs in the network, performs distributed rateless encoding of collected data packets (i.e., rateless coding is used at the application level). Each AP independently produces k_{AP} encoded packets, where the actual value of k_{AP} is chosen such that it is sufficient for successful data recovery by all MCs with high probability (w.h.p.). Specifically,

$$k_{AP} \geq \left(\frac{1+\epsilon^{(\max)}}{1-P_{PL}} \right) \cdot k$$

where $\epsilon^{(\max)}$ is the reception overhead that allows for decoding w.h.p. and which depends on the properties of the applied rateless codes, while P_{PL} is the estimated link-layer packet loss probability. For each encoded packet, AP draws degree d from the employed degree distribution and then randomly selects d data packets from the pool of all k data packets residing in the AP network during the *current* data refreshment period. In general case, most of the selected data packets are likely to be stored in other APs, so the AP has to request them using the known indexing scheme. After reception of the missing data packets, AP creates encoded packets by simple bitwise XOR of associated data packets.

Finally, in the download phase, each AP disseminates encoded packets by simply *broadcasting* them to MCs currently falling in its coverage area. This approach has been adopted in order to minimize the complexity and the power consumption of MC receiver by always keeping it in a receiving mode. Depending on the provided service, two kinds of dissemination are possible: broadcast (BC) and geocast (GC) ones. The BC service covers simultaneous *global* distribution of the most important data such as key traffic info to all the associated MCs, using the BC MAC address. The GC service could be used for an additional (e.g., traffic congestion and air quality) *local* delivering of uncoded packets containing information on the actual hot spot (e.g., context-aware information for navigation software enhanced services).

Due to its duration and broadcast nature, the BC service is capable of delivering significantly larger amount of data per superframe to its users as compared to the GC service, which is why the former is preferable for delay-sensitive real-time information delivery. The dissemination starts in the first superframe that follows the encoding phase and lasts until the next data collection phase (i.e., the next data refreshment period). For the purpose of BC service, the natural choice is to use CFP part of the superframe, as it guarantees delivery of traffic info updates to all subscribed MCs within the service area.

While traveling within the service area, each MC performs a channel sensing at periodic intervals (say, θ), dynamically selects the best carrier, and transparently roams among adjacent APs, while downloading encoded packets from APs, until it collects enough for sensor data recovery using the iterative BP algorithm. The number of excessive encoded packets compared to k sensor packets is measured by the reception overhead ϵ*; that is, for successful recovery, MC needs in total $k' = (1+\epsilon') \cdot k$ encoded packets, where ϵ' usually is a small positive number. Since each encoded packet is an innovative representation of the original data, *any* subset of $k' = (1+\epsilon') \cdot k$ taken from the set of *all* the encoded packets in the network allows for restoration of the whole original data. This property of rateless codes makes them a perfect candidate to be used at the application level for content delivery in vehicular networks, since packet losses caused by the varying link characteristics are compensated simply by reception of the new packets and there is no need for standard acknowledgment–retransmission mechanisms, which cannot be supported by a semiduplex architecture as the one adopted. In other words, the usage of connection-oriented transport protocols like TCP can be avoided, as UDP-like transport provides a satisfactory functionality. Moreover, the loosing of packets caused by channel error or by the receiver deafness during the selection of a different AP does not impact on BC scheme, as MC continues downloading data *without* any need for (de/re)association, session management, or content reconciliation.

16.4 SIMULATION RESULTS

The simulation setup assumes that the urban area is covered by a regular hexagonal lattice, where each nonoverlapping hexagon represents the coverage area of a single AP and the hexagon side length is equal to the AP transmission range. MCs move throughout the lattice using the rectangular grid that models urban road infrastructure, associating with the nearest AP. The overlay hexagonal AP lattice is independent and arbitrarily aligned with the underlying rectangular road grid. The MCs move according to the Manhattan mobility model (Bai et al., 2003), a model commonly used for metropolitan traffic. In brief, Manhattan mobility model assumes a regular grid consisting of horizontal and vertical (bidirectional) streets; at each intersection, MC continues in the same direction with probability 0.5 or turns left/right with probability 0.25 in each case. The MC speed is uniformly chosen from a predefined interval and changes on a time-slot basis (time-slot duration is a model

* This takes into account both the decoding overhead and the redundancy needed in the presence of erasure channel.

parameter), with the speed in the current time slot being dependent on the value in the previous time slot. Besides temporal dependencies, Manhattan mobility model also includes spatial dependencies, since the velocity of an MC depends on the velocity of other MCs moving in the same road segment and in the same direction; as we are interested only in I2V communications from the perspective of a single user (i.e., a single MC), spatial dependencies are omitted in our implementation.

The purpose of the simulations is to estimate the duration of the download phase as the most important and the lengthiest phase of the data refreshment period. In each simulation run, while moving on the road grid, the MC starts receiving the encoded data from the AP in whose coverage zone it is currently located. The reception of the encoded packets continues until the MC collects enough to successfully decode all the original data. If during this process MC happens to move to another AP zone, it simply associates to a new local AP (i.e., handover takes place) and starts to receive its encoded packets. Also, if the AP has transmitted all of its encoded packets to the MC, but it failed to decode the data (e.g., due to link-layer packet losses), the MC suspends data reception until it enters the new AP coverage zone. The simulation run ends when the decoding is finished and all the original data packets are retrieved. All the presented results are obtained by performing 1000 simulation runs for each set of parameters.

Table 16.1 summarizes the values for the communication and mobility model parameters used in simulations. The number of APs is chosen such that it provides a coverage area that is approximately equal to a medium-sized city area. The data packet length is estimated in such way that is sufficient to accommodate single sensor readings and additional headers (i.e., IEEE 802.11 MAC and LLC, network, and

TABLE 16.1
Simulation Parameters

System Parameter	Value
AP transmission range	400 m
N_{AP} (no. of APs in the system)	40
N_s (no. of SNs per AP)	50
k (no. of data packets)	2000
L (data packet length)	250 byte
$k \cdot L$ (total amount of original data)	4 Mbit ≈ 0.48 Mbyte
c, δ (rateless code parameters)	0.03, 0.5
k_{AP} (no. of encoded packets per AP)	3600
R (bit rate)	6, 11, 12, 24 Mbit/s
T_{SF} (superframe duration)	100 ms
τ_{HO} (handover time)	0.5 s
P_{PL} (packet-loss probability)	0.3
Road-segment length	150 m
Velocity	4–17 m/s
Acceleration	±0.6 m/s^2
Mobility model time-slot duration	2 s

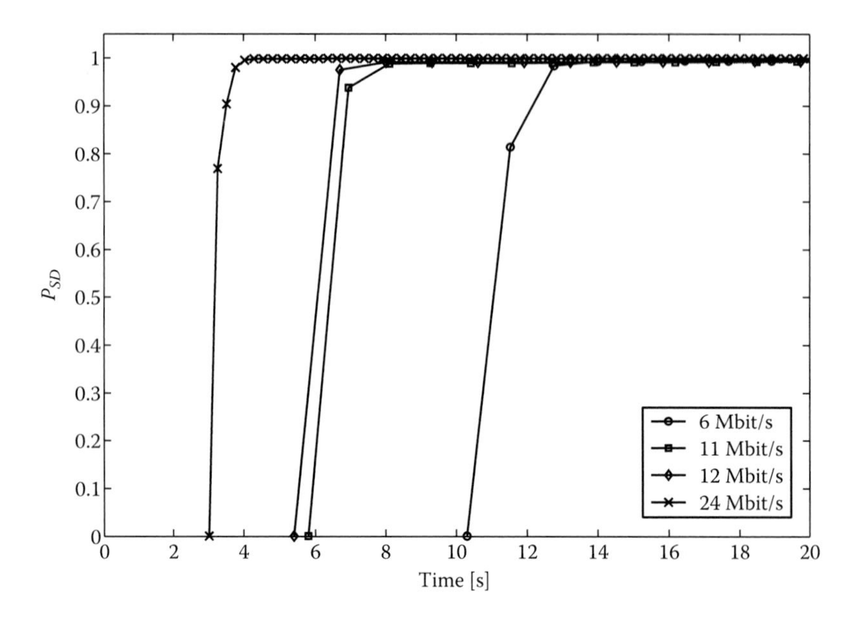

FIGURE 16.4 Probability of successful decoding P_{SD} for BC service, $T_{SF}^{(BC)} = 0.1 \cdot T_{SF}$.

transport layer). The values for bit rate and superframe duration are selected as suggested in Bohm and Jonsson (2008) and Eriksson et al. (2008), pessimistic assumption on packet-loss rate and estimate of the mean MC handover time were taken from Bychkovsky et al. (2006), and the average road-segment length (i.e., average distance between two intersections) was from Peponis et al. (2007). The number of encoded packets per AP k_{AP} is chosen such that an MC could decode all original data with probability of 0.99, when downloading from a single AP and considering employed rateless code properties and assumed link-layer packet-loss rate. In other words, $k_{AP} > (1+ \epsilon^{(\max)}) \cdot k \cdot L / (1-P_{PL})$.

Figure 16.4 presents the probability P_{SD} that the MC successfully decodes the sensor data as a function of time, for the BC service and $T_{SF}^{(BC)} = 0.1 \cdot T_{SF}$. The value for T_{SF} is selected such that it leaves enough room for the GC service and other usual best-effort services. As it can be observed from the figure, for higher bit rates (i.e., $R > 6$ Mbit/s), the MCN is able to successfully decode w.h.p. all the data in the time span of several seconds. The positive effect of rateless coding is inherent in the fact that, even in the worst case, the data refreshment period is below 15 s, a value that still allows for real-time information updates and that could be decreased further by assigning a larger superframe fraction to the BC service. As opposed to rateless encoded data delivery, the uncoded data delivery would result in retransmission feedback implosion for BC service, overwhelming the sender (i.e., AP) with unwanted traffic.

The probability of successful decoding for GC service is presented in Figure 16.5, where the fraction of the superframe assigned to a single user is assumed to be $T_{SF}^{(GC)} / N_{MN} = 0.01 \cdot T_{SF}$; the values for $T_{SF}^{(UC)}$ and N_{MN} are taken from the realistic analysis given in Bohm and Jonsson (2008). Figure 16.5 demonstrates that for the

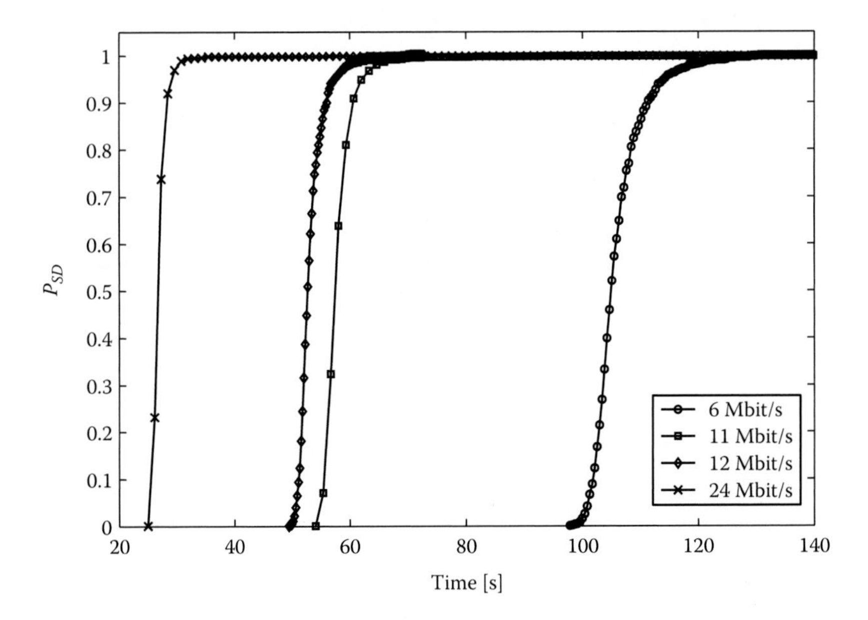

FIGURE 16.5 Probability of successful decoding P_{SD} for GC service, $T_{SF}^{(GC)}/N_{MN} = 0.01 \cdot T_{SF}$.

standard GC service, the data refreshment period is of the order of minutes rather than seconds, which limits its usage for the applications that tolerate larger update periods. However, this period would be significantly longer if rateless coding was not used, since the link layer retransmissions would make the data delivery process considerably less efficient. Finally, it can be observed that, for the GC service, the differences in transmission bit rate have a significant impact on the download delay, which makes higher bit rates desirable.

Figure 16.6 presents the duration of the time interval $T_{0.99}$ for which an MN, using the GC service, decodes all the original data with probability P_{SD}=0.99, as a function of the number of users N_{MN} and for the fixed $T_{SF}^{(GC)} = 0.8 \cdot T_{SF}$. The figure shows a linear increase in $T_{0.99}$ as the rate decreases or N_{MN} increases, verifying that the content reconciliation phase is indeed unnecessary, since the change of the AP does not introduce additional delays apart from the handover time. In other words, after a handover, MC seamlessly advances both with the receiving and decoding processes.

Finally, Figure 16.7 shows the cumulative distribution function F_T of the number of transmitted packets using the GC service from an AP to any MC within a single AP domain. The $T_{SF}^{(GC)}/N_{MN}$ ratio of the UC service is set to $0.005 \cdot T_{SF}$ or $0.01 \cdot T_{SF}$. As it can be observed, the number of transmitted packets to an MC reaches the threshold value k_{AP} equal to 3600 for the selected parameter values (Table 16.1), in all cases but for R=6 Mbit/s and $T_{SF}^{(GC)}/N_{MN} = 0.005 \cdot T_{SF}$. This means that number of encoded packets per AP (i.e., k_{AP}) is properly dimensioned to allow a single user to collect enough of encoded packets to decode all the data w.h.p. while moving through a single AP coverage zone.

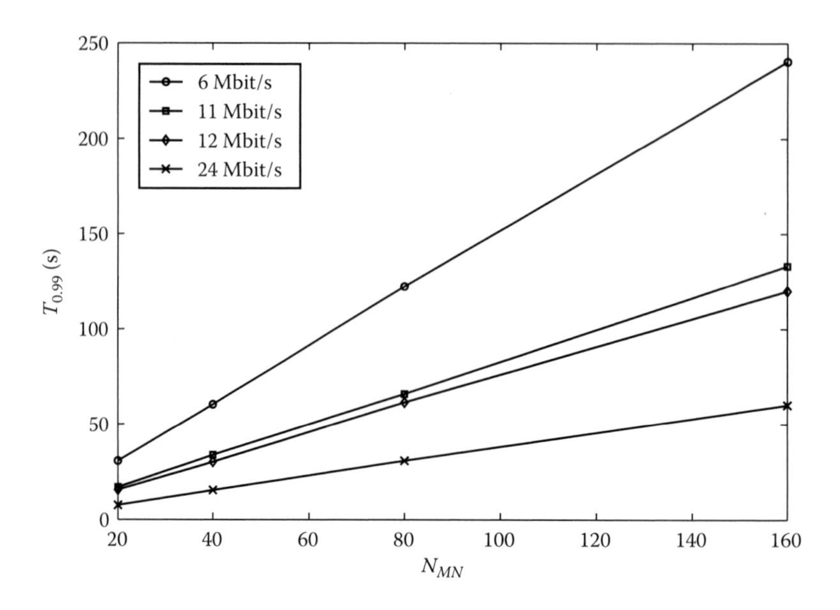

FIGURE 16.6 Duration of time interval $T_{0.99}$ for which MC decodes all data with $P_{SD}=0.99$ for GC service.

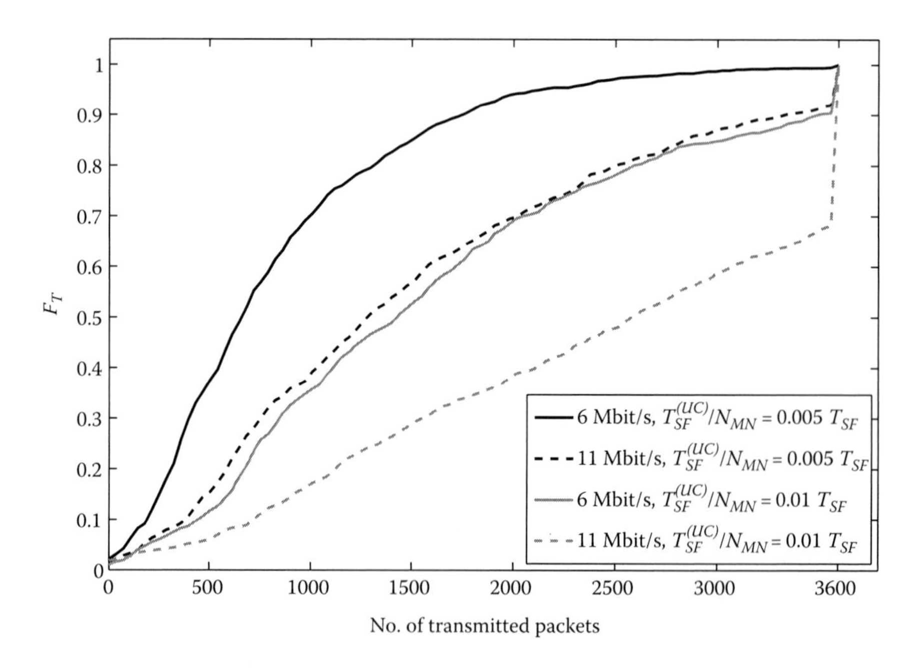

FIGURE 16.7 Cumulative distribution function F_T of a number of transmitted packets to MN in single AP cell for GC service.

To summarize the benefits provided by the proposed I2V data dissemination based on the rateless codes over traditional methods, it is worth noticing first of all that, by their design, rateless codes are tuned to the changing wireless link conditions and have a close-to-the-minimal reception overhead. Furthermore, each rateless coded packet is an equally important representation of the original data, which makes lengthy TCP-like reliability mechanisms unnecessary. These factors influence the time allocations within the superframe, allowing larger number of mobile nodes to be serviced during designated service-time portion of the superframe, or alternatively, service-time portion shortening, providing larger time allocations for best-effort traffic. Finally, while roaming through the network, mobile users can simply continue with data download from the new local AP after a handover, avoiding the redundant content reconciliation phase.

REFERENCES

Aoki, P. M., Honicky, R. J., Mainwaring, A., and Myers, C. (2009). A vehicle for research: Using street sweepers to explore the landscape of environmental community action, *ACM HFCS'09*, Boston, MA, pp. 375–384.

Bai, F., Sadagopan, N., and Helmy, A. (2003). IMPORTANT: A framework to systematically analyze the Impact of Mobility on Performance of RouTing protocols for Adhoc NeTworks, *Proceedings of IEEE INFOCOM 2003*, San Francisco, CA, pp. 825–835.

Baqer, M. (2010). Enabling collaboration and coordination of wireless sensor networks via social networks, *2010 Sixth IEEE International Conference on Distributed Computing in Sensor Systems Workshops (DCOSSW)*, Santa Barbara, CA, pp. 1–2.

Baqer, M. and Kamal, A. (2009). S-sensors: Integrating physical world input with social networks using wireless sensor networks, *IEEE ISSNIP'09*, Melbourne, Victoria, Australia, pp. 213–218.

Beach, A., Gartrell, M., Xing, X., Han, R., Mishra, Q. L. S., and Seada, K. (2010). Fusing mobile, sensor, and social data to fully enable context-aware computing, *ACM HotMobile'10*, Annapolis, MD, pp. 60–65.

Bohm, A. and Jonsson, M. (2008). Supporting real-time data traffic in safety-critical vehicle-to-infrastructure communication, *Proceedings of IEEE LCN 2008*, Montreal, Quebec, Canada, pp. 614–621.

Bollen, J., Mao, H., and Zeng, X. (2011). Twitter mood predicts the stock market, *Journal of Computational Science*, 1–8.

Bychkovsky, V., Hull, B., Miu, A., Balakrishnan, H., and Madden, S. (2006). A measurement study of vehicular Internet access using in situ Wi-Fi networks, *Proceedings of ACM MobiCom 2006*, Los Angeles, CA, pp. 50–61.

Byers, J., Considine, J., Mitzenmacher, M., and Rost, S. (2002a). Informed content delivery across adaptive overlay networks, *Proceedings of ACM SIGCOMM 2002*, Pittsburg, PA, pp. 767–780.

Byers, J., Luby, M., and Mitzenmacher, M. (2002b). A digital fountain approach to asynchronous reliable multicast, *IEEE Journal on Selected Areas in Communications* **20**(8): 1528–1540.

Chen, L., Peng, Y., and Tseng, Y. (2011). An infrastructure-less framework for preventing rear-end collision by vehicular sensor networks, *IEEE Communications Letters* **15**(3): 358–360.

Cordova-Lopez, L. E., Mason, A., Cullen, J. D., Shaw, A., and Al-Shamma'a, A. (2007). Online vehicle and atmospheric pollution monitoring using GIS and wireless sensor networks, *Proceedings of ACM International Conference on Embedded Networked Sensor Systems (SenSys)*, Sydney, New South Wales, Australia, pp. 87–101.

Eriksson, J., Balakrishnan, H., and Madden, S. (2008). Cabernet: Vehicular content delivery using WiFi, *Proceedings of ACM MobiCom 2008*, San Francisco, CA, pp. 199–210.

Gao, H., Utecht, S., Patrick, G., and Hsieh, G. (2010). High speed data routing in vehicular sensor network, *Journal of Communications* **5**(3): 181–188.

Gerla, M., Zhou, B., Lee, Y. Z., Soldo, F., Lee, U., and Marfia, G. (2006). Vehicular grid communications: The role of the Internet infrastructure, *Proceedings of WICON'06*, Boston, MA, pp. 199–210.

Guinard, D., Fischer, M., and Trifa, V. (2010a). Sharing using social networks in a composable web of things, *IEEE International Workshop on the Web of Things*, Mannheim, Germany, pp. 702–707.

Guinard, D., Trifa, V., and Wilde, E. (2010b). A resource oriented architecture for the web of things, *IEEE International Workshop on the Web of Things*, Mannheim, Germany, pp. 1–8.

IEEE (2007). Wireless LAN medium access control (MAC) and physical layer (PHY) specification, IEEE std. 802.11-1999 edition.

Internet of Things in 2020: Roadmap for the Future (2008). URL: http://www.smart-systems-integration.org/public/internet-of-things, accessed February 5, 2014.

Java, A., Song, X., Finin, T., and Tseng, B. (2007). Why we twitter: Understanding microblogging usage and communities, *ACM WebKDD/SNA-KDD'07*, San Jose, CA, pp. 56–65.

Jiang, D. and Delgrossi, L. (2008). IEEE 802.11p: Towards an international standard for wireless access in vehicular environments, *Proceedings of IEEE VTC2008-Spring*, Singapore, pp. 2036–2040.

Kansal, A., Nath, S., Liu, J., and Zhao, F. (2007). Senseweb: An infrastructure for shared sensing, *Multimedia* **14**(4): 8–13.

Kushalnagar, N., Montenegro, G., Hui, J., Culler, D. (2007). 6LoWPAN: Transmission of IPv6 Packets over IEEE 802.15.4 Networks. RFC 4944, September 2007.

Laisheng, X., Xiaohong, P., Zhengxia, W., Bing, X., and Pengzhi, H. (2009). Research on traffic monitoring network and its traffic flow forecast and congestion control model based on wireless sensor networks, *International Conference on Measuring Technology and Mechatronics Automation, 2009 (ICMTMA'09)*, Zhangjiajie, Hunan, China, Vol. 1, pp. 142–147.

Libelium: Interfacing the Sensor Networks with the Web 2.0 (2010). URL: http://www.libelium.com, accessed February 5, 2014.

Lovett, T., O'Neill, E., Irwin, J., and Pollington, D. (2010). The calendar as a sensor: Analysis and improvement using data fusion with social networks and location, *ACM Ubicomp*, Copenhagen, Denmark, Vol. 10, pp. 3–12.

Luby, M. (2002). LT codes, *Proceedings of IEEE FOCS 2002*, Vancouver, British Columbia, Canada, pp. 271–280.

Marcus, A., Bernstein, M. S., Badar, O., Karger, D. R., Madden, S., and Miller, R. C. (2011). Twitinfo: Aggregating and visualizing microblogs for event exploration, *ACM SIGCHI'11*, Vancouver, British Columbia, Canada, pp. 227–236.

Martinez, K., Hart, J., and Ong, R. (2004). Environmental sensor networks, *IEEE Computer Journal* **37**: 50–56.

Miluzzo, E., Lane, N. D., Fodor, K., Peterson, R., Lu, H., Musolesi, M., Eisenman, S. B., Zheng, X., and Campbell, A. T. (2008). Sensing meets mobile social networks: The design, implementation and evaluation of the CenceMe application, *ACM SenSys'08*, New York, NY, pp. 337–350.

Murty, R., Mainland, G., Rose, I., Chowdhury, A., Gosain, A., Bers, J., and Welsh, M. (2008). Citysense: An urban-scale wireless sensor network and testbed, *IEEE THS'08*, Waltham, MA, pp. 583–588.

Nagarajan, M., Sheth, A., and Velmurugan, S. (2011). Citizen sensor data mining, social media analytics and development centric web applications, *ACM WWW'11*, Hyderabad, India, pp. 289–290.

Ott, J. and Kutscher, D. (2004). Drive-thru Internet: IEEE 802.11b for automobile users, *Proceedings of IEEE Infocom 2004*, Hong Kong, China, pp. 362–373.

Pak, A. and Paroubek, P. (2010). Twitter as a corpus for sentiment analysis and opinion mining, *LREC 2010*, Valletta, Malta, pp. 1–7.

Palattella, M. R., Accettura, N., Vilajosana, X., Watteyne, T., Grieco, L. A., Boggia, G., and Dohler, M. (2012). Standardized protocol stack for the internet of (important) things, *Communications Surveys Tutorials, IEEE* **15**(99): 1–18.

Peponis, J., Allen, D., Haynie, D., Scoppa, M., and Zhang, Z. (2007). Measuring the configuration of street networks: The spatial profiles of 118 urban areas in the 12 most populated metropolitan regions in the US, *Proceedings of Sixth International Space Syntax Symposium*, Istanbul, Turkey, pp. 1–15.

Pinart, C., Calvo, J. C., Nicholson, L., and Villaverde, J. A. (2009). ECall-compliant early crash notification service for portable and nomadic devices, *Proceedings of IEEE VTC2009-Spring*, Barcelona, Spain, pp. 1–5.

Resch, B., Mittlboeck, M., Girardin, F., Britter, R., and Ratti, C. (2009). Real-time geo-awareness—Sensor data integration for environmental monitoring in the city. *Proceedings of the IARIA International Conference on Advanced Geographic Information Systems & Web Services—GEOWS2009*, February 1–7, 2009, Cancun, Mexico, pp. 92–97.

Rosi, A., Mamei, M., Zambonelli, F., Dobson, S., Stevenson, G., and Ye, J. (2011). Social sensors and pervasive services: Approaches and perspectives, *PERCOM Workshops*, Seattle, WA, pp. 525–530.

Sakaki, T., Okazaki, M., and Matsuo, Y. (2010). Earthquake shakes twitter users: Real-time event detection by social sensors, *ACM WWW'10*, Raleigh, NC, pp. 851–860.

Santini, S., Ostermaier, B., and Vitaletti, A. (2008). First experiences using wireless sensor network for noise pollution monitoring, *Proceedings of Third ACM Workshop on Real-World Wireless Sensor Networks (REALWSN'08)*, Glasgow, U.K., pp. 61–65.

Shu-Chiung, H., You-Chiun, W., Chiuan-Yu, H., and Yu-Chee, T. (2009). A vehicular wireless sensor network for CO_2 monitoring, *Proceedings of IEEE Sensors*, Christchurch, New Zealand, pp. 1498–1501.

Stefanovic, C., Vukobratovic, D., Chiti, F., Niccolai, L., Crnojevic, V., and Fantacci, R. (2011). Urban infrastructure-to-vehicle traffic data dissemination using rateless codes, *IEEE Journal on Selected Areas on Communications* **29**(1): 94,102

Tanner, J. C. (1957). The sampling of road traffic, *Journal of the Royal Statistical Society: Series C (Applied Statistics)* **6**(3): 161–170.

Thelwall, M., Buckley, K., and Paltoglou, G. (2011). Sentiment in twitter events, *Journal of the American Society for Information Science and Technology* **62**: 406–418.

Tse, R. T. S., Dawei, L., Hou, F., and Pau, G. (2011). Bridging vehicle sensor networks with social networks: Applications and challenges, *IET International Conference on Communication Technology and Application (ICCTA 2011)*, Beijing, China, pp. 684–688.

Tsugawa, S. and Kato, S. (2010). Energy its: Another application of vehicular communications, *IEEE Communications Magazine* **48**(11): 120–126.

Tubaishat, M., Zhuang, P., Qi, Q., and Shang, Y. (2009). Wireless sensor networks in intelligent transportation systems, *Wireless Communications and Mobile Computing 2009*, Leipzig, Germany, Vol. 9(3), pp. 287–302.

Vukobratovic, D., Stefanović, C., Crnojević, V., Chiti, F., and Fantacci, R. (2010). Rateless packet approach for data gathering in wireless sensor networks, *IEEE Journal on Selected Areas in Communications* **28**(7): 1169,1179.

Weng, J., Lee, F. (2011). Event detection in twitter, Proceedings of ICWSM 2011, Barcelona, Spain.

Wong, K., Chua, C., and Li, Q. (2009). Environmental monitoring using wireless vehicular sensor networks, *IEEE WCNMC'09*, Beijing, China, pp. 1–4.

Wu, C. and Li, B. (2007). Outburst: Efficient overlay content distribution with rateless codes, *Lecture Notes in Computer Science,* Springer Berlin Heidelberg, 2007, pp. 1208–1216.
Yerva, S., Jeung, H., and Aberer, K. (2012). Cloud based social and sensor data fusion, *2012 15th International Conference on Information Fusion (FUSION),* Singapore, pp. 2494–2501.
Yousefi, S., Mousavi, M. S., and Fathy, M. (2006). Vehicular Ad Hoc Networks (VANETs): Challenges and perspectives, *Proceedings of ITST 2006*, Chengdu, China, pp. 761–766.

17 Applying RFID Techniques for the Next-Generation Automotive Services

Peter Harliman, Joon Goo Lee,
Kyong Jin Jo, and Seon Wook Kim

CONTENTS

ABSTRACT

Recently, RFID has gained a lot of attention as a technology that facilitates ubiquitous computing. Nevertheless, it still requires improvement in some areas, particularly in its application development. In this chapter, we present ideas for new RFID applications in automotive fields. With these innovative ideas, we can generate new kinds of applications and services, instead of focusing on conventional item tracking applications. We then propose an architecture for integrating the RFID system into current automotive architectures. Finally, in order to show the feasibility of our ideas, we present our prototyped system and demonstrate how our ideas can be successfully implemented.

17.1 INTRODUCTION

Radio-frequency identification (RFID) technology has been used and developed for more than 50 years, first by Harry Stockman in 1948 [1], and its applications have been commonly used around us. Instead of the term RFID, we are popularly using the term wireless applications in their names. Due to this reason, RFID remains unpopular despite its long history. Today, most cars are equipped with a remote control to open and lock a door. In Korea, T-money [2] cards are used for public transportation payments. Although there is no RFID term in their names, both a car remote control and T-money are in the domain of RFID applications. RFID technology has become more and more widely used in real-world applications even without people realizing it.

RFID tags are often envisioned as a replacement for a barcode technology in the future. The reason for this is that RFID has many advantages over barcodes, such as a sight angle, a distance, a collision, and a tag size [3]. More detailed comparisons between RFID and barcodes are shown in Table 17.1.

Despite its numerous technological advantages over barcodes, the development of RFID technology did not occur as fast as it was expected, especially before the twenty-first century because of a high cost, lack of standards, privacy issues, lack of applications, etc. [4]:

1. *Cost*: RFID readers and tags use more advanced technology than barcode scanners and labels do, and it incurs a cost problem. Table 17.2 compares the prices for both technologies.
2. *Lack of standards*: Around 1020 years ago, when RFID applications began to be used in real-world applications, most developers used their own

TABLE 17.1
RFID and Barcode Comparison

System Parameters	Barcode	RFID
Data quantity (bytes)	1–100	16–64 k
Data density	Low	Very high
Machine readability	Good	Good
People readability	Limited	Impossible
Influence of dirt/dampness	Very high	No influence
Influence of covering	Total failure	Moderate
Data carrier cost	Very low	Medium
Reading electronic cost	Low	Medium
Unauthorized copying/modification	Slight	Almost impossible
Multiple reading	No	Yes
Reading speed	Low	Very fast
Maximum reading distance	50 cm	6 m

Source: CAEN RFID, About RFID, http://www.caen.it/rfid/about_rfid.php.

TABLE 17.2
Price Comparisons between RFID and Barcodes

	Barcode	RFID
Reader	US$100–300	US$700–2000
Passive tag	0–3 cent	50–100 cent
Active tag	—	US$10–100

versions of RFID systems. It resulted in lack of standards, which prevented the RFID markets from growing fast.

3. *Privacy issues*: Privacy is one major issue in RFID deployment. In RFID communication, unlike that of barcode, there is a wide gap in space between a reader and a tag. Some people think that this wide gap can be very vulnerable to data intruders. In February 2005, a team of researchers at Johns Hopkins University Information Security Institute and RSA Laboratories demonstrated that they could capture data from Texas Instruments RFID systems [5]. This issue forced RFID developers to reconsider their design, so it could guarantee enough safety to consumers.
4. *Lack of applications*: Currently, there are only a limited number of RFID applications that are commercially available. This makes it very difficult for RFID to attract various types of consumers.

Some of those problems have been resolved during the twenty-first century. It has been predicted that the cost of RFID readers and tags will drop quite significantly in the coming years, due to development of semiconductor technology and mass productions, as shown in Figure 17.1. In 2006, one of the major protocols, EPC Gen2,

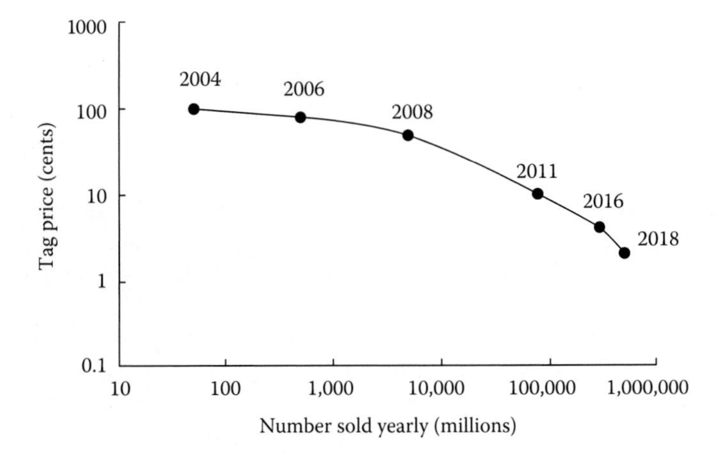

FIGURE 17.1 Tag's price prediction. (From Peter Harrop, The price-sensitivity curve for RFID, http://www.idtechex.com/products/en/articles/00000488.asp, 2006.)

has been already implemented by many commercial RFID reader and tag vendors, and it was adapted as ISO 18000-6 Type C standard [6]. Furthermore, current protocols, especially Gen2, are much safer than the previous ones, which had issues with illegal data capturing. For example, the protocol includes an option to kill a tag. After a tag is killed by a user, the tag will not respond to a reader anymore. This approach alleviates the concern about privacy.

We think that the slow growth problem of RFID comes from the initial concept to introduce the RFID technology as a barcode substitute. Due to this initial purpose, RFID developments always have been driven toward existing barcode applications. Barcode applications generally can be used only in a static environment, for example, by a cashier. In contrast, RFID technology can be used in a dynamic environment, because of its reading distance and reading speed. However, since RFID has so far only been used by the same static environment, we have failed to get the full advantage from RFID technology. Meanwhile, replacing barcode technology requires global effort to change many underlying systems. For this reason, most vendors are not willing to risk replacing a barcode system with RFID. One similar case with this RFID–barcode situation happened in PC–calculator transition around 30–40 years ago, as shown in Figure 17.2. Compared with calculators, PCs are much more powerful for calculations. But in order to be successful in a market, PC makers found many other applications like database and games, in addition to calculations. If PCs were only used for calculation, there would be no chance for them to grow through the market. We should not focus only on barcode's existing applications. Instead, we need to create new applications that obtain maximum benefits from RFID abilities.

In this chapter, we propose innovative ideas for these new applications, with more emphasis on automotive applications. We briefly review several RFID services in Section 17.2 and the currently existing automotive services in Section 17.3. We introduce innovative ideas for the next-generation RFID automotive services in Section 17.4 and discuss an architecture for the new services in Section 17.5. In Section 17.6, we introduce our prototyped system and finally conclude in Section 17.7.

17.2 BACKGROUND OF RFID SERVICES

Current RFID applications can be classified into several categories as follows:

1. *Transportation*. Most countries in the world have used RFID technology for a payment method in public transportations. Figure 17.3a shows T-money, an RFID smart card that is used for paying transportation fares in Seoul, Korea.
2. *Manufacturing control*. Some companies have applied RFID technology to track inventories inside their factory. In the application, an RFID system is used internally only by each company. This means that each company can use its own RFID standards without having concern about global standards. Hence, this application is easier to implement than other kinds of RFID applications. For this reason, a manufacturing control is one of the first successful RFID implementations in real-world applications.

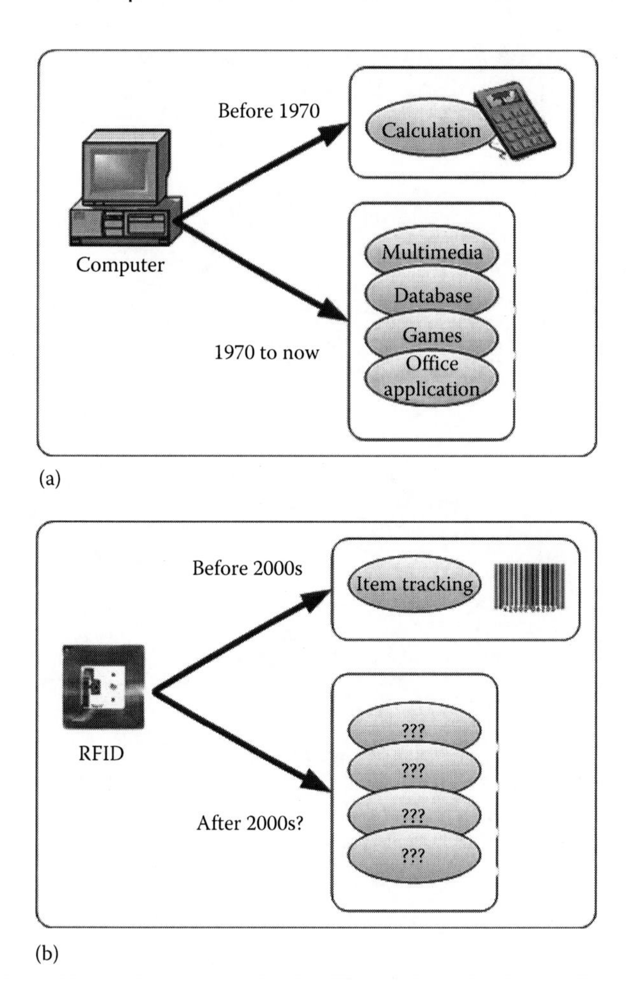

FIGURE 17.2 Analogy between PC–calculator and RFID–barcode cases. (a) PC calculator and (b) RFID barcode.

3. *People management.* Some companies and educational institutes have added RFID technology to their systems for managing their members. For example, RFID technology is commonly used to manage employee attendance in a company. In 1998, Malaysia became the first country to issue biometric passports, which are passports with RFID tags embedded inside [7]. Since that time, more than 20 countries followed to apply RFID technology in their passports. One disadvantage in this category of application was that IDs can be manipulated by some people. For example, in the case of employee attendance, an employee can easily lend his or her ID card to other employees. For this reason, RFID applications in this category cannot rely solely on RFID. Instead, they combine more than one identification technology inside them. For example, a biometric passport still has its usual

(a)

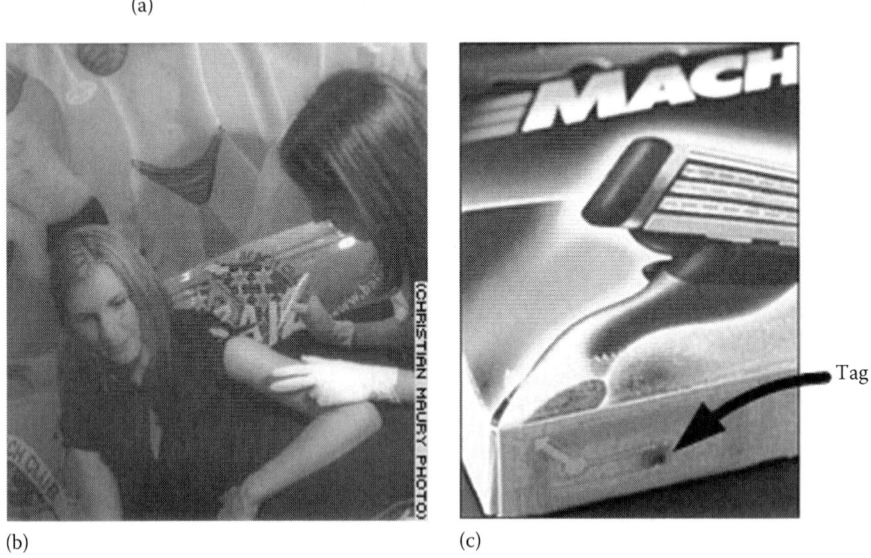

(b) (c)

FIGURE 17.3 Various RFID applications. (a) T-money RFID card for transports payment. (From Korea Smart Card Co. Ltd., Seoul's New Transportation System, http://www.t-money.co.kr/jsp/newpub/oversea/english/stories/S_story.jsp, 2006.) (b) RFID tag planted under the skin at Baja Beach Club. (From Robyn Curnow, The price to pay for VIP status, http://edition.cnn.com/2004/TECH/10/05/spark.bajabeach/, 2004.) (c) Gillette Mach3 embedded with RFID tag in its case. (From Claudia H. Deutsch and Barnaby J. Feder, A radio chip in every consumer product, 2003.)

physical appearance and a barcode label, in order to preserve the validity of its ID. Another way to overcome this problem is to treat humans like any other item. Rather than embed the tag in a card, it can be inserted beneath the skin. Thus, the validity of an ID is guaranteed because the tag cannot be switched. Baja Beach Club discotheque in Barcelona, Spain, has already planted RFID tags inside its VIP customers [8], as shown in Figure 17.3b. In this application, the RFID tag under a customer's skin is used for an automatic payment method inside the discotheque.

4. *Automatic payment method in stores.* At present, a barcode technology is used in many stores for identifying purchased items. By using RFID, people will not have to pay for them at a counter anymore. Instead, they just need to pass an RFID reader and the payment will be automatically charged online. Gillette moved one step ahead by adding RFID tags inside their Mach3 Turbo razor blade packaging [9,10], as shown in Figure 17.3c.

17.3 CURRENT AUTOMOTIVE SERVICES

In Section 17.1, we explained that in order for RFID to grow faster, new applications are required. The automotive industry is one of the best development targets for an RFID application.

The automotive world has changed greatly from the way it used to be 1020 years ago. Surprisingly, the most significant development is not happening in mechanical parts but in electronic parts [11]. From Figure 17.4, it can be seen that the proportion of electronics cost inside a car has been increasing constantly and will be increasing even more in the next 10 years. It has been estimated that it will rise to US$36.8 billion in 2005 and will reach US$52.1 billion by 2010 [12]. Some of these electronic parts are used for non-user-oriented applications (e.g., power train and chassis), but most of them are used for user-oriented applications. For example, cars are often equipped with global positioning system (GPS) services, sophisticated in-car multimedia systems and seating, various kinds of sensors, etc. In other words, there have been huge developments in automotive electronics toward user-oriented services.

There are various user-oriented services that can be found inside a modern car, as shown in Figure 17.5, and some are detailed as follows:

1. GPS was developed by the U.S. Department of Defense and came into use in 1993 [13]. Nowadays, GPS has been used around the world for many kinds of different applications. The most popular usage is in automotive navigation. Normally, such a GPS device has user-friendly controls

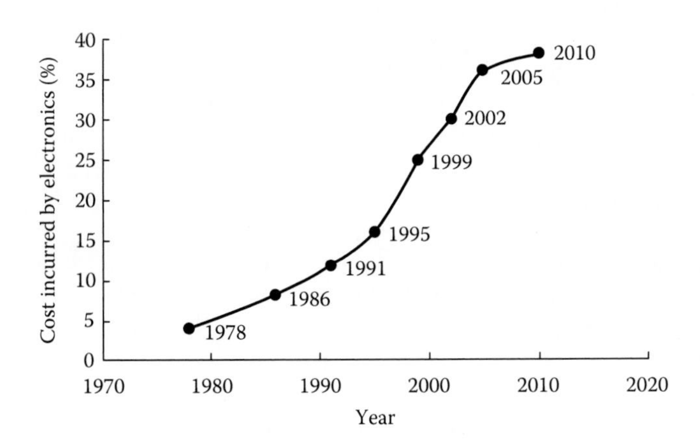

FIGURE 17.4 Proportion of electronics cost inside a car.

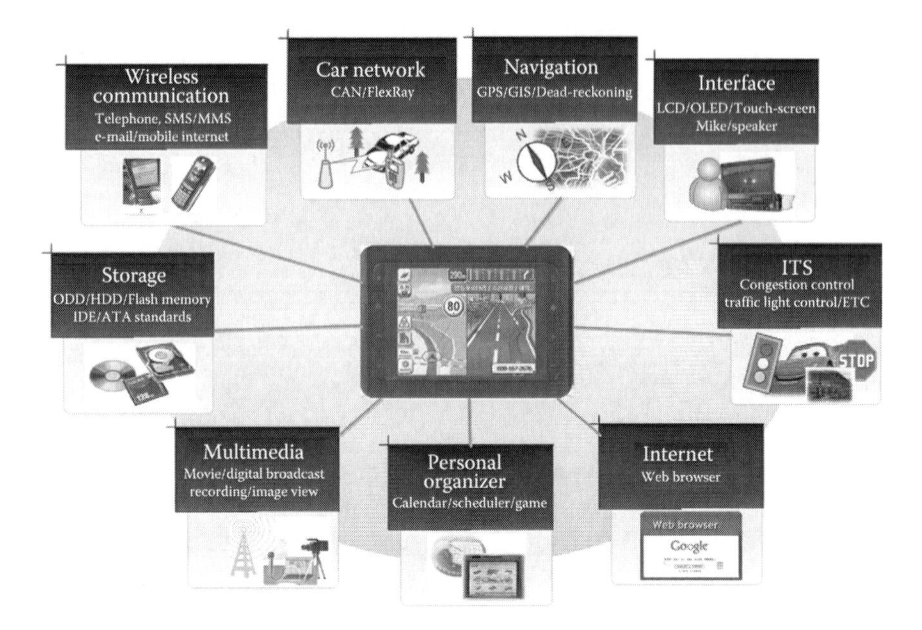

FIGURE 17.5 User-oriented applications in a modern telematics device.

and outputs and displays the car's current position on a map. Relying on this GPS navigation, other extended applications have also been developed, such as online road guide systems, emergency rescue systems, and car-tracking services.

2. Dedicated short-range communication (DSRC) is considered a subset of the RFID technology, since it also uses an RF to communicate [14]. Different from GPS and RFID, DSRC was specifically designed for automotive communication. DSRC technology is known to be fast, stable, and cheap. The DSRC system basically consists of two components, roadside equipment (RSE) and onboard equipment (OBE). The RSE communicates and provides useful information to the OBE. European countries, Japan, Korea, and the United States use the DSRC system for electronic toll collection, which automatically charges a toll payment as a car passes the RSE near a toll gate. This way, the toll gate queuing delay could be avoided. Other applications include an intersection collision avoidance, an emergency vehicle warning system, electronic parking payments, and an in-vehicle signing. Currently, DSRC applications are still very limited compared with GPS applications. However, due to its good performance and its economic efficiency, DSRC is predicted to occupy a great role in modern and future telematic services.
3. In-car Internet access. People are spending more time in their cars than ever before. For this reason, future cars must provide an environment such that a driver can process his or her own job inside a car. In other words, engineers need to expand their concept of a car from a transportation method to a moving office. Some modern cars are already installed with a multimedia

TABLE 17.3
Key RFID In-Car Appliances

RFID Combined with	Provided Feature
Ignition control	RKI system
Car lock system	RKE system
Seat and mirror setting	Automatic customizable seat and mirror setting
Speedometer, odometer, and ignition	Individual authority
Multimedia	Automatic profiled multimedia setting
Internet	Mobile advertising
Road information and multimedia	ARIS

tool that can connect to the Internet. This access will lend itself to infotainment (information + entertainment) services. One example of the related services is a multimedia player that can play mp3 files, DVD, and digital multimedia broadcasting (DMB). In order to connect to the Internet in a moving vehicle, a car uses WI-FI access points, cellular base stations, satellites, WiBro [15], or HSDPA [16]. Microsoft is developing a new foundry system, which is called Car.Net [17], which enables drivers to use the Internet in their vehicles. By using this system, drivers and passengers can check their e-mails, enjoy various kinds of entertainment, and search for stock market conditions simultaneously.

RFID technology could generate many new user-oriented automotive applications. Some possible automotive RFID applications are shown in Table 17.3, and the following section will discuss some of these ideas in detail.

17.4 INNOVATIVE IDEAS FOR AUTOMOTIVE RFID APPLICATIONS

17.4.1 Individual Authority

A car key is one of the most essential components of a car. It is used as a main identification tool that allows a car user to turn on an engine (ignition), open/close a door, open a glove compartment, open a car trunk, open a fuel tank, etc. One limitation in a car key system comes from the fact that a car keyhole is unique. Each car keyhole can only recognize one key, and each key can only be used in one keyhole. If a car is used by more than one driver, the key may be duplicated to give a copy of the key to each user. However, the keyhole is not able to recognize who a current driver is. Consider RFID readers and tags instead of conventional keys and keyholes. An RFID reader can read all tags in its reading zone and check the validity of the tag's ID. Thus, if we use an RFID tag as an identification tool for a car, we basically move the ID-checking process from a mechanical (like in a keyhole) to an electronic method.

In modern cars, keys have been replaced by remote keyless systems in many functions. More than 70% of currently manufactured cars come with a remote keyless entry (RKE) system, which allows doors to be locked and unlocked without using a key [18]. Meanwhile, some luxurious cars, such as Toyota Prius, Cadillac STS, and Audi A8, are also equipped with a remote keyless ignition (RKI) system, which allows car engines to be turned on without using a key. Instead of using a key, the RKE system uses an electronic key fob. In order to open a door, a user just needs to press a button on the fob within 10–20 m of the car. That typical range is possible since the fob communicates with the transmitter inside the car by using an RF signal. In an RKI system, in case of an ignition, a user just needs to press one button inside the car. However, due to security, most current RKI systems require the presence of the fob inside the car. This technique will prevent an engine from being started by unauthorized users. The car uses RF communication to check whether the fob is in the car or not. The RF transceiver inside the car will broadcast a signal, and if the fob is present, it will reply to the transceiver. It can actually be seen that RKE and RKI systems can be considered as RFID applications, since they use RF communication to check an ID.

Apart from the fact that RKE and RKI systems have more benefits than conventional car keys, there are still some issues that need to be discussed in these systems. For example, current RKE and RKI systems can only work for a single ID. In the case of a car that is shared by multiple users, all possible users need to have the same ID inside their fobs. Hence, the car will not be able to distinguish who the current driver is, since they use the same ID. In addition, the scope of current RKE and RKI systems is very limited. They only focus on services for a door entrance and an ignition. In this application, we can say that RFID technology has not been used effectively. For this reason, in this part, we propose one idea to expand the RKE and RKI systems to a broader scope of applications.

Consider if an RFID reader is put inside a car, and some electronic fobs with RFID tags are developed. Each of the fobs will have its own unique ID. Each possible car user will be given one of these fobs. When a user takes the car, the RFID reader will communicate with the user's fob. As already mentioned, in current RKE and RKI systems, there is only one ID that is recognized as an authorized user. Even if the fobs are duplicated, the IDs for each fob are still the same, as shown in Table 17.4. In our proposed idea, each fob will have a different ID, as shown in Table 17.5. Using these different sets of IDs, the car can recognize an individual driver.

The next step is to use a user recognition ability of a car for creating new applications. Since now the car is able to recognize a current driver, we could connect this

TABLE 17.4
Database Example of Single ID without a User Recognition Ability

Name	ID
User 1	10180604
User 2	10180604
User 3	10180604

TABLE 17.5
Database Example of Multiple IDs with a User Recognition Ability

Name	ID
User 1	10180604
User 2	10216154
User 3	12301240

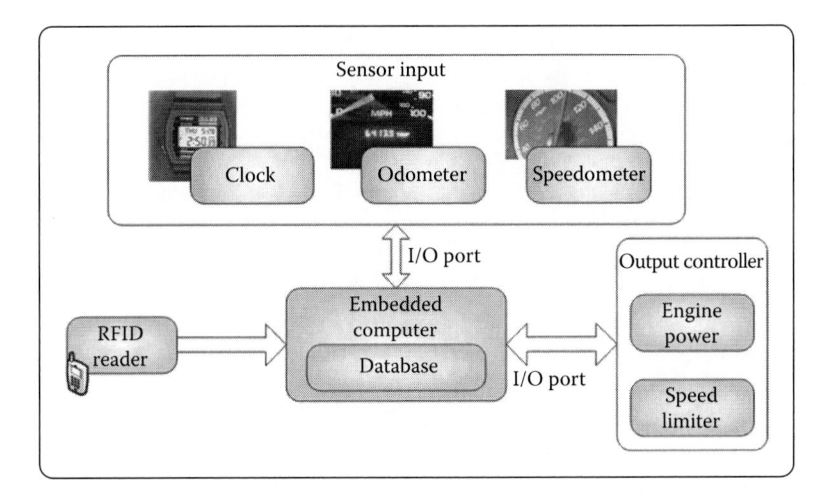

FIGURE 17.6 Individual authority architecture.

user recognition system with some hardware controllers inside the car. A simplified architecture for this idea is shown in Figure 17.6. The RFID reader is connected to a gateway. In this manner, the gateway is defined as an embedded computer that acts as a center of communication between the RFID reader and hardware controllers. This gateway is also the part that implements the user recognition system. It will have some amount of memory to hold a database for all users.

One possible application can be created by connecting a timer and an ignition controller to the gateway. This way, each user will have his own time limit when he drives a car. If the time limit expires, the driver will not be able to turn on the engine anymore. In this idea, the user recognition system will be expanded to keep settings for each user, as shown in Table 17.6. Some possible fields to be controlled are the following:

1. *Distance*. Each user has his own distance limitation. When a user drives a car more than the distance limit, the car will not be able to run anymore. In order to implement this application, the gateway is connected to a distance counter (e.g., an odometer).

TABLE 17.6
Database Example of Multiple IDs with a Customizable Authority

Name	ID	Speed Limit	Duration Limit	Distance Limit
User 1	10180604	15	60	120
User 2	10216154	—	—	50
User 3	12301240	50	30	—

2. *Time.* Each driver has his own time limitation, beyond which the car will not be able to run any farther. In order to apply this application, the gateway is connected to a timer.
3. *Speed.* Each user has his own speed limitation. A user cannot drive the car faster than his designated speed limit.

These approaches will have different kinds of benefits. For private cars, a user can limit authorities of other users. Parents who share a car with their children can give a low authority access to them. This will restrict the children to limited boundaries. For company cars, these approaches are also useful since their cars are occasionally used by different drivers. In car rental industries, these approaches can be used to limit use of rented cars.

17.4.2 Customizing a Vehicle for Multiple Drivers

We now expand the database from Table 17.6 to control more items. Here, the idea is not focused on a user's authority as in the previous idea. Instead, it focuses on any item inside the car that can be set for a user's preferences. The architecture for this idea is shown in Figure 17.7.

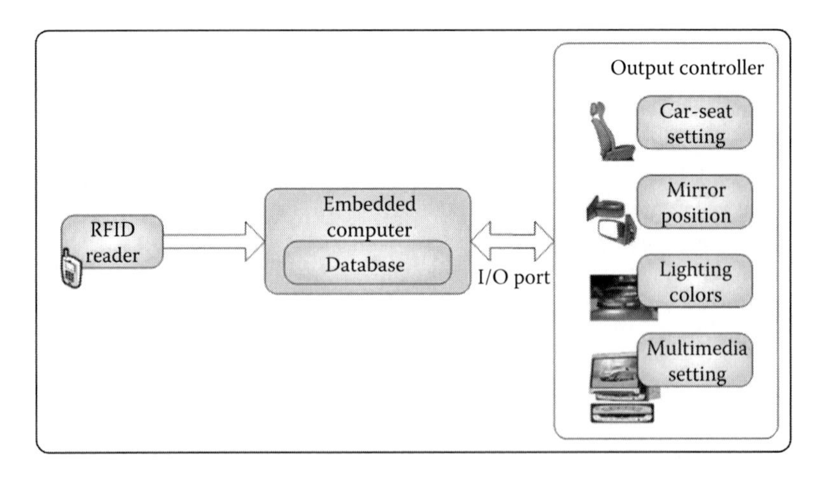

FIGURE 17.7 Customizable setting architecture.

One example of items to be controlled is car seats, which need to be set according to user's preferences. In conventional cars, the users need to set the seats manually. For a car that is shared by multiple users, each user will need to set the seats whenever there is a new driver or passenger. Some luxurious cars have already used a memory management for the car setting. For example, in the Saab 9–5 [19], seat controls have memory settings for three different drivers, including side-view mirror settings. However, they still need to choose among those three settings by pressing some buttons. In our idea, the users do not need to do any manual setting. Similar to the previous idea, the RFID reader will automatically recognize the user from his tag. The gateway then automatically sets the seat according to this user's preference. Other items that might be controlled by using this technique are the following:

1. *Mirror.* Car mirror settings are very critical for safety. Since they are dependent on driver's height, different drivers will usually require different settings. By using our RFID concept, the settings change automatically when a new driver takes a car.
2. *Multimedia player setting.* Drivers will likely want different radio channels and CD player settings. Furthermore, if the car is equipped with in-car Internet access, they also need different settings for Internet (web page, e-mail, etc.). With RFID, all settings will be changed automatically by the gateway.
3. *Future applications.* One of the biggest advantages of using RFID is that it triggers development of new applications. For instance, perfume vendors could develop some kinds of car perfumes that have different aromas, depending on the user's preference. If there is a new user, the perfume will automatically switch its aroma based on new user's preference. It is also possible to develop complex in-car lighting. Teenagers can choose varieties of colors for lighting, while general users can choose conventional lighting.

17.4.3 Automatic Vehicle Management

The condition of car tires needs to be monitored and replaced periodically. Goodyear has embedded RFID tags in their tires for NASCAR race use. Goodyear is the exclusive provider of race car tires for all NASCAR events, supplying about 200,000 tires to racers annually. Normally, participants have to buy their own tires. However, it is not uncommon for a team to use multiple sets of tires in a single race. Hence, the cost for tires could not be affordable for many potential racers. NASCAR came to Goodyear seeking a leasing alternative for drivers who are unable to afford the cost of buying tires for their vehicles. Goodyear then came with an idea to use RFID tags for managing the leasing of tires. Michelin went further than Goodyear by embedding tags in car tires for public use and more general purposes [20]. We basically extend the RFID tire service from Goodyear and Michelin by attaching tags to other parts of a car. Considering the fact that passive RFID tag costs are low, they can be attached to hundreds of car parts, as illustrated in Figure 17.8. Each tag should hold the part's information. An RFID reader can read information from a

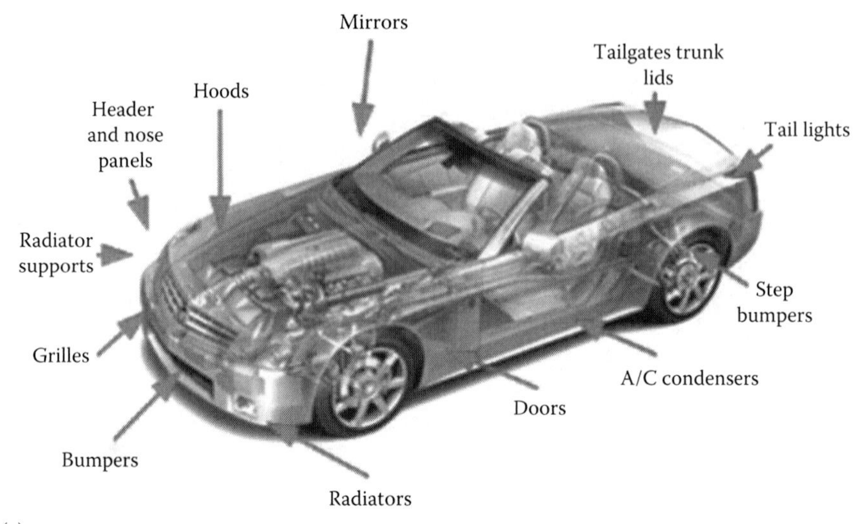

(a)

Devise name	Part number	Data type	Added by	Time added
Front left door	XXXXX	Production time	Manufacturer X	XXXXX
Front left door	XXXXX	Repair time	Car service Y	XXXXX
Front left door	XXXXX	Repair time	Car service Y	XXXXX
Front left door	XXXXX	Insurance checking	Company Z	XXXXX

Devise name	Part number	Data type	Added by	Time added
Front left door	XXXXX	Production time	Manufacturer X	XXXXX
Front left door	XXXXX	Repair time	Car service Y	XXXXX
Front left door	XXXXX	Repair time	Car service Y	XXXXX
Front left door	XXXXX	Insurance checking	Company Z	XXXXX

(b)

(c)

FIGURE 17.8 Tags are attached to various car parts for an automatic vehicle management. (a) RFID tags attached to various car parts. (b) Information such as historical time data is added to each tag's memory. (c) Allowed person can write/read the info from the tags for their purposes.

tag in milliseconds. Hence, the reader will have no difficulty in periodically reading information from all tags.

Conceptually, we exploit the fact that an RFID tag has some amount of user memory that can be written by an RFID reader. Although the memory size is not big, typically less than 1024 bit [21], it is enough to store a limited amount of information. One type of information that can be stored in tags is a history timetable for the car engine oil, which needs to be replaced periodically. Assuming that there is a tag attached to the car's oil tank, this tag can hold the projected time for oil replacement.

An RFID reader, controlled by a gateway inside the car, will read this information periodically and inform the user when oil needs to be replaced.

In addition, tags will be connected with various kinds of sensors for car management. Related applications include the following:

1. *Identification.* Tags are used for identification purposes, similar to conventional item tracking.
2. *Operating conditions.* Sensor devices are attached to tires. For example, a pressure detector is attached to each tire in order to detect when the tires' pressure exceeds a limit. The tag is connected with this sensor device, so that the gateway can read the pressure information through RF signals.
3. *Vehicle performance.* A sensor for detecting road condition is attached to each tire. The tag connected with this sensor device sends the information to the gateway. The gateway will adjust car's performance according to that road condition.

17.4.4 Advertising with RFID Tags

Many people have said that a multimedia system will be an essential component in cars of the future [22]. For this reason, we need to find new ideas that apply to the future automotive media system, which is predicted to be one of the best places for advertisement. Hence, we present an idea to use RFID technology for advertisement inside cars.

Eric Schmidt, Google's CEO, believes that when he is listening to a radio in his car, an advertisement should personally address his needs [23]. For example, if a person drives by a clothing store, a radio advertisement should remind him that he needs a pair of pants and instruct him to turn left at the upcoming clothing store. For this reason, Google has planned to target on GPS-based in-car personalized advertising in the near future. In October 2006, Viacom Outdoor in London added GPS technology to a number of public buses [24,25]. Each bus has a large digital LED advertising panel, as shown in Figure 17.9. Different from regular LED advertising, here the advertisement depends on the current location of the bus. This means we have a mobile advertising billboard. Furthermore, this kind of a billboard is customizable, which means that we can change the advertisement easily with little time and cost. In both of these cases, GPS is used to track the location of buses, since it is the only currently available technology to track the vehicle position.

The RFID system can be also used to track a vehicle's position. We put some tags at different roadside locations, and an RFID reader inside a bus reads the tag while in transit. Using this approach, we would be able to run a similar advertisement as with GPS. The difference from the GPS approach is that the location detection is done by putting tags in correct positions.

Advertising media can be implemented using in-car multimedia systems. It could be a monitor or a speaker, depending on the output form. Although some people might be willing to listen to any advertising channel, the best way to put an advertisement is by integrating it into some kind of entertainment. For example, the

FIGURE 17.9 Bus with GPS advertising. (From Dody Tsiantar, Getting on board, http://www.time.com/time/magazine/article/0,9171,901060424-1184037,00.html, 2006.)

advertisement could be integrated into some radio channels, so that it will play just like ordinary radio advertising. The only difference is that the advertisement will be different for each listener, depending on which location he currently is. Another method is through an Internet browser. In this method, an advertisement will pop up on a web page automatically when someone browses the Internet inside a car. The types of advertisement depend on which location the car currently is.

In addition to that, we can also implement a user filtering method for advertisement. In the current advertising method, a person will receive various kinds of advertisement. The problem is that sometimes a person only wants to receive some specific advertisement fields. In this RFID advertisement idea, each advertisement is treated as a single unit of data. For this reason, a driver can select which advertisement fields he is willing to receive. When his car is passing an advertisement tag, an RFID reader inside the car will check whether the tag's advertisement field matches the user preferences. If it does not match, the advertisement data will not be transmitted to the car.

Meanwhile, in order to implement the mobile advertisement source idea that was mentioned before, a tag is attached to a car. This car will then become a mobile billboard. If another car is located near enough to the advertising car, an RFID reader inside this car will communicate with the tag to get the corresponding advertisement.

The architecture for RFID advertising is shown in Figure 17.10. After a tag's ID is read by a reader inside a car, the ID will be sent to an advertisement database center using the Internet wireless access. Using that ID, the database center will search for a correct multimedia advertisement and then send it to the car by using Internet wireless access. Hence, in order to apply for this idea, we need a well-established in-car Internet infrastructure. Although this might sound unreasonable for the current time, in-car Internet infrastructures have been developed suddenly and recently. In the

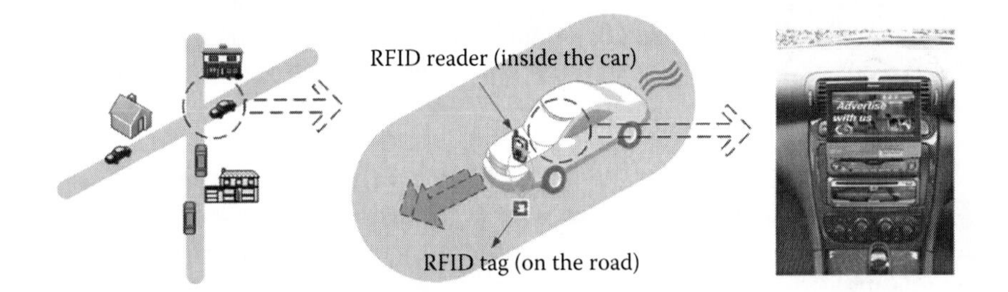

FIGURE 17.10 RFID advertising architecture.

United States, in-car Internet can be seen in some rental cars, although so far it has not been used by many people [26]. Currently, this kind of service is only worthy for a business person who likes to run his business from the car. However, a recent survey shows that the number of car Internet users is increasing [27]. Furthermore, there are many options to get an Internet connection inside a car (GPRS, EDGE, CDMA, WI-FI) [28]. For this reason, we believe that the well-established in-car Internet infrastructure, which is required for our idea, will be available in the very near future.

17.4.5 Advanced Road Information System

Some current telematics services have already provided a road information system. In this feature, while passing a road, a driver will be able to obtain information about the road, including speed limit and road conditions. One disadvantage of the currently used service is that in order to get the road information, the user needs to download and install it. If the road information is changing, the users will need to resynchronize the information. For this reason, a GPS-based road information system is not suitable for supplying such dynamic information as road conditions.

RFID can provide a better solution to implement this road information system. We name this service advanced road information system (ARIS) with its architecture shown in Figure 17.11. In the GPS approach, road information is read from a map data inside a GPS tool (located inside a car). In our approach, the road information will be read from a tag located on the road, assuming that all cars are installed with an RFID reader. In order to change road data, we change the tag's data. After the data has been updated, all cars that pass that tag can read the updated road

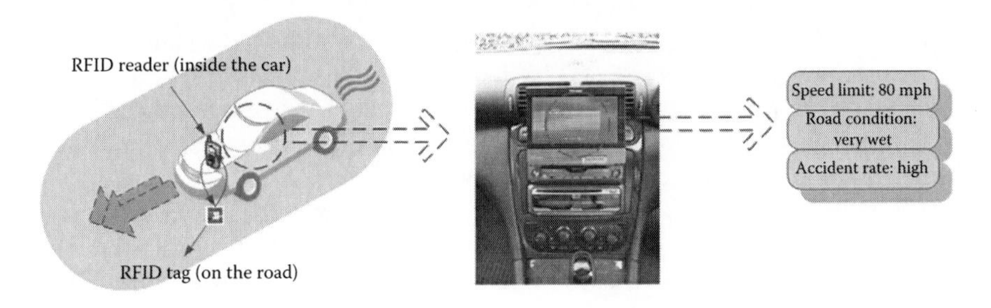

FIGURE 17.11 ARIS architecture.

information. Another option is to combine the existing GPS system with the RFID technology. Here, RFID tags are used as calibration tools to correct GPS errors. This technique will further improve the accuracy of GPS technology.

The main limitation for using RFID in this system is its limited data capacity. Compared with GPS, the data that can be held are relatively small (typically less than 1024 bits). However, this storage size is generally enough, considering two things. First, the amount of information to be sent to the car is small. Second, we could use some protocols to further compress data. Using this way, we could limit the required data to be transferred by focusing only on the information and leave the decoding process to the reader inside the car. Furthermore, if in-car Internet service is available, a combination with Internet can be used to get more information data. Similarly to RFID advertising, in this idea, the car will read the tag as it runs at a relatively high speed.

17.5 ARCHITECTURE FOR AUTOMOTIVE RFID SYSTEMS

In order to implement RFID systems with the currently existing automotive systems, the knowledge about the current car networking and telematics systems is needed. Fortunately, current car technologies have already provided flexible network infrastructures, such as controller area network (CAN) [29], intelligent transportation system (ITS) [30], and FlexRay [31]. For this reason, automotive RFID architecture can be easily attached onto the current car network systems. In addition, we can borrow some ideas from GPS systems, since GPS applications are built as separate add-on options for a vehicle.

17.5.1 Car Networking Systems

A currently available car contains thousands of circuits, sensors, and other electrical components, as shown in Figure 17.12. Around 30 years ago, communication among those components was handled by point-to-point wire connections. Adding more combinations inside the communication resulted in an enormous increment of wires. Adding more components will increase the complexity even more. This complexity increment would create damaging effects to the car itself, such as increasing its weight, weakening its performance, and reducing its reliability. For a normal car, every extra 50 kg of wiring will result in extra 100 W of power consumption; hence, it will increase fuel consumption by 0.2 L for each 100 km [32]. During the 1980s, centralized and distributed networks began to replace this point-to-point communication method, which provided more simple implementation [33].

CAN, which was developed in the 1980s, has been used to connect vehicle sensors and safety systems. Unlike a conventional point-to-point communication, CAN uses a central gateway that controls all communications among car's devices. Hence, this centralized networking system reduces wiring size, weight, and cost compared with conventional point-to-point communications. Furthermore, CAN also provides a standard serial bus that can be used to connect various kinds of devices. This means any vendor can synchronize their devices so that they could be hooked into CAN. This approach eliminated the portability problem that occurred in point-to-point

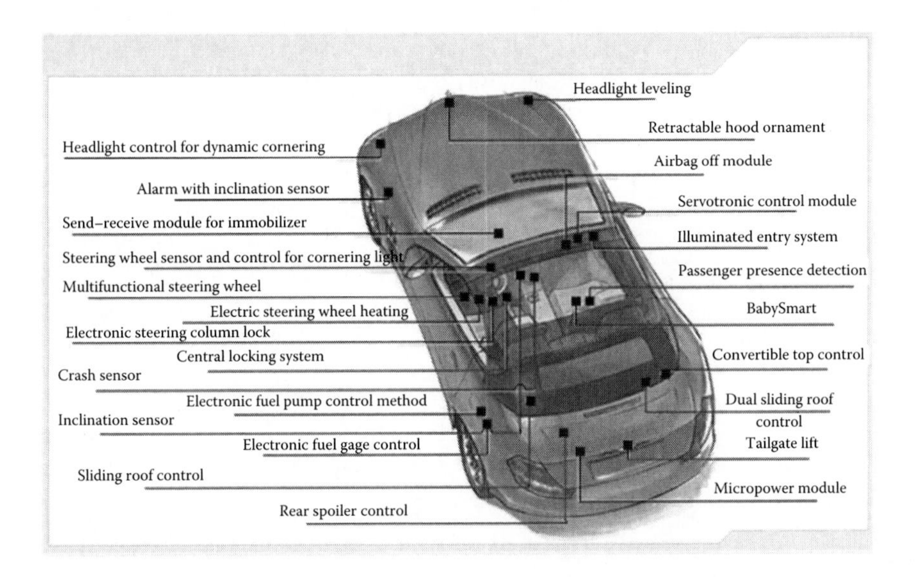

FIGURE 17.12 Electronics complexity inside current cars. (From Helbako, Details that make cars better, http://www.helbako.de/helbako/noflash/sprachen/eng/produkte.htm, 2006. With permission.)

connections and so increased the productivity of development. More than 100 million CAN nodes were sold in 2000, and currently it is the most widely used vehicle network [32].

Vehicle networks in CAN are commonly classified based on the Society of Automotive Engineers (SAE) standards. SAE formally classifies vehicle networks based on their bit transfer rate, as shown in Table 17.7. This classification relies on the fact that each application will need a different requirement of data bandwidth. Using this classification, an application that requires huge amount of data, such as multimedia applications, can use a bigger data bus in its communication with a gateway.

TABLE 17.7
Classification of Automotive Networks

Class	Speed	Example Application
A	Less than 10 kbit/s Low speed	Convenience features, e.g., trunk release, electric mirror Adjustment
B	10–125 kbit/s Medium speed	General information transfer, e.g., instruments, power window
C	125 kbit/s to 1 Mbit/s High speed	Real-time control, e.g., power train, vehicle dynamics
D	More than 1 Mbit/s	Multimedia applications, e.g., Internet, digital TV

Meanwhile, another application that does not require a huge amount of data can use a different data bus with smaller capability than the previous bus.

Although CAN includes many kinds of devices inside a car, ITS is focusing on factors that are visible to consumers. Examples of these factors are safety, transportation times, and fuel costs. In ITS, as in CAN, car networks are handled by a central gateway that controls all communications among various car devices.

FlexRay is one of the newest automotive network communication protocols. It is still under review for protocol specifications. According to its specification, FlexRay will give more benefits than CAN in terms of data rates (10 Mbps), redundancy, safety, fault tolerance, and price. Due to these reasons, FlexRay is predicted to replace other conventional automotive networking protocols in the near future. In 2006, the BMW X5 became the first vehicle to use FlexRay but limited to its pneumatic damping system. The full use of FlexRay is expected to be accomplished in 2008. Other recent car networking systems include local interconnect network (LIN) [34] and media-oriented system transport (MOST) [35].

17.5.2 Modern Telematics Devices

Recently developed modern telematics devices converge many technologies and services. Basically, a telematics device not only has one communication technology but also numerous communication methods, such as cell-based wireless communications for telephone services, in-car networking for controlling electronics of a car, GPS for location-based service (LBS), IP-based wireless communications for Internet services, DSRC for networking with other cars, and digital broadcast communications for a digital radio and TV. It also embeds a huge database for navigation and has many large storages such as optical disk drive (ODD), hard disk drive (HDD), and flash memory for saving files. A display and a voice/audio system have to be equipped to interface with a user. The device uses TFT-LCD or OLED with various sizes and resolutions for display, voice recognition, and noise cancellation methods for detecting user's voice commands and speakers with audio DSP or special ASIC for enhancing output sounds. Many brand new technologies are merged into one telematics device. Some major technologies and standards used in a telematics device are shown in Table 17.8.

Figure 17.13 shows that a telematics system includes numerous functional blocks and relevant modules. Typically, a host processor runs an operating system and executes many software applications for the technologies mentioned earlier.

17.5.3 Proposed Architecture for an Automotive RFID System

In order to provide that standard, the current architecture usually uses a layered system. The layer will act as a gate between two kinds of different networks. In a mobile RFID reader [36], the layer, which is called handset adaptation layer (HAL), acts as a gate between RFID readers and cell phones. By doing this, both RFID reader vendors and cell phone vendors can develop and produce their own part without having to depend on each other, which surely will accelerate the development for applications. The other advantage in a layered system is that a new system can be added easily.

TABLE 17.8
Technologies and Standards Used in a Recently Developed Telematics Device

Application	Technologies and Standards
Navigator	GPS, geographic information system (GIS), dead reckoning, NMEA 0183, NMEA 2000 protocol, SiRF, ISO 19100 series, GIS DB, route planning, route guidance, map matching
Interfaces with a user	TFT-LCD, OLED, touch screen, mike, speaker, camera, keypad, vision/audio enhancement, graphic user interface (GUI), voice recognition, noise cancellation, speech synthesis, speaker adaptation
Wireless communication	Telephone, SMS/MMS, e-mail, mobile Internet, video conference, SM/GPRS/EDGE, CDMA 1x/EV-DO, WCDMA
Internet-based services	Web browser, online market, wireless LAN (IEEE 802.11a/b/g), WiBro/WiMAX (IEEE 802.16), HSDPA
Storage	ODD, HDD, flash memory, IDE/ATA standards, file system, CD-DA, CD-ROM, DVD, DVD-R, DVD + R, DVD-RAM, DVD-RW, compact flash, multimedia card, Secure Digital, eXtreme Digital, memory stick, micro-SD (T-Flash), mini-SD, secure micro-SD, MMC-micro, RSMMC (reduced size multimedia card)
Car network	CAN, FlexRay, RIN, MOST, ISO 11898/11992/11783 series, SAE J1939 series, SAE J2411, SAE J2561
Multimedia	Movie/music play, digital broadcast, recording, image view, multimedia codec solutions (WMV, MPEG, Divx, QuickTime, WMA, MP3, AC3, JPEG), DSP (digital signal processing), DAB, T-DMB, S-DMB, digital radio, DVB-H, multitasking
ITS	Congestion control, traffic light control, electronic toll, collection, emergency warning/rescue, road information providing, DSRC, IEEE 1609 series, WAVE (IEEE 802.11p), LBS
Personal organizer	Calendar, scheduler, address book, word processor, spreadsheet, computer game, operating system, synchronization, software application

Figure 17.14 shows our proposed architecture for an automotive RFID system, which consists of three layers: physical layer, protocol layer, and application layer. The physical layer handles communication between reader RF circuits and tags through RF waves. The protocol layer controls the rule of how data transfer should be done, which component handles error checking and collision problems. Last, the application layer is the place where RFID applications should lie.

In the physical layer, more consideration should be put on performance such as the reading distance and the reading speed. The protocol layer is implemented inside a car central gateway. A local database, which holds a database for user IDs and a small amount of information, is also placed here. Next, the application layer should be implemented inside the gateway also. Object naming service (ONS) of the EPCglobal Network, which holds a global database for all applications, is also placed in this layer. This ONS will be used to handle requested information from a car. For example, in the case of RFID advertising, ONS will

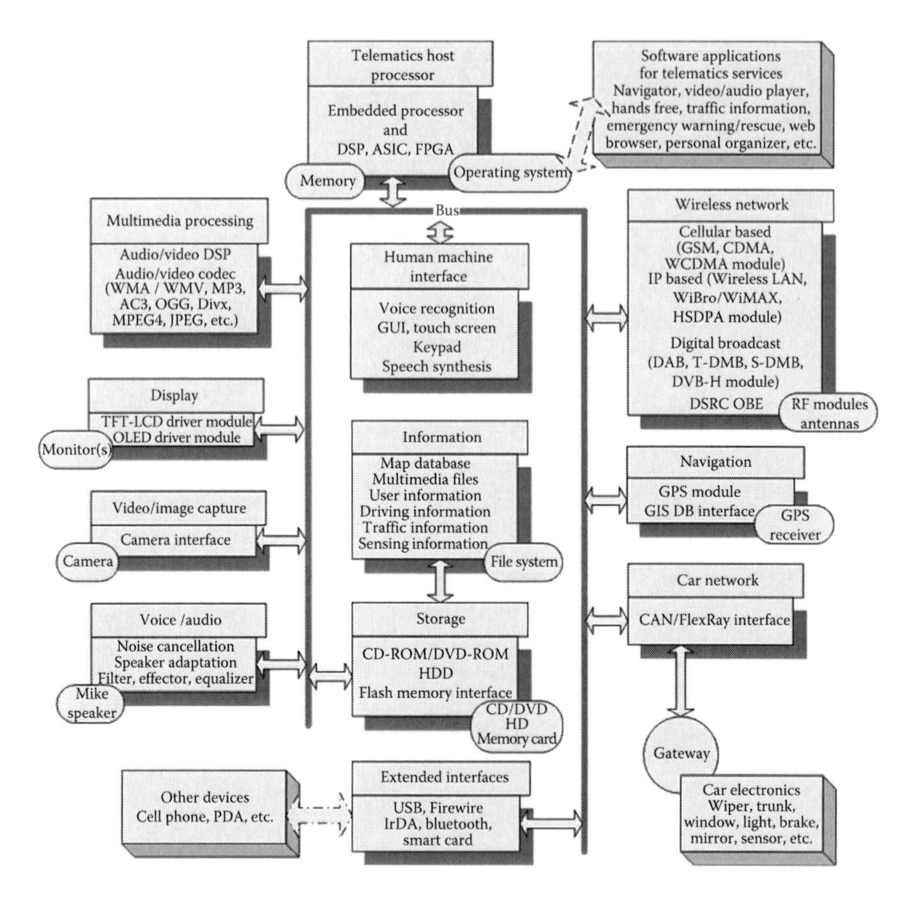

FIGURE 17.13 Typical architecture of a telematics system.

receive the advertisement's ID from a car, find a corresponding advertisement data based on this ID, and then transmit the data back to the car. There is another option to implement the protocol layer separately by using an additional baseband processor, which will require more resources but will improve the overall network performance.

17.6 EXPERIMENTAL DESIGN: TalusRFID

Most current RFID readers are implemented using a general processor. Some researchers have proposed new ideas for implementing an RFID baseband processor. One example can be seen in TalusRFID. In this approach, an RFID reader is designed using the Talus architecture [37], which uses Java as its programming language and FPGA technology as its target implementation.

Java has been evolved as one of the most preferable programming languages in the last decade. However, there has not been any RFID reader to be implemented using this language because Java execution speed is a lot slower than any other common programming language. The Talus architecture overcomes this slow-speed problem

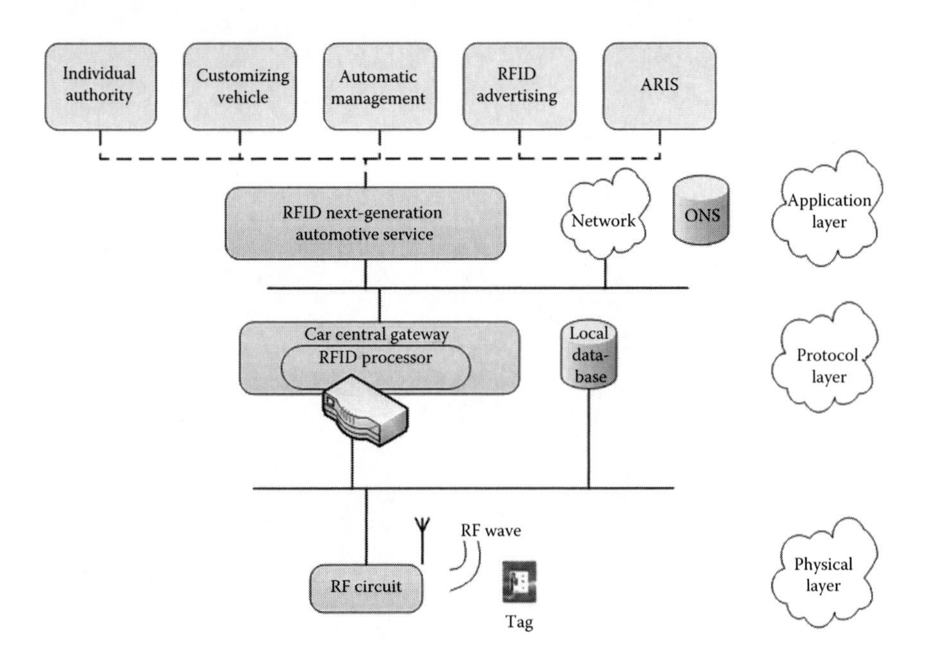

FIGURE 17.14 Proposed architecture for the next-generation automotive RFID system.

by using a coprocessor technique, which accelerates the execution on FPGA, hence making it able to implement an RFID baseband processor.

The TalusRFID reader consists of three different parts:

1. RF circuit
2. Processor part
3. PDA, as the user terminal I/O

As shown in Figure 17.15, the TalusRFID reader can be seen as a prototype for the RFID reader that will be used for the next-generation automotive RFID system. The physical layer discussed in Section 17.5.3 is the same as the physical layer in our proposed architecture. The protocol layer is implemented inside a baseband processor by the Java RFID software inside Talus architecture, and the application layer runs identification for users, implemented inside a PDA.

The details of the TalusRFID processor are shown in Figure 17.16, which supports ISO 18000-6 Type B and Type C (EPC Gen2) standards. The protocol layer was implemented in Java bytecodes, and lower layers, such as modulator, demodulator, and filters, are implemented in an FPGA reconfigurable hardware. TalusCore, which is a Java native processor, could only contain 3K instructions in its code memory due to resource limitation. For this reason, it will only execute manually selected time-critical methods, while the rest of the codes are executed on Java virtual machine (JVM), which in Talus is called Talus virtual machine (TVM). In order to provide the TVM environment, embedded Linux is installed in the ARM processor. In order to use hardware parts for RF circuits directly, a software programmer

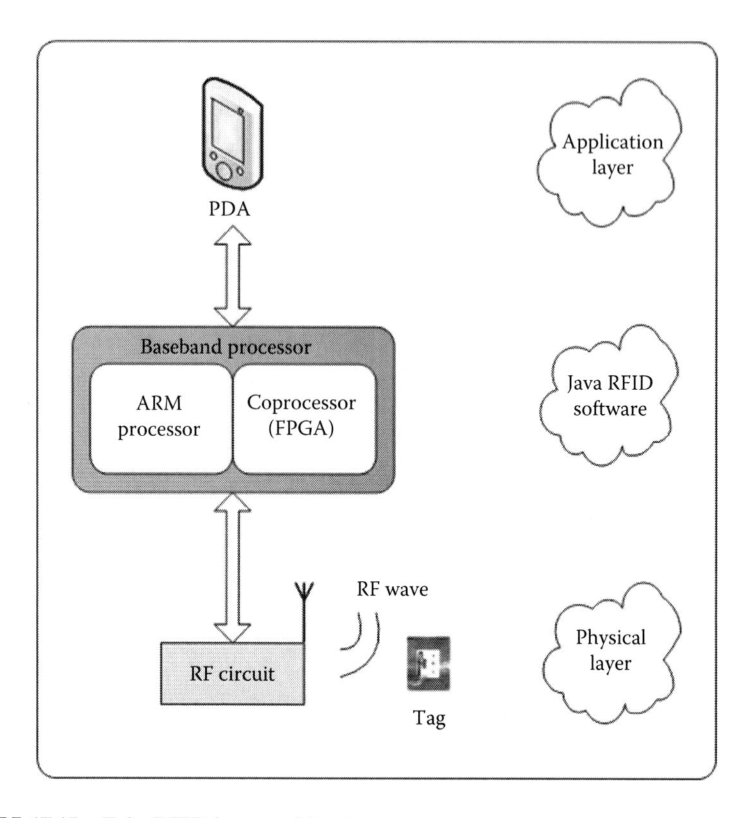

FIGURE 17.15 TalusRFID layer architecture.

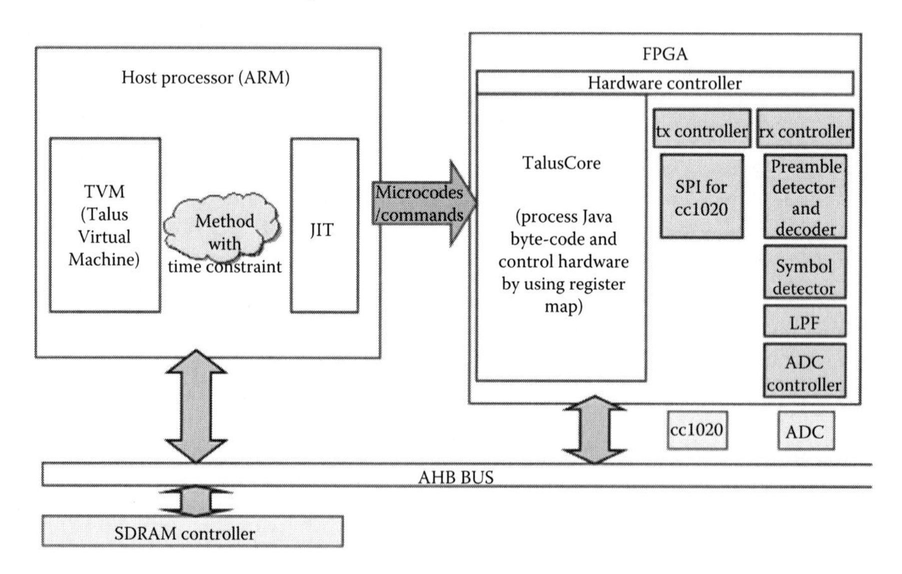

FIGURE 17.16 TalusRFID processor architecture. (From Lee, J.G. et al., TalusRFID: Java-based RFID baseband processor, *Proceedings of the 14th Korean Conference on Semiconductors*, pp. 187–188, 2007.)

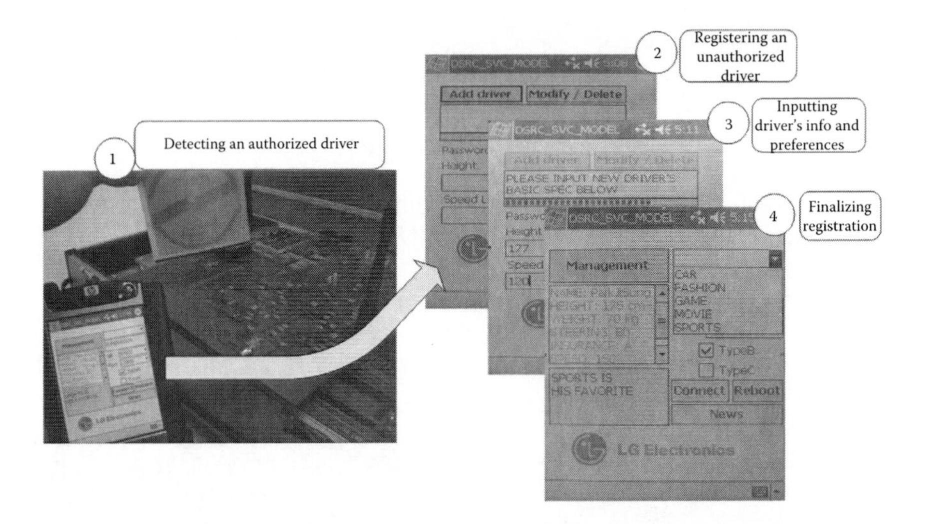

FIGURE 17.17 TalusRFID demo. (From Lee, J.G. et al., TalusRFID: Java-based RFID baseband processor, *Proceedings of the 14th Korean Conference on Semiconductors*, pp. 187–188, 2007.)

can use some special methods for external hardware control. The Java application can control hardware by invoking them. In this system, hardware demodulator and transmission controller are implemented in hardware. The demodulation of received signals in a software manner was almost impossible to satisfy a time constraint for tags' identification in ISO 18000-6 Type C due to overheads of OS. The details of the Talus architecture have been given in detail by Hwang et al. [37] and Lee et al. [38].

Figure 17.17 shows the TalusRFID demo with its PDA displays. The demo scenario was designed to illustrate our previous idea in Section 17.4.2 (customizing a vehicle for multiple drivers). In Step 1 of Figure 17.17, the reader detects an authorized tag and automatically sets car settings based on this tag owner's preference. In Steps 2–4, a car owner wants to add an additional user to the system. In this case, a new tag ID is registered in the system, including the new user's information and preferences. After the registration step is finalized, the system will be able to recognize the new driver as an authorized user.

17.7 CONCLUSION

Since it was introduced, RFID has been expected to replace barcode technology. For this reason, RFID is always compared with barcode technology in all kinds of parameters. Furthermore, the development of RFID applications has always focused on existing barcode applications. These factors limit the creation of innovative and creative applications for RFID systems. There are unlimited possibilities to generate new ideas for RFID applications. However, it will be easier to develop new ideas that have a visible effect on users. In other words, we should focus on user-oriented applications.

In this chapter, we proposed new ideas for user-oriented RFID automotive applications. For example, RFID technology can be used to provide some limits to drivers, depending on time, distance, speed parameter, etc. In addition, drivers can use RFID technology to automate settings in various devices such as car seats and mirrors. Since the cost of an RFID tag is predicted to drop in the near future, RFID tags can also be attached to numerous car parts in order to relay information to a car management system. Another possible innovation is to use RFID in advertising. Using RFID technology, a new technique for location-based advertising can be implemented inside a car. In ARIS, RFID is used to improve the accuracy of the existing GPS's road information system.

In order to implement those new applications, current car technologies have already provided flexible network infrastructures, such as CAN, ITS, and FlexRay. For this reason, automotive RFID architecture can be easily attached to the current existing car network systems. However, since automotive RFID applications combine many different types of applications, a fixed standard is required to accelerate the development of applications. With a fixed standard and networking protocol, different developers can focus on developing and producing their own parts without having to worry about compatibility with other parts. In this chapter, we have proposed an architecture for future automotive RFID applications. The architecture was designed so that only minimal modifications are required in integration with the current car networking architecture. Finally, we also discussed our prototyped system, called TalusRFID, which is an implementation of a Java-based RFID reader using the FPGA technology.

ACKNOWLEDGMENT

This work was supported in part by Digital Media Company, LG Electronics.

REFERENCES

1. Harry Stockman. Communications by means of reflected power. *Proceedings of the Institute of Radio Engineers* (*IRE*), 36, 1948, New York, pp. 1196–1204.
2. Korea Smart Card Co. Ltd. Seoul's new transportation system. http://www.t-money.co.kr/jsp/newpub/oversea/english/stories/S_story.jsp.
3. Institution of Electrical Engineers (IEE). Radio frequency identification device technology (RFID). http://www.rfidc.com/pdfs_downloads/IEE%20RFID%20Paper.pdf, 2005.
4. John R. Tuttle. Traditional and emerging technologies and applications in the radio frequency identification (RFID) industry. *IEEE Radio Frequency Integrated Circuit Symposium*, 1997, Denver, CO, pp. 5–8.
5. Steve Bono, Matthew Green, Adam Stubblefield, Ari Juels, Avi Rubin, and Michael Szydlo. Security analysis of a cryptographically-enabled RFID device. *Proceedings of the 14th USENIX Security Symposium*, 2005, Baltimore, MD, pp. 1–16.
6. Mary Catherine O'Connor. Gen 2 EPC protocol approved as ISO 18000-6C. http://www.rfidjournal.com/article/articleview/2481/, 2006.
7. Iris Corporation Berhad. Malaysia is the first country to use computer chips in passport. http://www.iris.com.my/News/new_detail.asp?id=12, 1998.
8. Robyn Curnow. The price to pay for VIP status. http://edition.cnn.com/2004/TECH/10/05/spark.bajabeach/, 2004.
9. David M. Ewalt and Mary Hayes. Gillette razors get new edge: RFID tags. http://www.informationweek.com/story/IWK20030110S0028, 2003.

10. Claudia H. Deutsch and Barnaby J. Feder. A radio chip in every consumer product, 2003.
11. Klaus Grimm. Software technology in an automotive company: Major challenges. *Proceedings of the 25th International Conference on Software Engineering* (*ICSE*), 2003, St. Louis, MO, pp. 498–503.
12. Ariz Scottsdale. Auto electronics market set to exceed US$50 billion by 2010. http://www.instat.com/press.asp?Sku=IN0603375RE&ID=1752, 2006.
13. United States Coast Guard's Navigation Center of Excellence. General information on GPS. http://www.navcen.uscg.gov/?pageName=gpsmain, 2014.
14. ITS Standards Program. Dedicated short range communications (DSRC). http://www.standards.its.dot.gov/Documents/advisories/dsrc_advisory.htm.
15. Seung-Que Lee, Namhun Park, Choongho Cho, Hyongwoo Lee, and Seungwan Ryu. The wireless broadband (WiBro) system for broadband wireless internet services. *Vehicular Technology Magazine*, 44, 106–112, 2006.
16. Javier Gozalvez. Mobile radio—HSDPA goes commercial. *Vehicular Technology Magazine*, 1, 45–53, 2006.
17. Microsoft. Microsoft Car.NET connects motorists to the wireless Internet. http://www.microsoft.com/Presspass/press/2000/oct00/carnetpr.mspx, 2000.
18. Maxim Integrated Products, Inc. Remote keyless entry systems overview. http://www.maxim-ic. com/appnotes.cfm/appnote_number/1774, 2002.
19. Svenska Aeroplan Aktiebolaget (SAAB). Saab 9–5 features and specifications. http://www2.saabusa.com/95s/features.asp?start=home.
20. Michelin. Intelligent tires: Michelin outlines new technology at industry conference. http://www.michelinman.com/difference/releases/pressrelease03092005a.html, 2005.
21. Melanie Rieback, Bruno Crispo, and Andrew Tanenbaum. Is your cat infected with a computer virus? *Proceedings of the 4th IEEE International Conference on Pervasive Computing and Communications*, 2006, Pisa, Italy, pp. 169–179.
22. Rajiv Mehrotra. Telematics might steer your car into the future. *IEEE Multimedia*, 9(3), 9–10, 2002.
23. Donna Bogatin. Google targets GPS-based in-car personalized advertising. http://blogs.zdnet.com/micro-markets/?p=131, 2006.
24. CBS Outdoor. Viacom outdoor launched GPS advertising. http://www.cbsoutdoor.co.uk/web/Current-news/Newspage-UK/Viacom-Outdoor-launches-GPS-advertising-a-global-first-with-Yell.com.htm, 2006.
25. Dody Tsiantar. Getting on board. http://www.time.com/time/magazine/article/0,9171,901060424-1184037,00.html, 2006.
26. Christopher Elliott. Wi-Fi is hitting the road in cars from Avis, but technical and legal bumps lie ahead. http://www.nytimes.com/2007/01/02/technology/02avis.html, 2007.
27. Jeff Goldman. A look at the future of in-car internet access. http://www.wireless-weblog.com/50226711/a_look_at_the_future_of_incar_internet_access.php, 2006.
28. Thomas Nolte, Hans Hansson, and Lucia Lo Bello. Wireless automotive communications. *Proceedings of the 4th International Workshop on Real-Time Networks (RTN'05) in conjunction with the 17th Euromicro International Conference on Real-Time Systems* (*ECRTS'05*), 2005, Palma de Mallorca, Balearic Islands, Spain, pp. 35–38.
29. Karl Henrik Johansson, Martin Torngren, and Lars Nielsen. Vehicle application of controller area network. *Handbook of Networked and Embedded Control Systems*, 741–766, 2005.
30. Fuqiang Liu and Fengzhong Li. Intelligent transportation system based on the next generation broadband wireless communication. *Proceedings of IEEE International Conference on Service Operations and Logistics, and Informatics*, 2005, Beijing, China, pp. 41–45.

31. FlexRay Consortium. FlexRay. http://www.flexray.com/.
32. Gabriel Leen and Donal Heffernan. Expanding automotive electronic systems. *IEEE Computer*, 35, 88–93, 2002.
33. Gabriel Leen, Donal Heffernan, and Alan Dunne. Digital networks in the automotive vehicle. *IEE Computing and Control Engineering Journal*, 10, 257–266, 1999.
34. LIN Consortium. Local interconnect network. http://www.lin-subbus.org/.
35. MOST Cooperation. Media oriented systems transport. http://www.mostcooperation.com/.
36. Joon Goo Lee, Seok Joong Hwang, Seon Wook Kim, Sunshin Ahn, KyungHo Park, Ji Hoon Koo, and Woo Shik Kang. Software architecture for a multi-protocol RFID reader on mobile devices. *Proceedings of the 2nd International Conference on Embedded Software and Systems* (*ICESS'05*), 2005, Xi'an, China, pp. 81–88.
37. Seok Joong Hwang, Peter Harliman, and Seon Wook Kim. Talus: Compiler-assisted Java accelerator. *Proceedings of the 4th International Workshop on SoC and MPSoC Design*, 2006, Yogyakarta, Indonesia, pp. 389–398.
38. Joon Goo Lee, Peter Harliman, Kyongjin Jo, Sungjea Ko, Seon Wook Kim, and Kwangjoo Choi. TalusRFID: Java-based RFID baseband processor. *Proceedings of the 14th Korean Conference on Semiconductors*, 2007, Jeju, Korea, pp. 187–188.

18 Intelligent Control Mechanisms for Vehicular Sensor Networks

Sami S. Alwakeel and Agung B. Prasetijo

CONTENTS

ABSTRACT

In recent years, there has been a trend of sensor networks being embedded into vehicular ad hoc networks (VANETs). For VANET, vehicles are equipped with sensors, and the output of such sensors is fed to various VANET devices. There are numerous benefits of VANETs: road safety, public service, driving experience, and vehicle business/entertainment.

Message delivery in mobile networks can be in the form of unicast, multicast, or broadcast. Broadcasting is preferable as it offers a simple technique to spread messages to all vehicles available in the VANET. Broadcasting messages may, however, lead to excessive redundancy, contention, and collision of messages, which are commonly known as *a broadcast storm* for VANET messages. Intelligent control of broadcast mechanisms may address such problems.

This chapter deals with various subjects related to VANET communication control and routing protocols. It discusses in detail various intelligence algorithms, known in the artificial intelligence field as classifiers, for controlling the VANET communication and solving the broadcast storm problem. The applications of Bayesian decision theory, Maximum-likelihood and Bayesian estimation, Nearest-neighbor estimation, support vector machine, and artificial neural network to improve rebroadcast selection problem are investigated. The chapter proves that the intelligent control of

VANET communication works very well in reducing message delays compared with flooding technique and significantly reduces the message redundancy of dense vehicular networks.

18.1 INTRODUCTION

Ad hoc networking exists to answer the need for communications without preexisting infrastructure such as routers in wired networks or access points in wireless environments such as in vehicular ad hoc network (VANET), battle fields, or natural disaster areas. Ad hoc networks tend to be wireless where nodes exchange messages directly between each other within their radio coverage. Classic routings in a wired local area network (LAN) are not applicable in an ad hoc network. Instead, broadcasting is used as the workforce for delivering messages within nodes. When a node broadcasts a message to its neighboring nodes, the message will be rebroadcast to other nodes by the neighboring nodes in such a way that the message will be transmitted over the network.

Wireless sensor networks (WSNs) are a kind of ad hoc network where every sensor cooperates with each other to deliver information to the destination. The information can be physical or environmental conditions: temperature, pressure, acidity, sound or light intensity, and image, to name a few. The sensor senses the condition of its surroundings and sends messages periodically to its neighboring sensors. Based on the routing algorithm applied, the selected neighboring sensors will rebroadcast the messages to other nodes until the messages reach the destination (this destination is often called the base station or sink). Figure 18.1 shows how sensor nodes scattered in a sensor field cooperate with each other in forwarding a message to the sink.

As sensor nodes are usually designed as being disposable and having a small form factor, they bear limited capabilities, such as short battery life, small memory, and low computational power and communication functionality [1–3]. Figure 18.2 shows the typical architecture of a sensor node. A typical architecture of a sensor

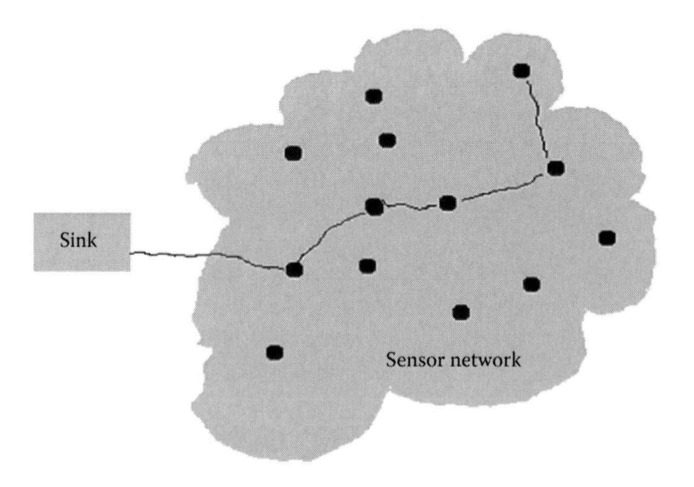

FIGURE 18.1 Sensor nodes scattered in a sensor field.

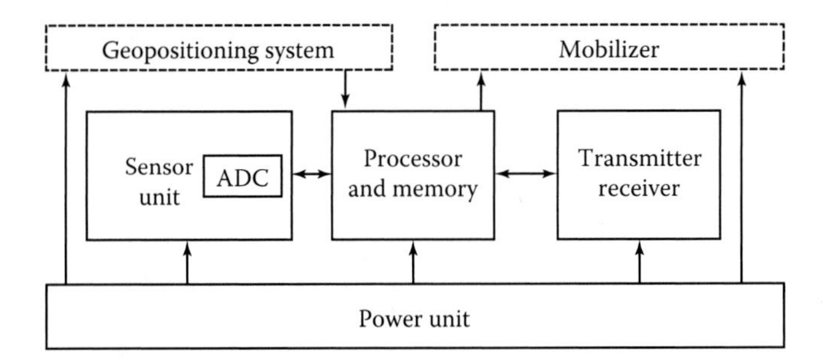

FIGURE 18.2 Typical architecture of a sensor node.

may have at least a power source, one or more sensor devices for gathering sensory information, a microcontroller, and a radio transmitter–receiver. An analog-to-digital converter (ADC) converts analog information obtained from the sensors, and its output is directly fed to the microcontroller for further processing. The microcontroller processes the information and decides when to send the information over a radio transmitter. The microcontroller also decides whether a message received from other nodes was received or rebroadcasted before. Information generated locally and messages received from other sensor nodes can be stored in the memory for further processing. The routing algorithm embedded in the microcontroller makes the decision mostly for the sake of energy conservation to prolong the life of the sensors, of course with a tolerable sacrifice on the communication performance (quality of service).

Some of the sensors, however, are equipped with a geopositioning device, and some with a mobilizer. The geopositioning information gives the user the ability to track the position of the sensor at all times. This is useful for a habitat monitoring scenario, for example. Geopositioning information also helps the routing algorithm to perform more efficiently in delivering messages to the sink. It helps to select nodes that are closer to the sink for conducting rebroadcast so that redundancy of messages over the network will further be reduced. Therefore, the energy consumption of nodes will also be minimized over time. Mobilizer is useful for gathering sensory information when the objects are not stationary. The mobilizer enables the sensor to approach such moving objects with the help of location finding unit. However, such nodes are slow in speed, and the movement is usually only for a short distance. The electromechanical devices of the sensor are also not designed for a fast and long-range movement.

Furthermore, besides the transmitter, the movement of a sensor also consumes extensive energy resource. Therefore, there should be a way for the sensor node to control its movement to be efficient. Power efficiency in transmission and movement are two well-known major issues for wireless sensor nodes. As such, a sensor network protocol stack is equipped with power management and mobility management planes. Figure 18.3 shows the sensor network protocol stack that has such planes.

With regard to mobility, sensor nodes in WSNs can be statically deployed over the sensor field or can move randomly within the sensor field's boundary. In the case of mobile nodes, WSNs are, therefore, classified as mobile ad hoc networks (MANETs).

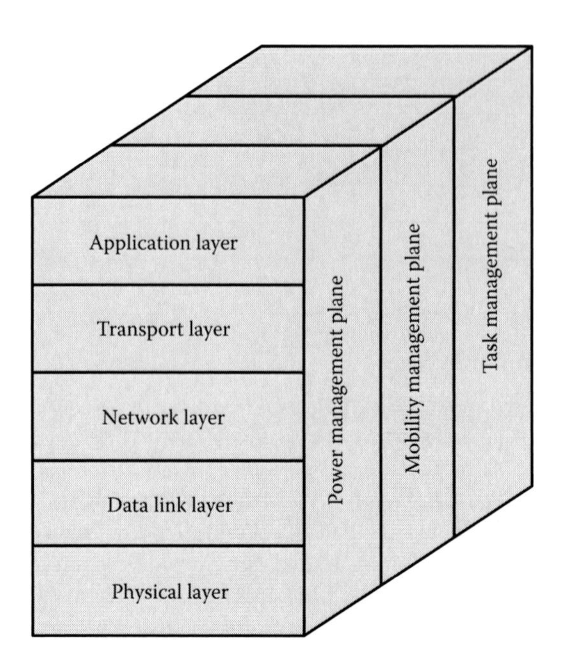

FIGURE 18.3 Sensor network protocol stack.

The characteristics of WSNs are similar to MANETs, and its routing strategies are, in general, applicable to WSNs, but of course with great care on the power consumption issue.

18.2 VEHICULAR SENSOR NETWORKS

In recent years, there has been a trend of sensor networks being embedded into VANETs, which is popularly called vehicular sensor networks (VSNs). Here, vehicles are equipped with sensors, and the output of such sensors is fed to VANET devices. Several benefits from such VSNs can be obtained for cases related to driving experience as well as road maintenance. VSNs may help drivers against car accident and improve their driving efficiency. A car having proximity sensors in front of it can give warning to the driver when the distance between the car and other objects, such as vehicles in front of it, is too close. Camera sensors located at the intersections may enable the use of traffic signal preemption (traffic prioritization) and may enhance traffic safety. Ultrasound sensor installed on cars can help the Department of Transportation conduct road maintenance in a fast and efficient way.

When sensors are integrated with VANETs, energy is no longer an issue. The movement of the sensors is also much faster compared with sensors that have an embedded electromechanical mobilizer. Power management and mobility management planes are not important in VSNs. The energy is available most of the time as cars have a generator installed under the bonnets. Likewise, the task management plane is also not required as energy preservation is not an issue.

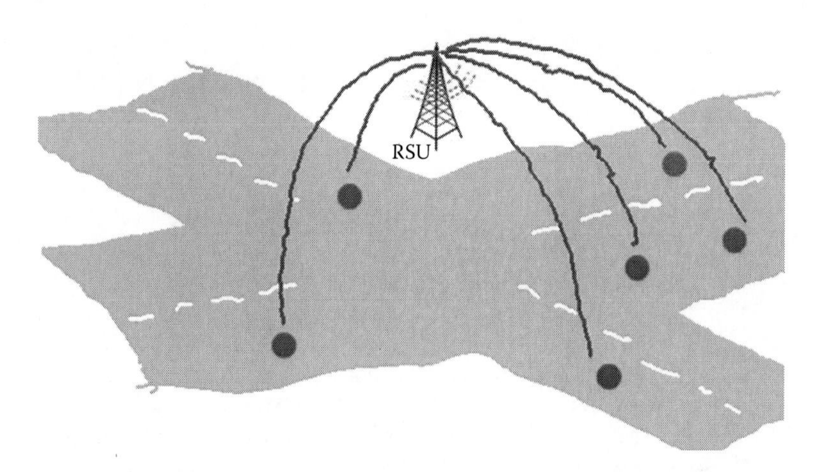

FIGURE 18.4 Roadside infrastructure communicates with vehicles in a road junction.

The communication between sensor nodes in VSNs is therefore different. In WSNs, sensor nodes communicate to each other in such a way that the main concern is to conserve energy. Routing control algorithms are mostly designed for such a goal so that WSNs have a minimal number of message broadcasts. Each node in WSNs cooperates with each other within the limited sensor field for routing such messages to a base station (sink).

In VSNs, however, communication between sensor nodes is greatly extended to the use of vehicles connected in an ad hoc fashion. The sink can now be more than one. Every roadside infrastructure (RSU) and every vehicular sensor node may become a sink. A camera sensor node in a road junction may deliver messages and vehicles reachable within a few hundred of meters will be the sink. In contrast, the RSU of the Department of Transportation can be the sink for information collected by sensor nodes that deliver information on road surface. A vehicular node can also be a sink of other vehicles when a car requests information of road traffic ahead. The scenario is very flexible, and it is open to the application software that is deployed for specific functions on the VSNs. Figure 18.4 depicts one scenario of VSN that the RSU communicates with cars in a road junction that can be a source of information or a sink to the surrounding vehicles.

Furthermore, as the energy conservation is not an issue, and the VSN's sensor mobility is greatly expanded. The way how sensor nodes are deployed will definitely change. The sensor field will be expanded greatly to the size of a big city, a suburban area, or an island. Thus, the geographical conditions will likely define the size of the sensor field. The node density will, therefore, be different among places and times, which makes the communication in VSNs a lot more challenging.

18.3 VSNs—CHARACTERISTICS AND MOBILITY

VSNs inherit the whole characteristics of VANETs. Road infrastructure makes vehicles in VSNs move in a more organized way compared with the movement of nodes in WSNs. Likewise, VANETs as a subclass of MANETs inherit most of the characteristics of MANETs that WSNs mostly deal with. However, there are typical characteristics in VANETs (so are there in VSNs) that are significantly different from MANETs. In literature, Liu et al. [4] described VANETs' characteristics as follows:

- It has a high mobility and rapidly changing topology as vehicles move very fast, particularly in highways.
- The availability of geographic position can be used to improve routing strategy.
- Mobility modeling and predication works in VANETs as vehicle movement is limited by prebuilt highways, roads, and streets.
- Hard delay constraints for safety applications must be carefully considered.
- There is no power constraint. VANETs can be utilized for a broad range of applications, covering both safety- and nonsafety-related applications.

All the characteristics mentioned before are applicable to VSNs as sensors are embedded in vehicles and their movements are no longer random, but according to the available road infrastructure. The sensor nodes are now scattered in a very large area and much faster in the movement compared with the typical nodes deployed within the WSNs. Therefore, connections and disconnections of nodes occur very frequently, and the nodes' density also change very rapidly. This definitely affects the message delay of VSNs that must be carefully considered, in particular for the safety-related applications.

Schoch et al. [5] identified the possible VANET applications and categorized them into the following: (1) active safety, (2) public service, (3) improved driving, and (4) business/entertainment. However, VSNs do not deal with business/entertainment applications, as such applications do not need sensors. Schoch et al. also identified five types of communication patterns in VANETs that also apply neatly into VSNs:

1. *Beaconing*: Periodic update of information among all neighboring nodes
2. *Geobroadcast*: Immediate distribution of information in a larger area, for example, to immediate neighbors
3. *Unicast routing*: Transport of data through the ad hoc network to a certain destination (node or RSU)
4. *Advanced information dissemination*: Disseminate information among vehicles during a certain time capable of bridging network partitions and prioritizing information
5. *Information aggregation*: Data are processed and merged by network nodes and not simply forwarded

As the energy resource is relatively available at anytime, beaconing can be made available in VSNs. This beaconing is very important for knowing the number of neighboring nodes to control message rebroadcast more efficiently and lower the message delay. With beaconing, information of nodes such as geolocation, speed, direction can be exchanged between each other. With such information, decision on whether a node is eligible for forwarding a message can be determined more accurately. In contrast to WSNs, beaconing should not be made available as this will consume extensive energy that leads to the shorter lifetime of the sensor nodes.

However, message redundancy to some extent can help VSNs to improve delay performance. For example, during midnight or in a suburban area, usually vehicles are very small in number. Forwarding a message to only a few of them can result in a longer delay when the message reaches the sink. Assigning more nodes to deliver such a message within such a scenario may improve delay performance. More message redundancy over the network increases the chance that the message reaches the destination faster. Therefore, selective control over nodes responsible for rebroadcasting must be carefully implemented in such situations. Moreover, a store-and-forward mechanism may also be used for partitioned connectivity in suburban areas that most of the nodes will be selected to increase message delivery performance.

Geobroadcasting and unicast routing are also useful in VSNs. An RSU may broadcast messages that dedicated to only one vehicle somewhere on the network or to more vehicles located in a specific geographic area. Geobroadcasting is also applicable when a car, for example, gives a sudden-brake warning to other vehicles at few hundred meters behind to avoid collision. A car sending a request for traffic in a specific way is also a kind of geobroadcasting.

VSNs also get benefits from aggregation functionality. For example, an RSU may aggregate information of messages sent by VSN nodes periodically measuring road surface. At the end, comprehensive information about road surface in a city is gathered that help that Department of Transportation conduct road maintenance in a smarter way. Information of road density from time to time can also be aggregated so that a complete picture of traffic density gives the idea of road support capabilities for future development.

With regard to mobile network's mobility, macroscopic and microscopic approaches can be used to design a mobility model for real conditions of vehicular networks [6,7]. The macroscopic approach sees vehicle traffic as if it is a flowing fluid with a given density and direction. On the other hand, the microscopic approach deals mostly with individual characteristics of vehicles related to its spatial and temporal dependency [4] such as road speed constrains, attraction and repulsion points, and vehicle's acceleration/deceleration [7]. Thus, the ability to realistically model the vehicle's movement is an important factor for routing protocol design during deployment.

Mobility models intended to generate realistic vehicular motion patterns should include accurate and realistic topological maps, obstacles, attraction/repulsion points, vehicles characteristics, trip motion, path motion, smooth deceleration and acceleration, human driving patterns, intersection management, time patterns, and external influences. However, most studies tend to take more simplistic assumptions and neglect several related mobility subjects. Harri et al. [7] noted on the development of the mobility model that can be classified into four classes: synthetic model (mathematical model), survey-based model (extracting from surveys), trace-based model (pattern from real mobility traces), and traffic simulator–based model (mobility traces from a detailed traffic simulator). As real mobility traces are prohibitively expensive, approaches on researches will be based on the other three classes.

18.4 BROADCAST CONTROL MECHANISMS

Message delivery in mobile networks can be in the form of unicast, multicast, or broadcast. Unicasting message is when a mobile node sends messages or data to a destination node, either within its network or its neighboring network. This action is required when a mobile node wants to communicate with another node, such as in intermittent connectivity (partitioned network) scenario that uses a store-and-forward mechanism. Multicasting (geocasting included) is when messages from a node are delivered to a limited number of mobile nodes within a network. Message broadcasting delivers messages or control data to all nodes within the network for message spreading or finding explicit routes [4].

Multicasting and broadcasting are usually employed to disseminate messages originating from safety applications such as collision avoidance, lane change, and accident information [8]. Both broadcasting and multicasting must, therefore, be conducted efficiently with minimum delay as bandwidth is limited [4]. Broadcasting is a simple technique to spread messages to all mobile nodes available in mobile networks within tolerable delays. It is a way for a node to deliver messages to its surrounding nodes, within or beyond its radio range. Broadcasting is essential, particularly for safety applications. A vehicle can spread information on various road conditions (e.g., foggy weather, traffic jam, or accidents) so that other vehicles are aware. General information, news updates, or advertisement can be spread with such a broadcast mechanism. Besides, broadcasting is also useful for discovering neighborhood nodes and helps multicasting decisions [9].

A simple technique for broadcasting in mobile networks is blind flooding. Blind flooding depicts sending messages to every surrounding node, and the recipient nodes respond by rebroadcasting the messages. However, broadcasting messages blindly may lead to excessive redundancy, contention, and collision of messages that trigger a broadcast storm [9,10]. Networks will be congested as massive redundant messages are populated and start taking a lot of bandwidth [4,11]. Ni et al. [10] clearly discussed how broadcast storm problem appears in MANETs.

A broadcast storm occurs when every vehicle (node) attempts to rebroadcast messages received from its neighbors. For instance, when a message reaches n nodes, the message will be rebroadcast n times. Massive message redundancy will build up in the network that leads to network congestion. In turn, contention could not be avoided when many nodes in close proximity use the communication medium to rebroadcast such a message. Finally, if this situation cannot be managed, massive message collision will occur, and the network will soon be disrupted.

There are two possible solutions to eliminate the broadcast storm problem: to reduce the possibility of rebroadcasts or to differentiate the timing of rebroadcasts [10]. However, the time differentiation works only when network density is considerably low. Therefore, most of the previous studies focused on how to reduce the possibility of rebroadcasts but continue to maintain the performance. Several studies proposed to lessen the broadcast storm through various mechanisms: from involving only a simple probabilistic scheme to the use of an artificial intelligence engine. In a simple probabilistic scheme, a random number is generated when a message is received. Nodes with a greater probability number than the threshold number set beforehand have a privilege to rebroadcast the message. A threshold number can be chosen based on the density of the nodes within a network. Higher threshold number means a smaller number of candidates that conduct rebroadcasting.

With such threshold mechanism, broadcast algorithms will not adapt to the density change in a network. The effectiveness of the algorithms can only be achieved for a specific node density. Adaptive mechanisms are then offered for broadcast algorithms to work under varying density situations. However, for providing adaptiveness, each node must have knowledge on its local information, such as the number of neighboring nodes within a network. Every node will either periodically advertise its availability to others or advertise by request using beacons prior to deciding a rebroadcast. With this scheme, the threshold number can be adaptively changed according to the network density.

As the density of vehicles in VSNs greatly varies according to places and times, an efficient control mechanism for forwarding messages must be considered very carefully to foster higher message delivery performance. Partitioned connections must also be addressed by the algorithms adaptively, as messages must be stored and then transported prior to joining a new network. This surely complicates the forwarding methodology. Maximum additional coverage is achieved through a rebroadcast when a rebroadcast node is at the farthest distance of the radio range. Theoretically, with ideal omnidirectional antennas, the extended coverage of rebroadcast is up to 61% [10] (see Figure 18.5). The closer the distance between the sender node and the rebroadcast node, the slower the message will be spread as more hops are required for messages reaching the sink. Therefore, the location of VANET nodes relative to the sender is an important factor for faster message dissemination. The distance of the sender can be excerpted from the received signal strength [12]. Alternatively, if a geographic positioning device is available, the sender may send its current geographic coordinate along with the message. Worth noting, having a different antenna direction pattern such as bidirectional or unidirectional may lead to different values of the maximum additional coverage achieved.

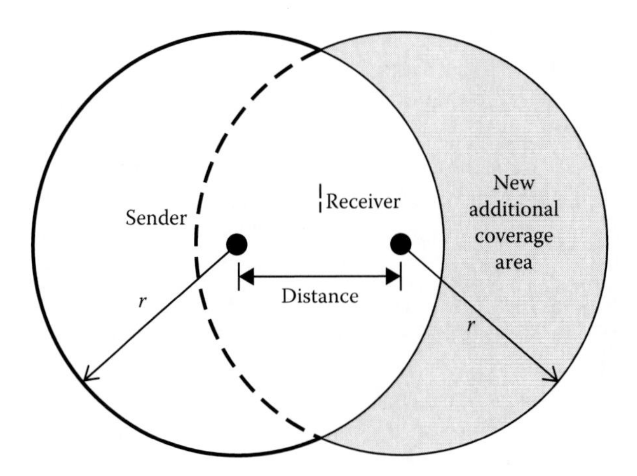

FIGURE 18.5 A maximum of 61% additional coverage for a neighboring node rebroadcast.

In addition, different vehicle density leads to a different number of rebroadcast nodes to select. Lower-density network may need a bigger portion of rebroadcast nodes to provide a higher delivery assurance as far as no congestion and only low contention occurs in the network. However, such a treatment does not work for a high-density network as redundant packets will rapidly fill up the bandwidth. A smaller number of rebroadcast nodes must be selected to reduce potential network congestion. Therefore, redundancy, to some extent, correlates to effectiveness of message broadcasting, but this also has a negative impact on the delay. In other words, higher redundancy is acceptable when there is a reduction in the expected average message delay.

Adaptive control mechanism is of great importance in coping with network density changes on VSNs. Instead of having a fixed threshold number, dynamic threshold is used. For example, one can reduce or expand the radio range to control network density. In fact, to provide adaptively, every node must advertise itself to other nodes periodically. This advertisement is conducted through the sending of control messages (beacons). Hence, some amount of bandwidth will be used for such node advertisements.

Several routing control schemes have been defined to cope with those issues we discussed. There is one category of routing protocols in mobile networks: topology-based routing protocol. This type of routing can be either proactive (table-driven) or reactive (on demand) [13]. Both protocols have their own advantages and drawbacks related to network density and road topology. Proactive routing protocols create more redundant packets compared with reactive protocols. The protocols maintain routing information in each node within a network as they have to send control messages to maintain current network connectivity information regularly.

Reactive routing protocols have its downside on the delay of routing messages as they need to broadcast small packets (control messages) to detect network configuration prior to sending messages. So, in highly dense networks, the proactive protocols can be less efficient compared with reactive protocols as most of the bandwidth is used only for sending such control messages. Therefore, some hybrid approaches are proposed as a trade-off. In addition, the use of GPS and occasionally road map information in routing protocol design create a separate routing class called geographic routing [13]. With such information, the occurrence of node disconnection and reconnection can be predicted more accurately, and hence, it increases the efficiency of VSN. Safety applications tend to disseminate safety messages to nodes within a similar geographical area. In other words, the importance of safety messages to an area far from the location of interest will diminish [8]. Hence, this geographic routing will significantly reduce communication traffic over VSNs.

Basic broadcast techniques in mobile networks follow either a 1-persistence or a *p*-persistence scheme. The 1-persistence scheme has the advantages of low complexity and high penetration rate but creates massive redundancy. While *p*-persistence scheme reduces the packet loss ratio at the expense of increased total delay and reduced penetration rate. However, such schemes do not satisfy the need of efficient broadcast as bandwidth consumption is significantly high. Initial study by Ni et al. [10] proposed four schemes dealing with message broadcasting: counter-based, distance-based, location-based, and cluster-based schemes. Not all of these schemes are adaptive. Instead, threshold value in every node should adapt to the density of the network. Tseng et al. [9] proposed adaptive approaches to address the changing threshold value, namely, adaptive counter-based, adaptive location-based, and neighbor coverage schemes.

Other studies on message broadcasting for specific situations/algorithms include the followings: Multihop Vehicular Broadcast (MVHB) [14]—a flooding protocol with congestion detection algorithm that suppresses unnecessary packets due to congested traffic and backfire algorithm for selecting adequate nodes for forwarding; Binary Partition–Assisted Emergency Broadcast Protocol (BPAB) [15]—a location-based protocol that adopts a repetitive 2-partitions method to divide the area inside the transmission range to find nodes in the farthest segment; Distributed Vehicular Broadcast (DV-CAST) [16]—a distributed broadcast protocol that relies on local topology information; and Distributed Probabilistic Broadcasting (DPB) [17]—based on probabilistic broadcasting, vehicles compute their own relay probability according to their own states.

Li et al. [18] suggested that VANETs and MANETs share a similar principle; that is, both do not rely on a fixed infrastructure for communication, and both have similarities such as self-organized, self-management, low bandwidth, and short radio transmission range. Hence, ad hoc routing protocols are still applicable to be implemented for VSN scenario. Jaap et al. [19] evaluated several MANETs routing protocols applied to a realistic highway mobility model: Ad hoc On-Demand Distance Vector (AODV), Dynamic Source Routing (DSR), Fisheye State Routing (FSR), and Temporally Ordered Routing Algorithm (TORA). Among these protocols, AODV has the best performance for such scenario while TORA is the worst. This proves that even though a realistic mobility model has been applied, not every MANET protocols can perform well in VANETs as well as in VSN scenarios.

18.5 VSNs' INTELLIGENT CONTROL MECHANISMS

Many mobile network routing control protocols are designed based on a specific road scenario. Therefore, the structure of nodes such as trees, clustering, and grids is usually investigated for typical areas. This situation contradicts the reality, as, for example, both dense and sparse traffic in urban areas is available at different places and times. There can be only few vehicles found on roads at midnight, and this situation might be similar to the rural area settings. Vehicular networks must, by themselves, adapt to every network change, from fully disconnected to a highly dense situation [4]. Routing control algorithms must work seamlessly and reliably in all of these scenarios. Therefore, mobile networks have a challenging issue with the scalability.

Besides, the decision for a node being a member of a network is the key to better performance. Local parameters such as location, speed, movement direction, and the number of neighbors can be used to determine network clusters. A network cluster is normally dedicated to vehicles located at the same region. However, vehicles will have short connection time to vehicles in the opposite direction. On the one hand, the inclusion of such oppositely moving vehicles for sitting in the same cluster is indisputable. On the other hand, the vehicles moving in the opposite direction are good candidates to spread messages faster to vehicles in backward direction, to warn drivers that there has been an accident ahead, for example. Therefore, nodes selection for a message rebroadcast is also affected by the characteristic of the message: its type and designation. Urgent messages such as sudden brake are important only for the vehicles right at the back and at a few hundred meters away moving in the same direction. Meanwhile, weather sensor unit may inform all vehicles in a city for giving drivers a warning on bad climate such as foggy weather and heavy raining. The delay requirements from both situations just described are different. Sudden-braking notification must have a low delay in less than 500 ms, while weather notification concerns with its spread rather than the delay itself.

The rebroadcast decision is therefore very complex as many parameters are involved and connections between each parameter to other parameters are so specific, depending on the messages to be delivered. Following is the list of parameters that affect the selection of nodes responsible for a rebroadcast:

- Communication distance
- Vehicle's density
- Vehicle's speed
- Vehicle's movement direction
- Message type
- Message designation

To have maximum additional coverage, rebroadcast nodes selected must be at the farthest possible distance from the sender but still within its radio coverage. However, different speed and movement direction between the sender and the receiver complicate the decision. A sender and a receiver having similar direction but different speed may lead to a critical situation. If the receiver is faster than the sender, the message may not reach the receiver, as the message is sent after a few milliseconds from beacons received. Therefore, a node may miss the potential candidates for a rebroadcast. If the message is about to be rebroadcast forward, in a geobroadcast scenario, leading vehicles having similar direction must be chosen. Faster vehicles are better candidates for conducting rebroadcast to lower the delay.

Message density as already discussed is so dynamic in mobile networks, and this is another burden for an effective nodes selection. An increase in the number of nodes participating in a rebroadcast may increase the chance of the message delivered. However, this must be conducted more carefully as a small increase in the number of participating nodes may lead to network congestion. Different to sparse networks such as in rural area, selecting all the neighboring vehicles to participate in message delivery will not lead to congestion but increase the communication performance.

All the mentioned situations make the relationships between performance target and the parameters mentioned earlier very hard to be represented in a mathematical model, if not impossible. Thus, intelligent control for message broadcasting via machine learning is a good candidate to solve such a situation. With machine learning, complex relationships between input parameters are observed and examined. The decisions will be based on past data experienced by learning algorithm employed during learning process. Such intelligence algorithms are commonly known in the artificial intelligence field as classifiers.

Classifiers are computer algorithms that specifically conduct learning from experience with respect to some class of tasks and performance measures. Classifiers categorize new data into more than one class (usually only two classes). Classifiers are commonly used when there are no exact input measurement or parameters (it has means, and its variances do not equal to zero) that can be used to classify one group among many others. In other words, the data to be solved may have specific probability distribution for every class so that the classification algorithm will classify data based on estimated means and variances [20]. Application of classifiers is almost everywhere. Classification algorithms can be used to recognize patterns such as face, shape of things, voice, and machine vision. In military field, classifiers can be used for radar image analysis, and in the economic field, classifiers help to predict the financial trends. Therefore, classification algorithms have now become widely accepted for problem solving in many fields.

Classification algorithms may reveal the hidden patterns of VSNs so that such algorithms will suggest the most promising nodes for rebroadcasting. One may select proper classifiers to solve complex relationships between input and the desired output from available data. Suitable classifiers will be highly related to the types of candidate models. Hence, classification is merely revealing the relationships between corresponding inputs and outputs based on the hidden information of the models such as the probability density of each group/category. If the exact distribution of data is known, decision for new data can directly be made based on the relevant inputs to the closest data distribution of a category. Figure 18.6 shows a classifier classifies new data based on the available classifier's knowledge/previous information. If possible, feature extraction can be employed to reduce the number of input parameters prior to processing the data. It is expected that with feature extraction, the classification performs better in class labeling.

Various categories of classifiers have been developed. Such classifiers classify new data through experience. These can be categorized as supervised, unsupervised, and reinforcement learning classifiers [20]. The use of supervised learning requires a sufficient number of labeled data, and it is known that having a large number of labeled data is considered as prohibitively expensive. Therefore, when no labeling is possible for the data, one must consider the other two types of classifiers since insufficient training data set can produce low-accuracy results. Unsupervised learning classifiers, do not need such a labeling, and the decision will be based upon hidden pattern/structure within the data. Reinforcement learning classifiers bridge the gap between supervised and unsupervised learning. They focus on goal-directed learning, which means every action of a classifier will be measured on a reward–punishment basis.

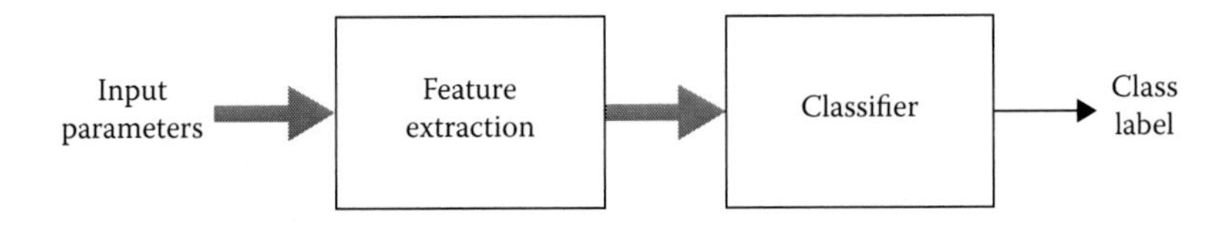

FIGURE 18.6 A classifier outputs a binary decision in response to the new input values.

The following are the popular types of the classification algorithms, but not limited to

- Bayesian decision theory
- Maximum-likelihood and Bayesian estimation
- Nearest-neighbor (NN) estimation
- Support vector machine
- Artificial neural network (ANN)

Bayesian decision theory is a fundamental statistical approach and can be used when the probability structures underlying the categories are exactly known [21]. Bayesian decision theory treats inputs using probabilistic terms such as probability distribution function (pdf) for every known input parameter. If the pdf is known, Bayesian decision theory is excellent for solving such classification problems. However, this situation is rarely found in reality. In most cases, the distribution of each category is not known. Maximum-likelihood and Bayesian estimation can be used when the class probabilities and the class probability density are unknown but models are clearly defined. Sufficient samples are taken to represent the probability distribution of every parameter. Bayes decision theory, maximum-likelihood, and Bayesian estimation are fitted only for problems dealing with pattern classification decision that can be directly or indirectly excerpted from the available parameter data distributions.

Unfortunately, VSNs do not have such definite pdfs for the sensor nodes and do not have data to be sampled for getting the predicted pdfs. Vehicles do not have a constant speed at all times. Likewise, their movement, of course, is not necessarily toward one specific direction, and the number of neighboring vehicles is always changing over time. Therefore, no specific distribution can be inferred from these parameters to judge which vehicles are eligible to conduct the broadcast. Above all, there are no specific probabilities from the parameters to differentiate among the nodes as every parameter has no fixed value. The only available information that can be excerpted for getting the pdfs is the vehicles' average speed, average number of neighbors, and average distance from the sender, as well as their standard deviations. Therefore, classification of nodes can only be based on this past information and a threshold value for those parameters in order to decide the node eligibility for rebroadcasting messages.

The downside of using such a method is that this cannot cope with network density changes. A vehicle commuting in a rural area is mostly a good candidate for conducting rebroadcast even though it has a small number of neighbors recorded in the past time. However, a vehicle commuting mostly in the urban area, even though the average number of neighbors recorded is much higher than that of the previously mentioned one, is not necessarily a better candidate for rebroadcasting messages. This is because that message redundancies in rural areas are highly likely, requiring a higher assurance of message delivery, but for urban areas, avoiding network congestion is the main issue.

NN estimation algorithm categorizes new data based on the closest training examples. This algorithm is considered as a lazy learning or instance-based learning algorithm as the classification is conducted by only measuring the distance between

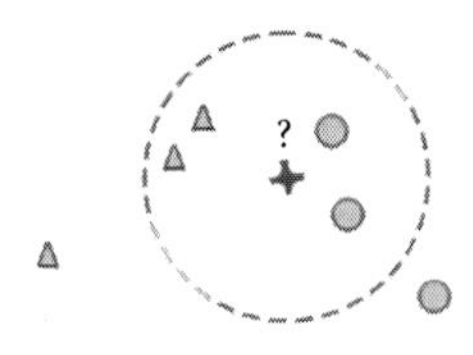

FIGURE 18.7 *K*-NN algorithm to classify new data plotted in a space domain.

the closest neighbors. If the closest neighborhood is categorized into class A, then the data will also be categorized into it. In an attempt to provide more accuracy to the selection, *K*-NN algorithm can be used instead. *K*-NN algorithm will conduct selection based on the majority of class members from the nearest *K* neighbors. Setting *K* equals 1 means that this is the same as NN algorithm. There should be at least *K* training samples that are already available for conducting new classifications. Figure 18.7 shows how new data will be classified in a given space domain. The new data will be plotted, and the node under question will count the closest neighbors until getting *K* data in the similar class.

VSNs may use this approach. Suppose that the radio coverage is of radius 300 m and we want to classify vehicles based on the distance between the sender and the receiver. At first time, we set two settings prior to selections: for example, 20 m to be not eligible for broadcasting and 250 to be eligible for broadcasting are considered as valid training samples. From the receiver's side, all the neighboring nodes recognized from the recent beacons are plotted for their distance from the receiver and classified based on these two values for the eligibility of rebroadcasting messages. At last, the receiving node's distance is plotted and is examined for its NN's class. This algorithm can be expanded to incorporate more than one parameters: distance of receiver from sender, vehicle's speed and direction, and the number of neighboring nodes, creating *N*-dimensions of classification plane.

Using current information of vehicle's movement direction is helpful in geocasting messages. The information on direction will be extracted as x and y vectors relative to the geographic location of both the node itself and the destination location. As the direction vectors x and y are not limited by their values, we need to create an assumption prior to setting two initial values for having the two classes. We can set the direction vectors to have values of maximum 1 and of minimum -1 for an inline direction toward the destination and an inline direction backward. The initial values to set can be 1 and -1.

Support vector machine (SVM) is a nonprobabilistic classification algorithm that groups the data based on spatial characteristic. This algorithm falls into supervised learning algorithms that analyze data and recognize the pattern from the training conducted prior to exercising with new data. SVM is trained using labeled data/examples that have already been classified correctly so that the SVM can infer the behavior of the system given the associated inputs. The knowledge gathered from the training is then used to classify the new inputs (data). The algorithm separates the data into two categories using a hyperplane—an n-dimensional plane as a generalization of a plane. If there are one-dimensional data, the hyperplane is in the form of a point. If there are two-dimensional data, the hyperplane will be a line and so on.

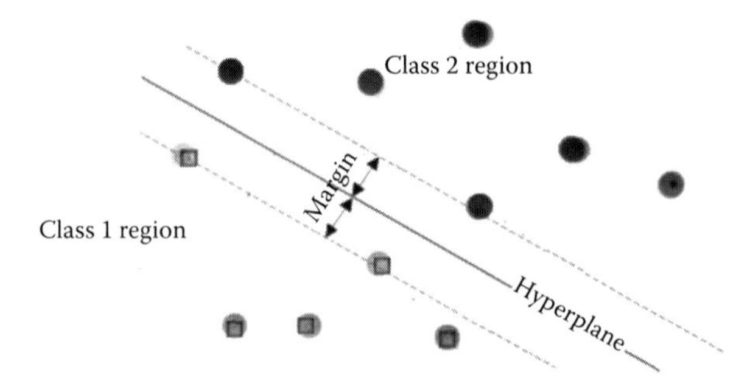

FIGURE 18.8 A hyperplane separating two classes in a maximum equal distance of *d*1 and *d*2.

If the data are linearly separable, the hyperplane can be straightforward, located in between the two classes of data (see Figure 18.8). However, for data that are not linearly separable, a kernel-trick is needed. A kernel-trick is a way of mapping data in a higher dimensionality feature space, in such a way that a hyperplane can separate the data with minimal error. The well-known nonlinear kernels in SVM are radial basis function, polynomial, and sigmoid function.

SVM can be implemented for solving the selection problem. In a very simplistic scenario, if the sender–receiver distance will be the only parameter for classifying nodes for a rebroadcast, the hyperplane will be a point located in the middle of the outer data of the two classes. Suppose that there are several available data that have already been labeled for their class. If the outer node in *no-rebroadcast* class having the sender–receiver distance equals 80 m and the outer node in *allow rebroadcast* class equals 140 m, then the hyperplane will be at 110 m. Note that there will be two outer nodes in every class, but what are mentioned here are the closest nodes to the other nodes in the class. Therefore, when new data arrive and the distance from the sender is 120 m, the node will be considered eligible to rebroadcast the message. Extension to the parameters employed in the SVM will complicate the decision. Separation may no longer be linearly established, so a nonlinear kernel selection can be the solution to recognize the VSN pattern.

ANN model tries to mimic the biological neural networks. Simple processors called perceptrons are connected to each other to have a complex and more powerful parallel processing intelligence. The ANN is often used to model complex relationships between inputs and desired outputs with the use of the perceptrons. The connection structure of ANN is very general, but it can be applied to any different problems. The intelligence results from the training of the ANN. This general-purpose intelligence makes the ANN superior over a conventional von Neumann's centralized computer architecture.

The ANNs have two types of connections between the perceptrons that are employed: feedforward and recurrent/feedback. According to its name, feedforward networks mean that the perceptrons are unidirectionally connected to each other, starting from the inputs and ending at the outputs. Figure 18.9 shows a feedforward network and a recurrent network. With feedforward connections, the corrections of

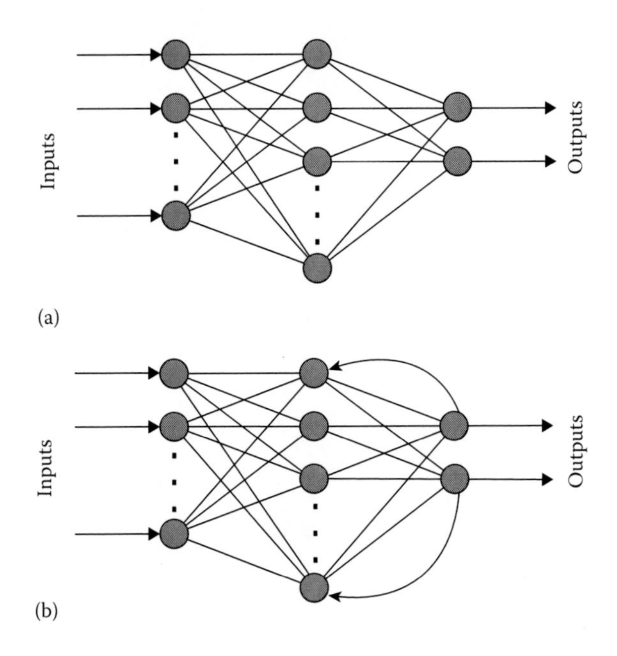

FIGURE 18.9 Two models of ANN: (a) feedforward network and (b) recurrent network.

weights during the training must be manually conducted based on the difference between the result and the desired target. When processing a single datum from a training data set, if the result is higher than the desired target, the weights are adjusted to a lower value. Conversely, if a lower result is achieved, the weights will be increased. The repetitive adjustment of weights for every record of training data will finally make the ANN intelligent. In feedback networks, the weights can be adjusted automatically as the networks have information going back to the input direction.

A hidden perceptron receives input signals, which are multiplied by an adjustable weight for every input. The multiplication of weights and inputs will be summed and inputted into an activation function for getting the decision. Figure 18.10 shows

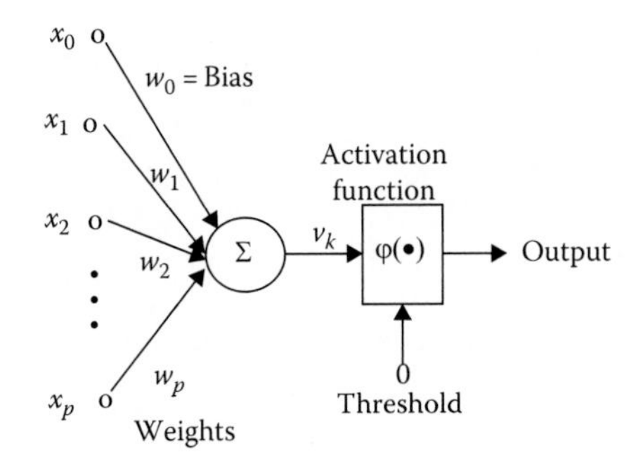

FIGURE 18.10 A mathematical representation of a perceptron with an activation function.

the mathematical approach for a perceptron. The weights are the only variables for memorizing the connections between input and output during the training. Initially, the weights are set to small number values (the suggested value is 0.5 for every weight), and during training, these weights are repetitively adjusted so that the output is getting closer to the target. At the end of the training phase, the weights are expected to make the output value very close to the training target value. An activation function works as a squashing function so that the output can only be in two conditions: either 0 or 1 or –1 and 1, for example.

The output of the summing function is of the form

$$x_0w_0 + x_1w_1 + x_2w_2 + \cdots + x_pw_p$$

While the activation function is a step function, the output will be

$$0 \quad \text{for } v_k < \text{threshold}$$

$$1 \quad \text{for } v_k \geq \text{threshold}$$

which represent the two classes.

VSNs can also implement the ANN to conduct the selections. All the parameters in the VSNs can be the inputs for this ANN, and a supervised training is conducted to program the weights. It is worth to note that the ANN can have more than one output. So we can get VSNs' performance parameters such as throughput, message duplicates, and delay as the desired targets. For such outputs, labeling the training data will consider that *to-broadcast* class will have a good message throughput, reasonable message redundancy, and an acceptable message delay. A detailed approach will be discussed in the section that follows.

As already mentioned, the NN algorithm, support vector machine, and ANN may solve the VSNs' rebroadcast selection problem. However, learning process is important for those algorithms prior to conducting the classification of new data, as probability distribution of each category will be memorized by the classifier. Sufficient data labeling, therefore, is the key to a successful new data classification. Correct selection of the classifier with a good match to its relevant attributes leads to efficient computation and excellent results.

18.6 VSNs CLASSIFICATION EXAMPLE: USING PERCEPTRONS

Node rebroadcast decision in VSN is a very complex task. Expressing the relationships among VSNs' parameters using either mathematical or logical expressions is very challenging. A good candidate solution is the ANN algorithm as it will automatically find the underlying relationships between those VSNs parameters during the training. However, ANN will give the expected output if the training data are sufficiently provided covering all scenarios of the VSNs, and the ANN engine itself is properly designed to solve the classification problem.

With regard to the ANN engine, in literature such as in [22], we find that using a single hidden layer of perceptrons cannot solve an Exclusive-OR problem

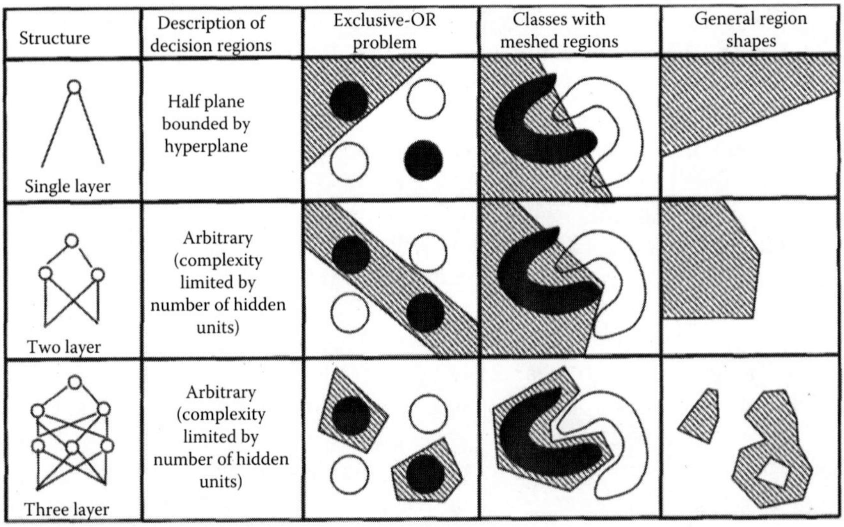

Structure	Description of decision regions	Exclusive-OR problem	Classes with meshed regions	General region shapes
Single layer	Half plane bounded by hyperplane			
Two layer	Arbitrary (complexity limited by number of hidden units)			
Three layer	Arbitrary (complexity limited by number of hidden units)			

FIGURE 18.11 The ability of ANN with different hidden layers to solve the separation problem in a two-dimensional input space represented geometrically. (Picture taken from Jain, A.K. et al., *IEEE Comput.*, 29(3), March 1996.)

(See Figure 18.11). The hyperplane cannot accommodate the data situated in such locations. It is noticed that the hyperplanes in the single-layer network are in the form of straight lines in the class space domain. Only with at least two-hidden-layer neural network (NN), the separation works well. However, the two-hidden-layer network cannot separate more complex problems such as problems that having classes with meshed regions. Here, three hidden layer network is required for solving such a situation. More hidden layers mean more powerful NN to deal with data complexity, but with a greater risk that the classification will cover a local minima—a value that is not the actual minimum value available from the solution space. The number of hidden layers employed in NN will rely much on

- The size of inputs and outputs
- The training case size and its complexity
- Noise in the targets
- The activation function of the hidden unit
- The type of training algorithm used

However, it is suggested in [23] that using single hidden layer is sufficient for most cases. Two-hidden-layer NN is for data having discontinuities such as square or saw-tooth waves.

There are many approaches to determine the number of perceptrons within a hidden layer. However, there are several rules of thumb on this issue that we can follow as suggested in [23]:

- The number of hidden perceptrons in a layer should be between the number of inputs and the number of outputs.
- The number of hidden perceptrons is two-third of the size of inputs added with the size of the outputs.
- The number of hidden perceptrons must be no more than twice the size of inputs.

These rules can be the best bet for the first phase on experimenting the ANN. Fine-tuning will be more likely treated on a trial-and-error basis. Too many perceptrons within the layers may cause overfitting. This means that the learning data do not sufficiently train the whole perceptrons for the problem. There should be sufficient training data when employing more perceptrons.

Following what are suggested in [23], the design for our problem is a single-layer NN as there are no discontinuity values in the parameters of VSNs. We define the whole parameters as the inputs, but with the vehicle's direction parameter will be split into two: direction toward X and directions toward Y; and the message regional designation will be represented as the regional center point X and regional center point Y as well as regional radius R. Message type will be in the form of priority values, say, 1 is for less urgent messages, and 2 is for urgent messages such as safety-related messages. Vehicle's relative speed to the sender's speed—that is, ratio between the receiver's speed and the sender's speed—is also introduced.

The extension of original parameters in VSNs is for explicitly representing each variable's role in the NN classification decision. For example, message regional designation can be for every node in the whole region, or can be only for a few hundred square meters in the north side. The location and its area will affect the desired performance. Likewise, the speed ratio between the sender and the receiver is an important parameter for the node to decide whether to pass the messages to either faster or slower neighboring vehicles.

So, after some modifications, the input of the VSN will be as follows:

- Communication distance
- Vehicle's density
- Vehicle's speed
- Vehicle's relative speed
- Vehicle's movement direction toward X
- Vehicle's movement direction toward Y
- Message's type
- Message's regional center point X
- Message's regional center point Y
- Message's regional radius R

For the output, we have the communication performance such as message throughput, delay, jitter, and the number of message duplicates. Having the most important performance is a reasonable assumption as accommodating every measure as an output requires using a very strict filter during classification. Message throughput and delay or message throughput and the number of message duplicates are a reasonable pair to select. Having 10 parameters as inputs and only 2 outputs and following [23], the number of perceptrons employed in the hidden layer will be $2/3 \times 10 + 2 = 9$.

Prior to using the NN, a data set is obtained with sufficient simulations, allowing several situations in the VSN to be represented, with different values to different parameters. We proceed by setting the scenario and defining the mobility model. Using wireless network simulator for reading the output of the mobility generator, we manually set the threshold for every parameter and measure the results: message throughput and the average delay, for example. Repetitive experiments with different threshold values for parameters and the associated outputs will build our required data. Such data can then be separated into two: training data and learning data. The generation of the training set will be considered good if the following are satisfied:

- Samples represent the general population/nodes.
- Each class contains sufficient data set.
- Each class contains a wide range of variations/scenarios.

The number of training set will be much related to the number of inputs, perceptrons to be employed, and the number of outputs. For the NN with 10 inputs, 9 perceptrons, and 2 outputs, there will be at least $(10+1) \times 9 + (9+1) \times 2 = 119$ training sets required. The addition of 1 in the equation is to accommodate biases employed in the system. This is actually analogous with having 119 unknown variables that need to be fulfilled by at least 119 equations. It is also commonly suggested that using training cases of 30 times higher than the number of weights may reduce the chance of overfitting. Overfitting will occur if the NN has no sufficient training data set. Therefore, using 300 test data is sufficient. The rest of the data set can be used as test data that can be 50% of the training data.

With the data set at hand, we then label every output to be in *to-broadcast* class or in the other class (the *no-broadcast* class). Categorization can be done, with the help of a spreadsheet processor, by sorting the data based on message throughput and the delay. Data set having good throughput with small delay will be categorized into the *to-broadcast* class. The rest will be classified into the *no-broadcast* class. Such data are then randomly separated into two: training data set and test data set. The training data set is used for training the NN. It is highly suggested to normalize all input parameters for both training and test data sets. Normalization will provide a common data range from 0 to 1. The training data set is fed one after the other to the classifier so that with the help of weight optimizer, such as back-propagation algorithm, the weights are adjusted iteratively. After the learning process is finished, the test data set will be used to justify the success of the training phase. Every test datum is inputted to the classifier, and the results will be observed. If the two outputs suggest a high throughput and an acceptable delay, the data will be categorized into *to-broadcast* class. Mean squared error (MSE) is one type of metrics available to grade the performance of the classifier. Once an expected MSE value is obtained, the NN is ready for a new classification task.

Flooding creates excessive message duplicates over network, and this drawback can be minimized with the use of such intelligent control algorithm. Figure 18.12 depicts the redundant message broadcasting suppression that can be achieved.

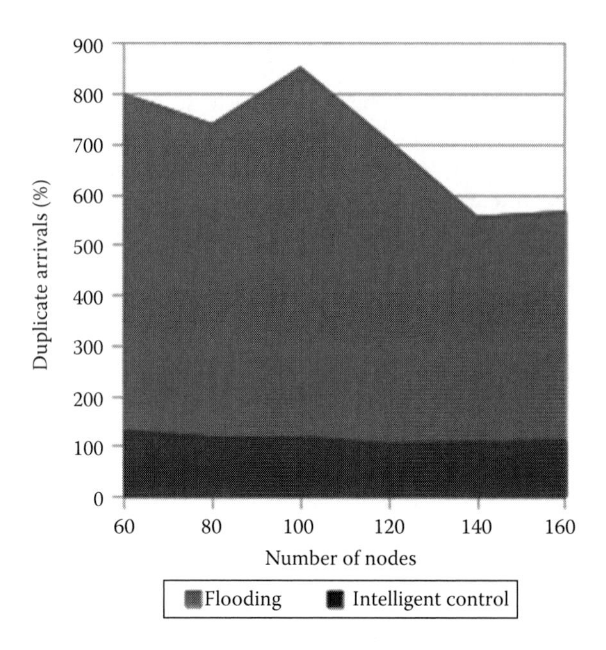

FIGURE 18.12 Number of duplicates arrived for different vehicle densities in a unicasting scenario.

The duplicate suppression measure is evaluated against different vehicle densities. As shown, there is about seven times reduction in average on message rebroadcasting redundancy when compared with flooding technique.

Besides the advantage of suppressing message redundancy, the intelligent control works very well on reducing the message transmission delay over flooding technique, particularly in a dense network scenario. There is a significant reduction of the delay when intelligent control mechanism is used for dense network (see Figure 18.13).

18.7 CONCLUSIONS

Broadcasting in ad hoc networks such as in VSNs must be treated carefully as flooding technique creates massive message redundancies. Therefore, decision on VSN nodes to rebroadcast messages must be related to the available input parameters and to accommodate every possible VSN scenario. The solution must not only be adaptive to network changes but also needs to maintain a predefined quality of service in terms of message throughput, delay, and message redundancy. The use of intelligent control for message broadcasting in VSNs is a very promising approach. It allows solving VSN problems that either do not have pdfs for the classes or the relationships between inputs and outputs are unknown and cannot be mathematically analyzed. The intelligent algorithms will gather the hidden relationships during the learning process and will easily adapt to changes of the network status over time.

Several artificial intelligence approaches can be used for VSNs' rebroadcast selection problem: *K*-NN, support vector machine, and ANN. The *K*-NN simply classifies

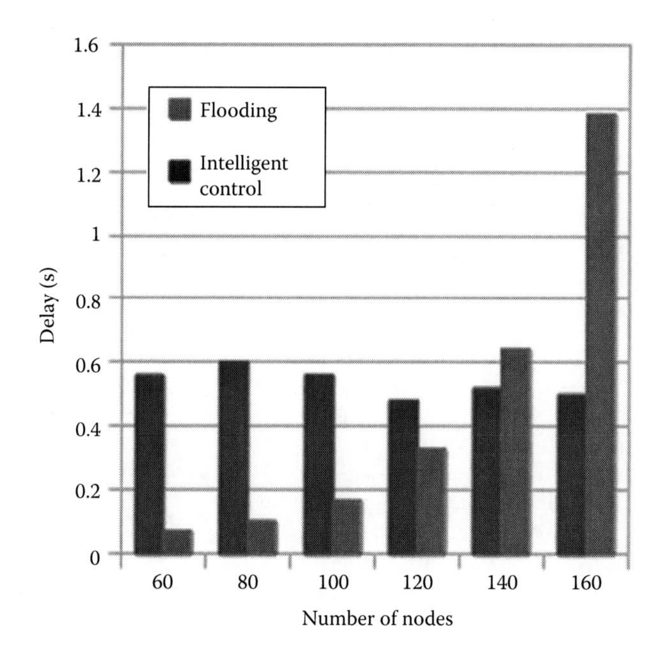

FIGURE 18.13 The delay over different vehicle densities in a unicasting scenario.

the data based on Euclidean distance of a small number of previously well-classified space domain data, while both SVM and ANN need a much larger set of training data. The dimension of the data can be 30 times higher of the number of weights employed in the algorithms. The performance is, therefore, highly related to the quality of the training data. A complete representation of all possible cases within the problem defined will be helpful for such artificial intelligence algorithm to record hidden distributions of classes.

ACKNOWLEDGMENT

This project work is sponsored by the Research Center for Sensor Network and Cellular Systems, the University of Tabuk, Kingdom of Saudi Arabia.

REFERENCES

1. J. Yick, B. Mukherjee, and D. Ghosal, Wireless sensor network survey, *Computer Networks*, 52, 2292–2330, 2008.
2. M. Cardei and J. Wu, Energy-efficient coverage problems in wireless ad-hoc sensor networks, *Computer Communications*, 29, 413–420, 2006.
3. K. Akkaya and M. Younis, A survey on routing protocols for wireless sensor networks, *Ad Hoc Networks*, 3, 325–349, 2005.
4. Y. Liu, J. Bi, and J. Yang, Research on vehicular ad hoc networks, in *2009 Chinese Control and Decision Conference*, Guilin, China, pp. 4430–4435, 2009.
5. E. Schoch, F. Kargl, M. Schoch, F. Weber, and T. Leinmuller, Communication patterns in VANETs, *IEEE Communications Magazine*, 46(11), 119–125, 2008.

6. N. Bellomo, M. Delitala, and V. Coscia, On the mathematical theory of vehicular traffic flow I. Fluid dynamic and kinetic modelling, *Mathematical Models and Methods in Applied Sciences*, 12(12), 1801–1843, 2002.
7. J. Harri, F. Filali, and C. Bonnet, Mobility models for vehicular ad hoc networks: A survey and taxonomy, *IEEE Communications Surveys & Tutorials*, 11(4), 19–41, 2009.
8. Y. Toor, P. Muhlethaler, A. Laouiti, and A. de la Fortelle, Vehicle ad hoc networks: Applications and related technical issues, *IEEE Communications Magazine*, 10(3), 74–88, 2008.
9. Y.-C. Tseng, S.-Y. Ni, and E.-Y. Shih, Adaptive approaches to relieving broadcast storms in a wireless multihop mobile ad hoc network, *IEEE Transactions on Computers*, 52(5), 545–557, 2003.
10. S.-Y. Ni, Y.-C. Tseng, Y.-S. Chen, and J.-P. Sheu, The broadcast storm problem in a mobile ad hoc network, in *Proceedings of the Fifth Annual ACM/IEEE International Conference on Mobile Computing and Networking—MobiCom'99*, Seattle, WA, pp. 151–162, 1999.
11. T. Kosch, C.J. Adler, S. Eichler, C. Schroth, and M. Strassberger, The scalability problem of vehicular ad hoc networks and how to solve it, *IEEE Wireless Communications*, 13(5), 22–28, October 2006.
12. C. Hu, Y. Hong, and J. Hou, On mitigating the broadcast storm problem with directional antennas, in *IEEE International Conference on Communications, ICC'03*, Anchorage, AK, vol. 1, pp. 104–110, 2003.
13. K.C. Lee, U. Lee, and M. Gerla, Survey of routing protocols in vehicular ad hoc networks, in *Advances in Vehicular Ad-Hoc Networks: Developments and Challenges, IGI Global*, IGI-Global, Hershey, PA, pp. 149–170, 2010.
14. T. Osafune, L. Lin, and M. Lenardi, Multi-hop vehicular broadcast (MVHB), in *Sixth International Conference on ITS Telecommunications Proceedings*, Chengdu, China, No. iv, pp. 757–760, 2006.
15. J. Sahoo, E.H.K. Wu, P.K. Sahu, and M. Gerla, BPAB: Binary partition assisted emergency broadcast protocol for vehicular ad hoc networks, in *2009 Proceedings of 18th International Conference on Computer Communications and Networks*, San Francisco, CA, pp. 1–6, August 2009.
16. O. Tonguz, N. Wisitpongphan, and F. Bai, DV-CAST: A distributed vehicular broadcast protocol for vehicular ad hoc networks, *IEEE Wireless Communications*, 17(2), 47–57, 2010.
17. Q. Yang, L. Shen, and W. Xia, Distributed probabilistic broadcasting for safety applications in vehicular ad hoc networks, in *2009 International Conference on Wireless Communications & Signal Processing*, Nanjing, China, pp. 1–5, November 2009.
18. F. Li, Y. Wang, and N. Carolina, Routing in vehicular ad hoc networks: A survey, *IEEE Vehicular Technology Magazine*, 2(2), 12–22, 2007.
19. S. Jaap, M. Bechler, and L. Wolf, Evaluation of routing protocols for vehicular ad hoc networks in typical road traffic scenarios, in *Proceedings of the 11th EUNICE Open European Summer School on Networked Applications*, Madrid, Spain, pp. 584–602, 2005.
20. R.O. Duda, P.E. Hart, and D.G. Stork, *Pattern Classification*, 2nd edn., Wiley-Interscience, New York, NY, 2000.
21. N. Friedman, D.A.N. Geigrr, and M. Goldszmidt, Bayesian network classifiers, *Machine Learning*, Kluwer Academic Publishers, Dordrecht, the Netherlands, Vol. 29, pp. 131–163, 1997.
22. A.K. Jain, J. Mao, and K.M. Mohiuddin, Artificial neural networks: A tutorial, *IEEE Computers*, 29(3), March 1996.
23. G. Panchal, A. Ganatra, Y.P. Kosta, and D. Panchal, Behaviour analysis of multilayer perceptrons with multiple hidden neurons and hidden layers, *International Journal of Computer Theory and Engineering*, 3(2), 332–337, April 2011.

19 Sensor Networks for Underwater Ecosystem Monitoring and Port Surveillance Systems

Ali Mansour, Isabelle Leblond, Denis Hamad, and Luis Felipe Artigas

CONTENTS

ABSTRACT

Providing a wide variety of the most up-to-date innovations in sensor technology and sensor network, our current project should achieve two major goals. The first goal covers some issues related to public safety and security, such as the coastal and port surveillance systems, while the second one will improve the capacity of public authorities to develop and implement smart environment policies by monitoring the shallow coastal water ecosystems. At this stage of our project, a surveillance platform has been already installed near the *Molène Island*, which is a small but the largest island of an archipelago of many islands located off the west coast of Brittany in France. Our final objective is to add various sensors as well as to design, develop, and implement new algorithms to extend the capacity of the existing platform and reach the goals of our project. This chapter describes the whole project by focusing on the variety of used sensors, and it will briefly introduce the most important required theoretical approaches such as blind signal processing, high-order statistics (HOS), classification algorithms, and data fusion methods, which will be applied to build up an original and reliable system able to perform a sustainable and long-term monitoring of coastal marine ecosystems and to enhance port surveillance capability. In addition, it discusses developed techniques and concepts to deal with several problems related to our project. The new system will address the shortcomings of traditional approaches based on measuring environmental parameters, which are expensive and fail to provide adequate large-scale monitoring. More efficient monitoring will also enable improved analysis of climate change and provide knowledge informing the civil authority's economic relationship with its coastal marine ecosystems. Some results are given and discussed.

19.1 INTRODUCTION

According to the National Oceanic and Atmospheric Administration (NOAA) of the US Department of Commerce, oceans cover 71% of the Earth's surface and more than 95% of underwater world remains unexplored. Evidently, oceans play an enormous role in sustaining life on Earth, in supporting international traffic, and in developing local, national, and international economies. Recently, oceanography, marine biosystems, and ecosystems have been the priority for many researchers all around the world. In addition, oceans' ecosystem has a major impact on the Earth ecosystem. Our project has two major targets to reach: underwater ecosystem monitoring and coast/port surveillance. According to the World Factbook published by the Central Intelligence Agency of United States, France has 4853 km of total coastlines. In addition, Metropolitan France is bordering the Bay of Biscay, the Celtic Sea, and English Channel, between Belgium and Spain, and bordering the Mediterranean Sea, between Italy and Spain. With more than 3400 km of coastlines in Metropolitan France, our study is obviously of great national interest and is, to some extent, representative of different marine ecosystems that can be encountered in Europe and in temperate countries all over the world. It is worth mentioning that only 23% of 194 independent countries recognized by the US state department don't have any sea coastline. Therefore, our developed systems, algorithms, and approaches can be of great benefit to most of the countries. In fact, the coastal marine ecosystem plays an important role in the future economy and sociology of any country. Monitoring coastal marine ecosystem leads us to analyze climate changes, to study ecological parameters, and to estimate economical factors. To study coastal marine ecosystem, various environmental parameters should actually be measured. This approach suffers from high cost and fails to provide adequate large-scale and high-resolution monitoring, needed to fully understand the dynamics of these highly dynamic ecosystems. In this project, we are proposing an original approach based on the use of a wide variety of sensors and sensor network. In addition, we will also analyze and classify marine microorganisms (MOs) (such as phytoplankton, zooplankton, and dinoflagellates symbionts of coral), which play a major role in coastal marine ecosystem. This project is a pioneer multidisciplinary project that its success requires a deep knowledge of various research axes such as information theory, advanced signal and image processing algorithms (blind identification and separation, wavelets, time–frequency–phase representations, shape extraction algorithms, adaptive filters, etc.), unsupervised classification, and pattern recognition approaches. In addition, data fusion approaches should be developed to achieve the final stage, which is the prediction and the decision making.

Topics and research studies related to image processing part of our project are outside the scope of this chapter. Hence, algorithm details and experimental results will be omitted. We will pay more attention to the active or passive acoustic sensors, as well as other specific sensors and sensor network. However, a model for an underwater acoustic transmission channel is briefly discussed; problems related to active and passive acoustics are also considered. Classification and features extraction techniques are introduced. Topics related to data fusion are also proposed. Finally, conclusions and future works are given.

19.2 PROJECT DESCRIPTION

In the last two decades, we have been involved in several research as well as industrial projects related to blind signal processing techniques (such as blind source separation, blind channel identification, inverse modeling, blind classification, and recognition algorithms), wireless communication, underwater acoustics, robotics, biomedical engineering, and electronic warfare. These applications motivated us to develop new algorithms, approaches, or systems, which have been published in several conference and journal papers along with many research reports. In more recent projects, we developed and implemented blind separation of sources (BSS) and classification algorithms along with feature extraction algorithms to deal with underwater passive acoustic applications. Underwater world fascinates all humankind by its life diversity and its natural resources. It was mentioned before that above 95% of this world is still unexplored. Besides, oceans play an enormous part in the regulation of climate, in the global as well as the local economical system (transportation, food reserve, etc.). In addition, oceans' ecosystem has a major impact on the Earth ecosystem and worldwide economy and sociology. Our project is related to France's coastal marine ecosystems. Therefore, it is necessary to undertake thorough monitoring of those ecosystems to ensure they are healthy and their economic use is sustainable.

The actual project can be mainly divided into three main distinctive phases:

1. In the first phase, an adequate prototype underwater sensor network platform should be validated and completely operational. Thus, the first phase focuses on the analysis and classification of variety of heterogeneous data collected using different sensors including but not limited to the following:
   a. Acoustic array, hydrophones: They will be used as passive acoustic sensors to monitor ambient noise, mammal sounds, anthropic activities, etc. In the following sections, the importance of this kind of sensors is highlighted. It worth mentioning that such sensors are crucial to perform port surveillance or electronic and antielectronic warfare. In fact, these sensors coupling with powerful signal processing and classification algorithms can quietly detect, distinguish, and identify acoustic signals generated by waves, weather conditions, dolphin whistles, whale sounds, shells' noises, vessels and submarine noises, etc.
   b. Fisheries echo sounder: they will be deployed to accomplish schools of Fish characterization, water column monitoring, algae analyzing, bubbles quantifying or zooplankton quantification. In order to improve the capacity of these sensors, one should perform echo integration of backscattering signal and apply inverse modeling theory. A general framework is currently applied in acoustic fisheries in order to obtain an evaluation of fish stocks: estimation of volume backscattering strength on large distance or, by shoal using echo-integration methodologies, tracking isolated fishes using signal processing applied on target strength (TS) detected by split-beam sounders, inverse modeling, etc.
   c. Acoustic Doppler current profiler (ADCP): It helps us to achieve current monitoring and water column backscatter monitoring. This can be done by performing the current estimation and an echo integration of backscattering signal.
   d. Side-scan sonar: Side-scan sonar systems (standard or interferometric side scan sonar) are widely deployed by different institutes such as government agencies, navies, port authorities, and research centers.
   e. Multibeam sounders, frequently used to obtain acoustic characterization of the seafloor, are now able to provide acoustic backscattering of the entire water column. So, they should be used to obtain data of the seafloor or the water column in a larger scale.
   f. Subbottom profilers, using low-frequency acoustic waves, are able to penetrate sediments on few meters. So, they can provide an internal characterization of the structure of the seafloor.
   g. Cameras: They will be very helpful in monitoring near structure and making visual identification. That could achieve event detection and image segmentation.
   h. A wide range of various sensitive sensors should be also added in order to measure or monitor the following parameters: benthic ecosystem, seafloor characterization, temperature, pressure detection, salinity, turbidity,

dissolved oxygen concentration, concentration of nitrate, concentration of phosphate, concentration of silicon, etc.

2. The second phase of the project consists of analyzing living underwater MOs in order to predict the health conditions of the macroenvironment; further details are given hereinafter. In this phase, we are planning to develop a novel method of monitoring coastal marine ecosystems, based on the analysis and classification of marine MOs. In fact, MOs are affected by many parameters, including water temperature, water pollution, and salinity; such monitoring will assist the modeling of coastal climate changes, the study of the marine food chain, or the analysis of the conditions of coral reefs. This monitoring method will focus on processing cytometry signals and the outputs of other sensors to assess the role of MO in monitoring these ecosystems.
3. During the final phase of the project, we should set up a network of immersed platforms in order to monitor a large part of France's coastlines. The obtained results could be coupled to satellite surveillance, which is using higher and more expensive technologies. Satellite images will be only considered to achieve very large-scale monitoring. Unmanned aircraft systems (UAS) could be also deployed in monitoring reasonable areas. Satellites or UAS images could be purchased and should be preprocessed and analyzed. In fact, a natural extension to our project will be the application of satellite remote sensing, airborne hyperspectral remote sensing, hyperspectral *in situ* sampling, in-water hyperspectral profiling, sea gliders supported by spectral fluorometers or water sample analysis (e.g., cytometry signal analyzer, *in vivo* fluorescence to assess chlorophyll *a* approaching the distribution of phytoplankton total biomass) to study the flux of in-water pigments, colored dissolved organic matter, and sediments in estuarine systems and their movement, transformation, and destination as they move into adjacent coastal waters. Finally, data fusion approaches should be developed to achieve the final stage of our project, which is the prediction and the decision making.

The first part of the project is already activated but not completed by deploying an autonomous underwater platform to collect and analysis real data. The platform, *MeDON observatory*, is installed near the *Molène Island*, which is located at the west coastline of the *Finistère* Department in Brittany, France. Data have been recorded with the help of an underwater cabled observatory containing multitude of various sensors used in the study. Figure 19.1 shows a schematic view of an advanced version of the actual deployed surveillance system. In fact, the station is equipped with 2 hydrophones, an ADCP, a video camera, and a multi-parameter probe. The success of our project requires an automatic process of the huge database, to develop and apply new classification algorithms along with blind image and signal processing, as well as automatic feature extraction and classification tools in such a way that the developed *smart* algorithms should help us to achieve better performances and at the same time reduce the overall deployment and maintenance costs. It is worth mentioning that underwater environment is a very hostile environment that offers us many challenges to overcome. This point will be clarified in the following sections.

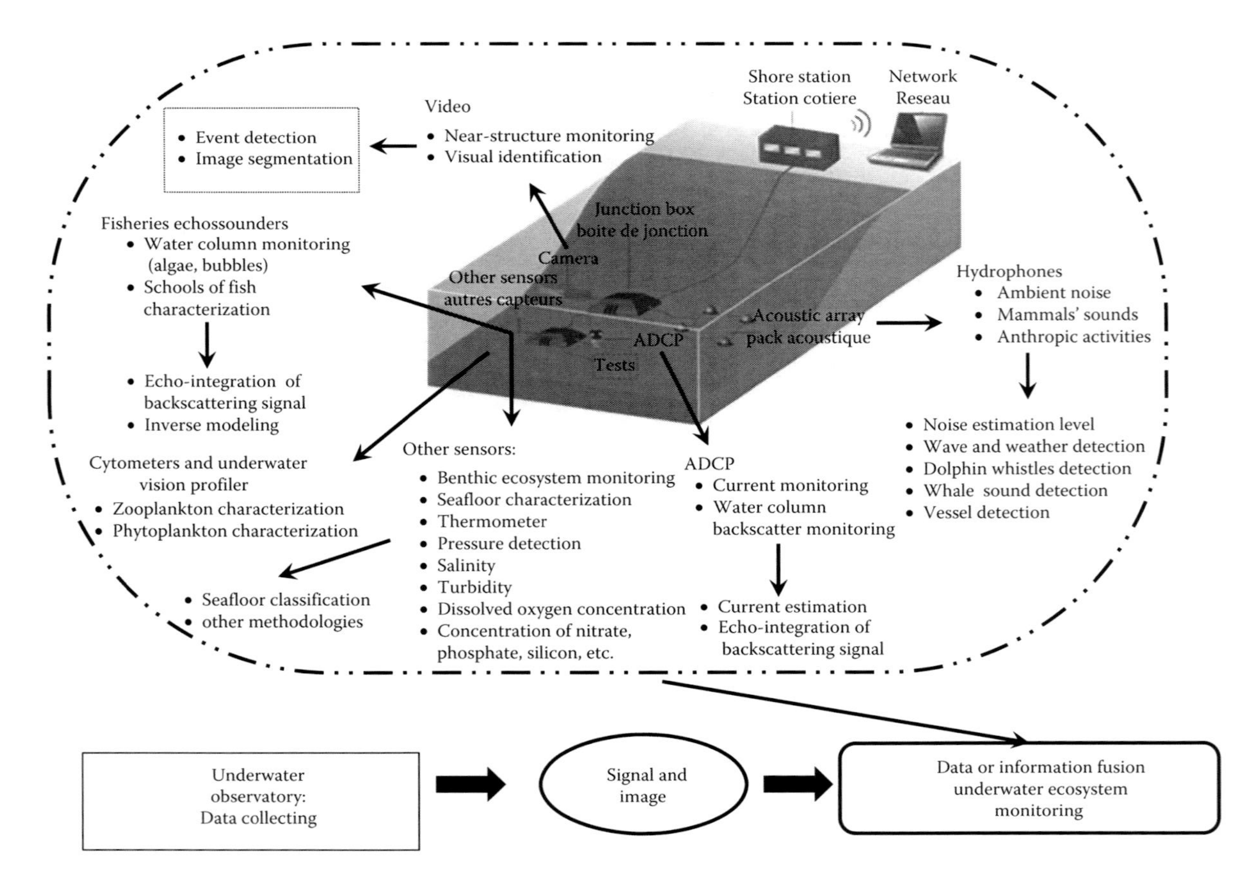

FIGURE 19.1 A schematic of a complete and future version of the underwater *MeDON observatory* platform.

This project is a pioneer multidisciplinary joint project (Electrical and Acoustic Departments at ENSTA Bretagne, LISIC of Université du Littoral Côte d'Opale [ULCO], and le Laboratoire d'Océanologie et Géosciences [LOG—UMR 8187 CNRS-ULCO-UL1] and l'Unité de Recherche *Recherches et Développements Technologiques [RDT]* of IFREMER). In fact, the success of our project requires a deep knowledge of various research axes such as information theory, advanced signal processing and image processing algorithms (blind identification and separation methods, wavelets, time–frequency–phase representations, shape extraction algorithms, adaptive filters, etc.), unsupervised classification and patter recognition approaches, and data fusion, without neglecting the essential part of bioengineering approaches and oceanography and oceanology methods. In addition, this project requires the collection and the process of huge amount of data. Many samples of marine MOs should be automatically collected, processed, analyzed, and classified. The analysis of various images and signals obtained from any ecosystem or biosystem is very challenging due to the complexity of the images and the nonstationarity and sparseness of the signals. The contribution of each research group is essential for the success of the whole project. In the following sections, major families of sensors are described and the essential approaches and methodology are briefly discussed.

It is worth mentioning here that conducting underwater experiments is mostly expensive and involves complex logistics and very large heavy items. In order to validate our approaches, techniques, and algorithms, real underwater experiments should be carried out. However, these experiments should be planned at an advanced stage of our project in order to save time and money. For these reasons, we developed a simulation model of an underwater acoustic transmission channels. Our model takes into consideration the acoustic underwater propagation properties, which are dependent on the depth of the water column, the number of reflection spots, the sound speed, the topology and the nature of the seabed, the surface of the water, the salinity, and underwater current, among other less important parameters. In addition, the sound speed is an increasing function of temperature, salinity, and pressure, which is a function of depth. The reflected acoustic waves on the seabed or on the water surface depend on many parameters (such as the composition and the topology of the bottom, the wind, the wave frequency, as well as the swell properties). This model will be briefly discussed.

19.3 HETEROGENEOUS SENSORS

Obviously, to achieve our goals of marine coast ecosystem monitoring and port surveillance for an entire country like France, one should develop and deploy very complex systems. The cost as well as the development time increases with the complexity of the designed and deployed systems. Therefore, one major task consists of reducing the total price of the project by improving the capacity and the performance of all types of sensors based on the development and the adaptation of powerful signal, image, classification, and fusion algorithms.

19.3.1 Active Acoustic Sensors

By their ease of propagation in the marine environment, the acoustic signals are often used as essential means of investigation in the underwater environment, whether for civil or for military: hydrography, cartography, meteorology, useful information to browsers, development concepts sonar, etc. (Baggeroer et al. 1993). The majority of the systems are active systems: sonars, single- and multibeam echo sounders, etc. Active mono- or multisounder beam, side-scan sonars, or ADCP mainly suffer from the following problems:

1. Problems related to underwater acoustic propagation (acoustic refraction, attenuation of acoustic signals, the reflections on the surface of the water or the seabed, fluctuating signals, Doppler effects, etc.).
2. Problems related to the distribution of acoustic signals by the reflectors present in the water column or on the seabed (dependence on the physical properties of signals, transmission medium, transmission frequency, multiple diffusion, interferences, speckle, etc.).
3. The formation of echoes or the backscattering signal, which depends on the frequency, target shapes and materials, incidence angles, etc.
4. Interpretation of recorded signals depends on each sensor; for example, a side-scan sonar uses grazing angles to insonify the seafloor when single beam sounder uses normal incidence and multibeam sounder several incidences. Many parameters could affect the interpretation, such as the frequency, the resolution, and the presence or not of secondary lobes. Figure 19.2 shows two images of the same water columns, insonified with two different sensors by two echo sounders at different frequencies (Simrad EA400 at 38 and 200 kHz): fishes, bubbles in surface, and scattering layers are more visible on the 38 kHz sounder (Figure 19.2a), but isolated targets are more detectable on the 200 kHz sounder (Figure 19.2b).

19.3.2 Passive Acoustic Sensors

The ocean observation using different acoustic passive sensors, that is, by using acoustic sources of opportunity, is relatively rare (Chapman and Lindsay 1996, Gaucher and Gervaise 2003, Gaucher et al. 2004, Gervaise et al. 2001, Gervaise et al. 2007, Hermand 1999, Munk 1995). However, this type of technology becomes more and more the center of attention of the scientific community for three main reasons:

1. Discrete system: A vital quality for any military application.
2. Passive systems don't require any transmission system, which is an essential part of active systems. Transmission systems are often sophisticated, bulky, and expensive. Marginalization and elimination, of such a system, are economical and logistical reasons.
3. Passive systems are ecological ones, as they produce no disturbance to the underwater ecosystem.

However, passive systems suffer from few drawbacks: the dependence of the system with respect to existing transmitters naturally in the environment and especially the

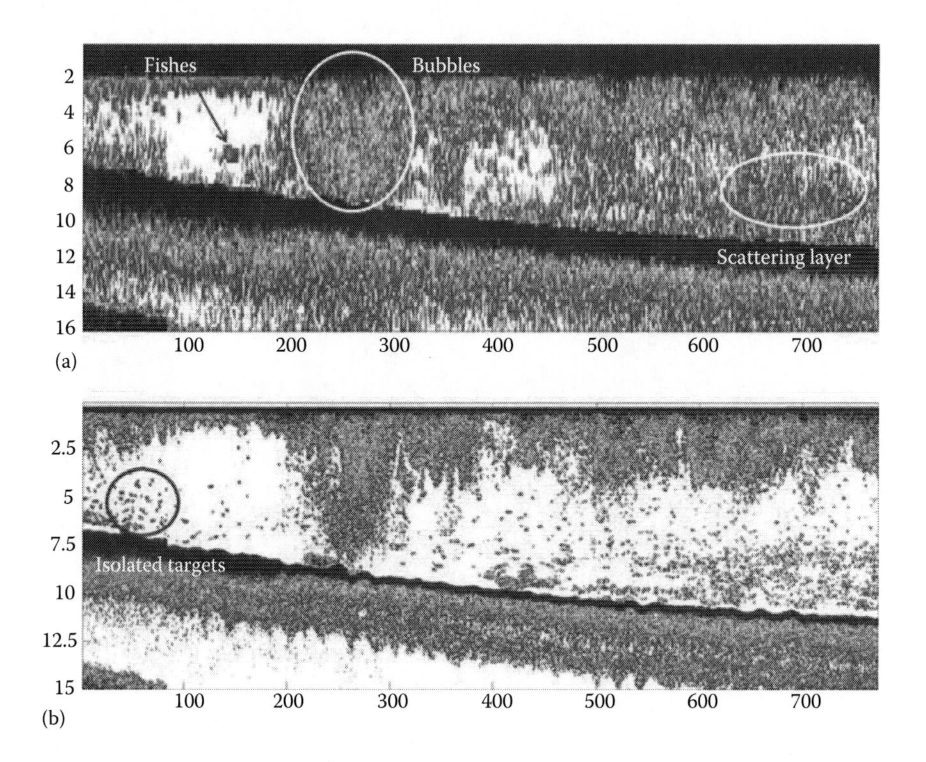

FIGURE 19.2 A comparison between two echograms of the same area insonified with diverse echo sounders at different frequencies (horizontal axis, ping number, and vertical axis, distance from sounder [m]).

lack of information on them (including mainly the number, positions, and types of issuers). In addition, most of existing acoustic processing algorithms assume the existence of a single issuer with a good signal to noise ratio (SNR). In real-life applications, this assumption can be hardly satisfied. Hence, several scenarios are possible:

1. Single source, single sensor: Consider the simplest case where we admit the existence of a single source in the channel. The traditional methods of identification are the most suitable, with respect to performance and in relation to their calculation time, to deal with this case.
2. Multisource, single sensor: Assume that for some reason (simplicity, strategy, or economics), the observation is made using only one sensor. For such a configuration, two cases are possible depending on waveform properties of the signal:
   a. Signals with different signatures of one dimension (i.e., straight or curved lines) in the plane of time–frequency or time–scale. In this case, algorithms based on various representation time–scale and time–frequency methods can be considered such as the short-term Fourier transform (STFT), the spectrogram, the scalogram, the Wigner–Ville distribution pseudo Wigner–Ville distribution page, and the

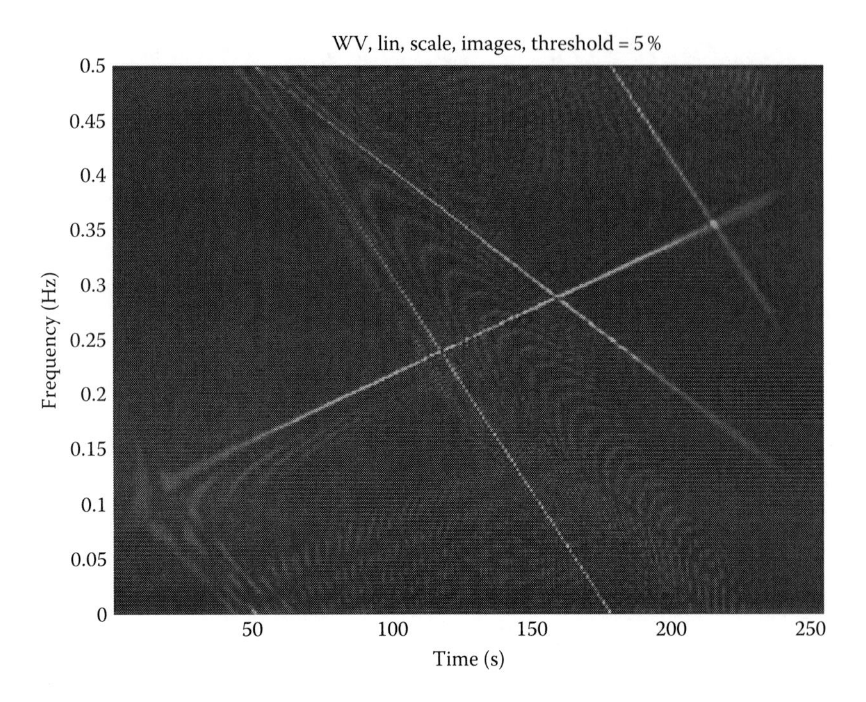

FIGURE 19.3 Pseudo Wigner–Ville transformation of the sum of different underwater acoustic natural signals.

spectrogram with signal reallocation, among others. Figure 19.3 shows the pseudo Wigner–Ville transformation of four different chirp signals. Chirp signals are frequency linearly modulated signals used in many advanced surveillance systems as radars or sonars. However, natural underwater sound signals don't belong to the family of frequency polynomial modulated signals (i.e., the generalization of frequency linearly modulated signals such as chirp signals).

b. Signals with the same signature or no exploitable geometrical signature (areas, volumes, dimensions higher than 1, irregular shapes, etc.) in the time–frequency or time–scale plane or space: in this case, there were no general or conventional solutions. Figure 19.4 shows the pseudo Wigner–Ville transformation of the sum of different underwater acoustic natural signals.

3. Multisource, multisensors: This scenario could represent the general case, that is, a channel with multiple sources and multiple sensors; this channel is also called multiple inputs, multiple outputs (MIMO) channel. We can see the importance of blind source separation algorithms to enhance existing passive acoustic. Being dependent on the nature of sources considered (Gaussian or not), independent component analysis algorithms can be used to separate different sources. Main inherited problems of passive systems are considered hereinafter.

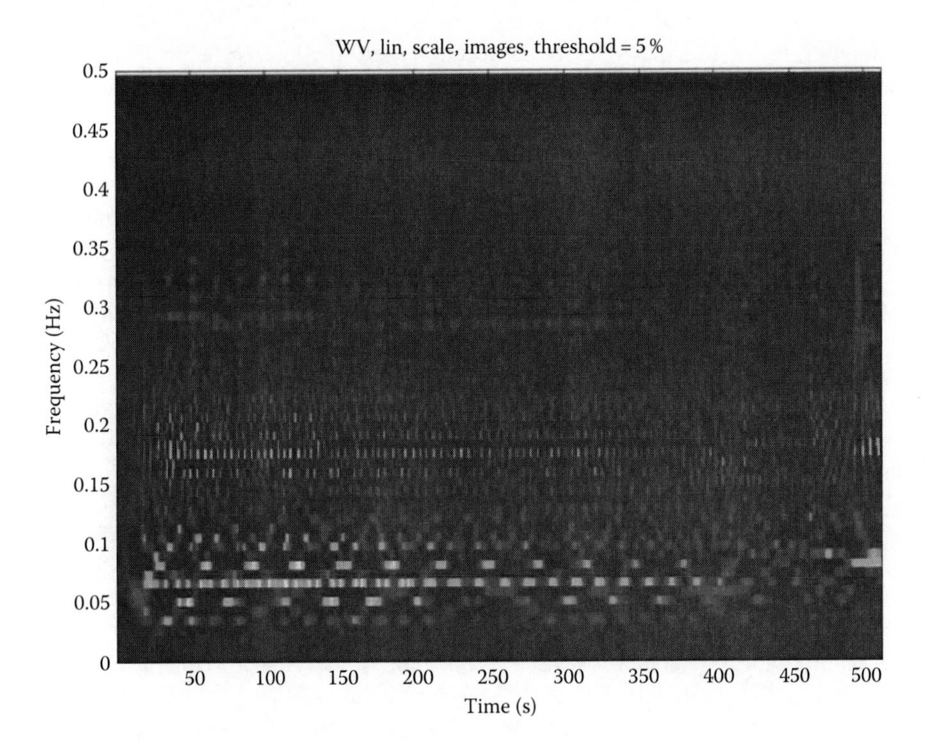

FIGURE 19.4 Pseudo Wigner–Ville transformation of four different chirp signals.

19.3.3 Flow Cytometers

Marine micro-organisms (i.e., MMOs such as phytoplankton and microzooplankton but also heterotrophic bacterioplankton) can be affected by many parameters such as water temperature, water pollution, and salinity. The study of these communities can help us to monitor the changes in marine ecosystem status and to detect both anthropogenic as well as long-term climate-driven changes and to study climate changes. In fact, MMO can adapt rapidly by changing their physiology and/or species composition, which can be monitored at high frequency with new sensors as *in situ* flow cytometers. In addition, their life depends on many parameters such as water pollutants, inorganic and organic dissolved compounds, temperature, and salinity (Alvain et al. 2008). By studying these tiny organisms, we could generate a precise image about the environmental conditions or ecological status of the marine environment. According to recent studies, as far as 70% of the oxygen in our atmosphere is produced or released by marine life. Most of that oxygen is directly produced by phytoplankton. For these reasons, the European Union classifies phytoplankton as a major indicator of the ecological status of coastal waters (WFD CE2000, MSFD CE2008). In fact, phytoplankton microorganisms represent the base of most marine food webs: All living resources depend on the plankton to survive. Moreover, some phytoplankton species can represent a threat (harmful algal blooms) to living organisms as well as to human health by their biomass accumulation and/or production of toxins which can damage marine organisms and/or accumulate in shellfish or fish consumed by humans. It is necessary to undertake thorough monitoring of coastal marine ecosystems to ensure they are healthy and their economic use is sustainable.

To study phytoplankton, researchers are using a combination of pulse-shape-recording flow cytometry (as the CytoBuoy machines) and other optical and chemical sensors deployed *in situ* (Olson et al. 2003, Thyssen et al. 2008, 2011). We should mention that technique is only performed by few laboratories around the world. One of these laboratories is the LOG which, in collaboration with the LISIC laboratory, are involved and coordinate a European INTERREGAIVA 2 Seas project involving French, English, and Dutch colleagues, on the intercalibration and application of innovative technologies for the monitoring of phytoplankton at high resolution (http://www.dymaphy.eu/?lang=en).

The researchers at LOG will provide the necessary data as well as the essential interpretation of the results from biological and environmental points of view. The analysis and classification of marine MOs will focus on processing cytometry signals and the outputs of other sensors to assess the role of MO in monitoring the coastal ecosystems. The new method will address the shortcomings of traditional approaches based on measuring environmental parameters, which are expensive and fail to provide adequate large-scale monitoring. More efficient monitoring will also enable improved analysis of climate change and provide knowledge informing the adequate authority's economic relationship with its coastal marine ecosystems.

In order to carry out our study, water samples will be automatically collected and analyzed. The analysis is based on pulse-shaped electrical-scatter and fluorescent signals and microscopic images. It is well known that the obtained electrical-fluorescent signals are random nonstationary and sparseness signals; see Figure 19.5. In this situation, classical signal processing algorithms can hardly give satisfactory results. Therefore, appropriate new signal and image processing algorithms should be developed (Caillault et al. 2009a,b, Wacquet et al. 2011).

19.3.4 Video Surveillance and Sensitive Cameras

Our vision and video surveillance systems will contain classical high-definition CCD cameras as well as infrared camera and microscopic cameras. Microscopic cameras are essential for the second stage of our project, that is, to monitor the health of MOs. The systems will as well contain different specter light projectors (white, red, infrared lights), which will mostly be switched off except when it becomes necessary in order to not disturb coastal wildlife. The white light helps us getting nice photos within short distance (<20 m); the red and the infrared lights are useful to get image without disturbing biological life. It is worth mentioning here that underwater images could suffer from the following six main limitations:

1. Light level
2. Water turbidity
3. Sampling and memory
4. The problem of image analysis
5. Data compression
6. Artifacts associated with sensors, movement, lighting control, etc.

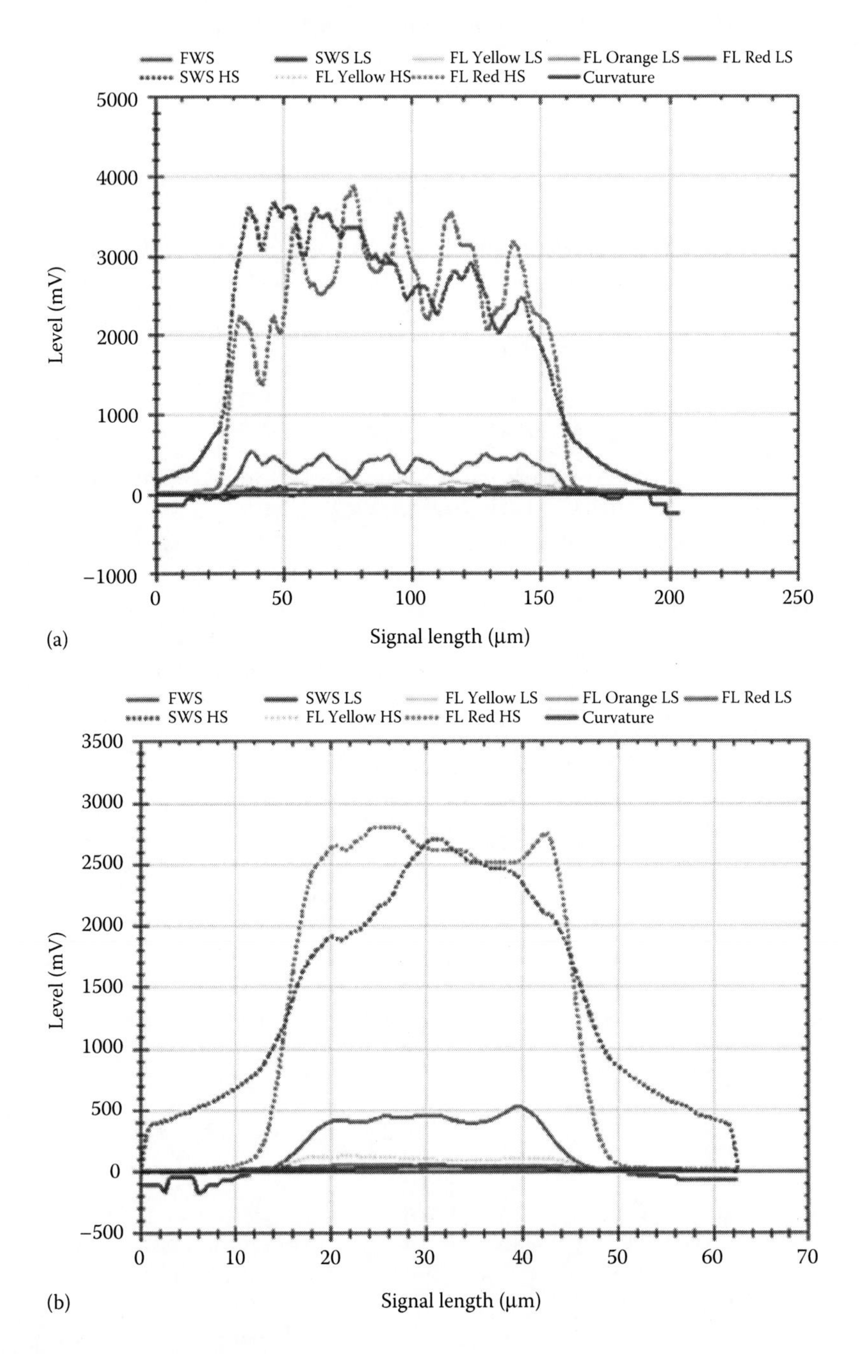

FIGURE 19.5 Two examples of typical cytometer electrical-fluorescent signal (Guiselin 2010, Guiselin et al. 2009). (a) Cytometer signals of a colony of *Lauderia Annulata* and (b) Cytometer signals of a single cell of *Lauderia Annulata*.

It has already been mentioned that in the third stage of the project, we are planning to deploy unmanned aerial vehicles (UAVs) equipped with stabilized camera, infrared camera, UV camera, and few other sensors (GPS, thermometer, barometer, air-pollution-monitoring sensor, etc.). Using such UAV to get several aerial photos of coastal regions, it can increase the efficiency of our approaches; aerial photos along with biosignals should be processed. At a later stage of our project, sensitive microscopic digital cameras (such as digital microscope camera ProgRes C14plus) could be introduced to our immersed platforms. Therefore, we will work on image processing and combining different proposed approaches. In addition, we should also merge and integrate the outlines of the whole system. The latest combination could be done using data fusion methods.

19.3.5 Multitude of Heterogeneous Auxiliary Sensors

At different stage of the project, different sensors and data acquisition systems will be deployed, such as sonars, ADCP, hydrophones, cameras, and sensitive microscopic digital cameras. At the final stage of our project, other sensors will be added such as cytometer, sensitive digital cameras, and satellite image data acquisition. As mentioned before, this project requires a considerable amount of complex and extensive experimentation, along with associated routine laboratory work, as well as detailed theoretical planning and interpretation. Data should be recorded with several possible platforms such as

- Vessels
- Autonomous underwater vehicles (AUVs) or remotely operated underwater vehicles (ROVs)
- Moorings
- Seafloor observatories

Therefore, we should conduct experiments, collect and process data, and implement various real-time processing approaches. We should also merge and integrate the outlines of the whole system. The latest combination could be done using

1. Information extraction
2. Harmonization of information
3. Data fusion

Figure 19.6 shows an example of hydrophone and ADCP data recorded on the MeDON observatory at the same time (06/20/2012 10:05–10:10 p.m.). The spectrogram shows the acoustic signature of dolphins since fish schools can be shown on the ADCP echogram.

All these platforms can be equipped with many different sensors: acoustic and video of course but also chemical sensors (oxygen or nitrate detectors, etc.), optical sensors as total and multiwavelength spectral fluorometers, etc. Especially, electrical field and magnetic sensors can be also considered as part of our future system. Our vision is to mimic natural underwater predators such as sharks (they are dotted with

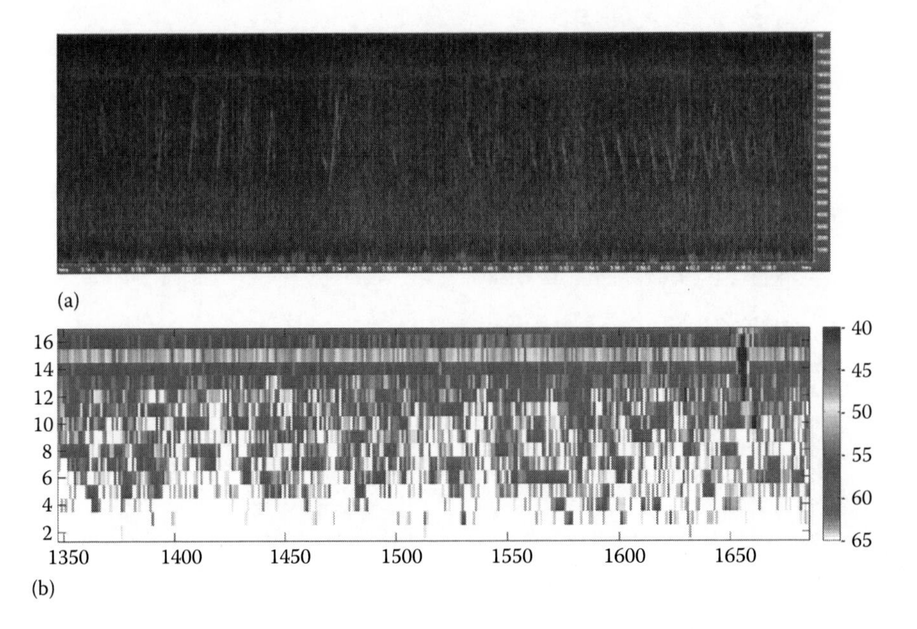

FIGURE 19.6 Spectrogram of hydrophone data and ADCP backscattering data recorded on MeDON observatory at the same time. (a) Hydrophone data; (b) ADCP backscattering data.

electrical sensors), dolphins (they are using advanced *sonar*), coral and shells (they use chemical detectors), or octopus (they are armed with big eyes).

19.4 APPROACHES AND METHODOLOGY

The analysis of various images and signals obtained from any ecosystem or biosystem is very challenging due to the complexity of the images and the nonstationarity and sparseness of the signals. Our main task will be the development of appropriate preprocessing signal and image algorithms as well as the elaboration of unsupervised classification methods.

19.4.1 Modeling of an Underwater Acoustic Transmission Channel

Underwater acoustic signals could be generated by mainly three types of sources: natural sources (waves, wind, rain, earthquake, etc.), animal sources (such as fish, shells, and dolphins), and human activities (such as sonars, the noise generated by moving boats, underwater activities, and fishing). These signals can't be easily classified or identified due to the wide diversity of their characteristics, some of which are Gaussian signals (wave noises); others are cyclostationary (boats' noises), low-band sparse signals (fish noise), high band signals (e.g., whale sounds), impulse signals (dolphin clicks, rain sounds, etc.), complex modulated chirp signals (dolphin whistles or, to a certain extent, sonars or echo sounders, etc.), and so on. Besides the complexity of signals of interest, one should take into consideration the supplementary difficulties related to the underwater acoustic transmission channels (Mansour 2012, Shulkin and Marsh 1962).

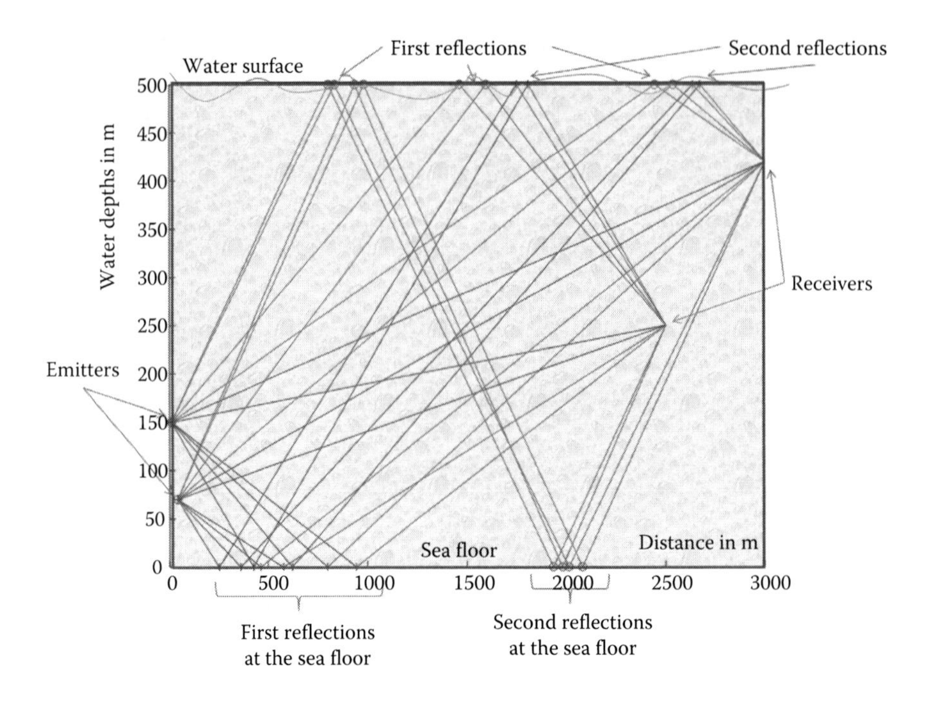

FIGURE 19.7 A simulated underwater acoustic transmission channel.

It is worth mentioning here that conducting underwater experiments is mostly expensive and involves complex logistics and very large heavy items. In order to validate our approaches, techniques and algorithms, real underwater experiments should be carried out. However, these experiments should be planned at an advanced stage of our project in order to save time and money. For these reasons, we developed a simulation model of an underwater acoustic transmission channels. In this section, this model is briefly discussed (Figure 19.7).

Underwater sounds propagate through water as a continuous change in the pressure. The propagation of the acoustic wave in a stationary medium is obtained by Helmholtz's equation. A general solution of the Helmholtz's equation is very difficult to obtain. Therefore, simplified propagation models have been widely used such as the ray theory, the mode theory, the parabolic model, and the hybrid model. For coastal ecosystem monitoring or port surveillance applications, the height of the water column may change from a couple of meters up to a few hundred meters, that is, the case of shallow water. In such scenario, the ray theory is the more appropriate propagation model. In the ray theory, the propagation channel is represented by finite impulse response (FIR) filters where the coefficients and the delays depend on the depth of the water column, the number of reflection spots, the sound speed, the topology and the nature of the seabed, the surface of the water, the salinity, and underwater current, among other less important parameters. The sound speed C (m/s), in oceans, is an increasing function of temperature T (°C), salinity S (parts per thousand, ppt), and pressure, which is a function of depth D (in m) (Etter 1991, 2001):

$$C = 1449 + 4.6T - 0.05T^2 + 23\times10^{-5}T^3 + 16\times10^{-8}D^2 + 0.02D$$
$$+ (1.34 - 0.01T)(S - 35) - 7\times10^{-13}TD^3$$

The preceding equation is an empirical relationship satisfied when $0 \le T \le 30$, $30 \le S \le 40$, and $D \le 8000$. The reflected acoustic waves on the seabed or on the water surface depend on many parameters (such as the composition and the topology of the bottom, the wind, the wave frequency, and the swell properties) (Lurton 2002, Brekhovskikh and Lysanov 2003). As water surface isn't a flat surface, the reflected acoustic waves are dispersed in the space. However, in average term, reflected acoustic waves can be considered as obtained by a flat surface with absorption coefficients (Lurton 2002). Ray trajectories and sound speed profile allow us to compute propagation times. In addition, ray trajectories, water attenuation, boundary roughness, and subbottom properties allow us to compute the signal magnitude. From a computational view point, ray trajectory is computed by solving the *Eikonal equation*, but signal magnitude is obtained as a result of *transport equation* (Jensen et al. 2000).

19.4.2 Estimation of the Number of Active Sources

In the context of underwater surveillance system, acoustic signals can be generated from various sources (natural, animal, or artificial). In most cases, the number of active sources cannot be exactly evaluated.* However, recorded acoustic signals can be practically clustered and attributed to few generic sources (a school of fish, a commercial or military boat, waves, dolphins, etc). In any surveillance system, a rough estimation of sources is crucial. In a previous work (Mansour 2012), the estimation of source number is done by implementing existing algorithms as well as by developing new methods (Figure 19.8) (Chen et al. 1996).

19.4.3 Source Localization

We are aiming to develop new underwater localization algorithms, which can be based on MIMO systems, modified MUSIC algorithms, HOS, channel equalization algorithms, time difference of arrival, triangulation, etc. Experiments realized in Canada show that using an array of hydrophones, it is possible to detect and give an estimation of the localization of marine mammals (Roy et al. 2010). Also, using only one hydrophone (Aubauer et al. 2000) proved that it was possible to find the distance and depth of marine mammals in particular conditions (shallow water, presence of multiple paths, in the acoustic signal of dolphin whistles, etc.). These algorithms should be tested on real data recorded with the fixed cabled observatory MeDON† located near Brest (France) at 20 m depth. This

---

* What is the number of sources that are generated by waves? A school of small fishes or sea animals, like shrimp, can generate similar sounds, which are dispersing in space. Even artificial noises, like the noise of moving boat, can be obtained from different sources located at various parts of the boats such as the boat propeller, the engine noise, sailors' activities, and boat's wake.

† Marine e-data observatory network (http://www.medon.info/).

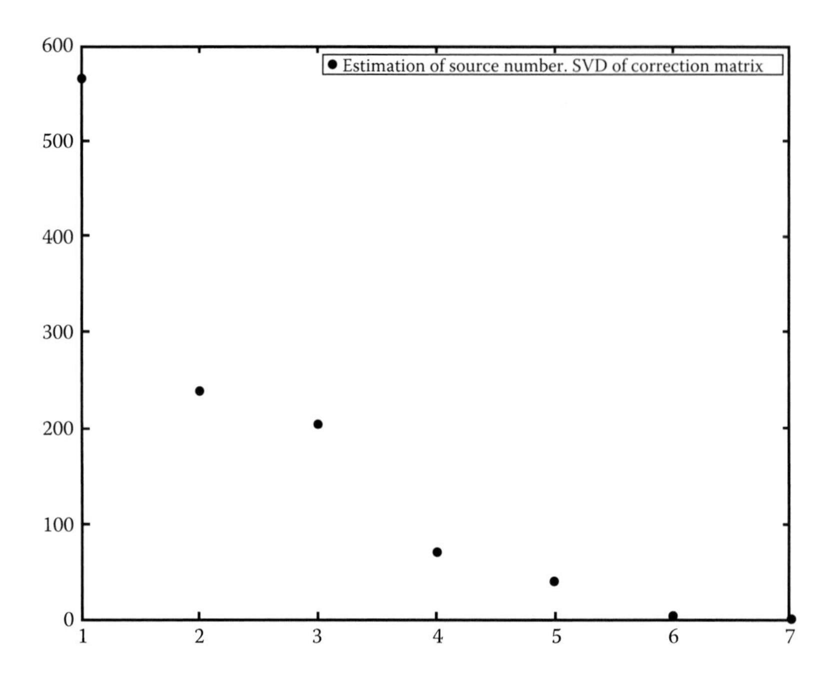

FIGURE 19.8 Estimation of the number of sources based on the eigenvalue of the autocovariance matrix.

observatory contains, among other sensors, three hydrophones, and localization of marine mammals should be then possible.

19.4.4 Estimation of High-Order Statistics

HOS are used in many localization, identification, and separation algorithms (Mansour and Jutten 1999, McCullagh 1987, Shiryaev 1984). In order to exploit spatial diversity, many blind or semiblind separation or identification algorithms use HOS, in time or frequency domain. The estimation of cross cumulants and moments, up to the fourth order, has been investigated in our previous studies. We proposed new adaptive HOS estimators for fourth-order cross cumulants. Many simulations were conducted to elaborate our unbiased estimators, and they showed that this estimator can be applied on underwater acoustic signals, which are nonstationary signals.

The nth-order moment, $\mu_n(X)$, of a random variable (RV), X, is the mathematical expectation of the nth degree monomial of X:

$$\mu_n(X) = E(X^n)$$

The nth-order cumulant, $Cum_n(X)$, is the mathematical expectation of an nth degree polynomial of X (Kendall and Stuart 1961). The moments (respectively, the cumulants) can be found in the Taylor expansion of the first (respectively, second) characteristic functions (FCFs) of X. Using Leonov–Shiryaev formula, the nth-order cumulant of X can be developed as a polynomial function of all X's moments of order less and equal to n:

$$Cum(X_1,\ldots,X_n)=\sum(-1)^{k-1}(k-1)!E\left(\prod_{i\in\gamma_1}X_i\right)E\left(\prod_{j\in\gamma_2}X_j\right)\cdots E\left(\prod_{k\in\gamma_m}X_k\right)$$

where the addition operation is over all the set of $(1\leq i\leq p\leq n)$ and composes a partition (Papoulis 1991) of $\{1, \ldots, n\}$. If the estimation of the moments can be done using unbiased estimators (low pass estimators, direct or adaptive ones), there are a few unbiased estimators of cumulants. For this reason, the estimation of cross cumulants and moments, up to the fourth order, has been investigated in our previous studies (Martin and Mansour 2004, Mansour 2012). In the latest study, we proposed a new adaptive HOS estimator for fourth-order cross cumulants between X and Y:

$$C_N=\frac{N-2}{N}\gamma C_{N-1}+\frac{1}{N}\gamma\mu_{13}(X,Y)+\frac{N+2}{N(N-1)}\gamma x_N y_N^3-3\gamma x_N y_N\mu_{02}(X,Y)$$
$$-3\gamma y_N^2\mu_{11}(X,Y)+(1-\gamma)x_N y_N^3-3(1-\gamma)x_N y_N\mu_{02}^2(X,Y)$$

where
the forgotten factor $0<\gamma<1$
N stands for the number of iterations
C_N is the $Cum(X, Y, Y, Y)$ at the Nth iteration
x_N and y_N are the Nth samples of X and Y
$(n+m)$th-order cross moment at the Nth iteration, $\mu_{nm}(N)=\hat{E}(X^nY^m)$, is given by

$$\mu_{nm}(N)=\frac{1}{\lambda^N}\left(\lambda(1-\lambda^{N-1})\mu_{nm}(N-1)+(1-\lambda)x_N^n y_N^m\right)$$

Many simulations were conducted to elaborate our fourth-order cross cumulant unbiased estimator. Our experimental studies showed that this estimator can be applied on underwater acoustic signals, which are nonstationary signals. In some cases, x_N and y_N in the cumulant equation have been replaced by their average over a small estimation window (10–50 samples). The previously proposed estimators can be improved by considering non-iid samples. The improvement of the preceding estimator will be considered in a future work (Figure 19.9).

19.4.5 Source Separation

In the last two decades, two research axes have been raised in the field of signal processing: blind identification and separation approaches (Cardoso 1992, Cardoso and Comon 1996, Cardoso and Laheld 1996, Mansour et al. 1996, 2000, 2002) and multiscale representations (time–frequency representations or wavelet analysis).

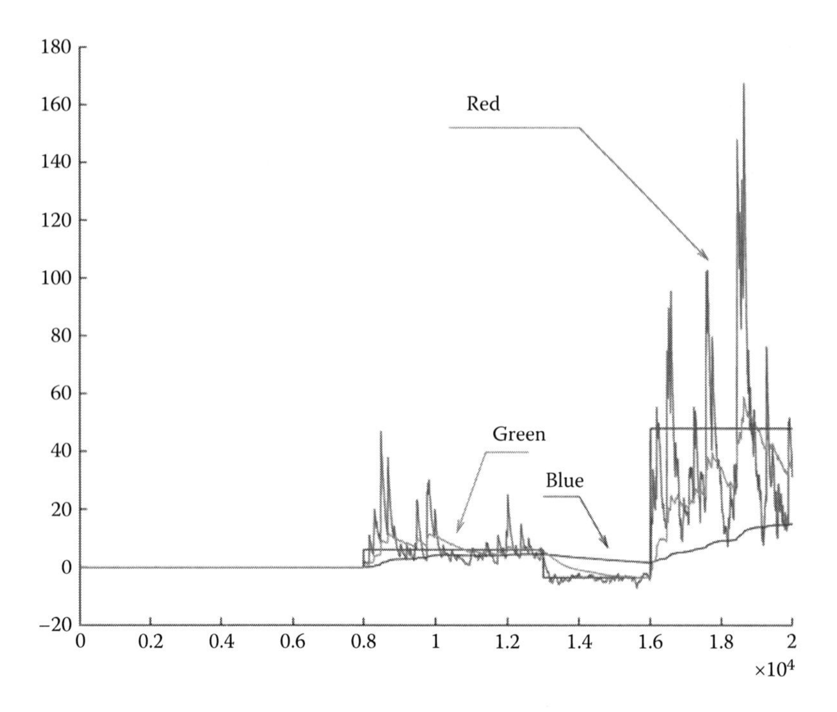

FIGURE 19.9 A comparison between three different estimators of fourth-order cumulates of a nonstationary signal, which contains four parts (two uniforms and Gaussians RV with different amplitudes).

The BSS problem consists of retrieving unknown sources from only observing a mixture of them. BSS was initially proposed to study biological phenomena (i.e., the central nervous system processes typically multidimensional signals). Recently, BSS can be found in various situations: radio communication, speech enhancement, separation of seismic signals, noise removal from biomedical signals, etc. (Cichocki et al. 1998, Comon 1994, Douglas et al. 1997, Jutten and Karhunen 2003, Kardec Barros et al. 1999, Karhunen et al. 1997, Kawamoto et al. 1997, 1998, 1999).

As mentioned earlier, the assumption about the existence of a single emitter is a kind of luxury that we cannot afford in any real underwater coastal activities, especially in passive mode. In this context, blind source separation algorithms can be of great interest to help achieve better results using passive mode of sensing. In underwater acoustics applications, the transmission channel can be modeled as a convolutive mixture (i.e., the transmission channel between the ith source and the jth sensor can be modeled using a FIR filter).

Few blind separation algorithms of convolutive mixtures are devoted to the treatment of nonstationary signals (a very important feature of underwater acoustic signals as natural vocalizations of whales and other marine animals or artificial sounds as boats, submarines, etc.) (Mansour et al. 1996, 2000, 2002, Matsuoka et al. 1995). None of these algorithms are dedicated to the processing of real acoustic underwater

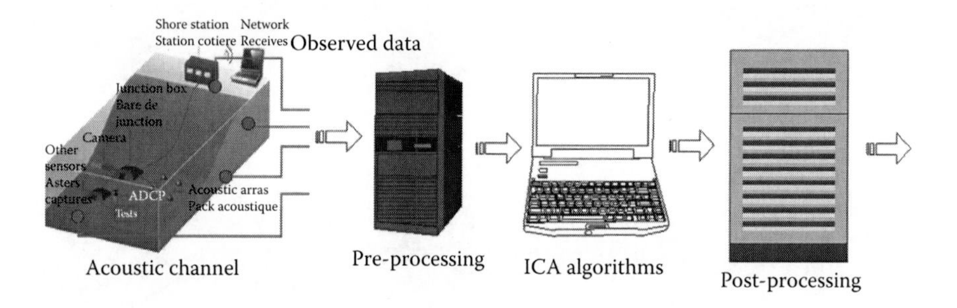

FIGURE 19.10 A proposed architecture to apply ICA algorithms on real underwater acoustic signals issued from passive acoustic sensors.

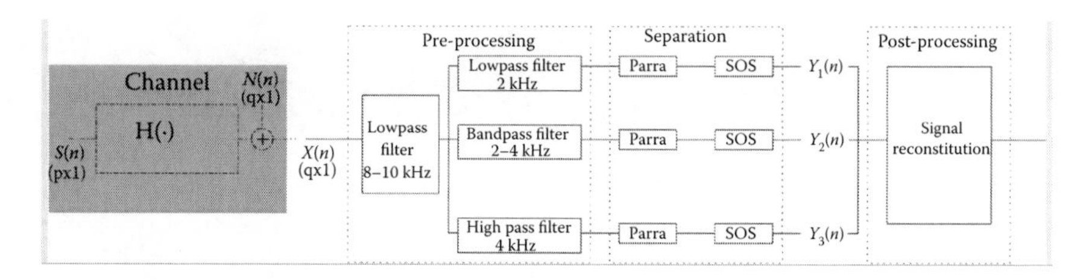

FIGURE 19.11 A schematic of a system that gives satisfactory results.

signals. Figure 19.10 shows a proposed architecture to apply ICA algorithms on real underwater acoustic signals issued from passive acoustic sensors. Generally, classic ICA algorithms are powerless in the face of the high nonstationarity, sparseness, and diversity of underwater acoustic signals (Amari 1996, Amari and Cardoso 1997, Babaie-Zadeh et al. 2006). Therefore, preprocessing and postprocessing stages are required. These stages should be adapted to deal with target signals and chosen ICA algorithms. Figure 19.11 shows a system schematic that gives satisfactory results; see Figure 19.12. Parra and SOS are two ICA algorithms proposed to a separate convolutive mixture of speech signals; further details can be found in Mansour (2001, 2012), Mansour and Gervaise (2004), Nguyen Thi et al. (1991, 1995), Parra and Alvino (2000), Puntonet et al. (2002), Rahbar and Reilly (2001, 2005).

19.4.6 Performance Indexes

Underwater acoustic signals are nonintelligible. This property should be taken into account in all stages of signal treatment. This can generate a problem related to measure the performance of our algorithms. To consider nonintelligible signals, one should investigate performance indexes. We conducted a preliminary study to propose or modify various performance indexes to deal with nonintelligible signals such as boat noises or marine animals.

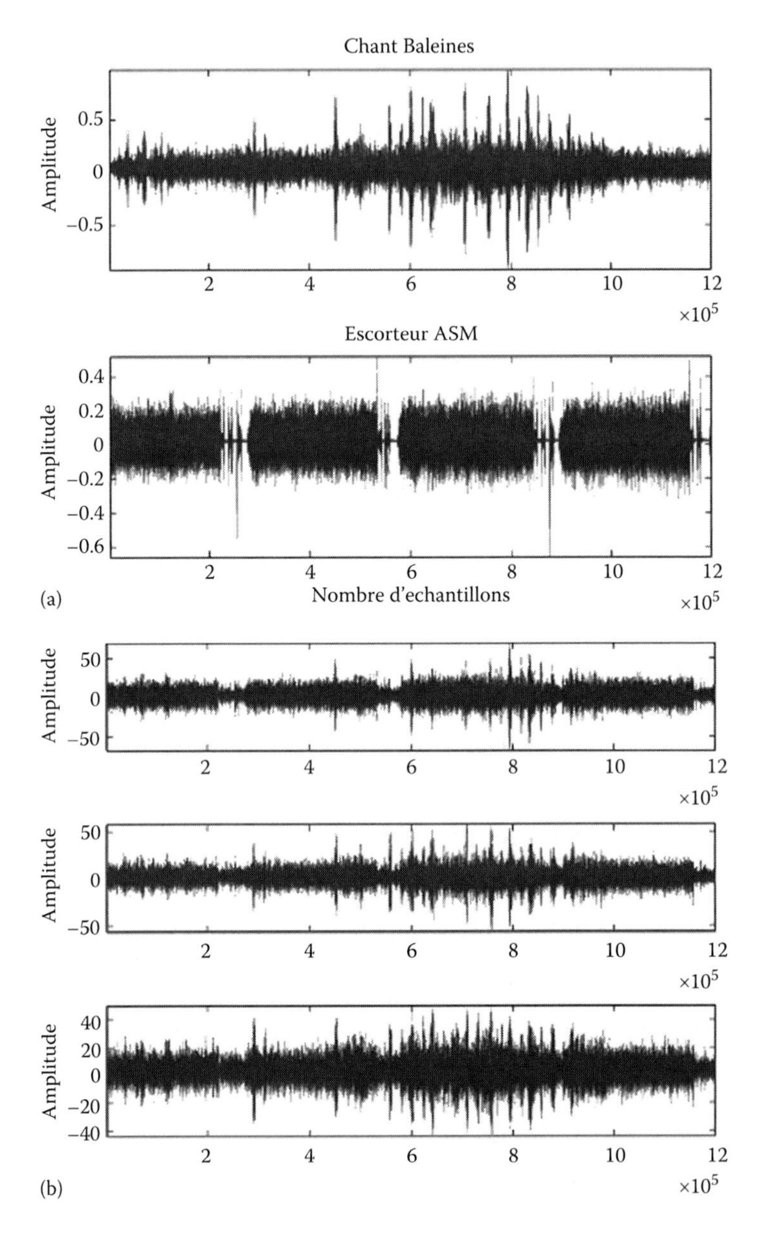

FIGURE 19.12 Blind separation of a mixture of two acoustic signals recorded by three acoustic sensors (whale sound and vessel): (a) original sources, (b) mixed signals.

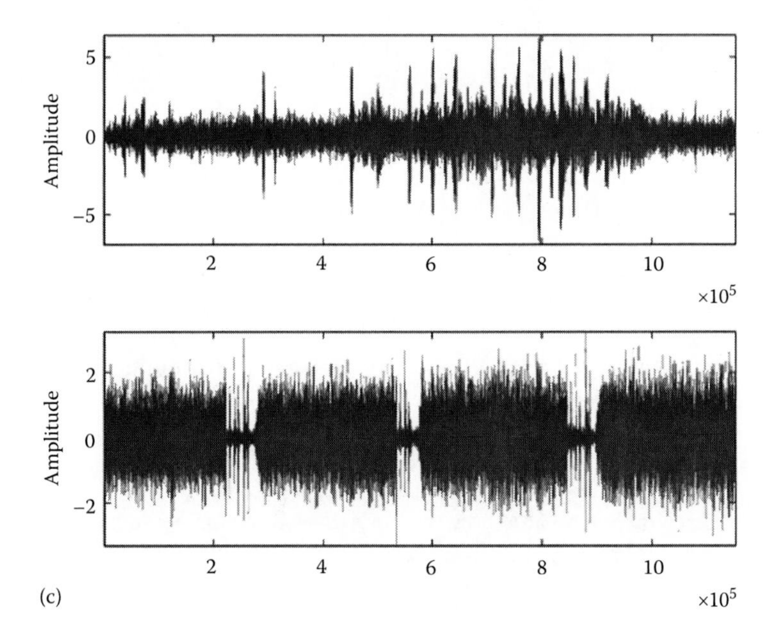

FIGURE 19.12 (continued) Blind separation of a mixture of two acoustic signals recorded by three acoustic sensors (whale sound and vessel): (c) separated signals.

It is well known that a blind separation of statistically independent sources of convolutive mixtures can lead us to the original sources up to a permutation and a scalar filter:

$$x(n) = a(z) * s(n) + b(z) * r(n)$$

where

$r(n)$ represents a residual mixture of noise and all sources except the one of interest $s(n)$, $a(z)$, and $b(z)$ are residual separation filters

* stands for the convolution product

The main objective of BSS algorithm is to increase the power ratio between the two terms of the estimated signal $x(n)$. In addition, the identification or classification of underwater acoustic signals is extraordinarily difficult step because these signals are nonstationary and nonintelligible sparse signals with low variable kurtosis.* In this context, the classification of ICA algorithms according to the separation quality becomes a difficult and important task.

The following discrimination criteria can be optimized to maximize the spatial diversity or the independence among estimated signals. At the same time, they can be very useful to quantify the separation achievement. In the last case, these criteria are called performance indexes:

---

* The kurtosis is a normalized fourth-order cumulant.

1. Modified cross talk: The cross talk is the inverse of SNR and it is widely used as a performance index for BSS algorithms of instantaneous mixture. To apply the cross talk, one should have original sources. Therefore, this performance index cannot be applied in real situation where sources are unknown. For this reason, we developed a modified definition for the cross talk as normalized distance between the estimated signal $x(n)$ and the original signal $s(n)$:

$$Mc(\hat{s}(n), s(n)) = 10\log_{10}\left(\frac{E(x(n)-h(z)*s(n))^2}{E(s^2(n))}\right)$$

where the estimated residual filter $h(z)$ is given as the minimum the least mean square (LMS) error ξ

$$h(z) = \min_h E(x(n)-h(z)*s(n))^2 = \min_h \xi$$

Our experimental results show that for a low-order channel filter (<20), this performance index can be used efficiently. When the order of channel is larger than 20, computing time becomes a big issue.
2. Mutual information: According to Tan et al. (2001), mutual information is one of the best independence indexes. In the context of BSS problem, the joint and the marginal probability density function (PDF) are unknown, but they can be estimated. To estimate the MI, we used a method proposed by Pham (2003). In his method, the integral is replaced by a discrete sum, and the PDF are estimated using kernel methods.
3. Quadratic dependence (Achard et al. 2003): To measure the independence among the components of a random vector X, the authors of Rosenblatt (1975) make a comparison between the joint PDF of the vector X and the marginal PDF product of its components x_i. Using similar approach, Kankainen (1995) proposed the quadratic dependence measure $D(X)$, which is a comparison between the joint FCF and the product of the marginal FCF. If the components of X are independent in their set, then the joint FCF is equal to the product of the marginal and $D(X)=0$. The main drawback of such performance index is the important computing time.
4. Nonlinear kernel decorrelation: Bach and Jordan (2003) propose an independence measure based on the concept of nonlinear decorrelation or the Φ-correlation function ρ_Φ:

$$\rho_\Phi = \max_{f,g\in\Phi}\left(\frac{Cov(f(X), g(Y))}{\sqrt{Var(f(X))Var(g(X))}}\right)$$

where

Cov and Var are, respectively, the covariance and the variance functions

Φ is a vectorial space of all functions applied from $\mathbb{R}$ to $\mathbb{R}$ that contents all Fourier transform basis (i.e., the exponential functions $\exp(j\omega x)$, with $\omega \in \mathbb{R}$)

$\rho_\Phi=0$ means the independence between X and Y.

According to Bach and Jordan (2003), the best choice of the two nonlinear functions f and g can be done using Mercer kernel functions.* We should notice that for acoustic signals, better results are obtained using polynomial kernel, which gives us a maximum difference between independent and correlated signals. Our experimental studies show that this performance index can be applied successfully in our project. However, computing time and needed memory become extremely important when the number of samples is over 500,000 samples. Finally, the difference between the NL-decorrelation of the sources and the mixed signals depends on original signals, the chosen kernel, and the mixing model and parameters.

5. Simplified nonlinear decorrelation: Using a similar approach to Bach and Jordan (2003), we proposed a simplified performance index based on the concept of a nonlinear covariance matrix $\Lambda=(\rho_{ij})$ defined by

$$\rho_{ij} = \max_{f,g\in\Phi}\left(\frac{E(\langle f(x_i)\rangle_C,\langle g(y_j)\rangle_C)}{\sqrt{E\left(\langle f(x_i)\rangle_C^2\right)E\left(\langle g(x_i)\rangle_C^2\right)}}\right)$$

where

$X=(x_i)$ is a random vector

$f(x)$ and $g(x)$ are two nonlinear functions and $\langle X\rangle_C=X-E(X)$

If the components of X are independent from each other, then Λ becomes a diagonal matrix. Using the last definition, we suggest the following performance index:

$$NLD = 20\log_{10}\left(\frac{\left\|Off(\Lambda)\right\|^2}{\left\|diag(\Lambda)\right\|^2}\right)$$

Here, $diag(\boldsymbol{M})$ is a diagonal matrix that has the same principal diagonal of matrix $\boldsymbol{M}$ and $Off(\boldsymbol{M})=\boldsymbol{M} - diag(\boldsymbol{M})$. Functions f and g are chosen from a set of functions (such as Gaussian kernel, sixth-order polynomial Kernel that its coefficients are the components of a unitary vector, saturation kernel using arc-tangent function, and saturation kernel using hyperbolic tangent function). Our experimental studies show the effectiveness of this performance index to deal with underwater acoustic signals and channels. The main drawback of this performance index is that obtained values depend on the kind and the number of original independent signals. Therefore, this performance index can only be used in simulations where the original sources are known.

* A bilinear function $K(X, Y)$ from a vectorial space X (e.g., $\mathbb{R}^m$) to $\mathbb{R}$ is said to be a Mercer kernel if and only if its Gram matrix is a semipositive matrix. By definition, the Gram matrix of basis vectors $(X_1, \ldots, X_m)$ of a m-dimensional vectorial space X with respect to a bilinear function $K(X, Y)$ is the matrix given by $G_{ij}=K(x_i, y_j)$. $K(X, Y)$ should, also, have the translation invariance, the convergence in $L^2(\mathbb{R}^m)$, and isotropic properties. A possible kernel is the Gaussian kernel.

6. Independence measure based on the FCF: The joint FCF of a random vector X is equal to the product of the marginal FCF of its components if and only if they are independent from each other. Using that property, Feuerverger (1993; see also Murata 2001) proposed the following independence measure:

$$T = \frac{\pi^2}{\eta^2}\sum_{ij} g(X'_j - X'_i)g(Y'_j - Y'_i) - \frac{2\pi^2}{\eta^3}\sum_{ijk} g(X'_j - X'_i)g(Y'_j - Y'_k)$$

$$+\frac{\pi^2}{\eta^4}\sum_{ijkl} g(X'_j - X'_i)g(Y'_l - Y'_k)$$

where

g is an adequately chosen function (Feuerverger 1993)

$X' = \Phi^{-1}(8X - 3/8\eta + 2)$ is the approximation of the score function of X

$\Phi(X)$ is the PDF of zero mean and unite variance Gaussian signal

Our experimental studies show that the computing time is the main drawback of this performance index. We should mention that for stationary signals, this performance index is consistent. Unfortunately, the last interesting property is useless in our application since the acoustic signals are nonstationary signals.

7. Cross cumulants: Previously described performance indexes cannot be applied in real situations, where original signals are unknown because the performance values depend on the sources. Therefore, we developed a new performance index based on the average of the fourth-order cross cumulant $Cum(1, 3)(X, Y)^2$, which is obtained using a sliding estimation window. Good results have been obtained using this performance index on instantaneous or convolutive mixtures of acoustic signals. However, the computing time is relatively important.

19.4.7 Source Classification and Recognition

Once the sources have been separated, another important step of our project will be followed. This important step consists of performing source classification and recognition in order to divide the received signals into several families (artificial signals, fishes, underwater mammals, natural noises, etc.) and then to specify the signal type (boat, submarines, dolphins, etc.) (Dufrenois et al. 2009, Kalakech et al. 2011a,b). To reach our goals, we are planning to apply feature extraction, hidden Markov models, classification algorithms, etc. Figure 19.13 gives an example of few indentified sources recorded with hydrophones on MeDON observatory.

19.4.8 Inverse Modeling Methodologies in Echo Sounder Data

Inverse modeling is widely used in the literature to estimate the size, the abundance, and in certain extent species of schools of fish. In fact, acoustic data, which are recorded using a calibrated single- or multibeam echo sounder, give echograms of volume backscattering strength (S_v in the formalism current used

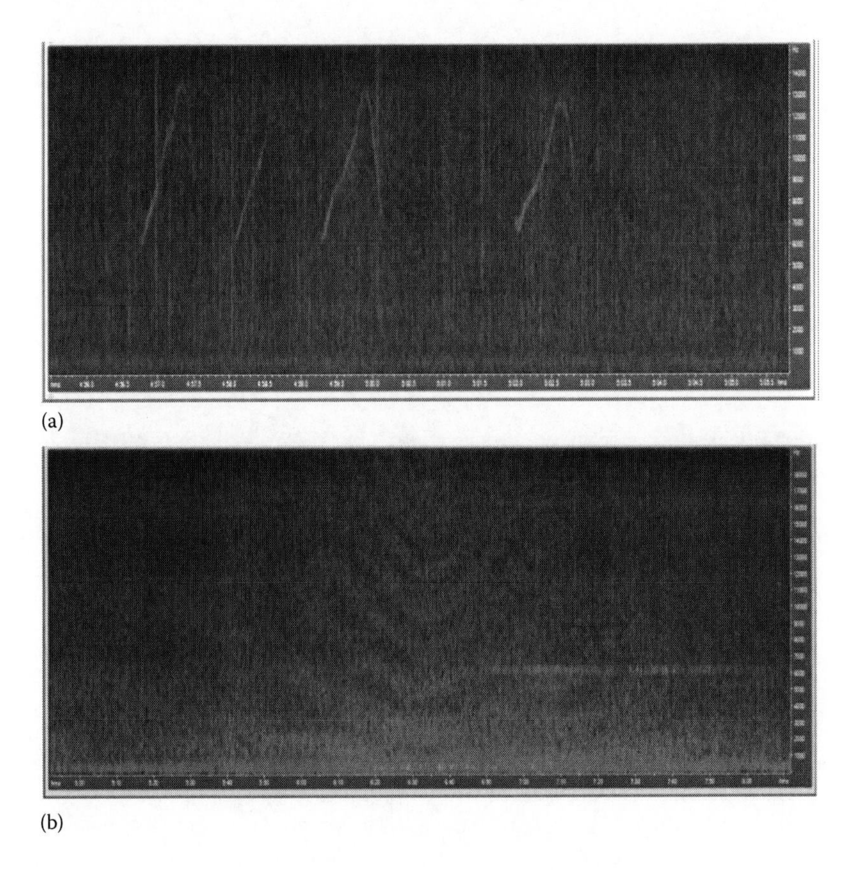

FIGURE 19.13 Spectrograms computed with hydrophone data recorded on MeDON observatory: (a) Dolphin whistles and (b) Lloyd's mirror coming from boat sound.

[MacLennan et al. 2002, Simmonds and MacLennan 2005]) in the water column and, if they are split beam, TSs of isolated targets.

If the concentration of targets (for example fishes) is not so important (i.e., the linear domain, this condition is generally satisfied), then the volume backscattering strength can be deduced linearly from the contribution of all insonified targets as shown in the following equation:

$$s_v = \int \sigma_{bs} n(a) da$$

where

s_v (in m^{-1}) is the volume backscattering strength
σ_{bs} (in m^2) is the acoustic backscattering cross section of one target of size a
$n(a)$ is the concentration of targets of size a
s_v is often given in dB, $S_v = 10\ \log10(s_v)$ (dB re 1 m^{-1}), as well as σ_{bs}, $TS = 10\ \log10(\sigma_{bs})$ (dB re 1 m^2).

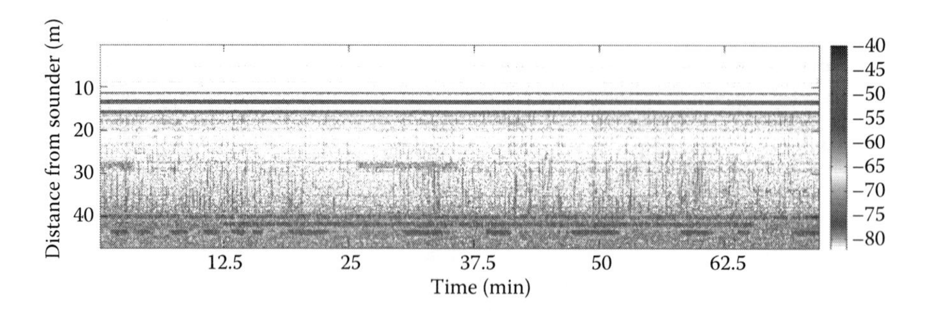

FIGURE 19.14 Echogram obtained on natural bubble seeps.

Traditionally, fisheries acoustics are used to obtain an evaluation of fish stocks. Nowadays, these techniques have been widely used to detect and track other targets such as the estimation of zooplankton (Stanton et al. 1994, Korneliussen et al. 2009), the monitoring of algae (Leblond et al. 2010), or a quantification of bubbles (Leblond et al. 2013). Figure 19.14 shows an example of echogram data of bubble seeps in Marmara Sea with an EK60 echo sounder at 120 kHz. In this experiment, the EK60 insonified horizontally.

19.4.9 Feature Extraction

In our project, a blind classification of MOs' images and signals will be a main step to help the experts in making decision. Each signal or image could be represented by a set of points in a multidimensional space. In this case, pertinent features of these images and signals should be firstly extracted. Later on, they should be projected in lower dimension space to reduce the computing time and efforts. To reduce the feature dimension, we will use nonlinear approaches (such as kernel PCA, ISOMAP, locally linear embedding, and independent component analysis), and we will propose similarity criteria. The project will be mainly approached along five research axes:

1. The extraction of relevant information often requires a deep knowledge of the measurement methods and the biological as well as the bio-optical characteristics of the aquatic MOs. This step could be very useful to optimally extract and represent significant features.
2. Once the signals and the images have been collected and relevant information has been well defined. The main part of the project will be the development of projection methods along with blind classification approaches.
3. To reach our final goal, new extraction and recognition and classification algorithms will be proposed using adaptive filtering theory, HOS, independent component analysis, wavelet, and time–frequency representations.
4. In order to better detect and treat the microscopic images, new image processing approaches will be developed to analyze the outcomes of a sensitive high-speed microcamera that will be added to the cytometer, and it will

provide us with microscopic image of phytoplankton cells along with the normal cytometry signals.

5. Sensitive satellite images of invisible lights could be also considered at a later stage in the project, to enhance the outcomes of our approaches and help us in monitoring a wide area. Specific focuses will be done in the following:
   a. Development of algorithms for improving the estimation of sediments and suspended sediments in the water column. This effort will require sample collection, filtering, and measurement of optical scattering properties.
   b. Development of algorithms that will enable the use of remotely sensed data.

19.4.10 Blind Classification

Recently, data visualization methods have received more attention from electrical and computer engineering community: image indexation and retrieval, data mining, surveillance, diagnosis, etc. To solve the latest problems, new projection methods in statistics and machine learning have been introduced. Most of these methods use assumptions based on geometrical or probabilistic characteristics of the data. On the other hand, pattern classification consists of representing data on computer screen. Expert feedbacks in the classification process have many advantages: visual representations could help making more confident decisions and feedbacks may improve the whole classification process. In our project, a blind classification of collected images and signals will be a vital step to help in making decision.

Using HOS, information theory along with multiscale analysis, we proposed various algorithms to classifying and recognizing noisy digital communication signals; see Figure 19.15. In the actual project, we are aiming to apply BSS and HOS to preprocess a variety of underwater acoustic signals.

In order to obtain a better knowledge of the nature of the seafloor, we also proposed a multiscale segmentation and classification algorithm on textured side-scan sonar image, using parameters coming from a wavelet analysis. Figure 19.16 shows an example of the results of the sonar classification (Klein 5000, data from Gesma), the segmentation (Leblond et al. 2005), and the classification (Leblond et al. 2008).

Multiview classification and fusion were used to improve the classification of sonar images of mines. The originality of this study was consisting in the automatic choice of the additional views using a predictive tool based on the estimation of the angle of view. Figure 19.17 gives a schematic view of the approach (Leblond and Scalabrin 2010).

In another study, we used data recorded in the water column by a split-beam echo sounder. Parameters computed on features automatically extracted from nautical area scattering coefficient were used to provide a segmentation and classification in order to obtain a prediction of algae abundance in a channel. Figure 19.18 shows an example of result of the prediction algorithm where the nautical area scattering

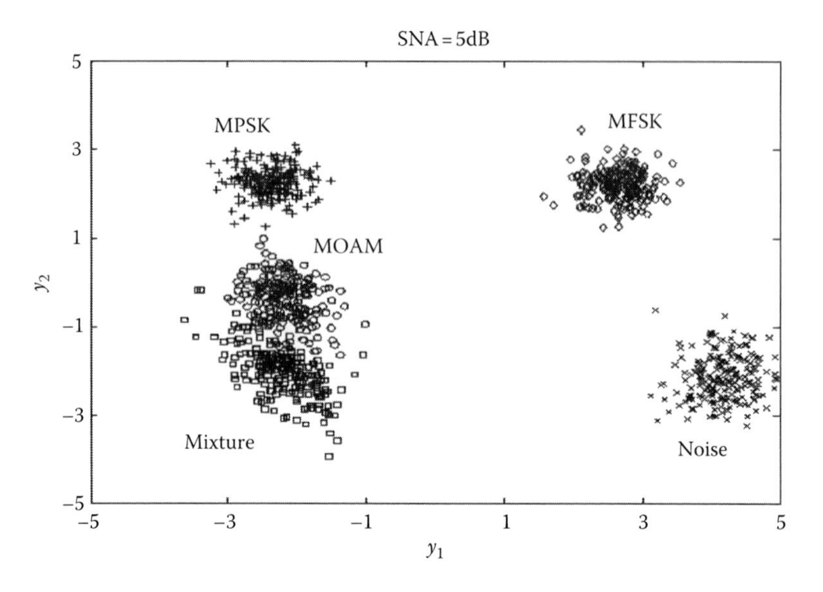

FIGURE 19.15 Classification of intercepted digital signals. (This figure extracted from our works published at ICA 2004.)

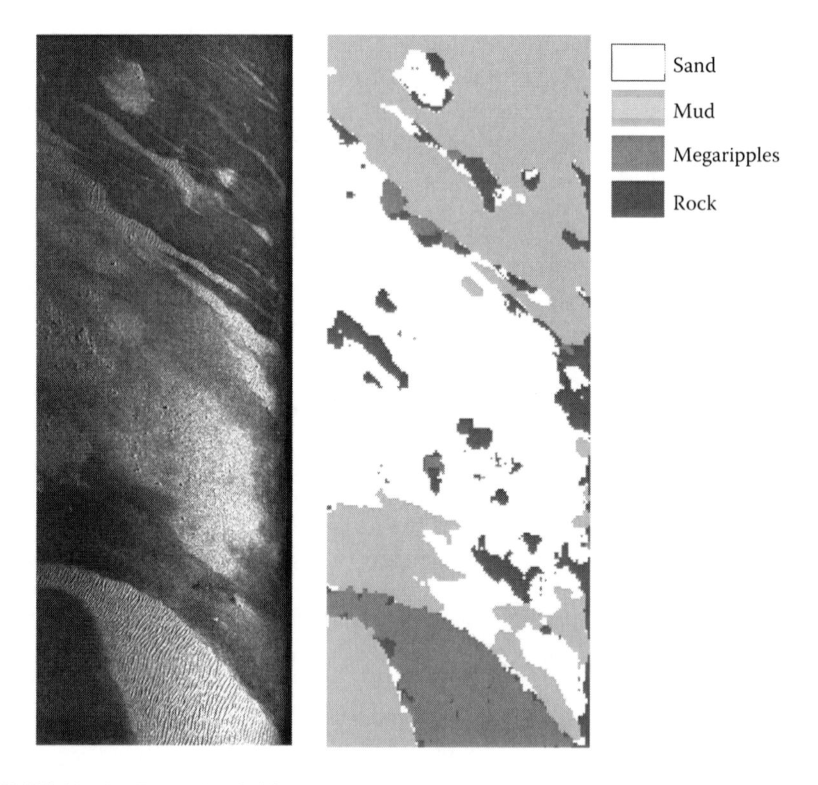

FIGURE 19.16 Example of side-scan sonar image.

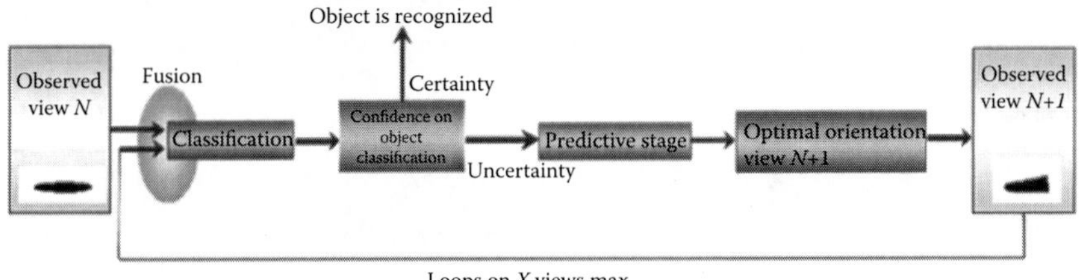

FIGURE 19.17 Schematic of the multiview classification methodology.

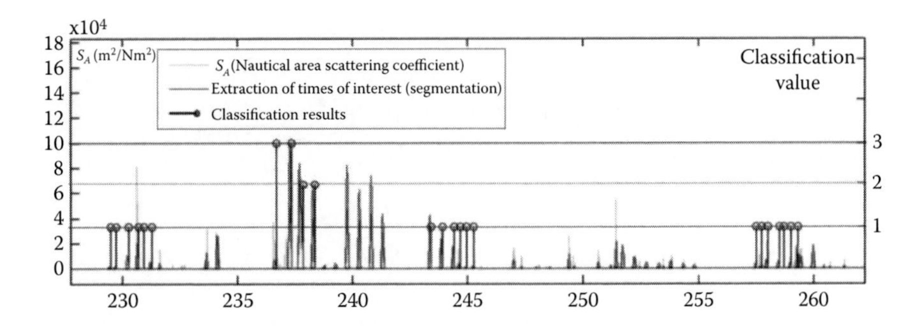

FIGURE 19.18 Example of result of the prediction algorithm (classification value = 1, no much algae; value = 2, more algae; and value = 3, critical presence of algae).

coefficients are computed on acoustic data of algae in the water column, the segmentation of interest periods, and classification results of algae abundance in a channel (Leblond et al. 2010).

19.4.11 Image Processing

Video and statistic images are precious data in order to achieve our goals. Through the three stages of the project, various images should be collected and processed. It was mentioned in previous sections that images acquired with the flow cytometer and sensitive satellite images of invisible lights will be also considered to enhance the outcomes of our approach and help us in monitoring a wide area. Our short-term and long-term studies should focus on specific research axes such as the following:

1. Development of algorithms for improving the estimation of sediments and suspended sediments in the water column. This effort will require sample collection, filtering, and measurement of optical scattering properties.
2. Development of algorithms that will enable the use of remotely sensed data to classify phytoplankton into subcategories, namely, phytoplankton

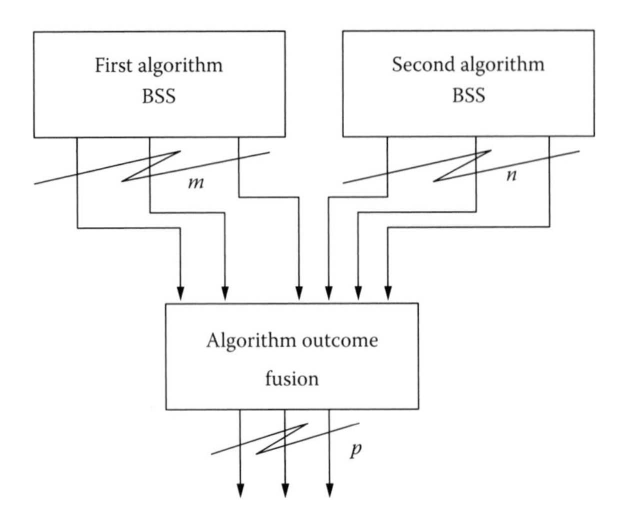

FIGURE 19.19 Fusion data.

functional types. The issue will be to determine how the phytoplankton mix varies regionally and interannually with climate and environmental variability (e.g., coastal water temperature, salinity, freshwater outflow from river systems onto the coast, etc.).

3. Understanding better the vertical distribution of marine pigments. Specifically, the value of remotely sensed observations may be limited in achieving that, especially when it comes to estimating in-water pigment concentrations. In this case, the sensor typically may only be observing the top 15 or 20 m of the ocean. Deep chlorophyll maximum (DCM) has frequently been observed at the base of the mixed layer where new and nutrient-rich water is entrained.

19.4.12 Data Fusion

As mentioned before, this project requires a considerable amount of complex and extensive experimentation, along with associated routine laboratory work, as well as detailed theoretical planning and interpretation. In addition, we should implement various developed approaches in real-time processing algorithms and maintain the experimental equipments. We should also merge and integrate the outlines of the whole process. The latest combination could be done using data fusion methods; see Figure 19.19.

19.5 CONCLUSIONS

Our project is a long-term and a complex project that should achieve port surveillance and costal underwater ecosystem monitoring. To reach our goals, we divided the project into three major stages. The first stage has already been started but far from completed. Up to now, an immerse observation platform is dedicated to our project. This platform contains sensor networks. Various sensors have been already deployed. Major sensors were discussed in this chapter. It is worth mentioning that it is tough to make general conclusions at this early stage of our project. However, as our team and our partners have been approaching the project from different academic and engineering points of view, partial conclusions can be made on every part. According to our studies, we can conclude that the advanced signal, image, and classification algorithms are essential to both simplify the design of the final observation platform and reduce the total price. In addition, once our typical platform has been carefully designed and our algorithms have been optimized, we are planning to deploy many similar platforms in sensitive and strategic places. All these immersed platforms will create a sophisticated and powerful surveillance and monitoring systems.

ACKNOWLEDGMENTS

A part of this work was supported by the French Military Center for Hydrographic and Oceanographic Studies (SHOM, i.e., Service Hydrographique et Océanographique de la Marine) and the French Research Institute for Exploration of the Sea (IFREMER, Institut Français de Recherche et d'Exploitation de la Mer). The exploratory phase of the work on innovative techniques for studying phytoplankton at high resolution is currently supported by the European Regional Development Funds (ERDFs) through the INTERREG IV A *2 Seas* program and the DYMAPHY project (2010–2013) involving partners from northern France (ULCO, CNRS, UL1, IFREMER), England (CEFAS), and the Netherlands (RWS). Thanks are due to all colleagues involved and also to colleagues from AMU-MIO in Marseilles who started the use of *in situ* flow cytometry (CytoBuoy) in France (M. Thyssen, G. Grégori, and M. Denis).

REFERENCES

Achard, S., D.-T. Pham, and C. Jutten, Quadratic dependence measure for nonlinear blind sources separation, in *Fourth International Workshop on Independent Component Analysis and blind Signal Separation, ICA2003*, Nara, Japan, pp. 263–268, April 1–4, 2003.

Alvain, S., C. Moulin, Y. Dandonneau, and H. Loisel, Seasonal distribution and succession of dominant phytoplankton groups in the global ocean: A satellite view. *Global Biogeochemical Cycles*, 22, GB3001, 2008.

Amari, S. I., Neural learning in structured parameter spaces: Natural Riemannian gradient, in *Neural Information Processing System—Natural and Synthetic*, San Diego, CO, December, 2–7, 1996.

Amari, S. I. and J. F. Cardoso, Blind source separation-semiparametric statistical approach, *IEEE Transactions on Signal Processing*, 45(11), 2692–2700, November 1997.

Aubauer, R., M. O. Lammers, and W. W. L. Au, One-hydrophone method of estimating distance and depth of phonating dolphins in shallow water, *Journal of the Acoustical Society of America* 107(5, Pt. 1), 2744–2749, May 2000.

Babaie-Zadeh, M., C. Jutten, and A. Mansour, Sparse ICA via cluster-wise PCA, *Neurocomputing*, 69(13–15), 1458–1466, August 2006.

Bach, F. R. and M. I. Jordan, Finding clusters in independent component analysis, in *Fourth International Workshop on Independent Component Analysis and Blind Signal Separation, ICA2003*, Nara, Japan, pp. 891–896, April 1–4, 2003.

Baggeroer, A. B., W. A. Kuperman, and P. N. Mikhalevsky, An overview of matched field methods in ocean acoustics, *IEEE Journal of Oceanic Engineering*, 18(4), 401–424, 1993.

Brekhovskikh, L. M. and Y. P. Lysanov, *Fundamentals of Ocean Acoustics*, Springer-Verlag, New York, 2003.

Caillault, E., P.-A. Hébert, N. Guiselin, and L.F. Artigas, Classification de cytogrammes par appariement élastique: vers la discrimination automatique du phytoplankton marin par cytométrie en flux. L'objet-8/2009. LMO'2009, pp. 1–15, 2009a.

Caillault Poisson, É., P. A. Hébert, and G. Wacquet, Dissimilarity-based classification of multidimensional signals by conjoint elastic matching: Application to phytoplanktonic species recognition, in *11th International Conference Engineering Applications of Neural Networks, EANN 2009*, London, U.K., Vol. 43, pp. 153–164, Springer, Berlin, Germany, August 27–29, 2009b.

Cardoso, J.-F., Iterative techniques for blind source separation using only fourth-order cumulants, in *European Signal Processing Conference*, Vandewalle, J., Boite, R., Moonen, M., and Oosterlinck, A. (eds.), Brussels, Belgium, pp. 739–742, Elsevier, August, 1992.

Cardoso, J.-F. and P. Comon, Independent component analysis, a survey of some algebraic methods, in *International Symposium on Circuits and Systems Conference*, Atlanta, GA, Vol. 2, pp. 93–96, May 1996.

Cardoso, J.-F. and B. Laheld, Equivariant adaptive source separation, *IEEE Transactions on Signal Processing*, 44(12), 3017–3030, December 1996.

Chapman, N. R. and C. E. Lindsay, Matched-field inversion for geoacoustic model parameters in shallow water, *IEEE Journal of Oceanic Engineering*, 21(4), 347–354, 1996.

Chen, W., J. P. Reilly, and K. M. Wong, Detection of the number of signals in noise with banded covariance matrices, *IEE Proceedings—Radar, Sonar and Navigation*, 143(5), 289–294, October 1996.

Cichocki, A., S. C. Douglas, and S. Amari, Robust techniques for independent component analysis (ICA) with noisy data, *NeuroComputating*, 22, 113–129, 1998.

Comon, P., Independent component analysis, a new concept? *Signal Processing*, 36(3), 287–314, April 1994.

Douglas, S. C., A. Cichocki, and S. I. Amari, Multichannel blind separation and deconvolution of sources with arbitrary distributions, in the book *Neural Networks for Signal Processing*, in *IEEE Workshop on Neural Networks for Signal Processing*, New York, pp. 436–445, September 1997.

Dufrenois, F., J. Colliez, and D. Hamad, Bounded influence Support Vector Regression for robust single model estimation, *IEEE Transactions on Neural Networks*, 20(11), 1689–1706, November 2009.

Etter, P., *Underwater Acoustic Modeling Principles, Techniques and Applications*, Elsevier, New York, 1991.

Etter, P., Recent advances in underwater acoustic modelling and simulation, *Journal of Sound and Vibration*, 240(2), 351–383, 2001.

Etter, P. C., *Underwater Acoustic Modeling and Simulation*, Spon Press Editor, London, U.K., 2003.

Feuerverger, A., A consistent test for bivariate dependence, *International Statistical Review*, 61(3), 419–433, 1993.

Gaucher, D. and C. Gervaise, Feasibility of passive oceanic acoustic tomography: A Cramer Rao bounds approach, in *Oceans 2003 Marine Technology and Ocean Science Conference*, San Diego, CA, pp. 56–60, September 22–26, 2003.

Gaucher, D., C. Gervaise, and H. Le Flock, Contributions to passive acoustic oceanic tomography, in *7me Journes d'Acoustique Sous-Marine*, Brest, France, October 19–20, 2004.

Gervaise, C., A. Quinquis, and N. Martins, Time frequency approach of blind study of acoustic submarine channel and source recognition, in *Physics in Signal and Image Processing, PSIP 2001*, Marseille, France, January 2001.

Gervaise, C., S. Vallez, O. Ioana, Y. Staphan, and Y. Simard, Passive acoustic tomography: Review, new concepts and application using marine mammals, *Journal of Marine Biology Association of United Kingdom*, 87, 5–10, 2007.

Guiselin, N., Etude de la dynamique des communautés phytoplanctoniques par microscopie et cytométrie en flux, en eaux côtières de la Manche orientale, PhD, ULCO, Calais, France, 2010.

Guiselin, N., L. Courcot, L. F. Artigas, A. Le Jéloux, and J.-M. Brylinski, An optimised protocol to prepare *Phaeocystis globosa* morphotypes for scanning electron microscopy observation, *Journal of Microbiological Methods*, 77, 119–123, 2009.

Hermand, J. P., Broad-band geoacoustic inversion in shallow water from waveguide impulse response measurements on a single hydrophone: Theory and experimental results, *IEEE Journal of Oceanic Engineering*, 24(1), 41–66, 1999.

Jensen, F. B., W. A. Kuperman, M. B. Porter, and H. Schmidt, *Computational Ocean Acoustics*, Springer-Verlag, New York, 2000.

Jutten, C. and J. Karhunen, Advances in nonlinear blind source separation, in *Fourth International Workshop on Independent Component Analysis and Blind Signal Separation, ICA2003*, Nara, Japan, pp. 245–256, April 1–4, 2003.

Kalakech, M., Ph. Biela, D. Hamad, and L. Macaire, Semi-supervised evaluation of constraint scores for feature selection, in *International Conference on Neural Computation Theory and Applications*, Paris, France, pp. 175–182, October 24–26, 2011a.

Kalakech, M., Ph. Biela, L. Macaire, D. Hamad, Constraint scores for semi-supervised feature selection: A comparative study, *Pattern Recognition Letters*, 32(5), 656–665, April 2011b.

Kankainen, A., Consistent testing of total independence based on empirical characteristic functions, PhD thesis, University of Jyvaskyla, Jyvaskyla, Finland, 1995.

Kardec Barros, A., A. Mansour, and N. Ohnishi, Removing artifacts from ECG signals using independent components analysis, *NeuroComputing*, 22, 173–186, 1999.

Karhunen, J., A. Cichocki, W. Kasprazak, and P. Pajunen, On neural blind source separation with noise suppression and redundancy reduction, *International Journal of Neural Systems*, 8(2), 219–237, April 1997.

Kawamoto, M., A. Kardec Barros, A. Mansour, K. Matsuoka, and N. Ohnishi, Real world blind separation of convolved non-stationary signals, in J. F. Cardoso, Ch. Jutten, and Ph. Loubaton (eds.), *First International Workshop on Independent Component Analysis and Signal Separation, ICA99*, Aussois, France, pp. 347–352, January 11–15, 1999.

Kawamoto, M., K. Matsuoka, and N. Ohnishi, A method of blind separation for convolved non-stationary signals, *Neurocomputing*, 22, 157–171, 1998.

Kawamoto, M., K. Matsuoka, and M. Oya, Blind separation of sources using temporal correlation of the observed signals, *IEICE Transactions on Fundamentals of Electronics, Communications and Computer Sciences*, E80-A(4), 111–116, April 1997.

Kendall, M. and A. Stuart, *The Advanced Theory of Statistics: Design and Analysis, and Time-Series*, Charles Griffin & Company Limited, London, U.K., 1961.

Korneliussen, R. J., Y. Heggelund, I. K. Eliassen, O. K. Øye, T. Knutsen, and J. Dalen, Combining multibeam-sonar and multifrequency-echosounder data: Examples of the analysis and imaging of large euphausiid schools, International Council for the Exploration of the Sea, Published by Oxford Journals, 2009.

Leblond, I., C. Berron, and I. Quidu, Mise en oeuvre de stratégies prédictives sur les vues à acquérir pour la classification multi-vue d'objets immergés à partir d'images SAS, in *Conférence RFIA'2010 Reconnaissance des Formes et Intelligence Artificielle*, Caen, France, Janvier du 19 au 22, 2010.

Leblond, I., M. Legris, and B. Solaiman, Use of classification and segmentation of sidescan sonar images for long term registration, in *IEEE OCEANS'05 EUROPE*, Brest, France, June 20–23, 2005.

Leblond, I., M. Legris, and B. Solaiman, Apport de la classification automatique d'images sonar pour le recalage à long terme, Revue Traitement du Signal, Vol. 25 numéro double 1–2, 2008 "Caractérisation des Milieux Marins".

Leblond, I. and C. Scalabrin, Etudes sur la détection d'algues dans la colonne d'eau par sondeurs halieutiques, in *Workshop MOQESM (MOnitoring Quantitatif de l'Environnement Sous-Marin)*, Sea Tech Week, Brest, France, June 23, 2010.

Leblond, I., C. Scalabrin, L. Géli, and L. Berger, Acoustic monitoring of gas emissions from the seafloor: Estimation of bubbles volumetric flows by inverse modeling, 2013, in preparation.

Lurton, X., *Introduction to Underwater Acoustics Principles and Applications*, Springer, London, U.K., 2002.

MacLennan, D. N., P. G. Fernandes, and J. Dalen, A consistent approach to definitions and symbols in fisheries acoustics, *ICES Journal of Marine Science*, 59, 365–369, 2002.

Mansour, A., A mutually referenced blind multiuser separation of convolutive mixture algorithm, *Signal Processing*, 81(11), 2253–2266, November 2001.

Mansour, A., Enhancement of acoustic tomography using spatial and frequency diversities, *EURASIP Journal on Advances in Signal Processing*, doi:10.1186/1687-6180-2012-225, Issue: On Line, 24 Oct 2012. The electronic version of this article is available at: http://asp.eurasipjournals.com/content/2012/1/225.

Mansour, A. and C. Gervaise, ICA applied to passive ocean acoustic tomography, *WSEAS Transactions on Acoustics and Music*, 1(2), 83–89, April 2004.

Mansour, A. and C. Jutten, What should we say about the kurtosis?, *IEEE Signal Processing Letters*, 6(12), 321–322, December 1999.

Mansour, A., C. Jutten, and Ph. Loubaton, Subspace method for blind separation of sources and for a convolutive mixture model, in *European Signal Processing Conference*, Triest, Italy, pp. 2081–2084, September 1996.

Mansour, A., A. Kardec Barros, and N. Ohnishi, Blind separation of sources: Methods, assumptions and applications, *IEICE Transactions on Fundamentals of Electronics, Communications and Computer Sciences*, E83-A(8), 1498–1512, August 2000.

Mansour, A., N. Ohnishi, and C. G. Puntonet, Blind multiuser separation of instantaneous mixture algorithm based on geometrical concepts, *Signal Processing*, 82(8), 1155–1175, 2002.

Martin, A. and A. Mansour, Comparative study of high order statistics estimators, in *International Conference on Software, Telecommunications and Computer Networks*, Split, Dubrovnik, Croatia; Venice, Italy, pp. 511–515, October 10–13, 2004.

Matsuoka, K., M. Oya, and M. Kawamoto, A neural net for blind separation of nonstationary signals, *Neural Networks*, 8(3), 411–419, 1995.

McCullagh, P., *Tensor Methods in Statistics*, Chapman & Hall, London, U.K., 1987.

Munk, W., P. Worcester, and C. Wunsch, *Ocean Acoustic Tomography*, Cambridge University Press, Cambridge, U.K., 1995.

Murata, N., Properties of the empirical characteristic function and its application to testing for independence, in *Third International Workshop on Independent Component Analysis and Signal Separation, ICA2001*, San Diego, CA, pp. 295–300, December 9–12, 2001.

Nguyen Thi, L. and C. Jutten, Blind sources separation for convolutive mixtures, *Signal Processing*, 45(2), 209–229, 1995.

Nguyen Thi, L., C. Jutten, and J. Caelen, Separation aveugle de parole et de bruit dans un mlange convolutif, in *Actes du XIIIème colloque GRETSI*, Juan-Les-Pins, France, pp. 737–740, September 1991.

Olson, R. J., A. Shalapyonok, and H. M. Sosik, An automated submersible flow cytometer for analyzing pico- and nanophytoplankton: FlowCytobot, *Deep Sea Research I*, 50, 301–315, 2003.

Papoulis, A., *Probability, Random Variables, and Stochastic Processes*, McGraw-Hill, New York, 1991.

Parra, L. and C. V. Alvino, Convolutive blind separation of non-stationary sources, *IEEE Transactions on Speech and Audio Processing*, 8(3), 320–327, May 2000.

Pham, D.-T., Fast algorithm for estimating mutual information, entropies and score functions, in *Fourth International Workshop on Independent Component Analysis and Blind Signal Separation, ICA2003*, Nara, Japan, pp. 17–22, April 1–4, 2003.

Puntonet, C. G., A. Mansour, C. Bauer, and E. Lang, Separation of sources using simulated annealing and competitive learning, *NeuroComputing*, 49(12), 39–60, 2002.

Rahbar, K. and J. Reilly, Blind separation of convolved sources by joint approximate diagonalization of cross-spectral density matrices, in *Proceedings of International Conference on Acoustics Speech and Signal Processing 2001, ICASSP 2001*, Salt Lake City, UT, May 7–11, 2001.

Rahbar, K. and J. Reilly, A frequency domain method for blind source separation of convolutive audio mixtures, *IEEE Transactions on Speech and Audio Processing*, 13(5), 832–844, 2005.

Rosenblatt, M., A quadratic measure of deviation of two-dimensional density estimates and a test of independence, *Annals of Statistics*, 3(1), 1–14, 1975.

Roy, N., Y. Simard, and C. Gervaise, 3D tracking of foraging belugas from their clicks: Experiment from a coastal hydrophone array, *Applied Acoustics*, 71, 1050–1056, 2010.

Shiryaev, A. N., *Probability*, Springer-Verlag, London, U.K., 1984.

Shulkin, M. and H. W. Marsh, Sound absorption in sea water, *Journal of the Acoustical Society of America*, 134, 864–865, 1962.

Simmonds, J. and D. MacLennan, *Fisheries Acoustics*, Blackwell Publishing, Oxford, U.K., 2005.

Stanton, T. S., P. H. Wiebe, D. Chu, M. C. Benfield, L. Scanrlon, L. Martin, and R. L. Eastwood, On acoustic estimates of zooplankton biomass, *ICES Journal of Marine Science*, 51, 505–512, 1994.

Tan, Y., J. Wang, and J. M. Zurada, Nonlinear blind source separation using a radial basis function network, *IEEE Transactions on Neural Networks*, 12(1), 124–134, January 2001.

Thyssen, M., B. Beker, D. Ediger, D. Yilmaz, N. Garcia, and M. Denis, Phytoplankton distribution during two contrasted summers in a Mediterranean harbor: Combining automated submersible flow cytométrie with conventional techniques, *Environmental Monitoring and Assessment*, 173, 1–16, 2011.

Thyssen, M., D. Mathieu, N. Garcia, and M. Denis, Short-term variation of phytoplankton assemblages in Méditerranean coastal waters recorded with an automated submerged flow cytometer, *Journal of Plankton Research*, 30, 1027–1040, 2008.

Wacquet, G., P. A. Hébert, É. Caillault Poisson, D. Hamad, Semi-supervised K-way spectral clustering using pairwise constraints, in *International Conference on Neural Computation Theory and Applications*, Paris, France, pp. 72–81, October 24–26, 2011.

Section VI

Disaster Management

20 Triage with RFID Tags for Massive Incidents

Sozo Inoue, Akihito Sonoda, and Hiroto Yasuura

CONTENTS

ABSTRACT

This chapter proposes a triage system using radio-frequency identification (RFID) and wireless systems that can be used to manage disaster situations. Triage is a procedure used by emergency personnel to manage limited medical resources to benefit a large number of injured people using triage tags. The proposed system of using RFID triage tags and wireless sensors has many advantages. Effectiveness of the proposed system is evaluated through an assumed situation/experiment involving a complex car crash of five cars and more than eighty injured people. The comparative results show significant improvement. The system can be applied to other disaster situations such as earthquakes, storms, or floods.

20.1 INTRODUCTION

In this chapter, a triage system using RFID and mobile devices with a wireless network is proposed, and its advantages are verified through an experiment assuming an incident of massive injured people.

Triage is a procedure used by emergency personnel to ration limited medical resources to massive injured people, in which triage tags are used to (1) classify and transport the injured people effectively and (2) obtain the information about the state and the scale of the casualty incident to publish to the masses or to utilize for decision making such as medical resource procurements. Figure 20.1 is a picture of a triage tag.

So far, triage is operated manually using paper triage tags, tallies, and radiophones. However, manually obtaining the information about the state and the scale of the casualty incident to publish to the masses or to utilize for decision making such as medical resource procurements leads to failure, inaccuracy, and delay in the information transmission while emergency personnel have priority over treatments for injuries and causes inefficiency in classification and transport of the injured people effectively, which is the essential goal of emergency medical services.

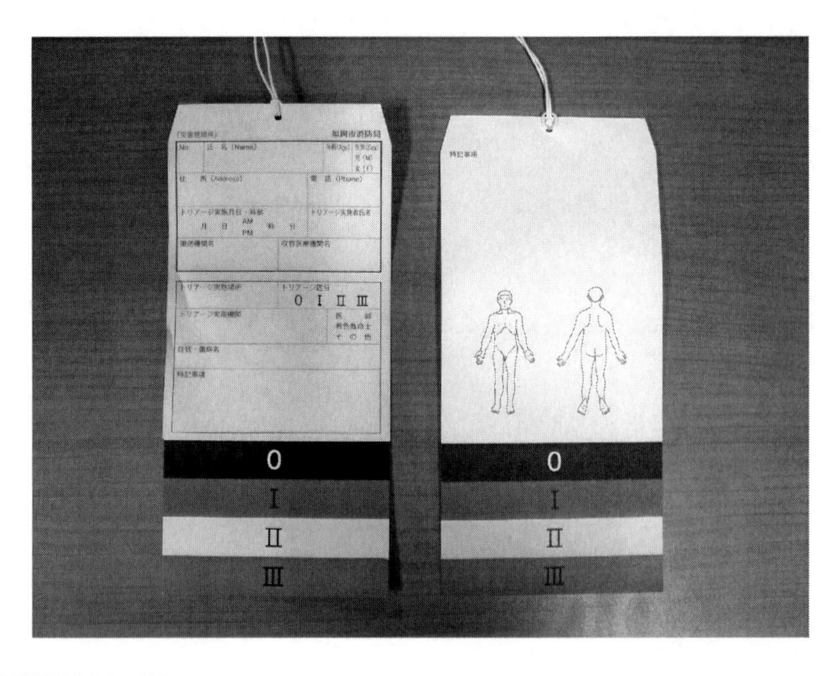

FIGURE 20.1 Triage tag.

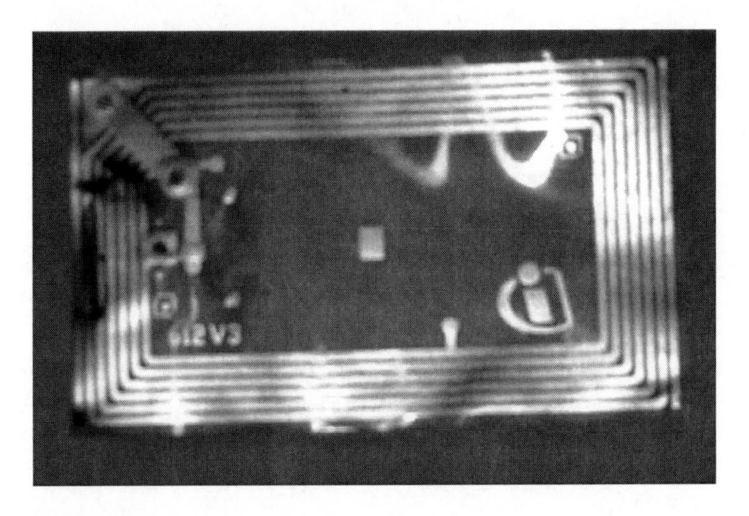

FIGURE 20.2 RFID tag.

In this chapter, we propose a triage system in which RFID tags, which are silicon chips with their IDs, radio-frequency functions, and some additional logic and memory [1,2], are attached to triage tags. Most of the RFID tags are passive, which means the power is supplied through radio-frequency communication from external readers. Employed RFID tags in this work are passive and have 1 kb of rewritable memories. Figure 20.2 is a picture of an RFID tag we employed.

Embedding an RFID tag to a triage tag has the following advantages:

1. A terminal that the emergency personnel use can identify the injured person by the unique ID value in each RFID tag.
2. Rewritable RFID tags provide the storage for the information of the injured person, and the emergency personnel can obtain the information when they are in a place where wireless communication is out of service, such as deep in a mountain or in the underground.
3. Time required to input an injured person's information can be independent of the network communication, since the information already input so far about the injured person, which is shown to the emergency personnel when they begin to add information, is not obtained through the network, but from the RFID tag.

This application addresses important challenges for pervasive computing: data integrity, the information that should not be lost after being input; input throughput, the time required to input the injured person's information should be as short as possible; availability, that is, emergency personnel should be able to use the system any time; and low latency of communication. These should be as independent of the network status as possible.

In this chapter, we show a realistic solution for the challenges by specializing the network usage in a way that only particular paths are used in particular stages of the workflow by analyzing the workflow and exploiting RFID tags to slim down the possible paths by the following approaches: input throughput and availability are assured by using RFID tags as local buffer; data integrity is assured and latency is improved by defining minimum wireless communication areas in the paths in triage workflow.

To evaluate the effectiveness of the system, we performed two experimental performances of triage assuming a complex car crash of 5 cars and about 82 injured people, where one of the performances is done in the current method and the other is done using the system. As a result of the experiment, the information collection was accelerated and made more complete by using the system, especially for the information that is utilized in early stages of the triage. The acceleration of the collection also resulted in the acceleration of the transportation of the injured people, which is the most important objective in triage. Moreover, we could estimate that the average acceleration of each input to the terminal by employing RFID tag in the experiment was 14.3 s.

The rest of this chapter is organized as follows: Section 20.2 describes triage and challenges for pervasive computing, Section 20.3 introduces the RFID triage system, Section 20.4 describes the related work, Section 20.5 shows the result of a preliminary experiment to complement the experiment, Section 20.6 shows the result of the experiment to confirm the effectiveness of the system, and Section 20.7 concludes this chapter.

20.2 TRIAGE AS A CHALLENGE FOR PERVASIVE COMPUTING

We position triage as not only a practical application that requires urgent and continuous improvement since we are facing massive casualty incidents every day but also a start point for new horizons in pervasive computing concerning unexpected human behavior, rapid deployment, and insufficient computing infrastructure. In this section, we describe the workflow of triage operated by emergency personnel, address technical requirements for triage regarding the paths of the information collected as that of network, and motivate the challenges for pervasive computing hidden in the requirements.

We target on incidents in which 10–100 people are injured in a regional place. Examples of such incidents are a crash/derailment/overturn of a train, a crash/fall of a car, a crash/fire of a plane, and terrorism in a subway/building.

20.2.1 TRIAGE

We call the people who are engaged in triage emergency personnel. Figure 20.3 shows an abstract workflow of triage.

Current triage is done in the following procedure without using information systems by the emergency personnel:

1. Emergency personnel who arrive to the incident site first establish a first aid area, which is a safe place for first aid close to the incident site, and an operation point, which is a place to command the triage.

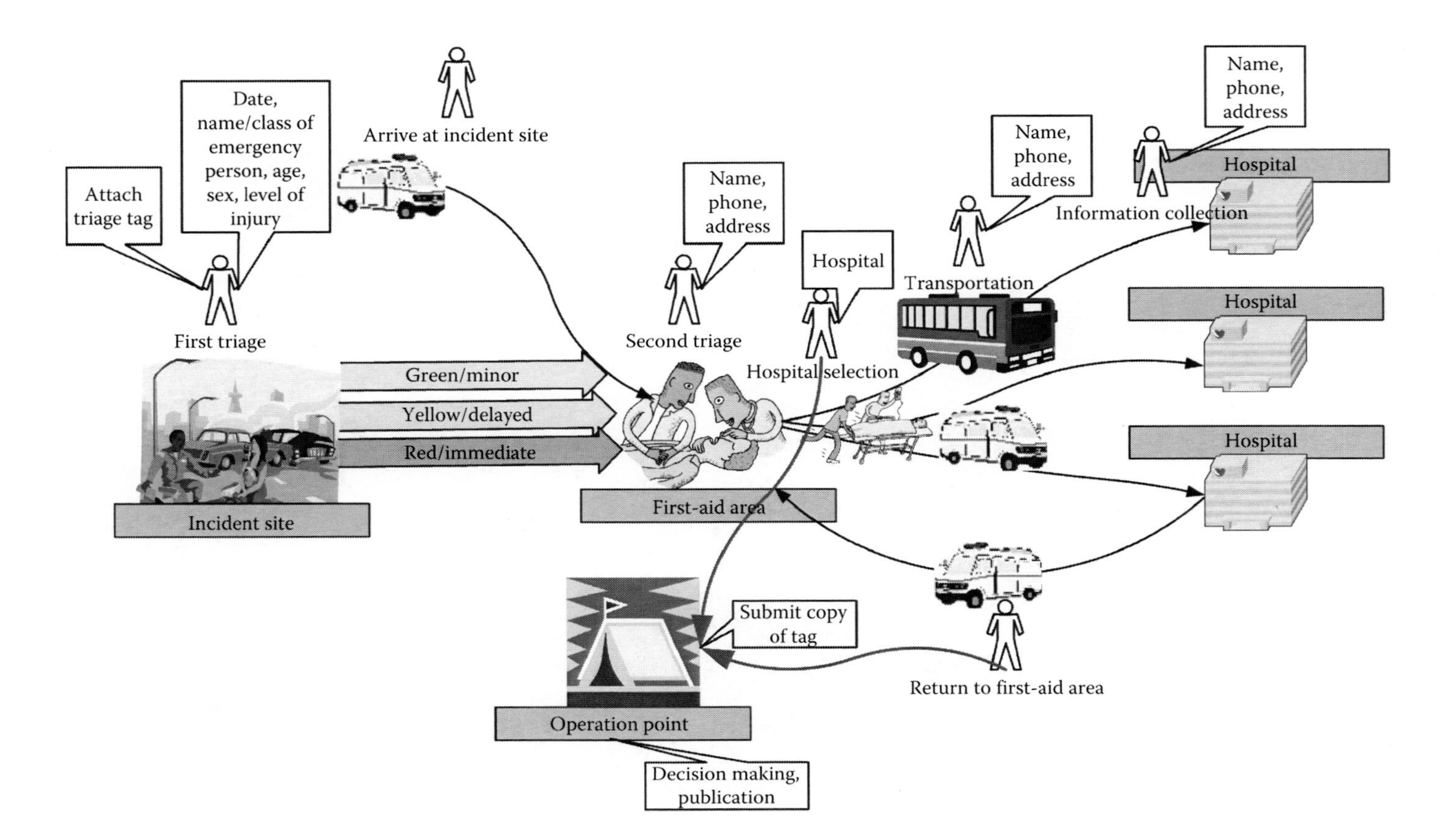

FIGURE 20.3 Workflow of triage.

2. *First triage*: After the evacuation of injured people once the incident site is secured by firemen, the emergency personnel enter the incident site and perform first triage, in which the personnel distinguish the injury level in about 30 s. During the first triage, they attach triage tags, each of which has a perforated colored label representing the injury level, as the following, to the injured people:
   a. *Black/deceased*: The person is severely injured to die and any hospital care does not help survive.
   b. *Red/immediate*: The person requires immediate surgery or other life-saving intervention; the person is unconscious and has first priority to be transported to hospitals.
   c. *Yellow/delayed*: The person requires hospital care and requires to be watched by trained persons, but the person remains conscious and the condition is stable for the moment.
   d. *Green/minor*: The person may require a doctor's care but not immediately, and the person is able to walk on her/his own.

   Simultaneously, the following information is written by the emergency personnel as much as possible:
      a. Date of input
      b. Name of the emergency personnel
      c. Category of the emergency personnel, such as doctor, emergency medical technician
      d. Rough age of the injured person
      e. Sex of the injured person
      f. Written version of the injury level
3. Injured people are moved from the incident site to the first aid area.
4. *Second triage*: In the first aid area, besides treatment of the injured people, second triage is performed if possible, in which the information of the injured people such as
   a. Name
   b. Phone number
   c. Address
   d. Updated information from first triage

   are collected as much as the emergency personnel and the injured people can afford to and written on the triage tags.
5. *Hospital selection*: At the exit of the first aid area, the hospital to which each injured person is transported is decided and written on the triage tag. A triage tag has carbon copies, and one of them is left to the emergency personnel of the operation point.
6. When an ambulance or a transport vehicle arrives at the first aid area, the injured people are transported from the first aid area to the hospital. In the ambulances, the information of the injured people are collected as much as possible and written on the triage tags.
7. Other emergency personnel stand by at each hospital and collect the information of the injured people transported. The ambulance, after transporting

injured people to hospitals, returns to the incident site again and repeats the transportation of injured people to hospitals. On returning to the incident site, the ambulance transports the carbon copy of the triage tag at the hospital to the operation point.
8. In the operation point, the information on the carbon copies are collected and reported to the fire department of the area, which is to be used for decision making, publication, etc.

20.2.2 Requirements

The first mission of triage targeted on in this chapter is to transport injured people as quickly as possible. As for the mission, information collection of injured people affects the latency of transportation in two ways: troublesome operations for information collection by an emergency personnel cause latency by disturbing her/his work, and particular types of input information should be collected in the early stages since they are used for decisions in the transportation. For the latter, information such as the level of injury and hospital to be transported have to be collected within the first or second triage before the transportation where these information are used, whereas in the other types, they have not been collected since they are used in the afterward. However, in the current triage, all of these information must be collected before transportation for the sake of avoiding loss and latency of the collection after the ambulance has left.

Thus, several technical requirements that can be identified as important challenges for pervasive computing arise regarding the paths through which the information of injured people is collected:

1. *Data integrity*: The information of injured people should not be lost after being input. However, in current triage, the complete information of injured people cannot be collected when triage tags fail to be cut off, are lost, or left in the pocket of an emergency person. Moreover, the information filled in a triage tag later than the first or second triage stages will be lost when the ambulance does not return to the incident site.
2. *Input throughput*: The time to input the injured person's information should be as short as possible. Especially, it should be so even if the latency of the network communication becomes large.
3. *Availability*: Emergency personnel should be able to use the system. Especially, input operations of the injured person's information should be available anytime in the triage, even when the network is unreachable.
4. *Low latency*: Several types of input information should be quickly collected and viewed from the operation point. In current triage, the information filled into a triage tag later than the first or second triage stage is not collected until the ambulance that transported the injured person returns to the incident site, even if they are filled in the ambulance or at the hospital, since additional manual communication from the operation point to hospitals or the ambulance increases the work in the operation point.

20.2.3 Challenges in Pervasive Computing

The requirements listed in the previous section are important challenges in pervasive computing, which aims one in harmony with real-world circumstance, including unexpected human behavior, rapid deployment, or insufficient computing infrastructure. In this chapter, we motivate these challenges and address an improvement in a special but realistic case of pervasive computing, in addition to the improvement of triage by introducing information technology.

Data integrity and latency will be improved by exploiting a wireless network, since the input information can be collected through the network, even if emergency personnel do not directly or indirectly give a triage tag to the person who is in charge of collecting the information. However, this stands on the assumption that information once input can be reached in a destination in the network for more than a while. This assumption does not always hold in pervasive computing. Hence, data integrity and low latency in the semireachable network are some of the challenges in pervasive computing.

Input throughput and availability against the network are further challenging requirements in pervasive computing. The real world is not of course composed as a set of packets like TCP/IP but behaves as the real-time system. The interaction between the real world and the network in the interface such as mobile devices, sensors, and RFID tags becomes incoherent if the gap is not absorbed in the network nodes in appropriate layers from physical to human workflows. In this sense, independence of network status for real-world interfaces is also one of the attractive challenges in pervasive computing.

In this chapter, we show a realistic solution for the challenges by specializing the network usage in a way that only particular paths are used in particular stages of the workflow by analyzing the workflow and exploiting RFID tags to slim down the possible paths by the following approaches:

1. Input throughput and availability are assured by using RFID tags as local buffers: users can input data by referring the RFID data already input so far, and the input device pushes the input data to the queue that is sent to the destination independent of the user's operation, as well as writing to the RFID tag.
2. Data integrity is assured and latency is improved by providing wireless communication areas at least to the final stages of the paths in triage workflow: hospitals and the operation point. Actually, latency will be more improved if the stages covered with wireless communication areas increase, but we believe it will be better than current triage, as experimented in Section 20.6.

The requirements in environments of insufficient network infrastructure including disasters are also discussed in mobile and ad hoc networks [3–5] with a requirement for quick deployment, but our approach of using RFID tags for input throughput and availability is unique, while most of the RFID applications in the literature use RFID tags as a device to identify objects or people by embedding unique IDs.

20.3 RFID TRIAGE SYSTEM

In this section, we describe the triage system using RFID tags. Figure 20.4 shows an abstract workflow in the RFID triage system.

Other than the advantages described in Section 20.2.3, we can employ the following advantages of the system including primitive ones:

1. Using mobile devices with wireless communication, the information of the injured people is collected quickly via the network.
2. Input method using mobile devices provides ease of reading compared with handwriting.
3. Input throughput is improved by automating the information of the emergency personnel and addresses from postal codes and by reducing the data types necessarily required to be input in the early stages only to the injury level and hospital.

20.3.1 System Architecture

The following are the system components, which are shown in Figure 20.5:

1. *Triage tag*: A tag with input forms for the information of the injured person and an RFID tag with rewritable memory of 1 kb and wireless communication in the frequency of 13.56 MHz. Each RFID tag has a unique ID in the system.
2. *First triage terminal*: A mobile terminal device that an emergency person in charge of the first triage stage uses. It is equipped with an RFID reader and a wireless communication interface. An emergency person can input injured peoples' information through the touch panel or buttons. After input, the information is saved to the RFID tag, as well as sent to the server through the wireless network. In case the network is disconnected, the information is stored on the terminal and resent when the network is connected. Static or automatic information such as the date and name/category of the emergency personnel are input by the terminal automatically. Figure 20.6 is the display of input terminal.
3. *Second triage terminal*: Similar to the first triage terminal, there is also a mobile terminal device that an emergency person in the second triage stage uses.
4. *Hospital selection terminal*: Similar to the first triage terminal, there is also a mobile terminal device that an emergency person in the hospital selection stage uses.
5. *Ambulance terminal*: A notebook PC equipped with a handy RFID reader and a wireless communication interface is placed in each ambulance. Emergency personnel can input the information of injured people using a keyboard and a mouse. First, several parts of an address can be converted from a postal code.

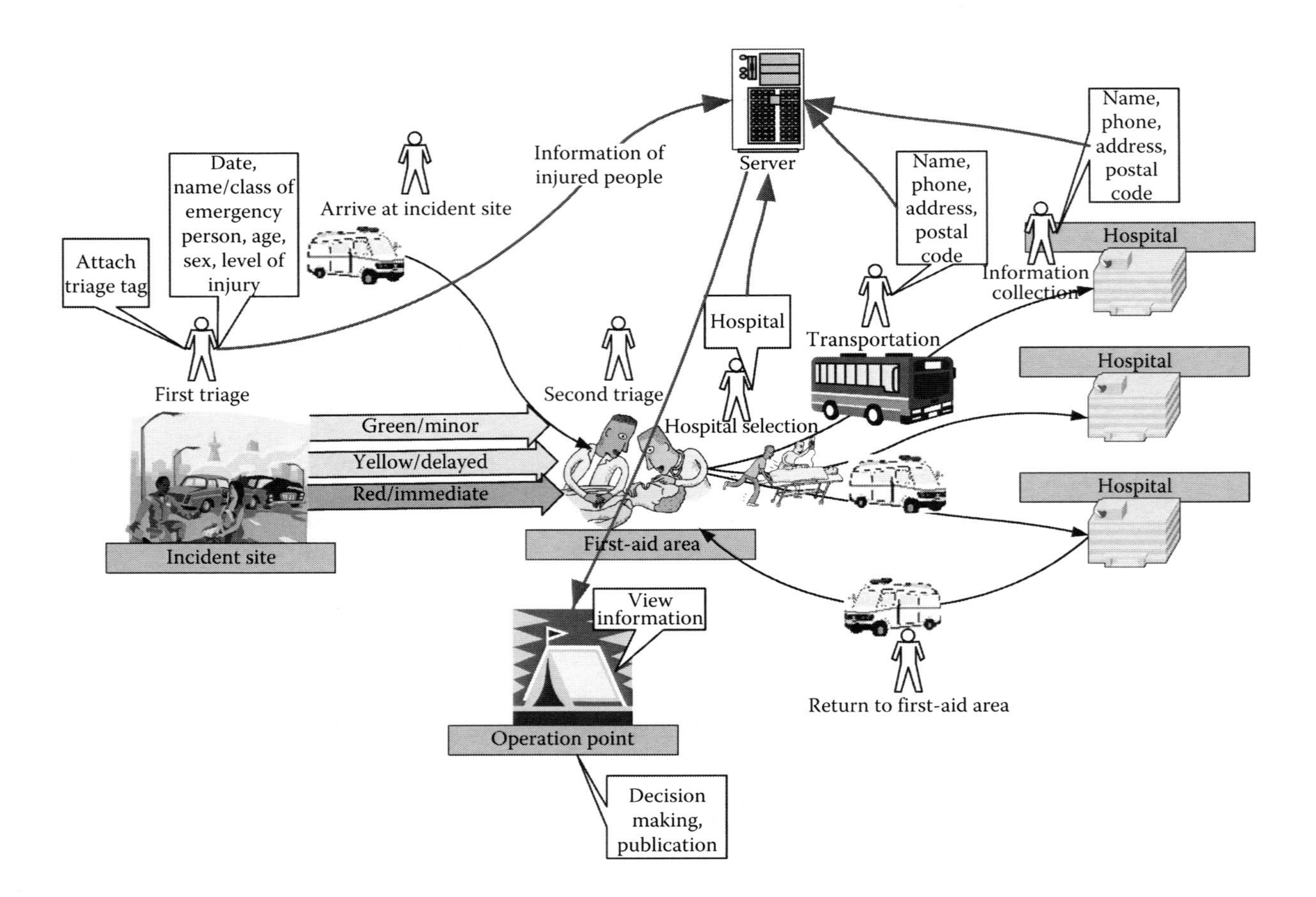

FIGURE 20.4 Workflow in RFID triage system.

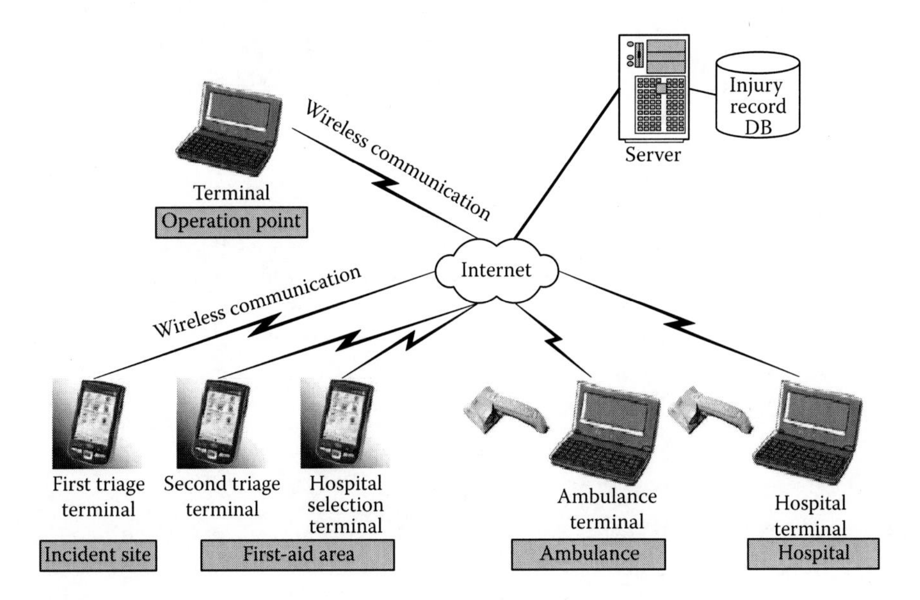

FIGURE 20.5 System and the network architecture.

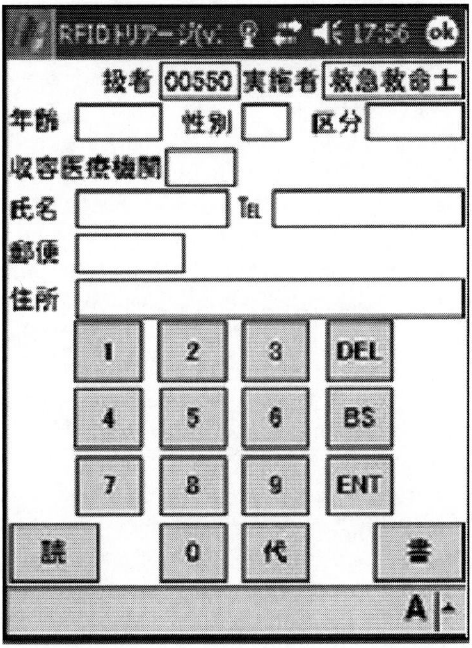

FIGURE 20.6 Display of input terminal in the first and second triage stages.

6. *Hospital terminal*: Similar to ambulance terminals, a notebook PC is also available that emergency personnel in each hospital use.
7. *Operation point terminal*: A notebook PC equipped with a wireless communication interface is placed in the operation point. Emergency personnel can browse the information collected to the server through a web browser software.
8. *Server*: Server equipment that has a database of the injured people is placed away from the incident site. The server stores the information sent from the terminals to a database and responds to the browsing request from the operation point terminals with HTTP protocols. Moreover, the server keeps the information of an injured person up to date from those sent by multiple terminals.

20.3.2 System Flow

Rough workflow of triage does not change when the system is applied. The difference is that instead of writing and collecting by hand, emergency personnel read the information of each injured person from the RFID tag, input the information of each injured person to the terminal, and then write it on the RFID tag. In the following, we describe the usage and the behavior of the system for each role of the emergency personnel.

An emergency person in charge of first triage inputs the information of each injured person as much as possible to the first triage terminal, and then writes it to the RFID tag. Simultaneously, she/he attaches the triage tag with the RFID tag to the injured person and cuts off the perforated label to the corresponding color. The information of the injured person written to the RFID tag is sent to the server as soon as the network becomes available.

An emergency person in charge of the second triage first reads the information of each injured person from the RFID tag through the second triage terminal and adds/updates the information to the RFID tag by interviewing the injured person. An emergency person in charge of hospital selection selects the hospital each injured person is transferred to and inputs it to the RFID tag through the hospital selection terminal.

An emergency person in an ambulance or in a hospital does the same operation to the injured person as the second triage but does it using a notebook PC and a handy RFID reader. Address is automatically complemented when a postal code is input. Additionally, the emergency person clicks the commit button on the display and records the time of carrying the injured person to the ambulance.

An emergency person in the operation point views the information of the injured people, which is collected to the server, on the operation point terminal using a web browser software and informs them to the emergency control center in the municipality as she/he needs. The server stores the information sent from the terminals and serves them to the operation point terminal through HTTP protocol. Multiple uploads from terminals for the same injured person are sorted in time order and kept up to date. This workflow is a natural extension from current triage except the time for inputs, which we evaluated to be trivial in Section 20.6.5.

Table 20.1 shows the types of collected information in each stage in the current workflow or in using the system. From the table, we observe that the information

TABLE 20.1
Types and Stages of Collected Information in Current Workflow or RFID Triage System

Type of Information Stage	Date	Emergency Person	Category of Emergency Person	Age	Sex	Level of Injury	Hospital	Name	Phone Number	Address
First triage	Auto	Auto	Auto	Y	Y	Y	—	—	—	—
Second triage	Auto	Auto	—	—	—	—	—	Y > N	Y > N	Y > N
Hospital selection	—	—	—	—	—	—	Y	—	—	—
Ambulance	—	—	—	—	—	—	Y	Y	Y	Y
Hospital	—	—	—	—	—	—	Y	Y	Y	Y

Notes: Y, input by emergency personnel both in current workflow and RFID triage system; Y > N, not input in RFID triage system, while input in current workflow; Auto, automatically or previously input in the system, while input in current workflow; —, not input in both cases.

used for the publication and which needs time to input, such as name, phone number, and address, is input after the hospital selection finishes, and the emergency personnel in charge of the stage not later than hospital selection only have to input the information that is simple to input, such as the level of injury, age, sex, and hospital. Moreover, the name and the category of the emergency personnel and the date can be input automatically or previously before the incident.

20.4 RELATED WORK

Triage is captured as an application that requires immediate improvement in the field of emergency medicine [6–11] and is realized as one of the important applications in pervasive computing [12].

A variety of information systems can be considered to be useful to support triage, such as wireless and mobile telemedicine systems [13]. Wireless network composition for such an application is discussed [14]. Tiny devices such as sensors to the network are tried to be used for vital sensors for injured people [12]. Active RFID tags are tried to be used for tracking the location of injured people [15].

On the other hand, information systems for supporting disaster management are proposed by featuring several key technologies in pervasive computing. Geographical information systems are used to locate and identify the context of people and objects [16]. Immediate deployment of wireless networks is discussed in [17]. RFID tags are tried to use for planning and guiding the evacuation [18].

Several works have exploited wireless networks, and several exploit RFID tags. However, these works along with networks do not address the requirements for data integrity, input throughput, availability, and latency as we have shown explicitly. Moreover, RFID tags are not used for improving the requirements. Our approach is unique in the sense that we use RFID tags for the alternative of current triage tags and for improving the requirements by alternating wireless network communication with the communication with RFID tags.

Similar to our approach, the US Navy tried to use rewritable RFID tags as triage tags, but they did not try to incorporate wireless network communication and RFID communication [19].

20.5 PRELIMINARY EXPERIMENT

Before the massive experiment, we evaluated several points with a small number of participants. We measured input operations five times in a room. Figures 20.7 through 20.9 are the results of the measurements.

Figure 20.7 is the comparison of input times of each part in first triage with and without the system. As shown in the figure, the time for an injured person is reduced to about 38 s from about 42 s. The detail shows that the most contribution is because the emergency personnel, with the system, do not have to input the level of injury, the name of the personnel, and the date. On the other hand, the time for preparing and cutting triage tags is increased by having input terminals in the case and about 2.5 s for writing on triage tags.

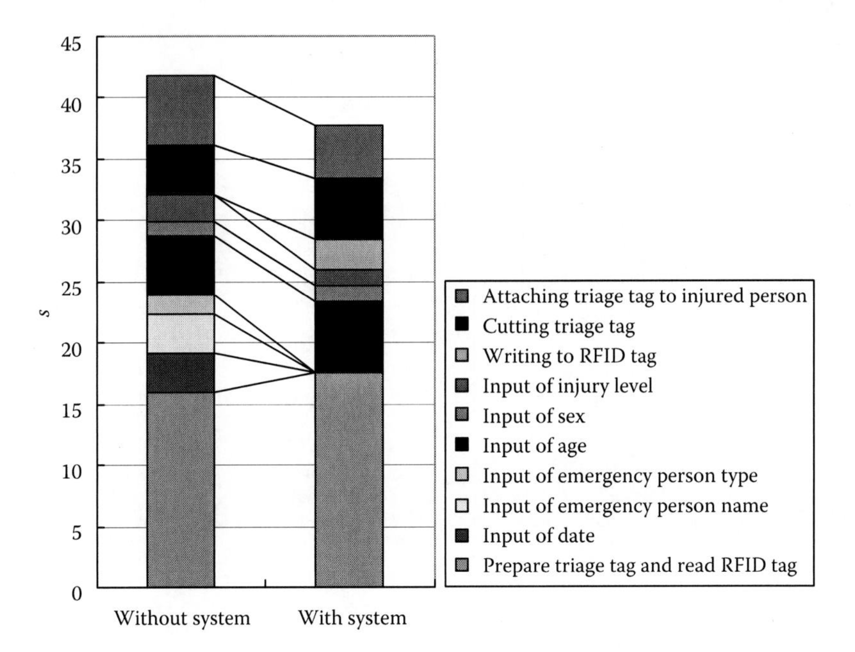

FIGURE 20.7 Input time in first triage.

Figure 20.8 is the comparison of input times of each part in ambulances or at hospitals with and without the system. As shown in the figure, the time for an injured person increased to about 43 s from about 35 s. The detail shows that the address, the postal code, the phone, and the name of the injured person take a longer time with the system than without the system. Although the system employs automatic address complements from postal codes, it is still longer than that without the system. Additionally, it needs about 3.5 s for writing on triage tags. It requires further work to reduce the input time for ambulances and hospitals, while there is a problem that, without the system, manual and hastened writing is often difficult to read.

Figure 20.9 is the comparison of input times of each part in hospital selections with and without the system. As shown in the figure, the time for an injured person was reduced to about 8 s from about 12 s. The detail shows that the operations without the system require the emergency personnel to pass the copy of the triage tags to other personnel nearby, whereas they do not with the system. On the other hand, the operations without the system can start the hospital information immediately, whereas in the case with the system, emergency personnel need first to read the information input so far from the RFID tags before input.

20.6 EXPERIMENT

In this section, we describe the experiment of using the RFID triage system. The experiment is done in a driving school in a day, assuming a complex car crash of 5 cars and about 82 injured people. On hearing from professionals of emergency

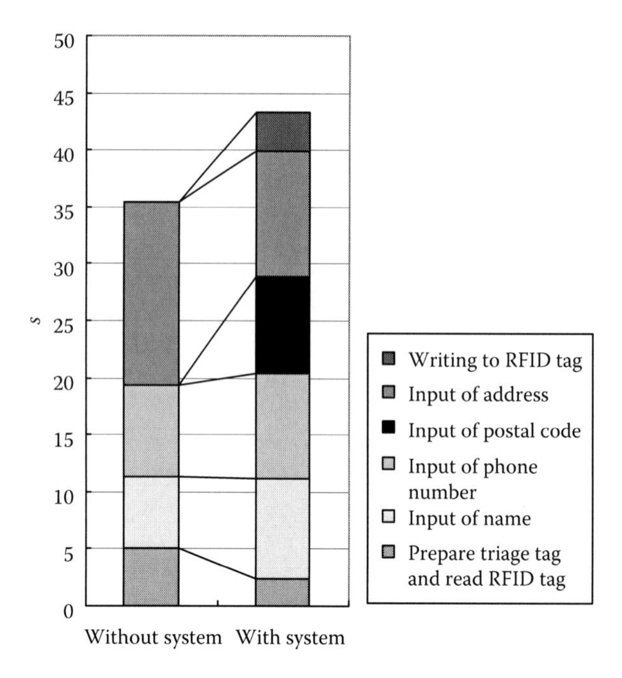

FIGURE 20.8 Input time in ambulance and hospital.

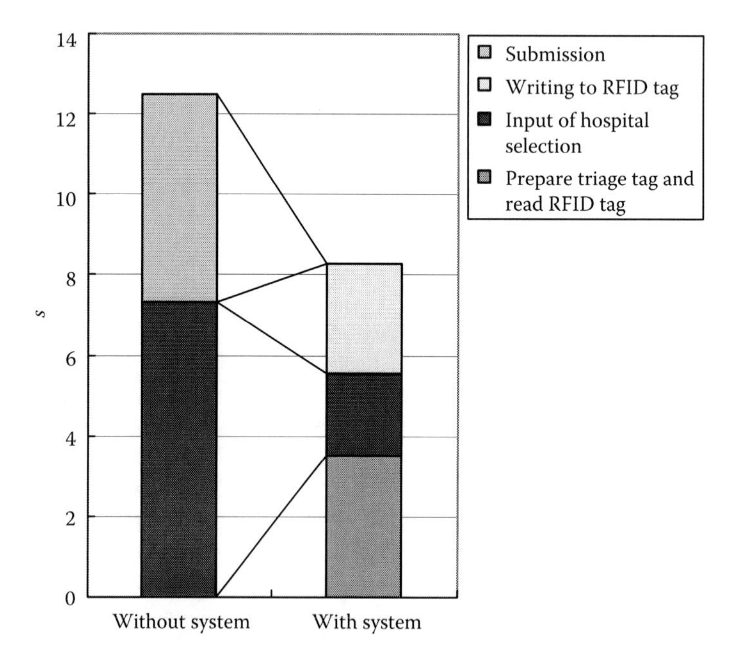

FIGURE 20.9 Input time in hospital selection.

operations, this population is considered to be about maximum where current triage is feasible. The goal of the experiment is to evaluate whether and how much the RFID triage system accelerates and completes (1) the transportation of the injured people and (2) the collection of the injured people's information.

20.6.1 Assumption

The experiment spot is a driving school of about 2250 m^2. In the experiment spot, we set a hospital area, which corresponds to the three points of hospitals, and set two hospitals about 700 m away from the spot.

We assume that two buses and three cars crash, where the injured people are located and classified as follows: 5 people as black/*deceased*, 18 as red/*immediate*, 31 as yellow/*delayed*, and 28 as green/*minor.*

In the experiment, 14 ambulances (3 of them are for mass transportation) and 16 other vehicles such as commander cars and machinery and materials cars are used. The population of emergency personnel is 85, where 4 (2 pairs) for first triage, 1 for second triage, 1 for hospital selection, 3 for each ambulance, and 1 for each hospital.

20.6.2 System Assignment

In the experiment, two first triage terminals, one second triage terminal, and one hospital selection terminal are used as mobile devices. Fourteen ambulance terminals are used, where six of them are equipped with RFID readers and wireless communication interfaces, three are equipped with RFID readers but wireless communication interface, and five are without both. The ambulance terminals with no RFID readers and wireless communication interfaces are introduced to assume the heterogeneous environment; here, not all the ambulances support the RFID triage system but only a usual notebook PC, such that a neighbor municipality helps the triage. The ambulances with no wireless communication interfaces are introduced to assume the situation where communication is not available.

One hospital terminal is placed at each hospital. One operation point terminal is placed at the operation point when the point is made during the experimental performance. The server is set in a data center about 1000 km away from the incident site.

20.6.3 Network

We prepared IEEE 802.11b wireless LAN interfaces for mobile device terminals and FOMA data communication interface that provides up to 385 kbps by NTT DoCoMo, Inc.

As access point facilities are necessary if we adopt wireless LAN, since otherwise it provides only several 10 m, we had to adopt them since there have been no mobile devices that will be able to be equipped with an RFID reader and any long-range data communication interface of several kilometers. In the future, this cost of access points will vanish when a mobile device with an RFID reader and a long-range data communication interface is used.

20.6.4 Process of Experiment

In the experiment, we first take about 1 h to perform the naive performance, which is the current triage, and then perform the advanced performance, which is the triage using the system, in the same condition.

Each performance starts with the departure of ambulances and fire engines having predefined time differences from the car parking. Each ambulance that transports the injured people to the hospitals inside the experiment spot arrives at the hospitals after driving around inside the spot, and those that transport to the hospitals outside the spot take the predefined route.

The injured people are predefined and informed with their level of injury and reply to the questions from the emergency personnel acting according to the defined level. The emergency personnel perform the triage as described in Sections 20.2.1 and 20.3.2.

In the two performances, seven persons measured the time of the flow of injured people and the flow of the collected information by stopwatches. Moreover, the collected information flow was also logged in the system in the advanced performance.

20.6.5 Evaluation

The result of the experiment is shown in Figures 20.11 through 20.15. The photos of the experiment are given in Figure 20.10.

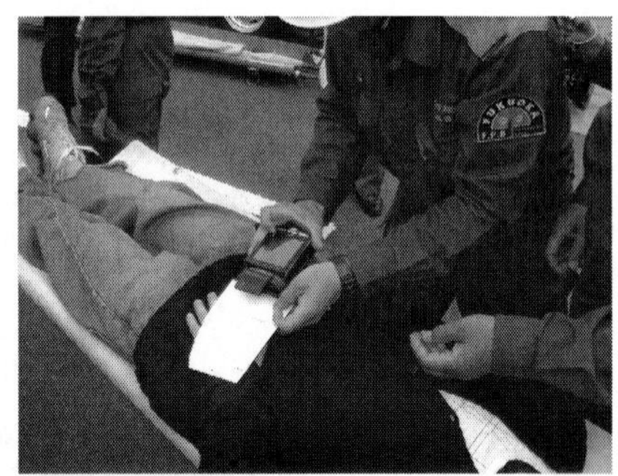

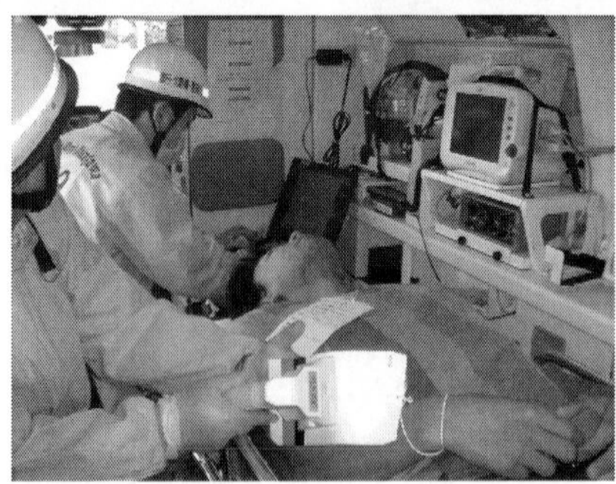

FIGURE 20.10 Photos of the experiment.

20.6.5.1 Transportation of Injured People

Figure 20.11 shows the progression of cumulative populations of the injured people who are carried from the incident site to the first aid area and carried out from the first aid area to the hospitals in each performance.

The population carried into the first aid area in the advanced performance always exceeds that of the naive performance (Figure 20.11). This demonstrates that, using the system, the injured people can be carried into the first aid area faster than the naive performance. This mainly resulted from the fact that multiple injured people could be carried in simultaneously using the system, whereas they must have been carried serially to manually count the people by emergency personnel in order not to lose the information of the injured people in the naive performance. Moreover, the population radically increases at around 2 min after the start. This is because it was possible to carry in those who are in green/*minor* level of injury and those who can walk simultaneously as early as possible, whereas they must wait for the people of other levels to be carried in first in the naive performance.

Moreover, the population carried out from the first aid area to the hospitals in the advanced performance (Figure 20.11) always exceeds that of the naive performance. This shows that, using the system, the injured people can be carried to the hospital faster than the naive performance. This is assumed to be because, in addition to the

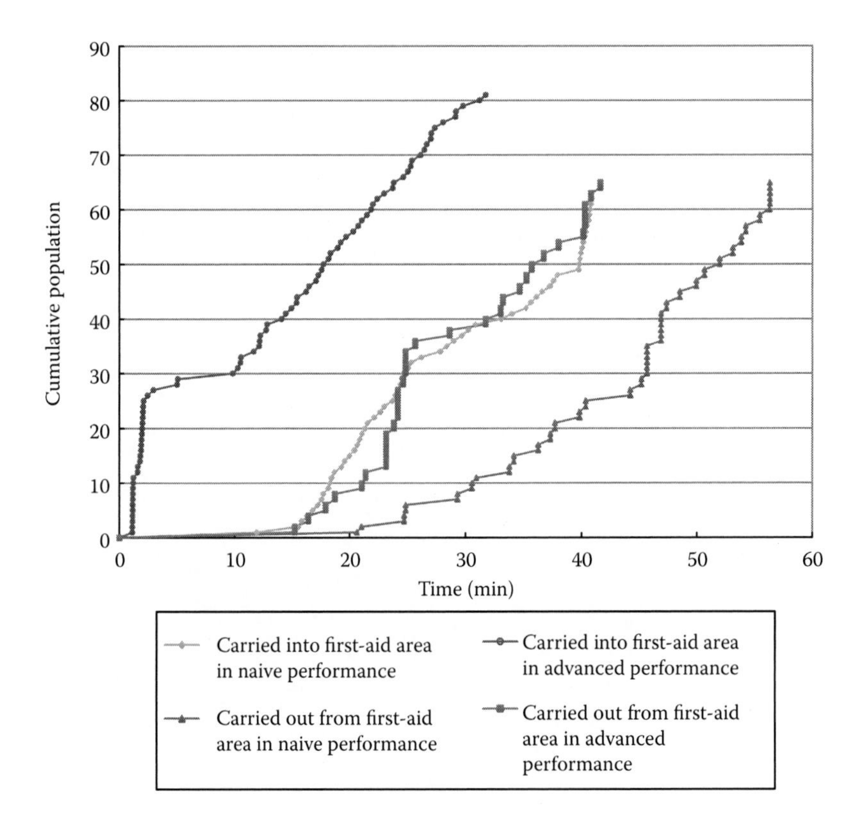

FIGURE 20.11 Transportation time.

acceleration of carrying into the first aid area, the hospital selection was effectively performed using a mobile device in the advanced performance. Moreover, the figure shows that plenty of injured people are carried out after 22 min. This is assumed to have resulted from the fact that the injured people of *green/minor* level could be carried out early followed by the acceleration of being carried in.

The population in the naive performance expires before it becomes the total population assumed because of the essential lack of the current workflow: several injured people in *green/minor* level were dismissed before going to the first aid area and could not be monitored; the *black/deceased* people who were not carried could not be monitored, whereas the advanced performance could. Moreover, the population carried out from the first aid area in the advanced performance exceeds that carried into there. This is because the vehicles for *green/minor* level could stop at different areas from the ambulances for other levels, and several people of *green/minor* level happened to be carried out without being carried into the first aid area.

20.6.5.2 Information Collection

Figure 20.12 shows the progression of the cumulative populations whose information about the injury level is collected in the manual copy at the operation point in the naive performance, in the RFID tag of the injured person in the advanced performance, and in the server in the advanced performance. Figure 20.13 is about the hospital selection, and Figure 20.14 is about the address. If there are multiple input records for a single type of information, we adopted the first period of the collection. The time in the server was measured in minutes, whereas the others were in seconds.

In the following, we describe the method for recording and adjusting the record in the naive performance. The level of injury, the sex, and the age were assumed to be collected when the entry in the triage tag is copied to a tally sheet before an injured person is transported from the first aid area toward a hospital. The name, the

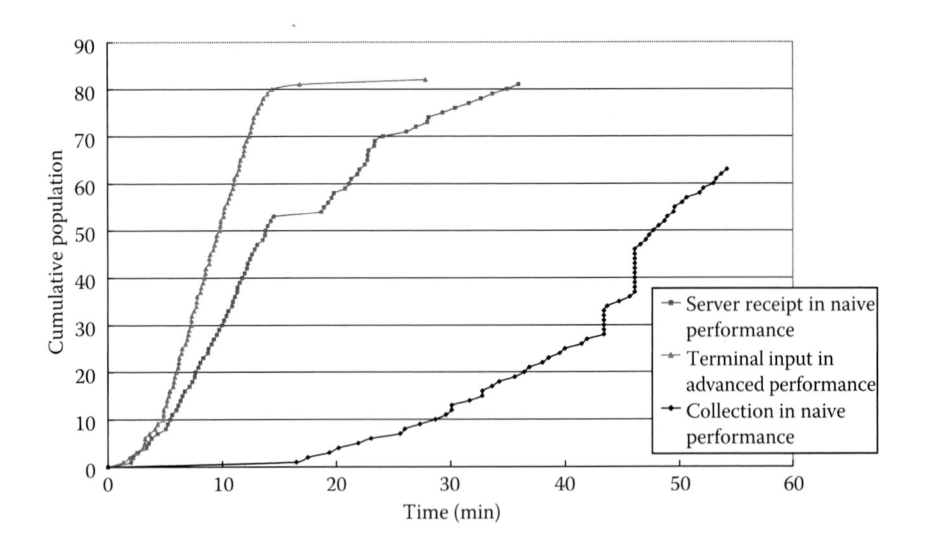

FIGURE 20.12 Time of information collection (level of injury).

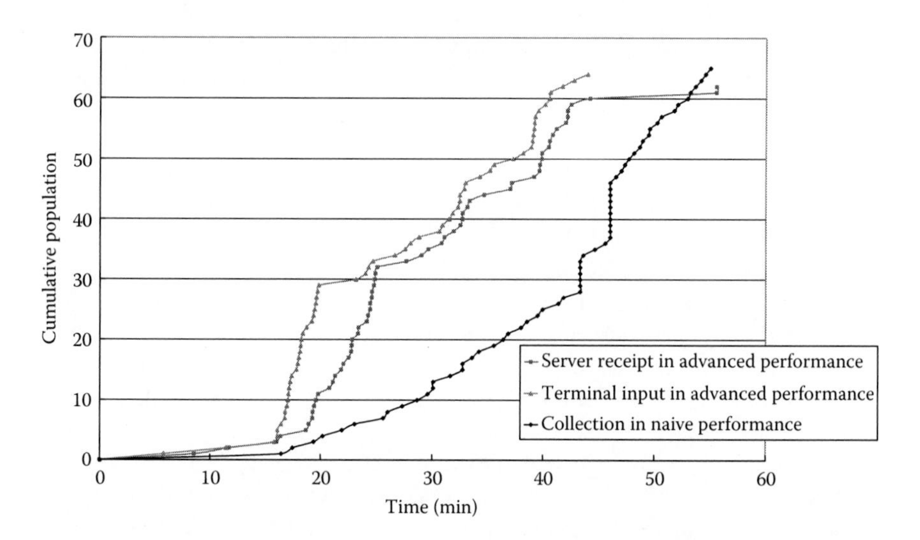

FIGURE 20.13 Time of information collection (hospital selection).

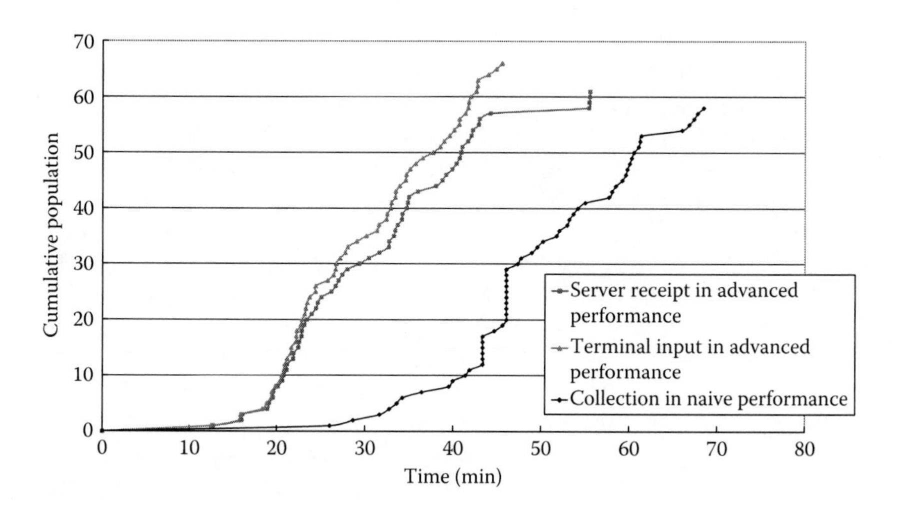

FIGURE 20.14 Time of information collection (address).

address, and the phone number were assumed to be so when they are copied after an ambulance returned to the operation point after the transportation. The records have been made slightly complementary to the deficit in the data obtained from the stopwatch measurement in the following ways:

1. We adopted the time of particular people for seven lacks of time where the subject injured people could be assumed to be in the same ambulance.
2. We adopted the time of the closest line in the tally sheet for six lacks of time where *green/minor* people could be assumed to be transported at a time.

3. For two time lacks of the manual copy to the tally sheet after the transportation when the times of collection are known, we adopted the collected times added by the averages of the periods from the collection to the manual copy of other people in the same hospital.
4. For one person whose collection time after the transportation and also that of manual copy lack, we adopted the average time of the times of manual copy for the same hospital.

It is not easy to compare the times of the naive and advanced performance since the methods of measurement are different, but we can see the times for the operation point to view the information for the first time, if we consider that they correspond to the manual copies at the operation point in the naive case and so to the data arrivals to the server in the advanced case.

From the figures, the information of the injured people using the system could be collected faster than the naive performance in any information. Although we omit the graphs for the other information, there was little difference in the shapes in the graphs of the level of injury, sex, and age. Similarly, so was in name, phone number, and address. These mainly owe to the timing of input shown in Table 20.1, and we can observe that the collection of the information in the early stages of the triage could radically be accelerated using the system.

Collecting information of all the injured people is essentially difficult, since that of the black/*deceased* people are not collected quickly and the green/*minor* people tend to walk out on their own. However, Figures 20.12 and 20.14 show that the number using the system is higher than the naive performance, which means that the system provides more accurate and complete information of injured people, especially in the early stages of triage. Although Figure 20.13 does not show the same, this can be assumed since the input time using the system was longer than the naive performance, which we confirmed in Section 20.5 in which the average input time of the hospital selection in the system was about 8.2 s, whereas it was 7.3 s without the system, with five trials in each case. Improving the time and complexity of input, there is a possibility of making the information collection of hospital selection accurate.

For data integrity in the advanced performance, all the peoples' data have not been collected in Figure 20.14 since the black/*deceased* and green/*minor* peoples' information have not been collected. However, the number of collection is more than that in the naive case. In Figure 20.13, the number of collection in the naive case is better than the advanced case. This can be assumed because the trial with the system needs, except for submission, much time than that without the system, as shown in Section 20.5.

20.6.5.3 Communication Time in the System

From Figures 20.12 through 20.14, we can observe that the network communication time in the system does not affect the triage to the critical delay. In addition, the receipt time on the server can be earlier if the time for communication is reduced, although the result in the advanced performance is better than the naive one. Figure 20.15 shows the distribution of the difference between the time

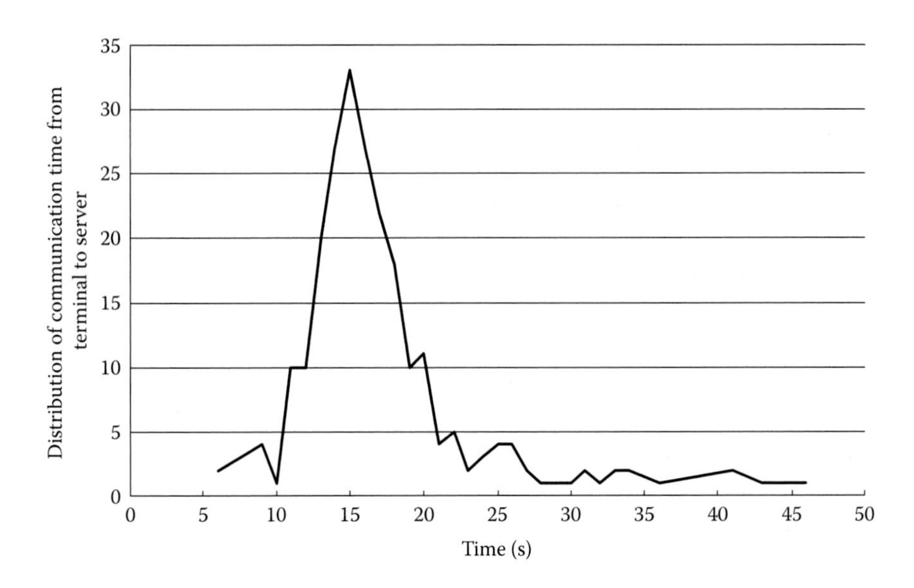

FIGURE 20.15 Distribution of communication time in the system.

received by the server and that input to the input terminal of each input of the injury information, where the differences of more than 60 s are omitted as a device without wireless communication functionality. The average time of the difference between the time received by the server and that input to the terminal was 17.3 s, and the variance was 35.39.

20.6.6 Discussion

We discuss the implications in network with the requirements discussed in Section 20.2.

Data integrity and availability are apparent to be improved without the experiment, as discussed in Section 20.2.3.

As to input throughput, note that the communication time does not affect the input time by the emergency personnel, since they first read the old information from an RFID tag and finish right after updating the information. The terminal automatically communicates with the server subsequently, but the emergency personnel do not have to wait for the network communication. If the system does not employ RFID tags, the first step of reading from the RFID tag must rely to the network communication. The throughput of an input by discarding RFID tags can be supposed to increase by 14.3 s, since the network communication time is 17.3 s in average and the user waiting time for the RFID communication is 3 s, as shown earlier.

Although latency needs further research to clarify the advantage of using RFID tags, we could decrease the latency, especially the required types of information compared with the naive performance, as shown in Figures 20.3 and 20.4, without affecting other factors such as input throughput in triage.

20.7 CONCLUSION

From the experiment, the RFID triage system is demonstrated to be effective in mass casualty incidents of about 100 injured people. In the future, the system will be more valuable when it becomes applicable for other incidents, such as with smaller number of injured people, or in broad area, such as earthquakes, storms, or floods.

ACKNOWLEDGMENTS

This work has been supported by the Grant-in-Aid for Young Scientists (A) No.18680009 of the Ministry of Education, Science, Sports and Culture (MEXT) from 2006 to 2008. We are grateful for the support of the experiment from Toppan Printing Co., Ltd., Fukuoka City; Mizuho Information & Research Institute, Inc.; Institute of Systems & Information Technologies/KYUSHU, NTT DoCoMo, Inc.; NTT West Corp.; and Kasuga-Ohnojo-Nakagawa fire department.

REFERENCES

1. P. Hewkin, Smart tags—The distributed memory revolution, *IEEE Review*, 35(6), 203–209, 1989.
2. R. Want, K.P. Fishkin, A. Gujar, and B.L. Harrison, Bridging physical and virtual worlds with electronic tags, *Proceedings of International Conference CHI 99*, Pittsburgh, PA, pp. 370–377, 1999.
3. M. Gerla and J. Tsai, Multicluster, mobile, multimedia radio network, *ACM/Baltzer Journal of Wireless Networks*, 1(3), 255–265, 1995.
4. G. Zussman and A. Segall, Energy efficient routing in ad hoc disaster recovery networks, *22nd Conference of the IEEE Computers and Communication Societies* (*INFOCOM*), San Francisco, CA, 10 p., 2003.
5. T. ElBatt, S. Krishnamurthy, D. Connors, and S. Dao, Power management for throughput enhancement in wireless ad-hoc networks, *Proceedings of IEEE ICC 2000*, New Orleans, LA, 9 p., 2000.
6. F. Subash, F. Dunn, B. McNicholl, and J. Marlow, Team triage improves emergency department efficiency, *Emergency Medicine Journal*, 21, 542–544, 2004.
7. T. Kilner, Triage decisions of prehospital emergency health care providers, using a multiple casualty scenario paper exercise, *Emergency Medicine Journal*, 19, 348–353, 2002.
8. S. Goodacre, F. Morris, B. Tesfayohannes, and G. Sutton, Should ambulant patients be directed to reception or triage first? *Emergency Medicine Journal*, 18, 441–443, 2001.
9. J. Terris, P. Leman, N. O'Connor, and R. Wood, Making an IMPACT on emergency department flow: Improving patient processing assisted by consultant at triage, *Emergency Medicine Journal*, 21, 537–541, 2004.
10. A. White, Change strategies make for smooth transitions, *Nursing Management*, 35(2), 49–52, 2004.
11. P. Parker, Move care to a higher level with emergency systems, *Nursing Management*, 35(9), 82–84, 2004.
12. K. Lorincz, D. Malan, T. Fulford-Jones, A. Nawoj, A. Clavel, V. Shnayder, G. Mainland, and M. Welsh, Sensor networks for emergency response: Challenges and opportunities, *IEEE Pervasive*, 3(4), 16–23, 2004.
13. S. Voskarides, C. Pattichis, R. Istepanian, E. Kyriacou, M. Pattichis, and C. Schizas, Mobile health systems: A brief overview, *Proceedings of SPIE AeroSense*, 2002, 124–131, 2002.

14. K. Banitsas, R. Istepanian, S. Tachakra, and T. Owens, Modelling issues of wireless LANs for accident and emergency departments, *IEEE EMBC Conference*, 4, 3540–3543, 2001.
15. http://www.activewaveinc.com/applications_hospitals.html.
16. P. Oosterom, S. Zlatanova, S. Fendel, and M. Elfriede (Eds.), *Geo-Information for Disaster Management*, Springer Verlag, Berlin, Germany, 1434, 2005.
17. S. Midkiff and C. Bostian, Rapidly-deployable broadband wireless networks for disaster and emergency response, *1st IEEE Workshop on Disaster Recover Networks* (*DIREN '02*), New York, NY, 10 p., June 2002.
18. S. Sharma and S. Gifford, Using RFID to evaluate evacuation behavior models, *North American Fuzzy Information Processing Society Conference*, 5 p., 2005.
19. J. Yoshida, Navy to use RFID technology in Iraq, *EE Times News*, May 2003, http://www. embedded.com/showArticle.jhtml?articleID=10700142.

21 Survey of Sybil Attacks in Networks

Wei Chang and Jie Wu

CONTENTS

ABSTRACT

Most peer-to-peer systems are vulnerable to Sybil attacks. The Sybil attack is an attack wherein a reputation system is subverted by a considerable number of forging identities in peer-to-peer networks. By illegitimately infusing false or biased information via the pseudonymous identities, an adversary can mislead a system into making decisions benefiting herself. For example, in a distributed voting system, an adversary can easily change the overall popularity of an option by providing plenty of false praise or bad-mouthing the option through these fake identities. In this chapter, we summarize the existing Sybil defense techniques and further provide some new research areas. Unlike traditional surveys about Sybil defense, we first categorize the Sybil defense methods, mainly according to their designed time, and then classify the methods by their approaches. We believe that by understanding the evolution of the solutions, readers could essentially have more insights on the problem. In a nutshell, the research on the Sybil defense technique has experienced four phases: (1) traditional security key-based approaches, (2) specific peer-to-peer system feature-based solutions, (3) social network-based methods, and (4) social community-based techniques. Besides all of these anti-Sybil methods, readers will also find some Sybil attack-related topics, such as sockpuppets in online discussion forums. By the end of the chapter, we will provide some predictions about directions for future research.

21.1 INTRODUCTION

We are entering a distributed computing era where a problem is cooperatively computed by many participating entities (also called nodes). Such cooperation mechanisms require that each participant trust one another. Just like in a team project, each person needs to trust his coworkers. In a typical (distributed) peer-to-peer system, the participators usually play three roles, simultaneously. First, they are the data owners, and they also share certain data with others. These data could be local raw data, such as sensor readings, or could also be partial computing results. Second, they are data processors: each participant locally processes the data according to some rules or algorithms. Third, they are also data transmitters. In a large-scale peer-to-peer system, a direct connection between each pair of nodes is impossible; therefore, the participating nodes usually build up a network, and a message is transmitted from one node to another via the relay operations of multiple intermediary nodes (transmitters). If attackers control one or more participating nodes, they could modify the local raw data, local computed results, or all of the transmitted data. Clearly, by such an attacking mechanism, the attackers can modify the overall computation results of a peer-to-peer system or even subvert the whole system. Therefore, security is a very important aspect of the research of peer-to-peer systems.

In this chapter, we investigate the Sybil attack, a particularly harmful attack in distributed peer-to-peer systems. Almost all distributed peer-to-peer systems are based on a common assumption that each participating entity controls exactly one identity. However, whenever the assumption cannot be satisfied, the system is subject to Sybil attacks. In a Sybil attack, an adversary creates a large number of forging/fake/pseudonymous identities (also named Sybil identities), and since all Sybil identities are controlled by the adversary, she can maliciously introduce a considerable number of false opinions into the system and subvert it, by making decisions benefiting herself. Essentially, Sybil attacks break and manipulate the trust mechanism behind peer-to-peer systems.

For a better understanding of what a Sybil attack is, here, we provide three examples. First, in some distributed systems, critical resources are assigned based on the voting results of participants: usually, only the node that has received the highest number of votes can access the resources. If an attacker illegitimately creates a large number of Sybil identities, then the adversary can proportion more resources by instructing the fake identities to vote in certain ways, such as always voting for her fake identities. Since votes are collected indirectly (recall that, instead of transmitting data through direct communication between remote nodes, most data are transmitted by the replay of other nodes), it is hard to detect the illegitimate votes. Another example comes from an application of sensor networks called *pervasive temperature monitoring*. In a large region, multiple sensors are randomly and uniformly deployed. Each sensor measures its surrounding temperature and further forwards the readings to a sink node, which collects the data. From the sink node, an average temperature can be computed. However, if the attackers launch Sybil attacks and let each Sybil identity report one more temperature degree, then the average temperature result will be incorrect. Our third example comes from a Facebook voting application. If an adversary maliciously creates many identities, she can easily change the overall popularity of an option by providing plenty of false praise or badmouthing of the option through Sybil IDs. Since the false opinions of the Sybils may essentially change the final decision of any distributed system, the research works on Sybil defense techniques hold the most important position.

The researches on Sybil attacks have passed three phases, and now, they are just entering the fourth phase. Note that our classification is based on the mainstream of research trends. Sybil attack is named after the subject of the book *Sybil*, a case study of a woman diagnosed with dissociative identity disorder. The name was suggested in or before 2002, by Brian Zill from Microsoft Research. The term *pseudo spoofing* had previously been coined by L. Detweiler on the cypherpunk's mailing list and was used in the literature on peer-to-peer systems for the same class of attacks prior to 2002. However, this term did not gain as much influence as *Sybil attack*. Phase I of the research on Sybil defense techniques began in 2002 and mainly ended in 2004. Within this phase, researchers tried to find some general mechanisms to defend themselves from all types of Sybil attacks in various networks or systems. In a nutshell, people tried to prevent Sybil attacks through redesigning system architectures and by involving secure mechanisms, such as digital signatures and identity authentication. However, the majority of approaches found at this phase faced a common problem: their schemes required a central authority for the verification of identities. Clearly, the trusted third party is the bottleneck of systems, which could easily become a target point. Moreover, it is impractical, since there is definitely no globe agency who can be trusted by the entire public.

Around 2004–2006, the research trend entered phase II, which included specific peer-to-peer system feature-based solutions. Within this period of time, researchers focused on designing a defense system for a specific peer-to-peer application system. Different application systems hold several unique features. By exploring these features, Sybil attacks can be detected, or the damage they cause can be bounded. For example, in sensor networks, nodes are static. By monitoring the received signal strength of each received message, Sybil nodes can be detected. However, readers need to be aware that such anti-Sybil systems are specially designed; an efficient solution for one application is typically not suitable for the others. Moreover, during the period from 2005 to 2006, a majority of researchers shifted to other secure problems, instead of studying Sybil attacks; the research on Sybil attacks was cooling down.

In 2006 Association for Computing Machinery (ACM) SIGCOMM, a novel paper [1] was presented, which led the research on Sybil defense to enter phase III. With goals dissimilar to those of phase I, the authors [1] aimed to adopt the concept of social networks. They wanted to detect Sybils based on a unique structure of friendships. Through observations, they found that although attackers can create plenty of Sybils and further create plenty of friendships (also known as social links) among the Sybils, the number of links between Sybils and honest users is limited. This is so, because the links/friendships on a social network are built based on a trust relationship, as well as physical interactions among real people. Based on this key observation, many creative and interesting Sybil defense approaches were proposed at phase II, and the Sybil attacks got the attention of the public once again: in any network or security-related conferences or journals, you can easily find several papers mentioning Sybil attacks. These types of Sybil defense systems are also called social network-based Sybil defense. Note that these solutions do not detect Sybil attacks in social networks. Instead, they explore the social networks behind a peer-to-peer system. More details can be found in journal [2].

In the year of 2010, also in ACM SIGCOMM, another paper [3] provided a new trend for Sybil defense, which argues that the Sybil nodes can be detected by the community structures of social networks, since there is a shortcut on the social graph of an attacked system. Community detection is a relatively mature topic in computer science, and plenty of useful techniques have been proposed. The paper [3] suggested that several community detection algorithms may be directly applied to the anti-Sybil problem. Note that partial methods in phase II essentially detect the honest community part of a social network in an implicit way, while the others just use other social features. We can also regard phase IV, social community-based Sybil detections, as an extension of phase III.

The remainder of this chapter is organized as follows. In Section 21.2, we formally introduce the Sybil attacks on peer-to-peer systems. The examples about typical vulnerable systems are given in Section 21.3. From Section 21.4 through 21.7, we provide the general description of the anti-Sybil approaches for each phase. In Section 21.8, we provide the prediction of future research, and Section 21.9 concludes the chapter.

21.2 TAXONOMY OF SYBIL ATTACKS

To better understand the Sybil attacks, in this section, we provide a taxonomy of different types of Sybil attacks. The capability of the attacker is determined by several characteristics: (1) insider vs. outsider, (2) selfish vs. malicious, (3) directed vs. indirected communications, (4) simultaneously vs. gradually obtained Sybil identities, (5) busy vs. idle, and (6) discarded or retained.

21.2.1 Insider vs. Outsider

Whether an attack is an insider or outsider directly determines the capability of the attacker and the hardness of launching a Sybil attack. For an insider, the attacker holds at least one legitimate identity and claims that she receives certain data from the other nodes, by using the fake identities. Usually, a distributed system assumes that each node is trustworthy, and therefore, the false data can be forwarded to the whole system. However, for an outsider, she is any illegitimate entity; before launching a Sybil attack, she must first access the system. However, distributed systems typically employ some kind of authentication to prevent illegitimated access, such as a password for entering or data encryption. The outsider needs to understand all the mechanisms of the system prior to launching Sybil attacks. Therefore, distributed systems are more vulnerable to inside attackers.

21.2.2 Selfish vs. Malicious

For security-related problems, there are two different types of attackers: selfish and malicious. Selfish attackers manipulate the false data just for their own benefit, while malicious attackers attempt to subvert a system. Whether an attacker is selfish or malicious is usually determined by the different types of targeted distributed system and final attacking effects. For example, in our critical resource accessing example, if the attacker has resource accessing rights all to herself, then she is a malicious attacker, since others cannot use the resource. However, if other users can also access the resource with less probability, then she is selfish. Since malicious attacks usually have more serious effects, it is of higher importance to defend against potentially malicious attacks than those that are potentially selfish.

21.2.3 Directed vs. Indirected Communications

How Sybil nodes communicate with honest nodes is also a significant consideration during the designing of Sybil defense mechanisms. The attacker can directly communicate with an honest node by using one of her Sybil identities, or she can use only her real identity to communicate with others and route the Sybil data via this real identity. For the attackers, the easiness of direct communication with honest nodes directly influences the success of attacking and whether honest users can see through the attack. In general, the attackers with more directed communications are harder to detect. However, for certain distributed systems, direct communication may be difficult to establish.

21.2.4 Simultaneously vs. Gradually Obtained Sybil Identities

The attacker can obtain all of her Sybil identities simultaneously, or she can gradually generate them one by one. For an intelligent attacker, the more diverse features the Sybil nodes have, the harder it is to identify Sybil nodes. Gradually creating Sybil nodes may potentially differentiate the first appearing time of the Sybils. However, the process may delay the attacking time and increase the explosion time of some Sybils: if a distribution randomly checks the authentication of some identities, previously generated identities have a higher chance of being caught.

21.2.5 Busy vs. Idle

All Sybil identities can participate in a distributed system simultaneously, or only some of them can work, while others are in an idle state. Essentially, the selection of these two schemes is determined by how cheap it is to obtain an identity. If the attacker can easily get plenty of fake identities, having some idle Sybil nodes could make them much more real, since an honest node may leave or rejoin the system multiple times. However, the power of Sybil attacks results from the quantity of the identities. If obtaining a large number of identities is very difficult, the attacker has to use all of them in order to launch a successful attack.

21.2.6 Discarded vs. Retained

For an attacker, how to manage the old Sybil identities is important. After finding a Sybil node, one can further (and gradually) identify the others by monitoring the claimed communication between a suspect node and the detected Sybil node. Since the attacker is not aware of whether the old identities have been detected yet, once in a while, she has to determine whether or not to discard them. Consider that generating Sybil identities has certain costs and the possible naming space is not infinite. The capacity of attacks is related with the naming costs and the mechanism of using old identities.

21.3 EXAMPLES OF VULNERABLE SYSTEMS

The Sybil attack is an attack wherein a distributed system is subverted by forging identities. Usually, peer-to-peer networks are vulnerable to Sybil attacks. In this section, we will provide several realistic examples of these vulnerable systems.

Moreover, we will also provide another two examples, which are very similar to the Sybil attacks.

21.3.1 Vehicular Ad Hoc Networks

A vehicular ad hoc network (VANET) is a technology that uses moving cars as nodes to create a special mobile network, which takes safety as its main purpose. In VANETs, each participating car can communicate with roadside base stations or other cars. However, this type of network is vulnerable to Sybil attack. For example, a selfish driver may launch a Sybil attack by claiming that many vehicles are traveling nearby. If this is the case, other cars may falsely believe that there is a traffic jam on the corresponding road and therefore pick up an alternative road. The selfish driver will enjoy better traffic, with others paying the cost. Moreover, the Sybil attacks can also cause serious safety threats: a malicious driver may drop the warning messages. In VANETs, when a crash happens or speed significantly reduces, a warning message for slowing speed will be generated and is further forwarded to the following vehicles, one by one. By claiming many fake identities, the warning messages may all be transmitted to the malicious driver's car. If she drops these messages, other following cars will be in danger.

21.3.2 Distributed Voting Applications in Peer-to-Peer Systems

Any distributed voting aggregation system is vulnerable to Sybil attacks. Usually, a distributed voting system consists of a collection of identities, which vote for different objects. Most voting systems assume that each user only holds one identity and each identity can provide only one vote. Based on this restriction, if attacks have many identities, then she can offer many votes. The vote can be in any form, from the simplest case, where each vote represents a positive or a negative opinion, to more complex cases, where the value of a vote can range within a given set of values. To rank objects, a ranking mechanism typically collects (or aggregates) the votes from distributed participants and further combines the votes in a certain method, such as the majority rule. By Sybil attack, the real users' major decision can be outvoted by the attacker: since the attacker can easily create many fake identities, the false opinions can be introduced into the system by these identities. Here, we need to claim that although the Sybil nodes may be held by different attackers in reality, for the ease of description, researchers always assume that the Sybil identities are owned by a single adversary. This is due to the idea that this assumption will not influence the effects of the attacks and will also not affect the results of defense approaches.

The example of Amazon's user feedback system in the introduction is essentially an aggregating voting system, since the reputation of each merchant is determined by the votes from customers. However, we also have to mention that the Amazon voting system is a centralized system, where all of the voting processes are controlled by a central server. However, generally, an aggregating voting system can also be a distributed system: each node can launch a vote, and the range of votes' values may be different.

21.3.3 Distributed Storage Applications in Peer-to-Peer Systems

Peer-to-peer storage systems adopt replication and fragmentation mechanisms, and usually the mapping from data to the corresponding stored nodes is performed by distributed hash tables. From the consideration of system stability and easy accessing, the mapping function is in the form of one to many. However, if the attacker is an insider, she can manipulate the values of her Sybil identities such that all the replicated data may actually be stored on the same malicious node, although the data seem to be stored at different nodes outwardly. Without multiple copies of data, the attacker can easily launch many followed attacks without being detected. For example, she can modify some data. Since she holds all of the data copies, no one can detect the modification of the data.

21.3.4 Routing in a Distributed Peer-to-Peer System

To improve the performance or fault tolerance, wireless networks usually adopt a concurrent multipath routing technique. Instead of using a single routing path, multipath routing has multiple alternative paths throughout a network. The computed multipaths may or may not be overlapped. This technique provides better load balancing and fault tolerance than traditional routing methods. However, in wireless sensor or ad hoc networks, Sybil attacks can easily invalidate the technique: a computed multipath routing, which seemingly consists of multiple disjoint paths, could in fact only go through the same malicious node, which holds several Sybil identities. Other wireless routing mechanisms, such as the decentralized object location and routing (DOLR) algorithm and the geographic routing algorithm, are also vulnerable to Sybil attacks. In peer-to-peer networks, nodes communicate with each other by relaying messages from one node to another, and the quality of the selected relaying paths directly influences the performance of a network system. In some extreme cases, Sybil attack may even isolate one part of a network from the other part.

21.3.5 Sockpuppets in Online Discussion Forums

In online discussion forums, in order to cheat people on the Internet, for instance, to believe that a product is a good buy or that a particular investment plan has an extremely high return and low risk, a common trick is to use different fake online identities pretending to be different people. This is done to praise or create the illusion of support for the product [4]. In the same forum, different online entities that belong to the same person are referred as *sockpuppets*. Note that sockpuppet does not belong to Sybil attack, since online discussion forums are not peer-to-peer systems. However, because sockpuppets have several features similar to Sybil attacks, we want to mention them. First, both attacks are based on the usage of multiple identities belonging to the same person. Second, their success is related to the same assumption that each user is associated with one, and only one, identity. Third, they all break the reputation mechanism behind a given system. Last, for some distributed peer-to-peer systems, such as mobile social networks, there are social features and friendships associated with each identity; this also applies to an online discussion forum. Due to these similarities, the solution to one attack may help the design of the other.

21.3.6 PageRank in Searching Engines

Another attack, which is similar to, but different from, Sybil attacks, is called spoofed PageRank [5]. For modern search engines, the ranking of a page is determined by the quality and quantity of referenced links. In order to promote a page ranking, the page owner (a selfish attacker) may create a lot of small and meaningless spoofing pages and let them link to the page. This type of attack is called spoofed PageRank. If we consider each page as an identity and its corresponding PageRank as reputation or trust, then spoofed PageRank is the same as Sybil attack; both of them promote the opinion or reputation of a malicious entity by using plenty of fake entities. Again, these similar features may help to design some new Sybil defenses. For example, Advogato trust metric [6] is a Sybil defense system, where users certify each other in a kind of peer-review process. As the author observed that his notion of a trust metric was fundamentally very similar to the PageRank algorithm, the solution of identity-spoofing-related schemes in a field may inspire the design of defense in another field. Since trust, PageRank, and Sybil attacks have delicate relationships, we present several details about them.

In a centralized ranking system (like Google PageRank), the rank of an entity is computed from a stationary probability distribution of a Markov chain, in which a random walker moves from one node to another by following the edges with a constant probability d (also named damping factor) or randomly jumping to another.

Let $\mathbf{P}' = [p'_{ij}]$ be a $m \times m$ transition matrix, which is a row-normalized link matrix whose value shows the probability of transmitting from v_i to v_j during random walks (by following the link structure of a network). If there is an edge $v_i \rightarrow v_j$, then $p'_{ij} = 1/D^{+}(v_i)$; otherwise, $p'_{ij} = 0$. $\vec{Q}$ is a $1 \times m$ vector, $\vec{Q} = [q(v_1), q(v_2), \ldots, q(v_m)]$, and it stands for a preference vector during a random jump, $q(v_i) \geq 0$ and $\sum_{i=1}^{m} q(v_i) = 1$. The computation of PageRank is an iteration process; we use $\vec{R}(t)$ to represent the PageRank at the tth computing round. The steady value of $\vec{R}(t)$ is the final PageRank $\vec{R}$, which is a $1 \times m$ vector, $\vec{R} = [r(v_1), r(v_2), \ldots, r(v_m)] \cdot r(v_i)$ means the PageRank of the node $v_i \cdot \vec{R}(t+1)$ is defined as follows:

$$\vec{R}(t+1) = d\vec{R}(t)\mathbf{P}' + (1-d)\vec{Q}$$

Realistically, when $|\vec{R}(t+1) - \vec{R}(t)| < \epsilon$, we let the value of $\vec{R}(t+1)$ be $\vec{R}$. For the ease of description, we let $\mathbf{P}$ be a probability transition matrix used for computing PageRank, $\mathbf{P} = [p_{ij}]_{m \times m}, p_{ij} = d \cdot p'_{ij} + (1-d)q(v_j)$. Algorithm 21.1 gives the procedure for computing PageRank. Take Figure 21.1a as an example. For v_2, $D^{+}(v_2) = 3$, $p'_{21} = p'_{23} = p'_{25} = 1/3$, and $p'_{22} = p'_{24} = p'_{26} = 0$. When the damping factor $d = 0.85$ and $q(v_1) = q(v_2) = \cdots = q(v_6) = 1/6$, p_{21} can be computed as follows: $p_{21} = d \cdot p'_{21} + (1-d) \cdot q(v_1) = 0.85 \times (1/3) + (1-0.85) \times (1/6) = 0.308$

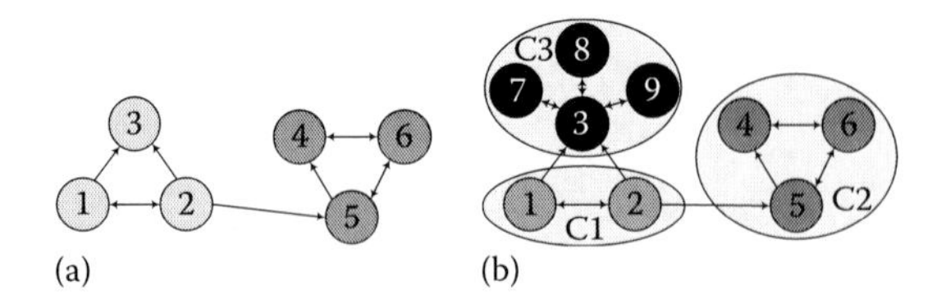

FIGURE 21.1 The change of a network's structure before and after having a spoofing page attack. (a) Shown is the network structure without the spoofing pages. (b) v_3 employs v_7–v_9 as spoofed pages.

$$\mathbf{P} = \begin{bmatrix} 0.025 & 0.450 & 0.450 & 0.025 & 0.025 & 0.025 \\ 0.308 & 0.025 & 0.308 & 0.025 & 0.308 & 0.025 \\ 0.167 & 0.167 & 0.167 & 0.167 & 0.167 & 0.167 \\ 0.025 & 0.025 & 0.025 & 0.025 & 0.025 & 0.875 \\ 0.025 & 0.025 & 0.025 & 0.450 & 0.025 & 0.450 \\ 0.025 & 0.025 & 0.025 & 0.450 & 0.450 & 0.025 \end{bmatrix}$$

After obtaining **P**, we used Algorithm 21.1 to compute the PageRank of each node. Initially, we let each node's PageRank equal $1/m$. Algorithm 21.1 iteratively computes the PageRank via using $\vec{R}(t+1) = \vec{R}(t)\mathbf{P}$, and the algorithm stops when $\vec{R}(t)$ is stable. For Figure 21.1a, the final PageRank is $\vec{R} = [0.0517\ 0.0574\ 0.0737\ 0.2686\ 0.1999\ 0.3487]$. Clearly, the webpage v_3 ranks as number 4 among these nodes. However, in Figure 21.1b, through adding three spoofing pages (alternatively, we called these fake identities as Sybils in a peer-to-peer system), the attacker, v_3, ranks as number 1, since those fake pages grab some ranking values from others and uniquely support v_3. Obviously, Google PageRank is vulnerable to spoofed PageRank attacks. Similarly, the other famous web page ranking algorithm, HITS, also has the same weak point.

Algorithm 21.1 PageRank

1: Assign initial values for entities: $\vec{R}(0) = [1/m, 1/m, \ldots, 1/m]$
2: $\vec{R}(1) = \vec{R}(0)\mathbf{P}$, $t=0$
3: **while** $(|R(t+1) - R(t)|) > \varepsilon$ **do**
4: $t=t+1$, $\vec{R}(t+1) = \vec{R}(t)\mathbf{P}$
5: Return $\vec{R}(t+1)$ as $\vec{R}$

21.4 SYBIL DEFENSE IN PHASE I: SECURE KEY-BASED ANTI-SYBIL TECHNIQUES

According to information theory, in order to detect Sybil nodes, one must possess asymmetric knowledge, which means the detecting algorithms must hold more information about either the Sybil part or the honest part. For the techniques in phase I, they usually only provide valid keys to the honest nodes. Here, the *key* is a general concept; in reality, these could be symmetric or antisymmetric keys. Also, the keys could be session keys or some permanent keys.

21.4.1 Trusted Certification (Centralized Solutions)

Sybil attacks can be avoided by using trusted certification. This type of method assumes that there is a special trusted third party or central authority who can verify the validity of each participant and further issues a certification for the honest one. In reality, such certification can be a special hardware device [7] or a digital number [8,9]. Note that essentially both of them are a series of digits but are stored on different media. Before a participant joins a peer-to-peer system, provides votes, or obtains services from the system, his identity must first be verified. Actually, this method is the most commonly used Sybil defense in our daily lives. For example, when we are applying for a credit card, we need to provide our social security number for verification; when we are voting in election years, we also need our official ID card for getting a ballot.

When a malicious user launches Sybil attacks, defense mechanisms usually require that a message be sent together with a signature, which could be used for authenticating the validity of the sender or the data. Actually, according to a paper [10], trusted certification is the only approach that has the potential to *completely* eliminate Sybil attacks. Since almost all authentication steps require the participation of the central server, we categorize this type of solution as a centralized trusted certification.

21.4.2 Trusted Certification (Semicentralized Solutions)

Centralized trusted certification approaches are often implemented by asymmetric (such as public/private keys) cryptography. However, the computational cost is a big problem. Some researchers proposed another type of solution, where partial identity verifications can be done without using the public/private keys. We named these kinds of solutions *semicentralized solutions*. Paper [11] provided a solution by using symmetric keys. They assumed that each node shares a unique symmetric key with a trusted base station. After verifying the validity of each other via a Needham–Schroeder-like protocol, a pair of nodes can establish a shared key. During data transmission between neighboring nodes, they can use the shared key for mutual authentication and can also encrypt the data.

21.4.3 Common Problems with Techniques in Phase I

Trusted certification usually relies on a centralized trusted authority for assigning and verifying identities. The authority must ensure that each node is assigned exactly one identity and that the identity list (also called a registration list) is well protected. In real life, the process of assigning identities is usually performed by human beings, which is costly and becomes the bottleneck of systems. Moreover, the central authority also needs to deal with lost identities and updates. The performance of real systems obstructs the usage of these solutions.

We summarize several obvious shortages of central authority-based methods as follows:

1. Single point of attack: In these schemes, the central authority can easily become a target. Besides Sybil attacks, attackers can also launch plenty of other attacks, such as denial of service, to crash the server.
2. Performance bottleneck: If several users access a central authority simultaneously, the authority may crash due to the huge workloads.
3. Communication cost: In this type of method, the authority is often required during the data transmission. For example, two strange nodes need the help of the authority for verifying each other. There is a considerable amount of extra communication between nodes and the authority.
4. Scaling: It is hard to construct an authority, which can be trusted by all participants, especially when a peer-to-peer system wants to include more users from diverse places.

21.4.4 Other Defenses in Phase I

21.4.4.1 Identity Fee

Unlike the trusted certification-based approaches, some other papers [12–15] add an economical *fee* with each certification. They argue that the attackers cannot easily subvert a peer-to-peer system unless they spend a lot of money. Indeed, they intend to build a system letting the cost of an attack outweigh the benefits of the attack.

21.4.4.2 Secure Hardware

As we have mentioned in the beginning of this section that the certification key is a general concept, a Sybil defense system can also be built based on the usage of some secure hardwares. Usually, these types of hardware periodically generate some time-sensitive token. Whenever a participant wants to verify the validation of his encounters, they just determine whether the token is valid or not. Note that there is an assumption behind the scheme that only honest users could get the secure hardwares and each valid user can only get one. Although the verifying process does not need the central authority, it is still responsible for dispensing the hardwares to valid users. Moreover, before giving out hardware to a user, the authority must verify the user's identity.

21.4.4.3 Resource Testing

Usually, each user owns only one identity, and each identity works on a single machine. However, when Sybil attacks are launched, the Sybil identities work on a single computer. When we give some time- or resource-consuming tasks to a group of identities, if they can finish the work within a threshold, then it is highly possible that they are honest users; otherwise, part of them may be Sybil. In general, the goal of resource testing [16–19] is to determine whether the selected identities have a reasonable amount of resources. The tests, which are adopted for these kinds of approaches, include (1) checking computing ability, (2) checking storage ability, and (3) checking network bandwidth. Readers must be aware that resource testing is not an efficient approach [10], but we still adopt them for deterring or discouraging the attackers.

21.5 SYBIL DEFENSE IN PHASE II: SPECIFIC SYSTEM FEATURE-BASED ANTI-SYBIL TECHNIQUES

Since attackers have a limited number of real devices in wireless ad hoc networks or sensor networks, a group of Sybils actually shares one device, and therefore, Sybils can be detected by letting honest users monitor signals' features or the moving patterns of coexisting identities. Consider that there are channel conflicts during the communication of honest users, while Sybil identities never have real data transmission. Paper [20] proposed a Sybil detection method by monitoring the neighbors' channel conflict rate. They assume that there is a central server that records the rate of each identity. Whenever a channel conflict happens, certain nodes should report the event to the server. If some identities have an abnormally low rate, then the server will regard them as Sybil identities. However, the readers should understand that this method will be inefficacious if the attackers are aware of the defense mechanism and will further purposely send certain signals for conflicts.

Considering the fact that Sybils usually appear together, paper [16] adopts moving patterns for Sybil detections. In order to increase the accuracy, they also introduce a signal collision-based improvement, based on the observation that the collision rates inside the Sybil groups are lower than that of the normal groups. However, when the density of honest users is not large enough, the accuracy of this type of Sybil defense will not be guaranteed.

Sensor networks are static. By exploring this feature, paper [19] proposed a received signal strength indicator (RSSI)-based scheme for Sybil detection. They assumed that each node has the ability to measure signal strength. When receiving a message, the receiver node will associate the RSSI with the identity of the sender, which is included in the message. For attackers, a Sybil node is able to send messages with different sender identities. Consider that all sensors cannot move and multiple Sybil identities are essentially sent from the same attacking sensor. Later, if the messages of another sender possess the same RSSI, then the receiver can treat these senders as Sybil nodes. This approach essentially verifies whether several identities share the same physical locations, and other papers [7,21,22] also call this type of approach *position verification*.

VANETs are also vulnerable to Sybil attacks. Paper [23] proposed a Sybil detection protocol by using vehicles' historical geographic information. The core technique for this method is position verification of mobile nodes: the method measures a possible existing area of a car, based on the car's and its neighbor's historical positions.

21.6 SYBIL DEFENSE IN PHASE III: SOCIAL NETWORK-BASED ANTI-SYBIL TECHNIQUES

In 2006, ACM SIGCOMM, paper [1] presented a novel idea on Sybil defense, which explores social networks *behind a given peer-to-peer system*. The authors want to detect Sybils based on a unique structure: although attackers can create plenty of Sybil identities and further establish several links among them, the total number of links between the Sybil and the honest users is limited, since the trust relationship on a social network is built based on the trust relationship among real people.

In other words, the corresponding social graph of an attacked peer-to-peer system contains a small cut structure. Admittedly, in real online social networks, such as Facebook or Twitter, a user may accept the friend request of a stranger. However, by using interaction networks, which provide a closeness rate based on historical interactions, or by using some special Apps, which allow users to manually enter some trust degree, we still can guarantee the limited number of trusted relationships between the honest and the Sybil users. After the publishing of this paper, many researchers came back to the works of Sybil defense, and even until now, there are still a lot of researchers working on this idea. In this section, we will introduce several typical anti-Sybil approaches based on the usage of social networks.

21.6.1 SybilGuard and SybilLimit

SybilGuard [1] and SybilLimit [24] are two famous Sybil defenses that use social networks. Since their core techniques are similar, we will only introduce SybilGuard. SybilGuard defines two terms: (1) a trusted path and (2) a trusted node. Similarly, for breaking the symmetric information constriction, SybilGuard also assumes that there is a known trusted node. From this trusted node, there are K random paths with a fixed length l. For the ease of description, we call these paths verifiers. From a suspect node, SybilGuard also sends K random paths. If a path encounters a verifier once, then we call the path *been verified once*, as shown in Figure 21.2b. If a path has been verified S times, then the path is a trusted path. When the majority paths of a suspect node are trusted paths, the suspect node will be regarded as a trusted node; otherwise, the node is a Sybil. Essentially, these random walks measure how well a suspected node and a verifier node are connected. The reason for SybilGuard working well is that the number of attack edges is limited in social network, as shown in Figure 21.2a. A majority of verifiers and a majority of the random paths from suspect nodes will remain in their resident communities.

Now, consider the case that a verifier comes into a Sybil region. Although this verifier can encounter plenty of Sybil-initialized paths, most of the encountered Sybil paths cannot get enough verifications, since only a very limited number of

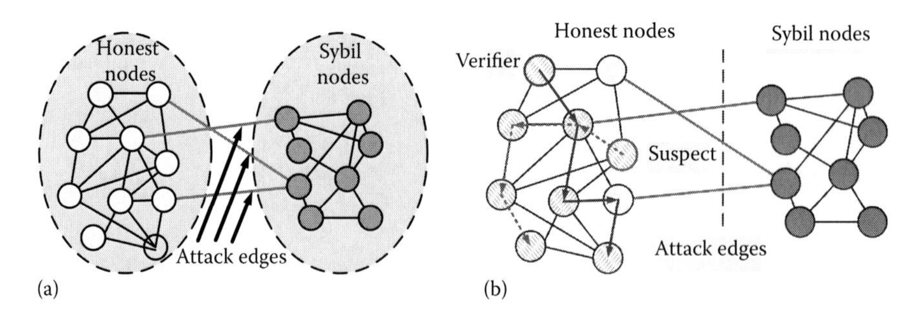

FIGURE 21.2 A social network-based Sybil defense scheme: SybilGuard. (a) Shown is the unique social network's structure for the distributed systems under Sybil attacks. Based on the fast-mixing feature of social networks and the unique structure (a shortcut), SybilGuard tries to identify Sybil nodes through random walks, as shown by (b).

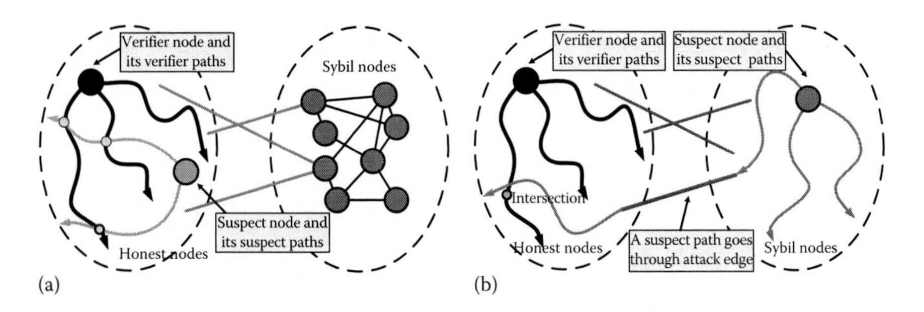

FIGURE 21.3 The difference between the verifications of the honest path and Sybil path in SybilGuard. (a) Shown is the condition that the verifier is checking an honest node. (b) Given is the case that a verifier is checking a Sybil node.

verifiers falsely enter the Sybil region. On the other hand, if a Sybil path enters the honest region through the attack edges, the path may intersect with verifiers many times and, therefore, become verified. However, because the number of attack edges is a limited number, the majority of Sybil-initialized random paths cannot be verified, as shown in Figure 21.3b. So, from the consideration of bounding the effects of Sybil attacks, SybilGuard works well.

21.6.2 SumUp

SumUp [25] is an anti-Sybil technique designed for a distributed voting system. Before we discuss the general idea of SumUp, we first need to understand the meaning of a credit network. Credit network [26] is a concept used in the electronic commerce field, and it is designed for building and measuring transitive trust among users. Note that, in the field of electronic commercial, trust is usually pairwise. Whenever a node (identity) trusts another node, a trust link will be established, together with certain trust value (credits). When a node gets services from others, the node can use the associated credits to pay for the services. Note that the credit network could also be used as a payment infrastructure between nodes that do not directly extend credit to each other [27]. Two remote nodes, which do not directly trust one another, can interact with each other when there exist credit paths between them. In some systems, such interactions will cost credits from the paths.

Formally, a credit network can be represented by a directed graph; each directed edge is associated with a credit value. In general, the credit is a dynamic value: each transaction (one time of interaction) consumes a fixed amount of credit. Note that if the credit of an edge becomes zero, then the corresponding two nodes are not able to trade any further. The interactions between two remote nodes are also allowed (suppose node v wants to get service from node w), if and only if there is at least one path from v to w and each hop on the path can *pay* a required credit. Moreover, such payment can also be split across multiple paths if they exist. Take Figure 21.4 as an example [27]. When node A gets service from a remote node E, we assume that the system requires four path credits in total. Node A adopts an equal splitting

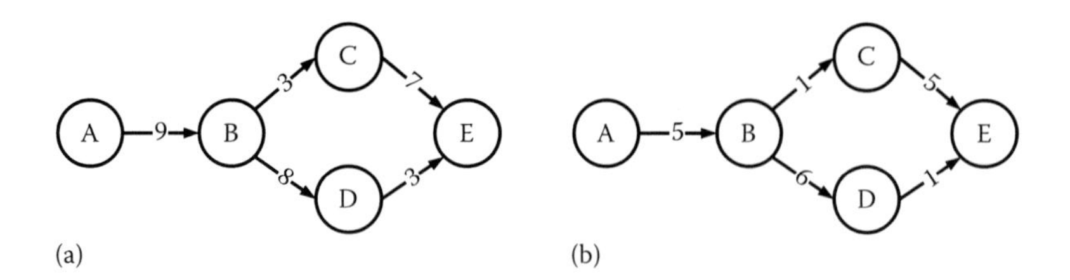

FIGURE 21.4 A credit network. Credit networks are directed and weighted graphs, where the weights of edges indicate the credits between two neighboring nodes. Interactions between two nodes are also called transactions; each transaction costs some credits on a trust path: (a) the credits of edges before transaction and (b) the credits of edges after transaction.

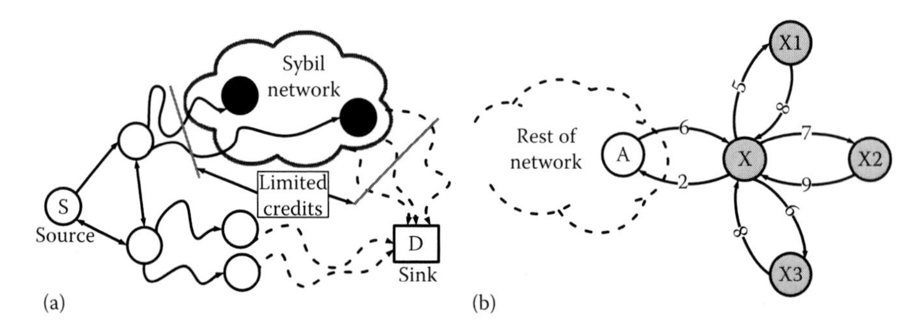

FIGURE 21.5 A Sybil defense named SumUp, which can be used in online distributed voting systems. (a) The idea of SumUp and (b) an example of SumUp. SumUp essentially restricts the damage of Sybil attacks by using credit networks. Note that although an attacker may create plenty of Sybils, the credits from the Sybils to the honest nodes are limited.

mechanism as follows: it first pays two credits along the path $A \to B \to C \to E$ (note that each hop costs two credits on the path) and then pays another two credits along the path $A \to B \to D \to E$.

The general idea of SumUp is that, in credit networks, the links between honest users and Sybil users are limited (as shown in Figure 21.5a) and the social networks in the honest users' part are fast mixing; most honest users can participate in a voting, since there are plenty of trusted paths from the honest user to a sinker node. However, since the number of credit links is limited, most Sybil nodes cannot provide their false opinions to the system. Figure 21.5b is an example. Although the attacker, node X, has three Sybil nodes, X_1, X_2, and X_3, the credit on the directed link from Sybil to the honest is only 2. Hence, if we assume that each action takes one credit, then no matter how many Sybils the attacker could create, he can only give two actions at most, which definitely will not change the voting result.

21.6.3 Canal

Canal [27] also adopts a credit network, and we can regard it as an extension of SumUp. In a credit network, each interaction between nodes always requires the

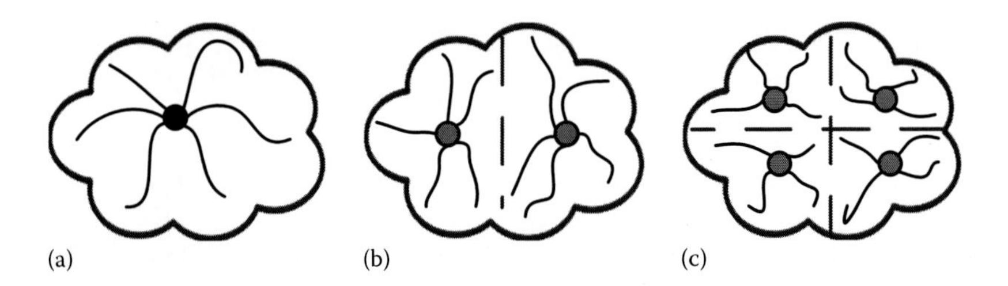

FIGURE 21.6 A extension of SumUp, named Canal system. The Canal system adopts landmarks to partition the whole network into layers. The trust routing between nodes is conducted by the routing between local landmarks: (a) level 0 landmark, (b) level 1 landmark, and (c) level 2 landmark.

system to first find at least one available credit path; clearly, such a process has a high computational cost. Essentially, the procedure of searching such paths is equivalent to the maximum flow problem. However, even the most efficient algorithms have the computation complexity in $O(V^2 \log(E))$. Instead of finding one or several *best* credit paths, Canal approximates the paths: they first partition the graph into several layers and regions and then adopt a landmark routing-based technique to find the paths. Landmark routing [28] is an old technique: in order to find a path between a pair of nodes, they first determine the paths from the nodes to several preselected nodes (also known as landmark nodes). Since the paths between landmarks are known, the resultant path will be a special path, which goes through at least one landmark. In Canal, as shown in Figure 21.6, the author uses multiple-level landmarks: if a credit needs to be transmitted to nearby nodes, the path will go through the lower-level landmark; when a credit is transmitting to far away nodes, the shortest path will pass a higher-level landmark. However, since the landmark absorbs the paths that may cause the credits of nearby paths to decrease quickly, the landmarks should be randomly generated and periodically updated.

21.7 SYBIL DEFENSE IN PHASE IV: SOCIAL COMMUNITY-BASED ANTI-SYBIL TECHNIQUES

Suppose that, in a peer-to-peer system, there are m honest nodes and n Sybil nodes. After the central authority obtains all of the data, based on the features of these data, he may predicate whether a node is honest or not. However, assume that the attackers somehow replicated all of the m honest users' data and built a network with the same structure as the honest one and then sent their replicated data to the collector. If this is the case, no matter how, the collector cannot discriminate a Sybil node from the others, since all of the information is exactly the same. For breaking such symmetric information, a Sybil defense system must build on some asymmetric information. Recall that social network-based Sybil defense algorithms always assume that the executants of the algorithms regard themselves as the verifier (each user at least knows that it is trustworthy, itself). The reason for having this assumption is just to break the symmetric information.

In 2010, paper [3] analyzed several classic social network-based anti-Sybil algorithms and found that those algorithms essentially detected the community structure of honest users. Following this discovery, the research community has begun to work on community-based Sybil defense approaches. Note that the exact concepts used by different papers may be different; some papers focus on that of traditional social networks, while some others work on signed social networks. The traditional communities are in a global view; however, it could also be in a local view. Since the concept of community-based Sybil defense is relatively new, we are not sure whether it will become a mainstream or if it is just an extension of social network-based solutions.

21.7.1 Pure Social Community-Based Sybil Detection

In 2010 ACM SIGCOMM, another novel paper [3], provided a new trend for Sybil defense, which argues that the Sybil nodes can be detected by their community structures. Since the research on community detection has been there for many years and there are plenty of useful techniques that have been proposed, the paper [3] may open up another option for anti-Sybil approaches. Since the paper is very fresh and, currently, we do not know how many followers there will be, we just say that the researches may, or may not, enter into the third phase.

Viswanath et al. [3] examine all the famous Sybil defenses in social networks and found out that, indeed, these methods partition a given network into communities; the preknown honest node's resident community is treated as an honest community, and all of the others' communities are regarded as Sybil nodes' resident regions. As shown in Figure 21.7, the authors found that all the existing Sybil defenses are essentially rating systems, which assign a value to each node, based on the distance towards preknown honest nodes. Different Sybil defenses may use different thresholds to partition the nodes. Hence, the authors proposed that the Sybil nodes may be detected through community detections.

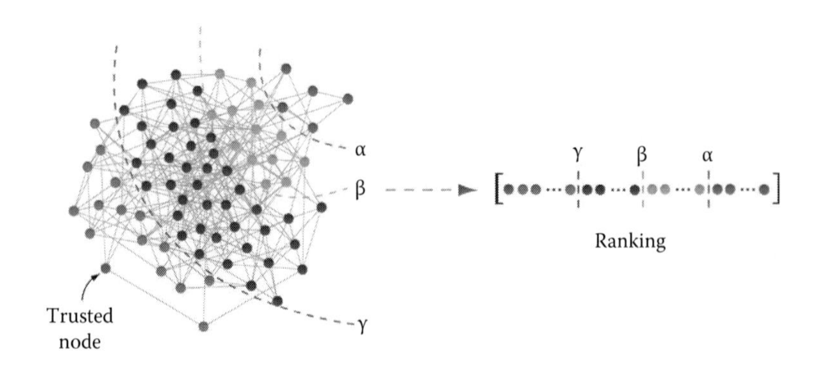

FIGURE 21.7 The essential approach of the existing Sybil defenses in social networks [3]. Most random walk-related Sybil defense algorithms essentially assign weights to other nodes based on their distance. By providing a threshold, the nodes with lower weights are regarded as Sybil.

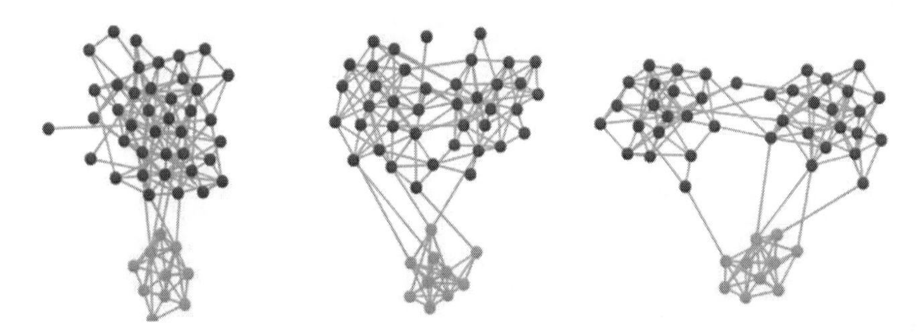

FIGURE 21.8 Future research directions [3]. In the experiment part of paper [3], the author shows that the accuracy of social network-based Sybil defense is related with the structure of honest nodes. When the honest nodes present multiple communities, most social network-based Sybil defense algorithms have high false-positive results. In this graph, the upper dark-gray nodes stand for honest nodes, and the lower light-gray nodes represent Sybils. From left to right, the figure shows the process that the structure of honest nodes changes from single community to two communities.

The authors of [3] proposed a Sybil defense by using conductance-based community detection. They regard all of the nodes in a social network as one of two types: the node resident in the honest community and node resident in the Sybil community, as shown in Figure 21.8. Through some simulations on synthetic data, they argue that their method can successfully detect Sybil nodes and their resident community.

21.7.2 SybilDefender

SybilDefender is another famous approach for anti-Sybil. It consists of two steps: Sybil node detection and Sybil community detection (Figure 21.9). For a suspect node, based on preknown honest nodes' statistical features, SybilDefender determines whether the suspect node is a Sybil or not. After finding a Sybil node, based on the assumption that Sybil nodes are more likely to connect with other Sybil nodes, the defense will detect the Sybil community in which the Sybil node resided. This defense is based on two assumptions: (1) the number of links between honest users and Sybils is limited and (2) the size of the Sybil community is smaller than that of the honest. The second assumption is realistic, since typical social networks contain millions of users; for the attacker, to register such a large number of identities is impossible. From an honest user, we can send a fixed number of random walkers to pass an l-length random path, assuming there are k walkers. At other nodes, we can compute the times that these random walkers passed through this node and call the times their *visiting frequency*. After that, we can calculate the statistic distribution of the visiting frequency. If the random walks from a suspect node do not follow some statistic distribution, then the suspect is a Sybil.

Figure 21.10 illustrates the SybilDefender. In Figure 21.10, suppose that we have already known an honest node. From this node, we send out k random walks with a fixed length l. Since social network (in the honest region) is fast mixing, which means that any pair of nodes can reach one another at an $O(\log n)$-length random

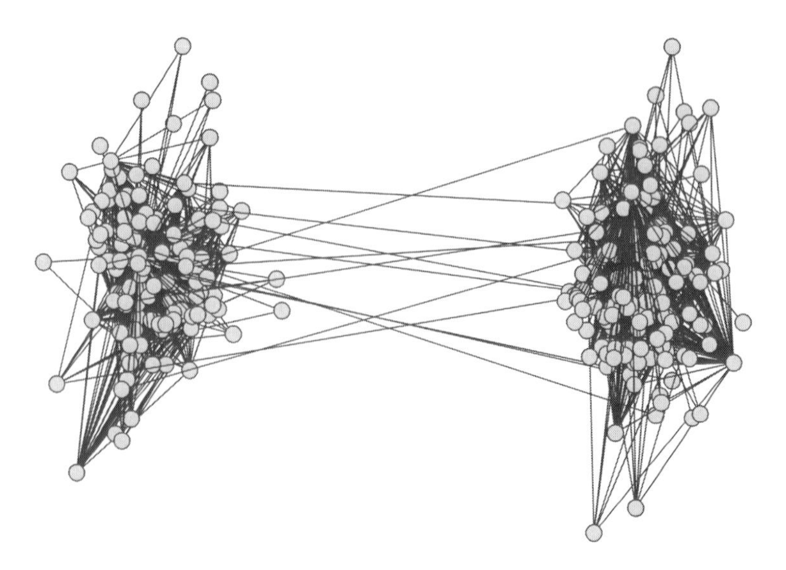

FIGURE 21.9 An honest community and a Sybil community. For the synthetic data, people usually create two communities; let one of them be the honest community and the other the Sybil community. The edges within each community are randomly generated, and the node degrees follow the power law distribution. By using a rewiring operation, a limited number of attack edges are added between the communities.

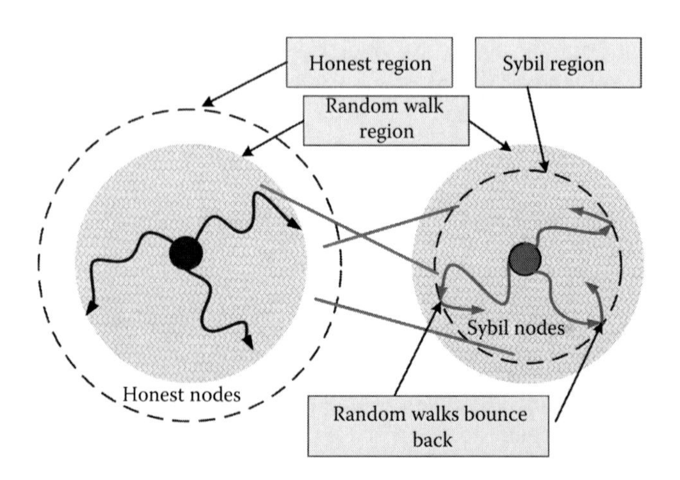

FIGURE 21.10 The idea of SybilDefender. One fundamental assumption of SybilDefender is that the size of Sybil community is much smaller than that of the honest community. When the length of random walks is appropriately set up, one can detect the existence of Sybil nodes via monitoring nodes' visiting frequency.

path, a circle region in the honest community will be covered by the random walk. However, because the size of the Sybil community is smaller than that of the honest one, the majority of random walks in the Sybil region will be reflected, which indicates that the distribution of the visiting frequency in the Sybil region is different from the honest one. By this way, a suspect node can be verified.

However, since we do not know the size of Sybil communities, how to determine the length of random walk is a problem. As shown in Figure 21.11a, if the length is longer than the radius of the honest community, or as shown in Figure 21.11b, if the size of a Sybil community is greater than a random walk's length, the distribution of visiting frequency may not show the difference. In other words, detecting by the nodes' visiting frequency may fail if the length of random walks is not set up appropriately. Hence, if the cases of Figure 21.11a and b happen, we have to try other random walk lengths.

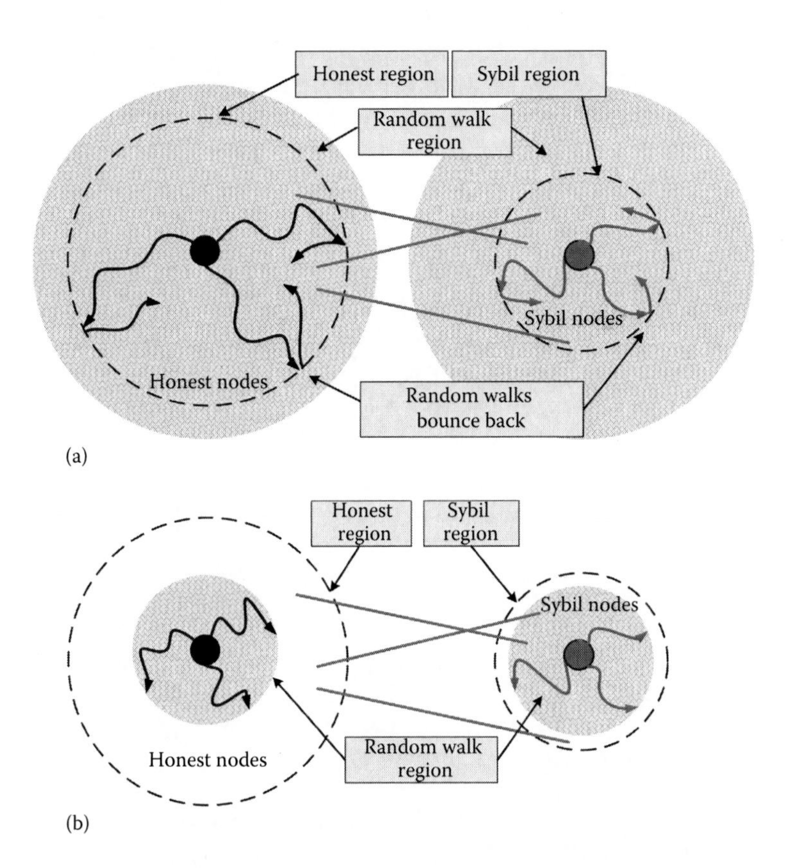

FIGURE 21.11 The appropriative length of random walk is related to the size of a Sybil community. The size of Sybil community is important to know for setting the length of random walks. In (a), the length of random walk is too long since the walks bounce back at both honest region and Sybil region, while in (b), the length is too short because the random walks do not bounce back at both regions.

Another challenge with the SybilDefender is that of how to extract the correct visiting frequency distribution from the honest region. Clearly, if we select an honest node that has been fooled by an attacker, the computed statistic feature of the honest node will definitely be different from that of other honest nodes, which locate at the core of an honest community. From this consideration, before computing the statistic feature of an honest node, SybilDefender first finds several K-hops neighbors. Considering that the links between honest and Sybil nodes are very limited, a majority of these K-hops neighbors will also be honest. Moreover, since the social networks are fast mixing, the statistic features of the nodes can correctly reflect that of the whole honest region.

Since SybilDefender does not know the size of the Sybil community, it initializes several groups of random walks with different lengths. Then, during the verification of a suspect node, several random walks (with different lengths) are conducted. If the distribution of the visiting frequency of the suspect node does not share the same distribution as the honest one, this suspect node will be regarded as a Sybil node; otherwise, SybilDefender will regard the suspect as an honest node.

After finding a Sybil node, the SybilDefender can also detect its resident Sybil community, based on the fact that Sybil nodes are more likely to connect with other Sybil nodes. The detection of a Sybil community can be done by using loop-free random walks. Consider that when a random walker passes the same node twice, it means the random walkers reach the boundary of the Sybil community. SybilDefender renders a random walker dead if it arrives at the same node twice. Similar to the process of verifying a suspect node, SybilDefender also initializes several random walks with different lengths. Again, the reason is that the size of the Sybil community is unavailable. If the dead ratio of the L-length group of random walks is greater than a predefined threshold, then all the passed nodes will be regarded as members of a Sybil community.

21.7.3 Signed Social Network-Based Sybil Defense [29]

Most social network-based Sybil defenses adopt the assumptions that the honest region is a fast-mixing network and that Sybil entities can only fool a limited number of honest entities. However, more and more evidence shows that some real social networks are not fast mixing, especially when only strong trust relations are considered. Moreover, the accuracy of all existing solutions is related to the number of attack edges that the adversary can build. For addressing these two important problems, we propose a local ranking system for estimating the trust level between users in mobile social networks.

Unlike traditional Sybil defenses, our proposed scheme has three unique features. First, our system creates a signed social network, which contains both trust and distrust relations. Second, consider that each mobile phone only has a relatively small storage and frequently accessing a remote server increases the amount of data flow, which costs an extra fee. In our solution, instead of storing the entire social graph, each user carries a limited amount of information related to him. Last, but not least, our system weakens the impacts of attack edges by removing several suspicious edges with high centrality.

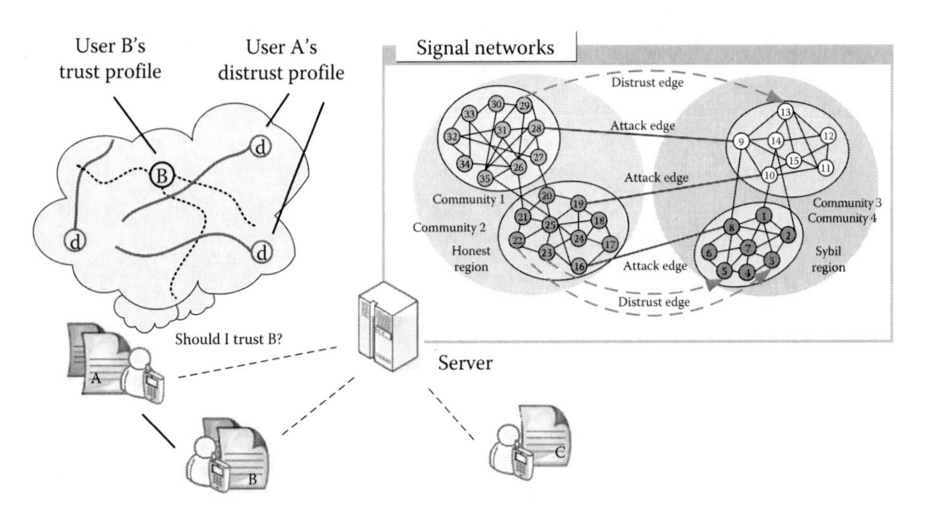

FIGURE 21.12 System model. The system consists of two parts as follows: (1) a central server, which stores signed social networks, and (2) users' smartphones, which detect the physical encounter of other users, report other users' misbehavior to the server, and estimate the trust level of the encounter.

In mobile social networks, our system consists of two parts: a remote server and several users, as shown in Figure 21.12. The server is responsible for two jobs: (1) storing and periodically pruning the created signed network graph and (2) assigning *randomly sampled* social profiles to users for computing the trust level between users. Note that now we are using two networks: a mobile social network and a created signed social network. The mobile social network is the network formed by physical interactions of phone users, while the signed network is created for Sybil detection. The positive edges on the created signed social network represent trusted social relationships, which could be obtained from an online-social network. The negative edges are generated based on users' physical interactions with each other. We assume that each honest user has one mobile phone, which is associated with a single real identity, while the attacker may hold more than one phone, and each phone runs multiple fake identities. Each identity is required to periodically send a special message to the server, and the server will return updated social profiles; otherwise, the identity will be deleted from the system. Unlike traditional social network–based Sybil defense models, we assume that the honest region of a social network may not be fast mixing. Exactly how many honest communities may be formed is determined by the social networks being considered. For instance, if we use a social network of political opinions, then the honest nodes may be gathered into two communities, for example, the Democratic and Republican parties. However, if we adopt Facebook, the honest users may only be clustered into one community.

Consider that multiple Sybils are sharing a single phone and that each Sybil identity needs to periodically report some message in order to keep itself valid. For some honest users, they may catch the instant that an attacker switches her Sybil identities. If that is the case, then the honest users will report this misbehavior to our server, and a distrust edge will be added from the reporter to the accused. Moreover, in mobile social networks, each identity used by a mobile phone is associated with a physical human, and some users may remember the appearance of others. When several honest users, who have been fooled by the same attacker, physically encounter the attacker at the same time, some of them may notice that the attacker is using a different identity; the honest one could report this event to our server. Besides these two options, any other neighbor monitoring techniques may also be adopted for the generation of negative edges.

The general procedure of our system is as follows. Each user locally stores two *randomly sampled* social profiles: a trust profile and a distrust profile, which are assigned and periodically updated by the server. Whenever two strangers encounter one another and want to have some cooperative service, each user's phone will exchange the trust profile and locally compute a trust and a distrust score to determine whether the other user is trustworthy. In order to increase the accuracy, a special pruning algorithm is running on the server.

When a new user V joins our system and provides his friend/foe lists, the server will generate a trust relation profile and a distrust relation profile. The generating procedure for the trust relation profile is as follows: the server first sends out K random walkers, and each of them will conduct an l-length random walk from V. The walkers only move along trust edges, and each path represents one possible way of trust propagation from V. As a result, there will be K random paths beginning with V, and the visited node list will be sent to V as a trust social profile. Obviously, the profile is a random sample of V's l-hop friendship. Consider that, in a mobile social network, a user may have different physical contacting frequencies to his directly trusted friends. User V is able to locally assign different weights to the paths, according to the frequencies.

In order to cheaply impersonate real users and to benefit from certain applications, Sybils always support each other by adding trust edges among themselves.* Therefore, friends of a distrusted node are likely to be distrustful. Moreover, the majority of random walks from a node will still reside in their own community. Based on these observations, the server creates V's distrust social profile by using the distrust relations of both V and his trusted friends. Before creating the profile, a distrust seed set needs to be generated: along trust edges, the server computes K short-length random paths from V, and nodes directly distrusted by the nodes on these paths form the seed set. Another l-length random walk will be produced from each seed, and the trails will be used as the distrust social profile of V. For instance, in Figure 21.13, the solid light-gray line p represents one of the short-length random paths from V, and there are three distrust edges (dashed lines) initiating from the nodes on p. The distrust seed set consists of three nodes (shadowed circles). From each of the seeds, the server conducts an l-length random walk, and the generated random paths compose V's distrust social profile; again, all random walkers are moving along trust edges.

When an honest user V encounters another user S, a trust level will be computed based on the similarity of their locally stored profiles. They first exchange their trust

* The conditions in which Sybil actively friends honest users are out of the scope of this chapter, since each Sybil normally interacts with honest users, instead of cheaply creating a fake identity.

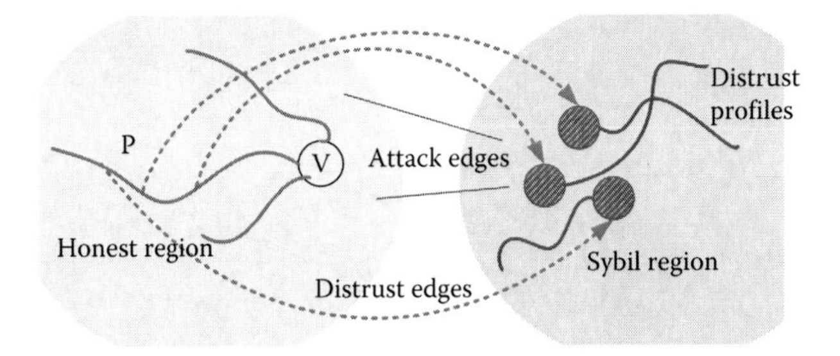

FIGURE 21.13 The generation of distrust social profiles. In order to obtain enough distrust relations, the server first creates three short-length random paths from *V*. Then, the server takes all the distrust relations of the nodes on the paths as a distrust seed set. In this figure, we use the shadowed dark-gray circles to indicate the seeds. From each seed, the server generates a random path along the trust edges, and these edges compose the distrust social profile. Here, we use solid dark-gray lines to represent the random paths belonging to *V*'s distrust profile.

social profiles, which are assigned by the server. Note that a user's identity (usually we adopt the user's public key as its identity), his trust (or distrust) social profiles, and the profile's valid time are signed together by the server's private key; the attacker cannot create or modify it. After obtaining the trust profile of user *S*, user *V* will verify it first, since the attackers may steal others' profiles. *V* generates a short random number and encrypts it by the public key of *S*; *S* needs to find out the random number, encrypt the random number by *V*'s public key, and send it back to *V*.

After the process of mutual verification, node *V* will locally compute a trust score and a distrust score for *S*. For the ease of description, the paths in *V*'s trust social profile are named as *verifier paths*, and the paths in the trust profile of *S* are called *suspect paths*, as shown by Figure 21.14. If there is a common node on both a verifier path and a suspect path, then we say that the suspect path is verified once; when a suspect path has been verified more than k_t times, where k_t is a constant, we say that

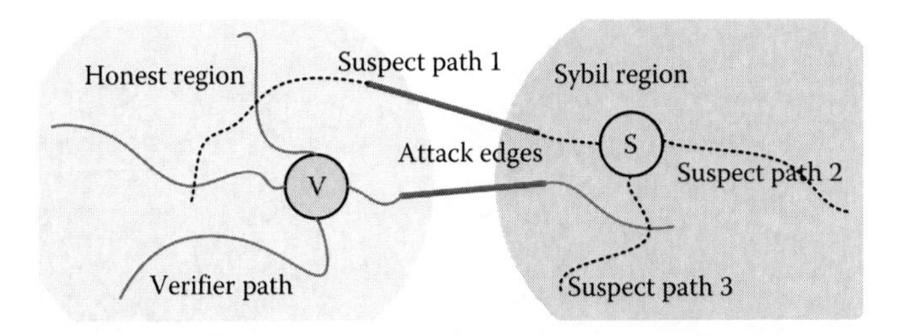

FIGURE 21.14 The computation of trust score. When the honest user *V* encounters a stranger *S*, they first exchange their social profile. In this figure, the solid gray lines indicate *V*'s trust social profile, and the dashed black lines represent the trust social profile of *S*. In *V*'s view, he regards himself as the verifier, and the paths in his trust profile are named as the verifier paths. Moreover, *V* regards the paths of *S* as suspect paths. *V* locally checks the connectivity between his profile and the others. A trust score will be determined based on the intensity of connectivity. In this figure, if *V* sets the verifier threshold $k_t = 2$, then only suspect path 1 will be fully verified. Therefore, $Ver(V, S) = 1$, $|K| = 3$, and $Trust(V, S) = 1/3$.

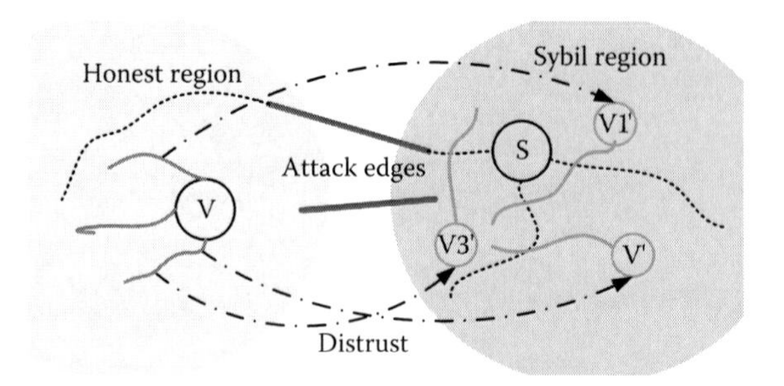

FIGURE 21.15 The computation of distrust score. In this example, the distrust social profile of V contains 3 distrust paths, which are represented by solid gray lines. V will compute a distrust score by using his distrust profile and S's trust profile. If we define a verified distrust path as a suspect path that comes across at least half of the distrust verifier paths, then, in this figure, only suspect path 3 is verified. As a result, $Dis(V, S) = 1$, $K' = 3$, and $DisTru(V, S) = 1/3$.

this suspect path is fully verified. Let $Ver(V, S)$ be the number of fully verified paths, and recall that there are a total of K random paths in a trust social profile. In regard to V, the trust score of S is given by

$$Trust(V, S) = \frac{Ver(V, S)}{K} \tag{21.1}$$

For the computation of distrust score, SNSD considers both the distrust social profile of V and the trust social profile of S, as shown by Figure 21.15. We name the paths from V's distrust social profile *distrust verifier paths*, and we use K' to represent the size of V's distrust social profile. Similar to the computation of trust score, when there are k_t distrust verifier paths having common nodes with a suspect path, this suspect path is a fully verified distrust path. Let $Dis(V, S)$ be the total number of fully verified distrust paths, and the distrust score of S in regard to V is given by

$$DisTru(V, S) = \frac{Dis(V, S)}{K'} \tag{21.2}$$

The final label of S, $L(V, S)$, is determined by the difference of the two scores: $z = Trust(V, S) - DisTru(V, S)$. Let α, β be two thresholds, $1 \geq \alpha > \beta \geq -1$:

$$L(V, S) = \begin{cases} \text{Trusted} & \text{for } z \geq \alpha \\ \text{Neutral} & \text{for } \alpha \geq z \geq \beta \\ \text{Distrusted} & \text{for } \beta \geq z \end{cases} \tag{21.3}$$

The accuracy of a majority of the existing Sybil defense systems is bounded by the number of attack edges. In order to improve the accuracy of our system, a special pruning algorithm called Sybil gateway-breaking algorithm (SGA) will run on the remote server once in a while. Essentially, the algorithm prunes some suspicious edges of the signed social network stored on the server. Consider that all of the paths connecting honest and Sybil nodes must go through the attack edges and that the number of attack edges is limited. So, the connectivity from an honest region (a group of honest nodes) to a Sybil region is bounded by the quantity of attack edges. If each honest node is able to locally check whether it is fooled by others based on the connectivity and deletes some attack edges, the accuracy of *any* social network-based Sybil defense will surely be enhanced.

The SGA consists of three subalgorithms, as shown by Algorithm 21.2: (1) suspicious edge selection algorithm, (2) gateway verification algorithm (GVA), and (3) attack edge detection algorithm. Note that a gateway is an edge that connects two communities together. SGA first selects high centrality nodes/edges as the suspicious ones and then verifies whether the associated edges are gateways. For gateway edges, the algorithm further detects whether they are Sybils. There are two options for finding the suspicious edges. The first one is based on node centrality, and the second one focuses on the edge centrality.

Algorithm 21.2 SGA

1: **Suspicious Edges Selection**:
2: Select edges with high local betweenness as suspicious edges.
3: Use signed network-based Sybil defense algorithm (Section 21.5) to determine honest and Sybil.
4: Find shortest paths from the honest nodes to Sybil nodes.
5: Compute the visiting frequency of edges.
6: Take the edges with high visiting frequency as suspicious.
7: **Gateway Verification**:
8: Generate the initial neighbor set, $\{u\}$, $\{v\}$, and $\{w\}$.
9: Compute the number of unique paths from $\{v\}$ to $\{u\}$, and from $\{v\}$ to $\{w\}$.
10: **for** Predefined times **do**
11: Respectively add Δk disjoint neighbors into $\{u\}$, $\{v\}$, $\{w\}$.
12: Compute the number of unique paths from $\{v\}$ to $\{u\}$, and from $\{v\}$ to $\{w\}$.
13: Compute the growing speeds of unique paths, S_{vu} and S_{vw}.
14: **if** $|S_{vu} - S_{vw}|$ is greater than a threshold **then**
15: E^{+}_{uw} is a gateway;
16: E^{+}_{uw} is not a gateway.
17: **Attack Edge Detection:**
18: Find attack edges from detected gateways by distrust relations; break them.

First, the attacker may adopt a target attack by densely adding attack edges to a target node and its neighbors; although the target region and Sybils are well connected, the nearby regions of the target may observe a weak connectivity from themselves to Sybils. Therefore, the server can first generate a local map (G) of each node and then compute each node's centrality on G. The GVA is only applied to the edges, which are connected to the high centrality nodes. The criteria of centrality we used is called the betweenness centrality [30], which is defined as the number of shortest paths passing a node out of the total number of shortest paths within a given network. $B(w) = \sum_{u,v \in G, u \neq v} g_{uv}(w)/g_{uv}$, where g_{uv} is the total number of shortest indirect paths linking nodes u and v and $g_{uv}(w)$ is the number of those indirect paths that include node w.

The second option is that since all of the paths between a Sybil node and an honest node must traverse the same set of attack edges, the suspicious edges could be the edges passed by the majority of random paths, which connect the nodes with antagonistic relations (at least one directly distrusted edge between the nodes). Therefore, the server randomly selects several pairs of antagonistic nodes, creates random paths between each pair of antagonistic nodes, and counts the visiting frequency of the transited edges. For the edges with high frequency, the GVA will be adopted. The aforementioned procedure substantially examines the centrality of edges based on partial nodes' relationships.

Nodes within the same community share certain characteristics [31]. Whether two nodes are located at the same community can be verified by their connectivity to other nodes. Intuitively, if one node's connectivity to the third node is much larger than that of the other node, it is very possible that the two nodes reside at different communities. We use the number of unique paths to measure the connectivity feature.

Definition 21.1

Unique paths indicate a group of paths connecting two distinct nodes or regions without sharing a common edge.

Since the amount of unique paths connecting two nodes is bounded by the node degrees of the ends, we compute the unique paths from a region to another region. Assume that the ends of a given edge are nodes u and w. Let v be the third node for checking the connectivity. $UP(u, w)$ represents the number of unique paths from u to w.

Generally speaking, based on the community structure of these three nodes, there are three possible cases that can be observed on connectivity:

1. Either u or w shares the same community as v.
2. Both u and w are located in the same community as v.
3. None of them come from v's community.

For the first case, assuming that node u resides in the same community as v, $UP(v, u)$ is much greater than $UP(v, w)$, since all the paths from v to w must go through the gateways between communities, which are limited. As we mentioned earlier, we count the amount of unique paths from region to region. If we gradually increase the size of regions by Δk, the number of unique paths from w's region will stop growing much earlier than those from u's region. However, for the second and third cases, we will not observe such differences. Based on this feature, we design a GVA for checking whether a given edge is a gateway or not.

GVA, respectively, calculates the growing speeds of unique paths from v to u and from v to w. The procedure is as follows. GVA gradually adds Δk disjoint neighbors into both regions, which, respectively, contain u and v (or w and v), and examines the amount of unique paths between the regions. Since we only care whether the edge E^+_{uw} is a gateway or not, we only need to check the existence of the growing speeds' difference for a given node v.

Whether a gateway is an attack edge is determined by the distrust relationships between the communities. If either one of them, or both of them, highly distrusts the other, it is very likely that the gateway is an attack edge. However, since the server does not know the exact community structures of the created social graph, the scheme of counting the total amount of distrust links from one community to another is infeasible. Consider that the majority of random paths are trapped inside their own communities; instead of all of the nodes within a community, we adopt random sampling to estimate the intensity of distrust relations between communities.

Our attack edge detection algorithm works as follows. First, for each gateway, the server temporarily breaks it. Then, from both its ends, the server sends out k random walkers along the trust edges, respectively. The length of the random walks is a small fixed number, and all of the visited nodes form a sampling set. The server also creates another set, called a distrust sampling set, which consists of nodes directly distrusted by the sampling set's members. The intersection of these two sets indicates the intensity of distrust of the communities. The larger the intersection set is, the more likely the gateway is an attack edge. Finally, the server keeps the gateways with large intersection sets being broken and restores other gateways.

21.7.4 Multiple Local Community-Based Sybil Defense [32]

Based on real data, paper [33] finds out that the probability of contagion is tightly controlled by the number of connected components (local communities) in an individual contact neighborhood, rather than by the actual size of the neighborhood. Based on this observation, we wonder whether there is a better way to measure the trust values among nodes in a peer-to-peer system. Consider that traditional social network-based Sybil defenses, such as SybilGuard, estimate the trust value by measuring the strength of trust paths between any pair of nodes. Since trust and contagion are closely related, instead of computing the trust paths at the node level, we should measure the strength of the paths at the local community level. Moreover, in traditional social network-based Sybil defense algorithms, each node must locally determine whether to accept all others' nodes. Therefore, the overall computation cost of these algorithms is a problem, especially in some large systems. Consider that, in a social graph, the neighbors of an honest user are still honest. Once similar users are clustered into groups, a Sybil detection algorithm only needs to determine whether a group (clump) is Sybil. Here, we propose three social clump-based Sybil defense algorithms, which can save plenty of computation costs, while maintaining the same level of accuracy as the traditional solution.

In social networks, all friendships between honest users are established based on their physical interactions, and therefore, the edges between users indicate credibility between them. In order to impersonate real users, the attacker has to create social profiles for each Sybil identity and manipulate the friendship edges among them. Consider an extreme case in which the attacker copies whole social networks of real users and replaces the real identities with the Sybil ones. It is impossible to discriminate Sybil IDs from the real ones by simply checking the friendship structures. However, since friendship is a mutual relation and is built by physical interactions between users, attackers cannot tamper with the friendships of real users, and therefore, the total number of friendships between Sybils and honest users is few; we use g to represent the number. In most cases, g should be smaller than the average number of honest users' friends. Suppose that we have previous knowledge about the value of g and that the attacker can arbitrarily add these g attack edges on a social graph.

The strength of a friendship is measured by the number of common friends. A pair of friends usually has several common friends. $N(u)$ is adopted to represent the friend set of node u, and let $|\cdot|$ stand for the cardinality of a given set. If the cardinality of the common friend set is large enough, then both users should be clustered into the same clump.

Theorem 21.1

When a pair of friends, u and v, shares more than g common friends $|N(u) \cap N(v)| > g|$, the users u and v must be both honest or both Sybil.

However, the requirement for having more than g common friends causes some honest nodes to have no clump to participate. Intuitively, in our solution, the more nodes that are clustered into clumps, the more computational costs that can be saved. Consider that an honest clump is a subgraph of the honest part. It should be cohesive inside and have low betweenness from the global view. Based on this observation, we propose a pair of rules for finding the members of a clump. A collection of nodes can locally form into a clump if and only if the majority of their friends reside in the same clump and the nearby nonclump members are still reachable within several hops after removing the clump. In detail, the members of constructed clump C must satisfy the following two rules:

Rule 1: if $u \in C$, then $|N(u) \cap C| \geq \alpha |N(u)|$.
Rule 2: if $u \in C$, then there exists $g - |N(u) \cap C| + 1$ unique paths between C and $N(u) \cap \bar{C}$, whose length is less than β.

Here, $\bar{C}$ represents the complementary set of C. α, β are two parameters.

Definition 21.2

Unique paths indicate a group of paths connecting two distinct nodes or regions without sharing a common edge.

We adopt Rule 1 to guarantee that nodes are well connected inside each clump. Rule 2 is set up in order to keep the bridge structure unchanged after grouping. However, for the honest nodes with lower node degree, they cannot be accepted by any clump by using Rules 1 and 2. We provide an additional rule:

Rule 3: if $|N(u)| < g$ and $\forall v \in N(u), v \in C$, then $u \in C$.

Note that Rules 1 and 2 work together and they only apply on the nodes with a large node degree. The additional rule, Rule 3, only suits nodes with a small friend set.

Algorithm 21.3 provides the procedure for clump generation. We associate one variable, called label, with each node. For the ease of description, we use l_u to represent the label of u. Initially, each node uses its own identity as the label. Based on Theorem 21.1, when two nodes, u and v, have more than g common friends, their resided clumps will merge into one (by adjusting the labels of both clumps' members into the least identity value). If the cardinality of the common neighbor set between a node u and its neighbor v satisfies Rule 1, then Rule 2 will be checked. When the rule is satisfied, then u's resided clump merges with v's. For a node, whose node degree is less than g, if all of its neighbors have the same label (Rule 3), the node will join its neighbors' clump.

Algorithm 21.3 Distributed Clump Generation Algorithm

1: Each node u sets up its label $l_u = u$
2: **if** $|N(u)| \geq g$ **then**
3: **if** $\exists v \in N(u)$ and $|N(v) \cap N(u)| > g$ **then**
4: Set the labels of l_u and l_v's clumps as $\min(l_u, l_v)$
5: **if** $\exists v \in N(u)$ and $|N(v) \cap N(u)| \geq \alpha |N(u)|$ **then**
6: Let $C = \{w | l_w = l_v\}$
7: **if** There are at least $g - |N(u) \cap C| + 1$ unique paths between C and $N(u) \cap \bar{C}$ **then**
8: Set the labels of l_u and l_v's clumps as $\min(l_u, l_v)$
9: **else**
10: **if** $\forall v, w \in N(u), l_v = l_w, l_u \neq l_v$ **then**
11: Set the label of u as $l_u = x$
12: When the labels of $N(u)$ are stable, run Algorithm 21.2

Note that line 5 of Algorithm 21.1 is not exactly the same as Rule 1. Since each node only regards itself as a clump in the very beginning, we need to put several similar nodes together. Admittedly, Theorem 21.1 can find the similar nodes, which definitely belong to the same clump. However, the number of nodes may not be large enough. So, we first adopt a modified Rule 1 as shown in line 5 of the algorithm, and then, after the process of labeling becomes stable, a special pruning algorithm will run on every node (except the nodes that satisfy Theorem 21.1), as shown by Algorithm 21.4. Essentially, the pruning algorithm checks whether the current label of a node satisfies Rule 1. If not, the node needs to redetermine its label by running Algorithm 21.1 again.

Algorithm 21.4 Clump Pruning Algorithm

1: Based on labels, find the members of u's clump C
2: **if** $|N(u) \cap C| < \alpha |N(u)|$ **then**
3: Set the label of u as $l_u = u$
4: Reconduct Algorithm 21.1 by replacing line 5 with Rule 1
5: Clump C will update its label based on connectivity

In traditional random walk-based Sybil defense algorithms, nodes need to launch a bunch of random walks since, initially, they only trust themselves. Moreover, from security concerns, these random walks are usually associated with secure signatures at each step. It is obvious that costs of the algorithms are too high. Consider the fact that most neighbors of an honest node are still honest. We propose three clump-based random walk (CRW) strategies. The proposed random walks can increase the efficiency of all existing random walk-based Sybil defense algorithms.

The strategy of CRWs is modified from traditional lazy random walk, where each walker randomly picks up one neighbor of its current location as the destination of the next hop. By using CRW, a random walker will transit from one clump to another, and the stationary distribution of each node depends on both the degree of the node and the stationary distribution of its resident clump. In order to keep the distribution of each node unchanged, we add a self-loop at each clump.

Given a clump $c = \langle V_c, E_c \rangle$, $c \subseteq G$, the transition probability $p(c, c')$ from clumps c to c' is defined as follows:

$$p(c,c') = \frac{\sum_{i \in c, j \in c'} a_{ij}}{\sum_{i \in c} deg(i)}$$

More specifically, when $c = c'$, $p(c,c) = 2|E|_c \Big/ \sum_{i \in c} deg(i)$.

Theorem 21.2

In CRW, clump C's stationary distribution π_C equals the summation of stationary distributions of its members (π_i) in traditional random walk. $\pi_C = \sum_{i \in C} \pi_i = \sum_{i \in C} \frac{|N(i)|}{2|E|}$, where $|E|$ is the total amount of edges in G.

A majority of existing social network-based Sybil defense algorithms adopt short-length random walks. As was introduced in the background section, there is a time gap between the first mixing time at the honest region and the mixing time of the whole graph. Essentially, the Sybil defense algorithms explore the features related with this time gap. The length of the short-length random walk usually is less than or equal to the first mixing of the honest region. Since we have already known that the honest region is fast mixing and that the mixing time of a fast-mixing network is proportional to the cover time of the network, we will discuss the variance of the cover time by using clumps.

Theorem 21.3

The cover time on honest region will be reduced in CRWs.

In a majority of random walk-based Sybil defense algorithms, the walkers are always easily spread to intensively connected nodes. Moreover, the similarity between social nodes in social networks can be used for measuring the strength of social links [34]. Based on the observations, we claim that by giving random walkers the ability to jump among clumps with high similarity, the mixing time at honest regions can be reduced. In other words, by assigning a higher probability of being visited, the speed of spreading trust scores to far away nodes can be accelerated.

For two given clumps c_i and c_j, which are not directly connected, the similarity between them is defined as $S(c_i, c_j)$:

$$S(c_i, c_j) = \frac{N(c_i) \cap N(c_j)}{N(c_i) \cup N(c_j)}$$

c_j of both c_i and c_j, if the cardinality of the percentage of the common neighbor sets, $|S(c_i, c_j)|$, is greater than a predefined threshold γ, the random walkers are allowed to jump between c_i and c_j. Essentially, clump-based random walk with groups jumping (CRWGJ) shrinks the length of jumping in intensively connected regions from two hops to single hop.

Before conducting any random walks, every clump in the given network needs to locally measure the common neighbor sets with direct neighbors and compute similarity scores with their two-hop neighboring clumps. If the result score is greater than the threshold, a virtual edge will be generated between the corresponding clumps. Suppose that the graph consisting of all virtual edges is presented by E_v. Clearly, $E_v \subseteq A^2(A^2 = [a_{ij}^2])$, since a virtual edge indicates that its two ends intensively connect with each other in two hops. During the computing of CRWGJ, a random walker will equally select an edge from its direct neighbors or virtual edges to be the destination of the next hop.

Based on the preceding strategy, the nodes with a large neighbor set or that are intensively connected with surrounding neighbors will have higher stationary distribution values. In the case that attack edges are randomly attached to honest nodes, CRWGJ can reduce the mixing time of the honest part and the false-positive rate* of Sybil detection results.

---

* False-positive rate indicates the percentage that an honest node is regarded as Sybil.

Theorem 21.4

The mixing time of CRWGJ at the honest region is less than that of CRW.

However, considering target attacks, where the attacker may do everything she can to establish social relations with a target node and its neighbors, CRWGJ may also result in the appearance of extra virtual edges between honest and Sybil parts; this is the trade-off between speed and accuracy.

Consider the fact that the position of a random walker will be independent from the initial location after several steps of transition. Most traditional random walk-based Sybil defense algorithms have two problems. First, although the probability for random walkers transiting from honest regions to Sybil regions is small, once a walker enters the Sybil regions, it is hard to get out. Second, when the number of random walkers is not large enough for a given graph, a node with a high degree may get unfair trust results. This may be due to the node trusting some relatively far away nodes instead of believing in some nearby nodes, which are not visited by the random walkers.

Here, we propose a new strategy called clump-based random walks with limited memory (CRWLM). In CRWLM, each random walker keeps a previous location list $l = \langle l_0, l_1, \ldots, l_{h-1} \rangle$ with length h. During transition, the walker can either move to one of its one-hop neighbors or jump back to the nodes in the list. Note that random walk with restart (RWR) is a special case of CRWLM, where the size of memory is 1, and each clump contains only one member. Assume that a random walker locates at clump c. The transition probability is given by the following:

$$p(c,c') = \begin{cases} \dfrac{W_{cc'}}{R+W_{c\bar{c}}} + \dfrac{R}{h(R+W_{c\bar{c}})} & \text{for} \quad c' \in N(c), c' \in l \\ \dfrac{1}{R+W_{c\bar{c}}} & \text{for} \quad c' \in N(c), c' \notin l \\ \dfrac{R}{h(R+W_{c\bar{c}})} & \text{for} \quad c' \notin N(c), c' \in l \\ 0 & \text{for} \quad \text{otherwise} \end{cases}$$

where $W_{cc'} = \sum_{i \in c, j \in c'} a_{ij}$ and $W_{c\bar{c}} = \sum_{i \in c, j \notin c} a_{ij}$.

The structure of traditional random walk's trajectory is a line, while CRWLM presents a dendriform structure. Moreover, such a tree contains an h-length trunk and several branches. One problem with traditional RWR is that when a node i sends out k random walkers, $k \gg deg(i)$, the walkers will repeatedly visit i's nearby nodes, which means the walkers do not spread out well. When adopting CRWLM, each walker first randomly creates a trunk and then expands several relatively short branches from it. Clearly, in CRWLM, a random walker has a better chance of visiting nearby nodes of the walk initiator and possesses more opportunities for returning to honest regions after having gotten into a Sybil region.

21.8 PREDICTION

Here, we provide several directions for further research. First, since there are certain types of friendships that are private, what will happen if we combine the privacy and Sybil attack problems together? Second, can we use the community detection method to deter Sybil attacks in a directed network? Third, the community detections are always time-consuming; can we find some other light algorithm that can detect the community structures quickly and accurately? Fourth, social networks contain multiple dimensions; how are we to combine this feature with traditional approaches? Based on the number of potential research directions, we believe that the community-based Sybil defenses may become the fourth phase of anti-Sybil techniques.

21.9 CONCLUSION

Peer-to-peer systems play an ever-increasingly important part of our daily lives. However, most of the peer-to-peer systems are vulnerable to Sybil attacks. In order to design more efficient and practical Sybil defenses, we write this survey. This article is the first survey focusing on the developments of Sybil defenses. We first give the definition of Sybil attacks and provide the classification of Sybil attacks. Then, we give several realistic systems that are vulnerable to Sybil attacks. After that, defense mechanisms and their corresponding strengths and weaknesses were discussed. Unlike other surveys, we describe these mechanisms according to anti-Sybil approaches' developing stages. By the end of this survey, we provide some directions for future research.

REFERENCES

1. H. Yu, M. Kaminsky, P. Gibbons, and A. Flaxman, Sybilguard: Defending against Sybil attacks via social networks, in *Proceedings of ACM SIGCOMM*, Pisa, Italy, 2006, Vol. 36(4), pp. 267–278.
2. H. Yu, Sybil defenses via social networks: A tutorial and survey, *SIGACT News*, 42(3), 80–101, 2011.
3. B. Viswanath, A. Post, K. Gummadi, and A. Mislove, An analysis of social network-based Sybil defenses, in *Proceedings of ACM SIGCOMM*, New Delhi, India, 2010, Vol. 40(4), pp. 363–374.
4. X. Zheng, Y. Lai, K. Chow, L. Hui, and S. Yiu, Sockpuppet detection in online discussion forums, in *Proceedings of IEEE IIH-MSP*, Dalian, China, 2011, pp. 374–377.
5. L. Page, S. Brin, R. Motwani, and T. Winograd, The PageRank citation ranking: Bringing order to the web, Technical Report, Stanford University, Stanford, CA, 1999.
6. J. Golbeck, B. Parsia, and J. Hendler, Trust networks on the semantic web, in *Cooperative Information Agents VII*, Helsinki, Finland, 2003, pp. 238–249.
7. J. Newsome, E. Shi, D. Song, and A. Perrig, The Sybil attack in sensor networks: Analysis & defenses, in *Proceedings of ACM IPSN*, Berkeley, CA, 2004, pp. 259–268.
8. J. Ledlie and M. Seltzer, Distributed, secure load balancing with skew, heterogeneity and churn, in *Proceedings of IEEE INFOCOM*, Miami, FL, 2005, Vol. 2, pp. 1419–1430.

9. G. Mathur, V. Padmanabhan, and D. Simon, Securing routing in open networks using secure traceroute, Technical Report MSR-TR-2004-66, Microsoft Research, 2004.
10. J. R. Douceur, The sybil attack, in *Peer-to-Peer Systems*, pp. 251–260, Springer, 2002.
11. C. Karlof and D. Wagner, Secure routing in wireless sensor networks: Attacks and countermeasures, *Ad hoc Networks*, 1(2), 293–315, 2003.
12. B. Awerbuch and C. Scheideler, Group spreading: A protocol for provably secure distributed name service, in *Automata, Languages and Programming*, Turku, Finland. Springer (Springer Science+Business Media, USA), 2004, pp. 183–195.
13. R. Gatti, S. Lewis, A. Ozment, T. Rayna, and A. Serjantov, Sufficiently secure peer-to-peer networks, *Workshop on the Economics of Information Security*, Minneapolis, MN, 2004.
14. N. Margolin, B. Levine, N. Margolin, and B. Levine, Quantifying and discouraging Sybil attacks, UMass Amherst Computer Science Technical Report, Vol. 67, 2005.
15. Y. Reddy, A game theory approach to detect malicious nodes in wireless sensor networks, in *Proceedings of IEEE SENSORCOMM*, Athens, Greece, 2009, pp. 462–468.
16. C. Piro, C. Shields, and B. Levine, Detecting the Sybil attack in mobile ad hoc networks, in *Proceedings of IEEE Securecomm*, Baltimore, MD, 2006, pp. 1–11.
17. B. Xiao, B. Yu, and C. Gao, Detection and localization of Sybil nodes in VANETs, in *Proceedings of ACM DWANS*, Los Angeles, CA, 2006, pp. 1–8.
18. Q. Zhang, P. Wang, D. Reeves, and P. Ning, Defending against Sybil attacks in sensor networks, in *Proceedings of IEEE ICDCS*, Columbus, OH, 2005, pp. 185–191.
19. M. Demirbas and Y. Song, An RSSI-based scheme for Sybil attack detection in wireless sensor networks, in *Proceedings of IEEE WoWMoM*, Buffalo, New York, 2006, pp. 564–570.
20. C. Zheng and D. S. Gilbert, Thwarting Sybil attacks and malicious disruption in wireless networks, 2012, http://www.comp.nus.edu.sg/~zheng-10/talk/grp-2012-09-14-paper-v3.pdf, accessed on March 2013.
21. P. Bahl and V. Padmanabhan, RADAR: An in-building RF-based user location and tracking system, in *Proceedings of IEEE CCS*, Athens, Greece, Vol. 2, 2000, pp. 775–784.
22. N. Priyantha, A. Chakraborty, and H. Balakrishnan, The cricket location-support system, in *Proceedings of ACM MobiCom*, Boston, MA, 2000, pp. 32–43.
23. Y. Hao, J. Tang, and Y. Cheng, Cooperative Sybil attack detection for position based applications in privacy preserved VANETs, in *Proceedings of IEEE GLOBECOM*, Houston, TX, 2011, pp. 1–5.
24. H. Yu, P. Gibbons, M. Kaminsky, and F. Xiao, Sybillimit: A near-optimal social network defense against Sybil attacks, in *Proceedings of IEEE Symposium on Security and Privacy*, Oakland, CA, 2008, pp. 3–17.
25. N. Tran, B. Min, J. Li, and L. Subramanian, Sybil-resilient online content voting, in *Proceedings of USENIX NSDI*, Boston, MA, 2009, pp. 15–28.
26. P. Dandekar, A. Goel, R. Govindan, and I. Post, Liquidity in credit networks: A little trust goes a long way, arXiv:1007.0515, 2010.
27. B. Viswanath, M. Mondal, K. Gummadi, A. Mislove, and A. Post, Canal: Scaling social network-based Sybil tolerance schemes, in *Proceedings of ACM EuroSys*, Bern, Switzerland, 2012, pp. 309–322.
28. P. Tsuchiya, The landmark hierarchy: A new hierarchy for routing in very large networks, in *Proceedings of ACM SIGCOMM,* Stanford, CA, 1988.
29. W. Chang, J. Wu, C. Tan, and F. Li, Sybil defenses in mobile social networks, in *Proceedings of IEEE Globecom 2013*, Atlanta, GA, 2013.
30. P. Marsden, Egocentric and sociocentric measures of network centrality, *Social Networks*, 24(4), 407–422, 2002.
31. S. Chan, I. Leung, and P. Liò, Fast centrality approximation in modular networks, in *Proceedings of ACM CNIKM*, Hong Kong, China, 2009, pp. 31–38.

32. W. Chang and J. Wu, Clump-based Sybil defense in crowdsensing, submitted for publication.
33. J. Ugander, L. Backstrom, C. Marlow, and J. Kleinberg, Structural diversity in social contagion, *Proceedings of the National Academy of Sciences*, 109(16), 5962–5966, 2012.
34. A. Mohaisen, N. Hopper, and Y. Kim, Keep your friends close: Incorporating trust into social network-based Sybil defenses, in *Proceedings of IEEE INFOCOM*, Shanghai, China, 2011, pp. 1943–1951.

Index

A

B

C

D

E

F

G

H

I

K

L

M

N

T

U

Z

DISCARDED
CONCORDIA UNIV. LIBRARY
CONCORDIA UNIVERSITY LIBRARIES
MONTREAL